Plant Cell Monographs

Volume 12

Series Editor: David G. Robinson
Heidelberg, Germany

Plant Cell Monographs

Recently Published Titles

Root Hairs
Volume Editors:
Emons, A. M. C., Ketelaar, T.
Vol. 12, 2009

Plant Microtubules
Development and Flexibility
2nd Edition
Volume Editor: Nick, P.
Vol. 11, 2008

Plant Growth Signalling
Volume Editors: Bögre, L., Beemster, G.
Vol. 10, 2008

Cell Division Control in Plants
Volume Editors: Verma, D. P. S., Hong, Z.
Vol. 9, 2008

Endosperm
Volume Editor: Olsen, O.-A.
Vol. 8, 2007

Viral Transport in Plants
Volume Editors: Waigmann, E., Heinlein, M.
Vol. 7, 2007

Nitric Oxide in Plant Growth, Development and Stress Physiology
Volume Editors: Lamattina, L., Polacco, J.
Vol. 6, 2007

The Expanding Cell
Volume Editors: Verbelen, J.-P., Vissenberg, K.
Vol. 5, 2007

The Plant Endoplasmic Reticulum
Volume Editor: Robinson, D. G.
Vol. 4, 2006

The Pollen Tube
A Cellular and Molecular Perspective
Volume Editor: Malhó, R.
Vol. 3, 2006

Somatic Embryogenesis
Volume Editors: Mujib, A., Šamaj, J.
Vol. 2, 2006

Plant Endocytosis
Volume Editors:
Šamaj, J., Baluška, F., Menzel, D.
Vol. 1, 2005

Anne Mie C. Emons • Tijs Ketelaar
Editors

Root Hairs

Springer

Editors
Prof. Dr. Anne Mie C. Emons
and
Dr. Tijs Ketelaar

Wageningen University
Laboratory of Plant Cell Biology
Arboretumlaan 4
6703 BD Wageningen
the Netherlands

Series Editor
Prof. Dr. David G. Robinson
Ruprecht-Karls-University of Heidelberg
Heidelberger Institute for Plant Sciences(HIP)
Department Cell Biology
Im Neuenheimer Feld 230
D-69120 Heidelberg
Germany

ISBN 978-3-642-09828-4 e-ISBN 978-3-540-79405-9
DOI: 10.1007/978-3-540-79405-9

Library of Congress Control Number: 2008934036

Softcover reprint of the hardcover 1st edition 2009

Cover design: WMX Design GmbH, Heidelberg, Germany

Printed on acid-free paper

9 8 7 6 5 4 3 2 1 0

springer.com

Editors

Anne Mie C. Emons studied Biology (M.Sc. 1969, cum laude) at the Radboud University of Nijmegen, The Netherlands, and after a family break, did her Ph.D. at the same university in 1986 and postdoctorate at Wageningen University in 1987, and John Innes Centre, Norwich, UK, on EMBO grant in 1988 under Dr. Keith Roberts. From 1988, she was an Assistant, Associate and Full Professor at Wageningen University, laboratory of Plant Cell Biology and since 2002 Advisor Bio-organization FOM Institute AMOLF, Theory Biomolecular Matter group.

Her major research interests are plant cell biology, cell polymers: actin filaments, microtubules and cellulose microfibrils, plant cell cyto-architecture, tip growth, and legume root hair Rhizobium interaction

Tijs Ketelaar studied Plant Breeding (M.Sc. 1996) at Wageningen University, the Netherlands, which was followed by a Ph.D. at the laboratory of Plant Cell Biology of Wageningen University in 2002. From 2001 to 2004, he was a postdoctorate in the laboratory of Professor Patrick Hussey at the University of Durham, UK, after which he received a VENI fellowship from the Dutch Science Foundation (NWO) at the laboratory of Plant Cell Biology of Wageningen University (2004–2007). From 2007 he has been an Assistant Professor in the same laboratory.

His research interests are actin cytoskeleton, plant cell cyto-architecture, polar cell expansion, and root hairs.

Preface

Root hairs stand out from the root body, extending the surface area of the root to facilitate anchoring in the soil and absorption of ions and water. They grow at their tips, and are the first target cells in the symbiotic relations of seed plants with micro-organisms. These properties make them an excellent model system for plant cell elongation studies and many aspects of plant cell signaling to an increasingly large number of plant biologists. In addition, root hairs are easy to observe in vivo because the growth process takes place on the surface of the plant. Root hair mutants are simple to distinguish and generally fertile, and drugs and signal molecules can be easily applied, making them close to ideal cells for experimental manipulation.

Due to the polarized nature of tip-growing root hairs, a polarized cyto-architecture is required to supply the material for cell growth and cell wall production to the cell tip. This cyto-architecture is characterized by a number of general features. The central vacuole is absent from the apical region and the nucleus is at a constant distance from the tip. The actin cytoskeleton is organized such that exocytotic vesicles reach the plasma membrane specifically at the site of cell expansion, whereas the microtubule cytoskeleton determines and maintains growth direction. The vesicle membranes fuse with the plasma membrane through exocytosis and deliver their contents and membrane-embedded enzymes for cell wall production. The robust intracellular organization employed in the growth process requires feedback regulation with, amongst others, calcium ions, ROP proteins, and plant growth regulators (auxin). Signal molecules from symbiotic microbes tap into these signaling cascades, and redirect them to orchestrate plant cell processes for their own benefit.

Almost all life on earth needs nitrogen compounds to produce nucleic acids and proteins, and most organisms, including plants, cannot use atmospheric nitrogen for this purpose. Many plant species have set up symbioses of various sorts with nitrogen fixing fungi and bacteria, which supply the useful nitrogen compounds and receive energy compounds in return. The last chapters of this book cover recent advances in research on these symbioses of plants with rhizobia and mycorrhizal fungi. The well-studied interaction between legume plants and *Rhizobium* bacteria, particularly, has provided insight into the early signaling events taking place in root hairs.

This book not only reviews recent advances in the molecular cell biology of root hair research, but most chapters also contain detailed explanations of techniques that were successfully used to study several aspects of root hairs, making the book useful in daily laboratory practice. Some chapters deal with these techniques in the form of detailed comparative "Methods" sections, while other chapters give straightforward protocols.

We enjoyed working on this book in collaboration with the contributing authors. It was especially gratifying to see the clear progress that has been made since the molecular cell biology book on root hairs, edited by Ridge and Emons, (Springer, 2000) appeared.

Wageningen, March 2008

Anne Mie C. Emons
Tijs Ketelaar

Contents

Contributors

Takashi Aoyama
Institute for Chemical Research, Kyoto University, Uji, Kyoto 611-0011, Japan
aoyama@scl.kyoto-u.ac.jp

Farhah F. Assaad
Technische Universität München, Botanik, Am Hochanger 4, 85354 Freising, Germany
farhah.assaad@wzw.tum.de

Tatiana Bibikova
Biology Department, The Pennsylvania State University, 208 Mueller Lab, University Park, PA, 16802, USA
tbibiko1@swarthmore.edu

Ton Bisseling
Wageningen University and Research Center, Graduate school of Experimental Plant Sciences (EPS), Laboratory of Molecular Biology, 6703 HA Wageningen, The Netherlands
ton.bisseling@wur.nl

Paola Bonfante
Dipartimento di Biologia Vegetale dell'Università di Torino and Istituto per la Protezione delle Piante-CNR, Viale Mattioli 25, 10125 Torino (Italy)
p.bonfante@ipp.cnr.it

Hyung-Taeg Cho
School of Biological Sciences, Seoul National University, 599 Gwanak-ro, Gwanak-gu, Seoul 151-742, Korea
htcho@snu.ac.kr

Anne Mie C. Emons
Laboratory of Plant Cell Biology, Wageningen University, Arboretumlaan 4, 6703BD Wageningen, The Netherlands
annemie.emons@wur.nl

John Fowler
Department of Botany & Plant Pathology, Oregon State University,
3021 Agr. Life Sci. Bldg., Corvallis, OR, 97331-7303, USA
fowlerj@science.oregonstate.edu

Nancy A. Fujishige
Department of Molecular, Cell and Developmental Biology, University
of California, Los Angeles, CA 90095-1606 USA
nfujishi@ucla.edu

Daniel J. Gage
University of Connecticut, Department of Molecular and Cell Biology,
91 N. Eagleville Rd., U-3125, Storrs, CT 06269-3125, USA
daniel.gage@uconn.edu

Andrea Genre
Dipartimento di Biologia Vegetale dell'Università di, Torino and Istituto per la
Protezione delle Piante-CNR, Viale Mattioli 25, 10125 Torino (Italy)
andrea.genre@unito.it

Simon Gilroy
Department of Botany, University of Wisconsin, Birge Hall, 430 Lincoln Drive,
Madison, USA
sgilroy@wisc.edu

Claire Grierson
School of Biological Sciences, University of Bristol, Bristol, UK
Claire.Grierson@bristol.ac.uk

Ann M. Hirsch
Department of Molecular, Cell and Developmental Biology and Molecular
Biology Institute, University of California, Los Angeles, CA 90095-1606, USA
ahirsch@ucla.edu

Tijs Ketelaar
Wageningen University, Laboratory of Plant Cell Biology, Arboretumlaan 4,
6703 BD Wageningen, The Netherlands
tijs.ketelaar@wur.nl

Sang Ho Lee
Department of Plant Biology, University of Minnesota, 250 BioSci. Center,
1445 Gortner Ave., St. Paul, MN 55108, USA
leex3704@umn.edu

Roger R. Lew
Department of Biology, York University, Toronto, Canada
planters@yorku.ca

Erik Limpens
Wageningen University and Research Center, Graduate school of Experimental Plant Sciences (EPS), Laboratory of Molecular Biology, 6703 HA Wageningen, The Netherlands
erik.limpens@wur.nl

Michelle R. Lum
Biology Department, Loyola Marymount University, 1 LMU Drive, Los Angeles, CA 90045
mlum@ucla.Edu

Bela M. Mulder
FOM Institute AMOLF, Kruislaan 407, 1098SJ Amsterdam, The Netherlands, Laboratory of Plant Cell Biology, Wageningen University, Arboretumlaan 4, 6703BD Wageningen, The Netherlands
mulder@amolf.nl

M. Niels de Keijzer
FOM Institute AMOLF, Kruislaan 407, 1098SJ Amsterdam, The Netherlands
dekeijzer@amolf.nl

Erik Nielsen
Department of Molecular, Cellular & Developmental Biology, University of Michigan, 830 North University Avenue, Ann Arbor, MI 48109
nielsene@umich.edu

Mara Novero
Dipartimento di Biologia Vegetale dell'Università di, Torino and Istituto per la Protezione delle Piante-CNR, Viale Mattioli 25, 10125 Torino (Italy)
mara.novero@unito.it

John Schiefelbein
Department of Molecular, Cellular, and Developmental Biology, University of Michigan, Ann Arbor, MI, USA
schiefel@umich.edu

Björn J. Sieberer
Laboratory of Plant-Microorganism Interactions, CNRS INRA, UMR2594, 24 Chemin de Borde Rouge, BP 52627, F-31326 Castanet-Tolosan, France
bjorn.sieberer@toulouse.inra.fr

K. Szczyglowski
Agriculture and Agri-Food, Southern Crop Protection and Food Research Centre, London, Ontario N5V 4T3, Canada
szczyglowskik@agr.gc.ca

Antonius C.J. Timmers
Laboratory of Plant-Microorganism Interactions, CNRS INRA, UMR2594, 24 Chemin de Borde Rouge, BP 52627, F-31326 Castanet-Tolosan, France
ton.timmers@toulouse.inra.fr

Viktor Žárský
Department of Plant Physiology, Faculty of Sciences, Charles University, Vinicna 5, 128 44 Prague 2, Czech Republic, Institute of Experimental Botany, Academy of Sciences of the Czech Republic, Rozvojova 263, 165 00 Prague 6, Czech Republic
zarsky@ueb.cas.cz

Genetics of Root Hair Formation

C. Grierson and J. Schiefelbein (✉)

Abstract There has been a great deal of recent progress in our understanding of the genetic control of root hair development, particularly in *Arabidopsis thaliana*. This chapter summarizes the genes and gene products that have been identified using forward and reverse genetic approaches. The involvement of these genes at various stages of root hair development is described, including the specification of the root hair cell type, the initiation of the root hair outgrowth, and the elongation (tip growth) of the root hair.

1 Introduction

The formation of root hairs has been used for more than a century to study fundamental problems in plant biology, using physiological, cell biological, and developmental approaches (Cormack 1935; Cutter 1978; Haberlandt 1887; Leavitt 1904; Sinnot and Bloch 1939). In the past 20 years, a great deal of attention has been devoted to using genetics to study root hair formation, particularly in the model plant species *Arabidopsis*. *Arabidopsis* root hairs are amenable to genetic dissection because (1) they are easily visible on the root surface and appear rapidly (within 3 days) after seed germination, making them one of the most convenient postembryonic cells for phenotypic analysis; (2) the entire developmental history of the *Arabidopsis* epidermis has been defined, from its embryonic origin through its mature cell features (Dolan et al. 1993, 1994; Scheres et al. 1994); (3) the root epidermal cells are generated and differentiate in a file-specific manner, which enables all stages of development to be analyzed along the root at any time; (4) the root hair cells are not required for plant viability or fertility, and so any type of mutant can be isolated and analyzed; (5) large-scale genetic screens are feasible, since large numbers of seedlings can be analyzed for their root hair phenotype; and

J. Schiefelbein
Department of Molecular, Cellular, and Developmental Biology,
University of Michigan, 830 North University Avenue, Natural Sciences Building,
Ann Arbor MI 48109-1048, USA
e-mail: schiefel@umich.edu

Plant Cell Monogr, doi:10.1007/7089_2008_15

(6) molecular and genomic resources are available in *Arabidopsis* for the rapid analysis of new genes and proteins.

In this chapter, we describe the genetics of root hair cell specification, root hair initiation, and root hair growth (elongation). A list of the genes identified from *Arabidopsis* is presented in Table 1.

Table 1 Genes involved in root hair formation in *Arabidopsis*

Gene[a]	Root hair mutant phenotype	Predicted product	Reference
AGP30	None	Arabinogalactan protein	van Hengel et al. 2004
AIP1	Short root hairs	Actin-interacting protein	Ketelaar et al. 2007
AKT1	Long root hairs	Potassium transporter	Desbrosses et al. 2003
ARA6	None	Rab GTPase	Grebe et al. 2003
ARF1	Dominant mutants have no hairs, or have short or double root hairs	ADP-ribosylation factor GTPases	Xu and Scheres 2005
AtEXO70A1	Reduced root hair length	Exocyst subunit Exo70	Synek et al. 2006
AtIPK2α	Long root hairs	Inositol polyphosphate kinase	Xu et al. 2005b
AUX1	Short hairs	Auxin transport	Pitts et al. 1998
AXR1	Bulges form but do not elongate	Subunit of RUB1-activating enzyme	Cernac et al. 1997; del Pozo et al. 2002
AXR2/IAA7	Dominant mutants are hairless, except where root meets hypocotyl	Repressor of auxin-responsive transcription	Nagpal et al. 2000
AXR3/IAA17	Dominant mutants are hairless, except where root meets hypocotyl	Repressor of auxin-responsive transcription	Leyser et al. 1996
BHLH32	High phosphate does not suppress root hair development	bHLH transcription factor	Chen et al. 2007
BIG	Short hairs in high phosphate conditions	Calossin	Lopez-Bucio et al. 2005
BRISTLED1 (BST1)/DER4	Short hairs, sometimes branched	Unknown	Parker et al. 2000; Ringli et al. 2005
CAP1	Hairs short, bulbous, occasionally branched	Actin-binding protein	Deeks et al. 2007
CEN3/DER1/ACT2	Short, wide hairs, some with wide bases, some hairs curled and/or branched	Vegetative actin	Ringli et al. 2002
CENTIPEDE1 (CEN1)	Short, wide hairs, sometimes curled	Unknown	Parker et al. 2000
CENTIPEDE2 (CEN2)	Short, wide hairs, sometimes branched and/or curled	Unknown	Parker et al. 2000

(continued)

Table 1 (continued)

Gene[a]	Root hair mutant phenotype	Predicted product	Reference
COW1	Short, wide root hairs, often branched at the base	Phosphatidylinositol transfer protein	Bohme et al. 2004; Vincent et al. 2005
CPC	Reduced root hair number	R3 Myb transcription factor	Wada et al. 2002
DER2	Hairs stop growing after bulge forms	Unknown	Ringli et al. 2005
DER3/ENL7	Hairs have wide bases and sometimes branch	Unknown	Ringli et al. 2005
DER5	Short, distorted hairs	Unknown	Ringli et al. 2005
DER6	Very short hairs	Unknown	Ringli et al. 2005
DER7	Hairs often short, wide, and bulbous	Unknown	Ringli et al. 2005
DER8	Hairs often depolarized	Unknown	Ringli et al. 2005
EGL3	None	bHLH transcription factor	Bernhardt et al. 2005
EIN2	Root hairs form further from the end of the epidermal cell	Ethylene response	Fischer et al. 2006
ENL1	Wavy, branched hairs	Unknown	Diet et al. 2004
ENL5	Short, wide, curved hairs	Unknown	Diet et al. 2004
EPC1	Short root hairs	Glycosyltransferase	Bown et al. 2007
ETC1	None	R3 Myb transcription factor	Kirik et al. 2004a
ETO1-4	Increased root hair length and density	Ethylene overproducer	Cao et al. 1999
ETR1, ERS1, ERS2, ETR2	Reduced root hair length	Ethylene receptor	Cho and Cosgrove 2002
EXP18	–	Expansin protein	Cho and Cosgrove 2002
EXP7	None	Expansin protein	Cho and Cosgrove 2002
GL2	Excess root hairs	Homeodomain-leucine-zipper transcription factor	Masucci et al. 1996
GL3	Increased root hair number	bHLH transcription factor	Bernhardt et al. 2005
IAA14	Dominant mutants are hairless, except where root meets hypocotyl	Repressor of auxin-responsive transcription	Fukaki et al. 2002
ICR1	In loss-of-function mutant hairs form closer to the root tip. Ectopic expression produces short, bulbous root hairs	Coiled-coil domain scaffold protein regulated by ROPs	Lavy et al. 2007
IRE	Short root hairs	Serine/threonine kinase	Oyama et al. 2002
KEULE	Hairs absent or stunted and swollen	Sec1 protein	Assaad et al. 2001

(continued)

Table 1 (continued)

Gene[a]	Root hair mutant phenotype	Predicted product	Reference
KIP	Wide hairs	SABRE-like protein	Procissi et al. 2003
KJK/SHV1/AtCSLD3	Hairs burst after bulge forms	ER-localized cell wall polysaccharide synthase	Favery et al. 2001
LPI1	Root hair length and density less affected by high phosphate	Unknown	Sanchez-Calderon et al. 2006
LPI2	Root hair length less affected by high phosphate	Unknown	Sanchez-Calderon et al. 2006
LRX1	Hairs short, swollen, or branched	Leucine-rich repeat (LRR)/extensin	Baumberger et al. 2001
LRX2	Some hairs short, swollen, or branched	LRR/extensin	Baumberger et al. 2003
MRH1	Short hairs	LRR class of receptor-like kinase	Jones et al. 2006
MRH2	Wavy, branched hairs	Armadillo repeat containing kinesin-related protein	Jones et al. 2006
MRH3	Hairs have wide bases	Inositol 1,4,5-triphosphate 5-phosphatase	Jones et al. 2006
Myosin XIK	Hairs short and slow-growing	Myosin XI	Ojangu et al. 2007
OXI1/AGC2	Variable root hair length	Oxidative-burst-inducible kinase	Anthony et al. 2004; Rentel et al. 2004
PAX1	Suppressor of dominant mutations in AXR3, pax1 single mutant has extra root hairs that are sometimes branched	Unknown	Tanimoto et al. 2007
PFN1	Overexpressing root hairs twice as long as wild type	Profilin	Ramachandran et al. 2000
PGP4	Long hairs	Auxin transport	Santelia et al. 2005
PI-4Kβ1	Double mutant with PI-4Kβ2 has branched, wavy and bulged root hairs	Phosphatidylinositol 4-OH kinase	Preuss et al. 2006
PIP5K3	Short, wide hairs. Overexpressors have fat, curled or bulbous hairs	Phosphatidylinositol-4-phosphate 5-kinase	Stenzel et al. 2008
PLDζ1	None	Phospholipase D	Ohashi et al. 2003
PRP3	None	Proline-rich cell wall protein	Bernhardt and Tierney 2000
RABA4b	None	Rab GTPase	Preuss et al. 2004

(continued)

Table 1 (continued)

Gene[a]	Root hair mutant phenotype	Predicted product	Reference
RHD1/UGE4	Wide swellings	UDP-d-glucose 4-epimerase	Schiefelbein and Somerville 1990; Seifert et al. 2002
RHD2	Hairs stop growing after bulge forms	NADPH oxidase	Foreman et al. 2003
RHD3	Wavy, short hairs	GTP-binding protein involved in ER–Golgi transport	Wang et al. 1997; Zheng et al. 2004
RHD6	Hairless	Unknown	Menand et al. 2007
RHL1, RHL2, RHL3	Reduced root hair density, dwarf	Topoisomerase subunits	Sugimoto-Shirasu et al. 2002, 2005
RHM1/ROL1	Suppressor of lrx1	Pectic polysaccharide rhamnogalacturonan modifying enzyme	Diet et al. 2006
RPA/At2G35210	Short, bulbous and branched root hairs	ARF GTPase-activating protein	Song et al. 2006
SAR1	Suppressor of axr1	Nucleoporin	Cernac et al. 1997
SCN1	Short, wide, hairs, sometimes branched	ROPGDP dissociation inhibitor	Carol et al. 2005
SHV2/COBL9/DER9/MRH4	Hairs stop growing after bulge forms	COBRA-like protein	Jones et al. 2006
SHV3/MRH5	Hairs stop growing after bulge forms	Glycerophosphoryl diester phosphodiesterase (GPDP)-like protein	Jones et al. 2006
SIMK	Overexpression increases root hair length	Mitogen-activated protein kinase	Samaj et al. 2002
SIZ1	Long root hairs very near root tip	SUMO E3 ligase	Miura et al. 2005
SOS4	Bulges form but do not elongate	Pyridoxal kinase	Shi and Zhu 2002
TIP1	Wide swellings; short, wide hairs, branched at base	ANK protein *S*-acyl transferase	Hemsley et al. 2005
TRH1	Hairs stop growing after bulge forms. Some cells with multiple bulges	Potassium carrier required for auxin transport	Rigas et al. 2001
TRY	None	R3 Myb transcription factor	Schellmann et al. 2002
TTG	Excess root hairs	WD-repeat protein	Galway et al. 1994
WER	Excess root hairs	R2R3 Myb transcription factor	Lee and Schiefelbein 1999
WRKY75	Increased root hair density	Transcriptional repressor	Devaiah et al. 2007

[a]Multiple names for the same gene are given when these are used in the literature

2 Genetics of Root Hair Cell Specification

2.1 Variation in Root Hair Pattern

The first stage in root hair formation is the specification of a newly formed epidermal cell to differentiate as a root-hair-bearing cell, rather than a non-hair-bearing epidermal cell. Depending on the plant species, root hair cells are specified by an asymmetric division mechanism, a random mechanism, or a position-dependent mechanism (Clowes 2000). Many monocots use the asymmetric division mechanism, whereby the smaller of the daughter cells from an epidermal cell precursor adopts the hair cell fate. In the random mechanism, the fate of a given cell cannot be predicted by its history and it adopts its fate at a relatively later stage.

The position-dependent mechanism is used by several plant families, including the *Brassicaceae* (Bunning 1951; Cormack 1935), as represented by the well-characterized *Arabidopsis* root hair pattern (Dolan et al. 1994; Galway et al. 1994). In these plant species, root hair cells arise over the intercellular space between underlying cortical cells (the "H" cell position), whereas nonhair cells develop over a single cortical cell (the "N" position). This implies that positional cues play an important role in cell fate determination in this tissue. Prior to hair formation, immature *Arabidopsis* epidermal cells in the H position can be distinguished from those in the N position by a greater rate of cell division (Berger et al. 1998), reduced cell length (Dolan et al. 1994; Masucci et al. 1996), enhanced cytoplasmic density (Dolan et al. 1994; Galway et al. 1994), reduced vacuolation rate (Galway et al. 1994), cell surface features (Dolan et al. 1994; Freshour et al. 1996), and chromatin organization (Costa and Shaw 2005; Xu et al. 2005a). Furthermore, the analysis of reporter gene lines and in situ RNA hybridization show that cells in the two epidermal positions already exhibit differential gene expression during embryogenesis (Berger et al. 1998; Costa and Dolan 2003; Lin and Schiefelbein 2001). These observations indicate that epidermal cells begin to assess their position and adopt their appropriate fate at an early stage, both during embryogenesis and postembryonic root development. However, despite early differential cellular activities in H and N cells, a relatively late change in the position of an immature epidermal cell can induce a change in its developmental fate (Berger et al. 1998). Together, these results suggest that epidermal cell fate is not fixed at an early stage in *Arabidopsis*; rather, positional signaling apparently acts continuously during postembryonic root development to ensure appropriate cell-type patterning.

2.2 Arabidopsis Genes Controlling Root Hair Cell Specification

Forward and reverse genetic approaches have led to the identification of more than eight *Arabidopsis* genes that influence early events in root hair specification (Grierson and Schiefelbein 2002; Pesch and Hulskamp 2004; Xu et al. 2005a). Most of these encode transcription factors that influence the expression of cell-type-specific genes. The emerging picture is that these act early in a complex regulatory network involving lateral inhibition with feedback (Larkin et al. 2003a; Lee and Schiefelbein 2002; Schiefelbein 2003b). Four genes, *TRANSPARENT TESTA GLABRA* (*TTG*), *GLABRA 3* (*Gl3*), *ENHANCER OF GLABRA 3* (*EGL3*), and *WEREWOLF* (*WER*), are required to specify the nonhair fate because mutations in these (alone or in combination) lead to the formation of root hair cells in place of nonhair cells ("hairy" mutants) (Bernhardt et al. 2003; Galway et al. 1994; Lee and Schiefelbein 1999; Masucci et al. 1996). Three genes, *CAPRICE* (*CPC*), *TRIPTYCHON* (*TRY*), AND *ENHANCER OF TRY AND CPC* (*etc1*), help specify the hair cell fate, and mutations in them (alone or in combination) cause nonhair cells to develop in place of hair cells ("bald" mutants) (Kirik et al. 2004; Schellmann et al. 2002; Simon et al. 2007; Wada et al. 1997). Current models suggest that TTG (a small protein with WD40 repeats (Walker et al. 1999)), GL3 and EGL3 (related bHLH transcription factors (Bernhardt et al. 2003)), and WER (R2R3 MYB-domain transcription factor (Lee and Schiefelbein 1999)) act in a central transcriptional complex in the N cells to promote the nonhair cell fate (Pesch and Hulskamp 2004; Schiefelbein 2003a; Ueda et al. 2005). This complex positively regulates the expression of *GLABRA2* (*GL2*), which encodes a homeodomain-leucine-zipper transcription factor protein (Rerie et al. 1994) required for non-hair-cell differentiation (Costa and Shaw 2005; Di Cristina et al. 1996; Masucci et al. 1996). The GL2 appears to negatively regulate root-hair-specific genes and positively regulate non-hair-cell-specific genes (Lee and Schiefelbein 1999, 2002; Masucci et al. 1996), as described later.

The TTG-GL3/EGL3-WER transcription complex also mediates lateral inhibition by positively regulating the transcription of *CPC*, *TRY*, and *ETC1* in the N cells (Kirik et al. 2004; Koshino-Kimura et al. 2005; Lee and Schiefelbein 2002; Ryu et al. 2005; Simon et al. 2007; Wada et al. 2002). These encode small one-repeat MYB proteins that lack transcriptional activation domains and appear to move to the adjacent H cells (possibly via plasmodesmata) and inhibit the ability of the WER/MYB23 MYB proteins to participate in complex formation and/or promoter binding (Esch et al. 2003; Kirik et al. 2004; Kurata et al. 2005; Ryu et al. 2005; Schellmann et al. 2002; Wada et al. 1997, 2002). In addition, the central TTG-GL3/EGL3-WER complex represses expression of the *GL3/EGL3* bHLH genes, and as a result, the bHLH genes are preferentially active in the H cells. This implies that these bHLH proteins move from the H cell (where they are produced) to the N cell (where they are required for non-hair-cell specification) to complete a regulatory circuit involving two loops of communication between the N and H cells (Bernhardt et al. 2005). It has been proposed that these intercellular events result in mutual dependence of cell identities between neighboring cells, and thus help to ensure that distinct cell fates arise (Schiefelbein and Lee 2006).

The preferential accumulation of the central transcription complex in the N position has been proposed to result from the action of a positional signaling pathway (Larkin et al. 2003b; Schiefelbein and Lee 2006). A central question revolves around the molecular basis of this positional influence. A leucine-rich repeat receptor-like kinase (LRR-RLK), named SCRAMBLED (SCM), is likely to mediate this process.

Homozygous recessive *scm* mutations cause the developing root epidermis to exhibit a non-position-dependent distribution of *GL2*, *CPC*, and *WER* expression and epidermal cell types. Detailed genetic and expression studies indicate that SCM likely acts to inhibit *WER* transcription in the H cell position (Kwak and Schiefelbein 2006). The predicted SCM protein possesses the structural features of a typical LRR-receptor-like protein kinase, of which there are more than 200 in *Arabidopsis* (Initiative 2000; Shiu and Bleecker 2001, 2003; Torii 2004). It possesses a putative extracellular domain with six tandem copies of a 24-residue LRR, a single predicted transmembrane domain near its center, and a C-terminal putative intracellular kinase domain. Taken together, these findings suggest that SCM enables immature epidermal cells to detect a positional signal and adopt an appropriate fate.

Recent studies indicate that the positional cue may influence epidermal cell type differentiation by remodeling the chromatin around the transcription factor genes. Three-dimensional fluorescence in situ hybridization was used to show that the chromatin around the *GL2* gene is "open" in the differentiating N cells, whereas it is "closed" in the differentiating H cells (Costa and Shaw 2005). Further, the *GL2* chromatin status appears to be reset during mitosis, which makes the epidermal cells in the meristematic zone vulnerable to chromatin modification caused by a change in cell position (Costa and Shaw 2005). Two studies suggest that histone modification influences the cell fate decision. Inhibition of histone deacetylation, using trichostatin A, increased the expression of *CPC* and *GL2* genes, decreased the expression of *WER*, and led to an increase in root hair cell formation (Xu et al. 2005a). A protein that interacts with the cell division machinery, GEM1, influences histone H3 modifications at the *GL2* and *CPC* promoters (Caro et al. 2007). Together, these findings suggest a link between cell division, chromatin status, and cell fate decisions in the root epidermis.

2.3 A Model for *Arabidopsis* Root Hair Patterning

The molecular genetic results to date suggest a possible model for specification of root hair cells in *Arabidopsis* (Fig. 1). Essentially, this model proposes that SCM signaling influences, in a position-dependent manner, the accumulation of the central transcription factor complex (TTG-GL3/EGL3-WER) which promotes the non-hair cell fate (via GL2) as well as the hair cell fate (by lateral inhibition via CPC/TRY/ETC1). Specifically, the SCM pathway is proposed to negatively regulate *WER* transcription in the H cells, which causes these cells to preferentially succumb to the CPC/TRY/ETC1-dependent lateral inhibition and adopt the hair cell fate.

Several key aspects of this model are supported by experimental evidence. For example, a functional WER-GFP fusion protein localizes to the N cell nucleus (Ryu et al. 2005), whereas a CPC-GFP fusion protein accumulates in the nuclei of both N and H cells (Wada et al. 2002). This supports the differential mobility of the WER and CPC, which is crucial for the model's mechanism. Also, WER has been shown to directly regulate *CPC* expression, likely via distinct WER-binding sites

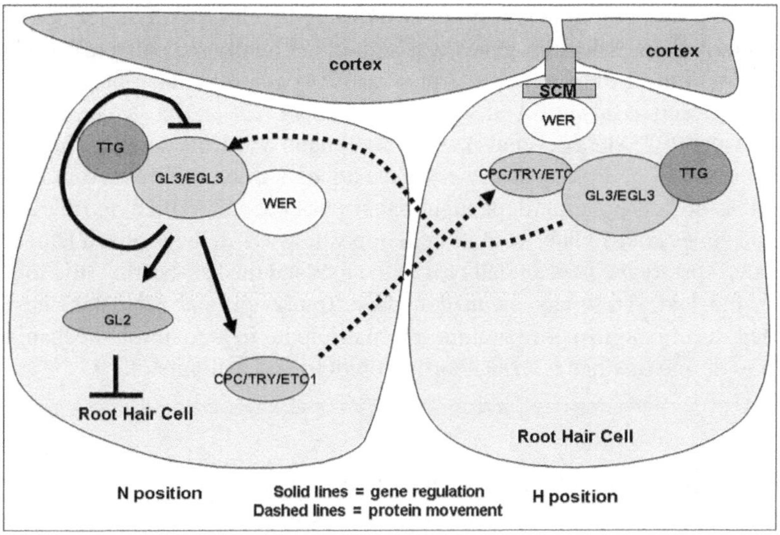

Fig. 1 A potential model to explain cell-type patterning in the *Arabidopsis* root epidermis. The figure illustrates a late stage after positional cues act through the SCM receptor

in the *CPC* promoter (Ryu et al. 2005), and the action of the WER-equivalent MYB protein GL1 is disrupted by TRY in yeast (Esch et al. 2003).

Although this model is consistent with the data to date, there are several unclear issues to resolve. For example, the precise way SCM alters the transcription factor network is not known. Although WER seems to be its ultimate target, we do not know the components or the action of the putative signal transduction chain. Also, other mechanisms that influence position-dependent epidermal patterning are not known, although they are likely to exist, because *scm* mutants do not completely abolish the cell pattern (Kwak and Schiefelbein 2006). Further, the putative SCM ligand that is hypothesized to be asymmetrically distributed in the epidermis is not known.

It is noteworthy that some aspects of this patterning mechanism appear to be present in the specification mechanisms for other epidermal cell types. Like the root, the *Arabidopsis* hypocotyl epidermis possesses two distinct files of cell types that arise in a position-dependent pattern (Berger et al. 1998; Gendreau et al. 1997; Hung et al. 1998; Wei et al. 1994). Cells in one file preferentially adopt the stomatal fate (located outside an anticlinal cortical cell wall, analogous to the H position of the root), whereas cells in the other file adopt the nonstomatal fate (located outside a periclinal cortical cell wall, analogous to the N position). With the notable exception of SCM, essentially all of the genes associated with root epidermal cell fate affect the hypocotyl epidermis in an analogous manner (Berger et al. 1998; Hung et al. 1998; Lee and Schiefelbein 1999). Considering that the root and hypocotyl derive from the embryonic epidermis, it appears that the same patterning mechanism is used throughout the embryonic axis to generate position-dependent specification of a common pattern of two different cell types. Further, some of the root epidermal specification genes (including *TTG* and *GL2*) also affect trichome (hair) formation in the shoot epidermis and seed coat mucilage production in the developing seed (Galway et al. 1994; Koornneef 1981; Koornneef et al. 1982; Larkin et al. 1997; Masucci et al. 1996; Schellmann and Hulskamp 2005). However, the patterning of cell types in these tissues appears to be different from the one in the root, as no cell-position-dependent pattern has been identified. Furthermore, the common genes control hair formation in opposite ways in the root and leaf; they are required to specify the nonhair cell type in the root and the hair-bearing (trichome) cell type in the leaf. This may mean that these transcriptional regulators have been recruited during angiosperm evolution to participate in a common mechanism that is able to define distinct cell fates during epidermis development.

3 Genetics of Root Hair Initiation

The first outward sign of root hair formation is the emergence of a bulge on the outside of an epidermal cell. This event, termed root hair initiation, is controlled by several known genes (initiation genes). In *Arabidopsis*, these are likely to act downstream of GL2. Likely targets of GL2 include *ROOTHAIRDEFECTIVE 6* (*RHD6*, required for hair initiation (Masucci and Schiefelbein 1994, 1996; Menand et al. 2007)) and *PLDζ1* (encoding a phospholipase D protein involved in hair outgrowth and elongation). GL2 binds to the *PLDζ1* promoter in gel shift assays and prevents expression of *PLDζ1 in planta*, supporting the idea that GL2 acts directly upstream of this gene in nonhair cells (Ohashi et al. 2003).

Root hair initiation is also dependent on auxin and ethylene signaling (see also Lee and Cho 2008). The relationship between auxin and ethylene signaling mechanisms in roots is intricate. These hormones can regulate each other's biosynthesis and response pathways, and sometimes independently regulate the same target genes (Stepanova et al. 2007). Ethylene acts, at least in part, through changes to auxin biosynthesis and transport (Ruzicka et al. 2007; Swarup et al. 2007). The role of auxin in root hair development is more clearly defined than that of ethylene. Auxin acts downstream of GL2, probably by controlling the abundance of AUX/IAA transcriptional repressors. The relative abundance of the AUX/IAA transcription factors SUPPRESSOROFHY 2 (SHY2) and AUXINRESISTANT 3 (AXR3) in a H cell controls whether or not initiation takes place. Mutant forms of AUXINRESISTANT 2 (AXR2) (Nagpal et al. 2000), IAA14 (Fukaki et al. 2002), and AXR3 (Leyser et al. 1996) that are poorly degraded in response to auxin prevent root hair initiation altogether, whereas similar mutants of the AUX/IAA SHY2 promote hair initiation closer to the root tip (Knox et al. 2003).

The first proteins to appear at a developing initiation site are Rho of plant (ROP) small GTPases, which arrive in a patch at the plasma membrane before the bulge starts to form (Jones et al. 2002; Molendijk et al. 2001). ROPs are unique to plants, but are related to the Rac, Cdc42, and Rho small GTPases that control the morphogenesis of animal and yeast cells. These proteins often act as intermediaries between receptors and downstream effects, such as cytoskeletal rearrangements, reactive oxygen generation, and mitogen-activated protein kinase cascades. Applying the ADP-ribosylation factor GTPase guanine-nucleotide exchange factor (ARF GEF) secretion inhibitor brefeldin A prevents ROP localization at the root hair initiation site. This suggests that either ROPs themselves or molecule(s) that localize ROPs are placed at the future site of hair formation by a targeted secretion mechanism that involves one or more ARF GTPases (Molendijk et al. 2001). However, although dominant mutants of ARF1 affect root hair initiation and tip growth, as well as ROP2 polar localization, they do so at a slower timescale than brefeldin A (Xu and Scheres 2005), suggesting that brefeldin A treatment may be acting through other, as yet unidentified, ARFs.

The location of a patch of ROPs, and hence of the bulge on each hair cell, is controlled by auxin. Externally applied or endogenously overproduced auxin or ethylene hyperpolarizes ROP localization and bulge formation so that it happens nearer to the apical end of the cell (towards the root tip). The auxin concentration gradient is virtually abolished in *aux1;ein2;gnomeb* triple mutant roots, and in these mutants locally applied auxin can organize hair positioning. Under these circumstances, hairs can even form at the basal (nearest the shoot) end of the cell (Fischer et al. 2006).

Within minutes of ROP localization, the root hair cell wall begins to bulge out. The size of the bulge is affected by the *ROOT HAIR DEFICIENT 1* (*RHD1*) and *TIP1* genes. Genetic evidence suggests that *RHD1* may act downstream of *TIP1*. *RHD1* encodes an isoform of UDP-d-glucose 4-epimerase (UGE4) that is thought to be involved in the biosynthesis of cell wall carbohydrates (Rosti et al. 2007; Seifert et al. 2002). *TIP1* encodes the first plant member of a class of transferases that *S*-acylate specific proteins, affecting their targeting and retention at different subcellular membrane domains (Hemsley et al. 2005). Target proteins that are *S*-acylated by TIP1 are still being identified.

As the cell wall begins to bulge out, the pH of the wall falls. The mechanism responsible for this pH change is uncertain; it may be due to local changes in wall polymer structure and ion exchange capacity, or to local activation of a proton ATPase or other proton transport activity (Bibikova et al. 1998). This pH change may activate expansin proteins that catalyze wall loosening (Baluska et al. 2000; Bibikova et al. 1998). Two root-hair-specific expansin genes have been identified whose promoters are induced in hair cells shortly before the bulge forms (Cho and Cosgrove 2002), and a *cis*-element responsible for this hair-cell-specific transcriptional response has been defined (Kim et al. 2006).

4 Genetics of Root Hair Elongation

4.1 Establishment of Tip Growth

Following initiation, the growth of the root hair proceeds by a highly oriented cell expansion mechanism known as tip growth. This type of growth is used by other tubular, walled cells, including pollen tubes. Numerous genes have been identified that are required for the correct direction and extent of root hair tip growth. These include genes involved in the mechanism of growth, as well as genes involved in regulating the location and amount of growth.

Mutants in many root hair genes fail to establish normal tip growth. They either fail to elongate a root hair from the bulge at all, or produce multiple hairs from each bulge. These results suggest that tip growth involves processes and/or cellular properties that are not essential for bulge formation. Several genes with mutants in this class encode actin or actin regulators. For example, *cap1 arp2-1* double mutants make bulges that fail to extend, implicating the actin regulators CYCLASE-ASSOCIATED PROTEIN 1 (CAP1) and the ACTIN-RELATED PROTEINS 2/3 (ARP2/3) complex in root hair tip-growth (Deeks et al. 2007). Loss-of-function mutants in the major vegetative actin, ACTIN 2, also usually fail to establish tip growth (Gilliland et al. 2002; Ringli et al. 2002), and the expression of the N-terminus of class Ie formin AtFH8 in root hairs can suppress root hair elongation, and sometimes, bulge formation, suggesting that formins may also play a role (Deeks et al. 2005). A second group of genes whose mutants fail to establish normal tip growth encodes proteins associated with cell wall properties, such as KJK, which contributes to cell wall synthesis (Favery et al. 2001), and COBL9, a COBRA-like protein (Jones et al. 2006). Many genes involved in intracellular signaling during tip growth also have mutants in this phenotypic class (e.g. *ROOTHAIR DEFECTIVE 2* (RHD2), MRH3, SUPERCENTIPEDE1 (SCN1)), and are discussed in more detail what follows.

There is accumulating evidence that an adequate auxin supply is required for tip growth. Mutants in some genes that affect auxin biosynthesis or transport, such as *SALT OVERLY SENSITIVE 4* (*SOS4*; Shi and Zhu 2002) and *TINY ROOT HAIR 1* (*TRH1*; Rigas et al. 2001), fail to establish tip growth. SOS4 is a pyridoxal kinase involved in the production of pyridoxal-5-phosphate. This probably acts via auxin (or possibly ethylene) synthesis, which may involve enzymes that use pyridoxal-5-phosphate as a cofactor (Shi and Zhu 2002). The TRH1 potassium carrier is essential for normal auxin transport, and Trh1− mutants make root hair bulges that do not elongate (Vicente-Agullo et al. 2004). Similarly, mutants of *AUXINRESISTANT 1* (*AXR1*) have root hairs that rarely elongate, so that most remain as small bumps on the surface of the root (Cernac et al. 1997). *AXR1* encodes a subunit of an RUB1-activating enzyme, which is required for auxin responses (del Pozo et al. 2002). Taken together, these results suggest that auxin transport and signaling are essential for root hair elongation. For the role of auxin in root hair elongation, see also Aoyama (2008).

4.2 Maintenance of Tip Growth

Our current understanding of root hair tip growth is summarized in Fig. 2. The transcription of at least 606 genes is upregulated during root hair tip growth (Jones et al. 2006). These encode proteins with roles in cell wall synthesis, assembly, and function, exocytosis and secretion, endocytosis, protein targeting, vesicle transport, cytoskeletal rearrangements, and protein phosphorylation. Many well-characterized

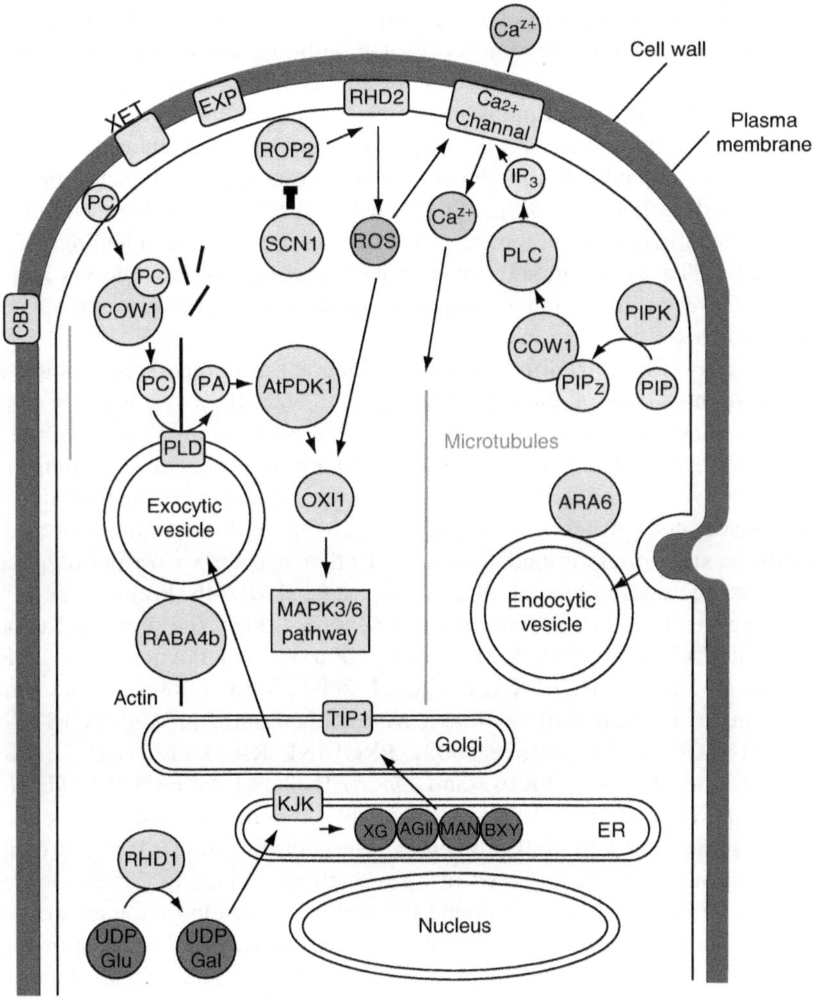

Fig. 2 A potential model of mechanisms involved in root hair tip growth. *UDP-Glu* UDP glucose, *UDP-Gal* UDP galactose, *XG* xylogucan, *AGII* arabinogalactan II, *Man* mannan, *bXY* b-xyloglucan backbone, *PIP* phosphatidylinositol monophosphate, *PIP2* phosphatidylinositol (4,5)-bisphosphate, IP_3 inositol triphosphate, *PC* phospholipase C, *PLD* phospholipase D {, *PIPK* phosphatidylinositol monophosphate kinase, *XET* xyloglucan endotransglycosylases, *EXP* expansins, *ROS* reactive oxygen species

examples of root hair genes are listed in Table 1. The resulting mRNAs are exported and translated, some presumably on cytosolic ribozomes, and others presumably at ribozomes on the surface of the endoplasmic reticulum (ER). In most cases the precise subcellular locations of gene products, and the mechanisms of protein targeting and movement within the hair cell are not known, but proteins probably either diffuse through the cytosol to their sites of action, move directly along microtubules or actin filaments, become associated with membrane-bound vesicles that bud from the ER, Golgi, and/or vacuole, or are produced on polysomes located in the subapex of the hair (see Emons and Ketelaar 2008). Some proteins, or the products of protein activity, such as cell wall polysaccharides, are contained within vesicles. Others are embedded in the inner or outer leaflet (e.g., GPI anchor proteins), or through both leaflets (e.g., some ion channels and cellulose synthase complexes), of vesicle membranes. Some proteins involved in intracellular signaling, such as ROP small GTPases, are thought to be bound directly, for example via attached lipids, or indirectly, for example via protein–protein interactions, to the outer surfaces of vesicles.

The vesicles are transported, probably along actin filaments, to specific destinations, in most cases at the tip of the root hair. Here they fuse with the plasma membrane such that the inner leaflet of each vesicle membrane becomes the outer leaflet of the plasma membrane, the outer leaflet becomes the inner leaflet, and the vesicle contents are delivered to the outer surface of the cell. Vesicle-associated proteins take part in growth and maturation processes, including vesicle docking and fusion, and other aspects of trafficking, secretion, membrane rearrangement, and cycling (e.g., RABA4b (Preuss et al. 2004), RPA (Song et al. 2006), ARF1 (Xu and Scheres 2005), the predicted exocyst component EXO70A1 (Synek et al. 2006; Zarsky and Fowler 2008), RHD3 (Zheng et al. 2004), KEULE (Assaad et al. 2001; Assaad 2008). Some protein components of the cell wall, such as EXPANSIN 7 and EXPANSIN 18 (EXP7 and EXP18; Cho and Cosgrove 2002), PROLINE RICH PROTEIN 3 (PRP3; Bernhardt and Tierney 2000), and ARABINOGALACTAN-PROTEIN 30 (AGP30; van Hengel et al. 2004), are probably also transported to the cell surface in vesicles. For cell wall molecules, see Nielsen (2008). It is not clear to what extent mechanisms involved in tip growth take place in the ER or Golgi, in vesicles while in transit, or outside the cell after vesicle fusion events.

Throughout tip growth new material is repeatedly delivered to the growing tip very precisely so that the root hair grows in a straight line, and with an almost constant diameter. The direction of growth is controlled by microtubules (Ketelaar et al. 2002; Sieberer et al. 2005; see Sieberer and Timmers 2008). These are influenced by calcium, which possibly marks the location at the tip where growth most recently took place (Bibikova et al. 1999, see Bibikova and Gilroy 2008). The Armadillo repeat containing kinesin MRH2 is somehow involved in setting up and/or interpreting microtubule structures that guide growth, as demonstrated by mrh2 mutants, which have very wavy hairs (Jones et al. 2006). Growth itself depends on actin dynamics (Ketelaar et al. 2002), and many of the actin regulators discussed earlier contribute to the elongation mechanism, which is also affected by profilin (Ramachandran et al. 2000). RNAi lines with reduced actin-interacting protein 1 (AIP1) levels have reduced turnover of F-actin, resulting in actin bundling near the root hair tip, and short root hairs. Myosin XIK is also required for normal root hair elongation (Ojangu et al. 2007).

The production of a mature hair takes several hours and involves very many reiterations of the precise export of new lipids, polysaccharide, and protein, and the import and recycling of used and left-over materials. A suite of signaling mechanisms coordinate this complex, dynamic system. These involve recurring interactions between small GTPase signaling, phosphoinositide signaling, calcium signaling, and reactive oxygen signaling pathways. The nature and sequence of these interactions are not yet clear, but components of the relevant pathways are being identified, and in some cases functional links between components have been found.

The small GTPase RabA4b acts, probably through the phosphatidylinositol 4-OH kinase *PI-4Kβ1*, to promote the organization of post-Golgi secretory compartments (Preuss et al. 2004, 2006). Other genes with roles in phosphoinositide signaling during root hair tip growth include the inositol 1,4,5-triphosphate 5-phosphatase MRH3 (Jones et al. 2006), COW1 (CAN OF WORMS 1) – a phosphatidylinositol transfer protein (Bohme et al. 2004; Vincent et al. 2005), inositol polyphosphate kinase (IPK2α; Xu et al. 2005b), and a phosphatidylinositol 4-phosphate 5-kinase (*PIP5K3*; Stenzel et al. 2008). An ARF GTPase-activating protein (ARF GAP), ROOT AND POLLEN ARFGAP (RPA), has been identified that plays an important role in maintaining tip growth. RPA loss-of-function mutants have short, isodiametrically expanded root hairs that sometimes branch. RPA is localized at the Golgi, and a possible target of RPA activity, ARF1, is seen at the Golgi and endocytic organelles (Xu and Scheres 2005). RPA activates ARF1 and the ARF1-like protein U5 in an in vitro assay, consistent with the idea that it acts as a regulator of ARF activity (Song et al. 2006). ARF1 activity has been shown to have a knock-on effect on ROP localization (Xu and Scheres 2005). ROPs are present at the tips of growing hairs, disappearing when growth ends (Jones et al. 2002; Molendijk et al. 2001). ROP overexpression produces root hairs much longer than normal (Jones et al. 2002), suggesting that ROP signaling can drive all of the processes required for tip growth. Actin dynamics (Jones et al. 2002) and reactive oxygen signaling via the NADPH oxidase RHD2 (Carol et al. 2005; Jones et al. 2007) are both regulated by ROP activity. RHD2 is required for the establishment of a calcium gradient (Foreman et al. 2003), and Ca^{2+} in turn activates ROS production by RHD2, providing a feedback mechanism that helps to maintain an active growth site (Takeda et al. 2008). ROP signaling also regulates *ICR* (*INTERACTOR OF CONSTITUTIVE ACTIVE ROPS 1*), which encodes a coiled-coil domain scaffold protein that affects root hair growth, but it is not clear which proteins assemble around ICR (Lavy et al. 2007). ROP2 is restricted to a single growth site by the negative ROP regulator, SCN1 (Carol et al. 2005; Parker et al. 2000). Protein phosphorylation cascades are also important for root hair growth. The OXI1/AGC2 oxidative-burst-inducible kinase affects root hair length (Anthony et al. 2004; Rentel et al. 2004), as does the serine/threonine kinase IRE (Oyama et al. 2002). There is also evidence from inhibitor and overexpression studies that mitogen-activated protein kinases play a role in root hair elongation (Samaj et al. 2002).

An intriguing set of proteins, including LRR/EXTENSIN1 (LRX1), LRX2, and COBL9, has extracellular domains that probably reach out into the cell wall. These are candidates for proteins involved in transducing signals between the intracellular and extracellular realms of the cell, but such a role is yet to be demonstrated experimentally. LRX1 and LRX2 encode LRR-extensin proteins (Baumberger et al. 2001, 2003). *lrx1* and *lrx2* mutant phenotypes are suppressed by mutations in the *RHAMNOSE BIOSYNTHESIS 1* (*RHM1*) gene, which encodes a UDP-rhamnose synthase, and affects pectin structure, suggesting that LRX proteins might function in a pectin-related process (Diet et al. 2006).

The length of root hairs is affected by auxin and ethylene, and many mutants in ethylene synthesis, perception, and response, and in auxin transport and response have altered root hair length (Table 1). Root hair length is used by some authors as an assay for auxin content (see Lee and Cho 2008), although the full complement of auxin transport and auxin response genes in root hairs is not known, and the molecular mechanism by which auxin regulates root hair length remains to be discovered.

5 Environmental Regulation of Root Hair Development

Root hairs are important for acquiring immobile nutrients such as phosphorous and potassium (Bates and Lynch 2000a,b; Lynch 2007), and probably play significant roles in the uptake of other nutrients, along with water. There is some evidence that root hairs play a role in detecting nutrient conditions (Shin et al. 2005). Because root hairs form most of the surface area of roots they provide a major interface between plants and the underground environment, where they encounter soil micro- and macrofauna, including root pathogens (such as club root) and symbionts (including nitrogen-fixing bacteria), which in some cases enter roots through root hairs. For *Rhizobium*–legume interaction, see Limpens and Bisseling (2008).

All of these environmental influences can affect root hair development (e.g., *Bacillus megaterium* can promote root hair elongation (Lopez-Bucio et al. 2007)), and both root hair patterning and root hair growth can be affected. The best understood effects of the environment on root hair development are responses to nutrient concentration. When nutrients are sparse the density and length of root hairs both increase. Hair development is regulated in response to many nutrients, including phosphate (Bates and Lynch 1996), iron (Schmidt et al. 2000), manganese, and zinc (Ma et al. 2001). Phosphate has the strongest and best characterized effect. Root hair density on Columbia roots grown in low phosphorus (1 mmol μ^{-3}) is five times greater than on roots grown in high phosphorus (1,000 mmol^{-3}). The number of hair-forming files is increased in low phosphorus from 8 to 12 files, and more of the cells in these files make hairs than on plants grown on high phosphorous (Ma et al. 2001). Hairs are also three times longer on low phosphorus than on high phosphorus (Bates and Lynch 1996). Different nutrients control root hair development by different mechanisms. For example, auxin and ethylene signaling are involved in some responses to iron deficiency but have little effect on responses to low phosphorus (Schmidt and Schikora 2001).

Most nutrient responses involve specific signaling pathways (Lopez-Bucio et al. 2003; Muller and Schmidt 2004; Perry et al. 2007; Zhang et al. 2003). Several genes with roles in responses to phosphate have been identified in genetic screens (e.g., *LOW PHOSPHORUS INSENSITIVE* (*LPI*); Sanchez-Calderon et al. 2006), and a few have been characterized in detail. BHLH32 is a transcription factor that negatively regulates phosphate-starvation-induced processes (Chen et al. 2007). The protein directly interacts with TTG1 and GL3, suggesting possible mechanisms for changes in root hair patterning.

We are beginning to understand how the genes that control epidermal patterning and root hair growth of *Arabidopsis* might function. New frontiers are opening at the interface between these genetic mechanisms and environmental factors. It is clear that, while some environmental responses involve familiar physiological mechanisms, such as auxin signaling, others use other pathways. It is likely to be some time before the full network of influences on root hair development comes into view.

6 Protocol: Rapid Preparation of Transverse Sections of Plant Roots

To analyze the pattern of root epidermal cells, it is typically necessary to obtain transverse sections. Compared with conventional methods, the protocol here is simple and rapid, and it enables one to determine the basic organization of root cell types.

Seedlings are grown on nutrient medium in petri plates. The seedlings are flooded with water, and then transferred to molten 3% agarose solution (made in distilled water). After the agarose hardens, the agarose block is removed from the mold and trimmed to a rectangular shape surrounding the root. A standard razor blade is used to cut thin cross-sections perpendicular to the length of the root. It is important to keep the agarose block wet throughout the procedure. The sections are placed in water as they are cut, and then stained briefly (e.g., 0.05% Toludine blue, pH 4.4). Following rinsing, the sections are mounted on glass slides under a cover slip and viewed under a microscope.

Acknowledgments We apologise to authors whose work could not be included because of space constraints. C.G. thanks J.S. and the editors for their patience, and Piers Hemsley for help with Fig. 2. Research in J.S.'s laboratory is supported by the U.S. National Science Foundation (IOS-0744599 and IOS-0723493).

References

Anthony RG, Henriques R, Helfer A, Meszaros T, Rios G, Testerink C, Munnik T, Deak M, Koncz C, Bogre L (2004) A protein kinase target of a PDK1 signalling pathway is involved in root hair growth in Arabidopsis. *Embo J* 23:572–581

Aoyama T (2008) Phospholipid signaling in root hair development. In: Emons AMC, Ketelaar T (eds) Root hairs: excellent tools for the study of plant molecular cell biology. Springer, Berlin Heidelberg New York. doi:10.1007/7089_2008_1

Arabidopsis Genome Initiative (2000) Analysis of the genome sequence of the flowering plant Arabidopsis thaliana. Nature 408:796–815

Assaad FF (2008) The membrane dynamics of root hair morphogenesis. In: Emons AMC, Ketelaar T (eds) Root hairs: excellent tools for the study of plant molecular cell biology. Springer, Berlin Heidelberg New York. doi:10.1007/7089_2008_2

Assaad FF, Huet Y, Mayer U, Jurgens G (2001) The cytokinesis gene KEULE encodes a Sec1 protein that binds the syntaxin KNOLLE. *J Cell Biol* 152:531–543

Baluska F, Salaj J, Mathur J, Braun M, Jasper F, Samaj J, Chua NH, Barlow PW, Volkmann D (2000) Root hair formation: F-actin-dependent tip growth is initiated by local assembly of profilin-supported F-actin meshworks accumulated within expansin-enriched bulges. *Dev Biol* 227:618–632

Bates TR, Lynch JP (1996) Stimulation of root hair elongation in Arabidopsis thaliana by low phosphorus availability. *Plant Cell Environ* 19:529–538

Bates TR, Lynch JP (2000a) The efficiency of Arabidopsis thaliana (Brassicaceae) root hairs in phosphorus acquisition. *Am J Bot* 87:964–970

Bates TR, Lynch JP (2000b) Plant growth and phosphorus accumulation of wild type and two root hair mutants of Arabidopsis thaliana (Brassicaceae). *Am J Bot* 87:958–963

Baumberger N, Ringli C, Keller B (2001) The chimeric leucine-rich repeat/extensin cell wall protein LRX1 is required for root hair morphogenesis in Arabidopsis thaliana. *Genes Dev* 15:1128–1139

Baumberger N, Steiner M, Ryser U, Keller B, Ringli C (2003) Synergistic interaction of the two paralogous Arabidopsis genes LRX1 and LRX2 in cell wall formation during root hair development. *Plant J* 35:71–81

Berger F, Haseloff J, Schiefelbein J, Dolan L (1998) Positional information in root epidermis is defined during embryogenesis and acts in domains with strict boundaries. *Curr Biol* 8:421–430

Bernhardt C, Tierney ML (2000) Expression of AtPRP3, a proline-rich structural cell wall protein from Arabidopsis, is regulated by cell-type-specific developmental pathways involved in root hair formation. *Plant Physiol* 122:705–714

Bernhardt C, Lee MM, Gonzalez A, Zhang F, Lloyd A, Schiefelbein J (2003) The bHLH genes GLABRA3 (GL3) and ENHANCER OF GLABRA3 (EGL3) specify epidermal cell fate in the Arabidopsis root. *Development* 130:6431–6439

Bernhardt C, Zhao M, Gonzalez A, Lloyd A, Schiefelbein J (2005) The bHLH genes GL3 and EGL3 participate in an intercellular regulatory circuit that controls cell patterning in the Arabidopsis root epidermis. *Development* 132:291–298

Bibikova T, Gilroy S (2008) Calcium in root hair growth. In: Emons AMC, Ketelaar T (eds) Root hairs: excellent tools for the study of plant molecular cell biology. Springer, Berlin Heidelberg New York. doi:10.1007/7089_2008_3

Bibikova TN, Jacob T, Dahse I, Gilroy S (1998) Localized changes in apoplastic and cytoplasmic pH are associated with root hair development in Arabidopsis thaliana. *Development* 125:2925–2934

Bibikova TN, Blancaflor EB, Gilroy S (1999) Microtubules regulate tip growth and orientation in root hairs of Arabidopsis thaliana. *Plant J* 17:657–665

Bohme K, Li Y, Charlot F, Grierson C, Marrocco K, Okada K, Laloue M, Nogue F (2004) The Arabidopsis COW1 gene encodes a phosphatidylinositol transfer protein essential for root hair tip growth. *Plant J* 40:686–698

Bown L, Kusaba S, Goubet F, Codrai L, Dale AG, Zhang Z, Yu X, Morris K, Ishii T, Evered C et al (2007) The ectopically parting cells 1–2 (epc1–2) mutant exhibits an exaggerated response to abscisic acid. *J Exp Bot* 58:1813–1823

Bunning E (1951) Uber die Differenzierungsvorgange in der Cruciferenwurzel. *Planta* 39:126–153

Cao XF, Linstead P, Berger F, Kieber J, Dolan L (1999) Differential ethylene sensitivity of epidermal cells is involved in the establishment of cell pattern in the Arabidopsis root. *Physiol Plant* 106:311–317

Caro E, Castellano MM, Gutierrez C (2007) A chromatin link that couples cell division to root epidermis patterning in Arabidopsis. *Nature* 447:213–217

Carol RJ, Takeda S, Linstead P, Durrant MC, Kakesova H, Derbyshire P, Drea S, Zarsky V, Dolan L (2005) A RhoGDP dissociation inhibitor spatially regulates growth in root hair cells. *Nature* 438:1013–1016

Cernac A, Lincoln C, Lammer D, Estelle M (1997) The SAR1 gene of Arabidopsis acts downstream of the AXR1 gene in auxin response. *Development* 124:1583–1591

Chen ZH, Nimmo GA, Jenkins GI, Nimmo HG (2007) BHLH32 modulates several biochemical and morphological processes that respond to Pi starvation in Arabidopsis. *Biochem J* 405:191–198

Cho HT, Cosgrove DJ (2002) Regulation of root hair initiation and expansin gene expression in Arabidopsis. *Plant Cell* 14:3237–3253

Clowes FAL (2000) Pattern in root meristem development in angiosperms. *New Phytol* 146:83–94

Cormack RGH (1935) Investigations on the development of root hairs. *New Phytol* 34:30–54

Costa S, Dolan L (2003) Epidermal patterning genes are active during embryogenesis in Arabidopsis. *Development* 130:2893–2901

Costa S, Shaw P (2005) Chromatin organization and cell fate switch respond to positional information in Arabidopsis. Nature 439:493–496

Cutter EG (1978). The epidermis. In: Plant anatomy. Clowes & Sons, London, pp 94–106

Deeks MJ, Cvrckova F, Machesky LM, Mikitova V, Ketelaar T, Zarsky V, Davies B, Hussey PJ (2005) Arabidopsis group Ie formins localize to specific cell membrane domains, interact with actin-binding proteins and cause defects in cell expansion upon aberrant expression. *New Phytol* 168:529–540

Deeks MJ, Rodrigues C, Dimmock S, Ketelaar T, Maciver SK, Malho R, Hussey PJ (2007) Arabidopsis CAP1 – a key regulator of actin organisation and development. *J Cell Sci* 120:2609–2618

del Pozo JC, Dharmasiri S, Hellmann H, Walker L, Gray WM, Estelle M (2002) AXR1-ECR1-dependent conjugation of RUB1 to the Arabidopsis Cullin AtCUL1 is required for auxin response. *Plant Cell* 14:421–433

Desbrosses G, Josefsson C, Rigas S, Hatzopoulos P, Dolan L (2003) AKT1 and TRH1 are required during root hair elongation in Arabidopsis. *J Exp Bot* 54:781–788

Devaiah BN, Karthikeyan AS, Raghothama KG (2007) WRKY75 transcription factor is a modulator of phosphate acquisition and root development in Arabidopsis. *Plant Physiol* 143:1789–1801

Di Cristina M, Sessa G, Dolan L, Linstead P, Baima S, Ruberti I, Morelli G (1996) The Arabidopsis Athb-10 (GLABRA2) is an HD-Zip protein required for regulation of root hair development. *Plant J* 10:393–402

Diet A, Brunner S, Ringli C (2004) The enl mutants enhance the lrx root hair mutant phenotype of Arabidopsis thaliana. *Plant Cell Physiol* 45:734–741

Diet A, Link B, Seifert GJ, Schellenberg B, Wagner U, Pauly M, Reiter WD, Ringli C (2006) The Arabidopsis root hair cell wall formation mutant lrx1 is suppressed by mutations in the RHM1 gene encoding a UDP-L-rhamnose synthase. *Plant Cell* 18:1630–1641

Dolan L, Janmaat K, Willemsen V, Linstead P, Poethig S, Roberts K, Scheres B (1993) Cellular organisation of the Arabidopsis thaliana root. *Development* 119:71–84

Dolan L, Duckett C, Grierson C, Linstead P, Schneider K, Lawson E, Dean C, Poethig RS, Roberts K (1994) Clonal relations and patterning in the root epidermis of Arabidopsis. *Development* 120:2465–2474

Emons AMC, Ketelaar T (2008) Intracellular organization: a prerequisite for root hair elongation and cell wall deposition. In: Emons AMC, Ketelaar T (eds) Root hairs: excellent tools for the study of plant molecular cell biology. Springer, Berlin Heidelberg New York. doi:10.1007/7089_2008_4

Esch JJ, Chen M, Sanders M, Hillestad M, Ndkium S, Idelkope B, Neizer J, Marks MD (2003) A contradictory GLABRA3 allele helps define gene interactions controlling trichome development in Arabidopsis. *Development* 130:5885–5894

Favery B, Ryan E, Foreman J, Linstead P, Boudonck K, Steer M, Shaw P, Dolan L (2001) KOJAK encodes a cellulose synthase-like protein required for root hair cell morphogenesis in Arabidopsis. *Genes Dev* 15:79–89

Fischer U, Ikeda Y, Ljung K, Serralbo O, Singh M, Heidstra R, Palme K, Scheres B, Grebe M (2006) Vectorial information for Arabidopsis planar polarity is mediated by combined AUX1, EIN2, and GNOM activity. *Curr Biol* 16:2143–2149

Foreman J, Demidchik V, Bothwell JH, Mylona P, Miedema H, Torres MA, Linstead P, Costa S, Brownlee C, Jones JD et al (2003) Reactive oxygen species produced by NADPH oxidase regulate plant cell growth. *Nature* 422:442–446

Freshour G, Clay RP, Fuller MS, Albersheim P, Darvill AG, Hahn MG (1996) Developmental and tissue-specific structural alterations of the cell-wall polysaccharides of *Arabidopsis thaliana* roots. *Plant Physiol* 110:1413–1429

Fukaki H, Tameda S, Masuda H, Tasaka M (2002) Lateral root formation is blocked by a gain-of-function mutation in the SOLITARY-ROOT/IAA14 gene of Arabidopsis. *Plant J* 29:153–168

Galway ME, Masucci JD, Lloyd AM, Walbot V, Davis RW, Schiefelbein JW (1994) The TTG gene is required to specify epidermal cell fate and cell patterning in the Arabidopsis root. *Dev Biol* 166:740–754

Gendreau E, Traas J, Desnos T, Grandjean O, Caboche M, Hofte H (1997) Cellular basis of hypocotyl growth in Arabidopsis thaliana. *Plant Physiol* 114:295–305

Gilliland LU, Kandasamy MK, Pawloski LC, Meagher RB (2002) Both vegetative and reproductive actin isovariants complement the stunted root hair phenotype of the Arabidopsis act2–1 mutation. *Plant Physiol* 130:2199–2209

Grebe M, Xu J, Mobius W, Ueda T, Nakano A, Geuze HJ, Rook MB, Scheres B (2003) Arabidopsis sterol endocytosis involves actin-mediated trafficking via ARA6-positive early endosomes. *Curr Biol* 13:1378–1387

Grierson C, Schiefelbein J (2002) Root hairs. In: Somerville C, Meyerowitz EM (eds) The Arabidopsis book. .American Society of Plant Biologists, Rockville, doi/10.1199/tab.0060. http://www.aspb.org/publications/arabidopsis/

Haberlandt G (1887) Ueber die Lage des Kernes in sich entwickelnden Pflanzenzellen. Berichte der deutschen botanischen. *Gesellschaft* 5:205–212

Hemsley PA, Kemp AC, Grierson CS (2005) The TIP GROWTH DEFECTIVE1 S-acyl transferase regulates plant cell growth in Arabidopsis. *Plant Cell* 17:2554–2563

Hung CY, Lin Y, Zhang M, Pollock S, Marks MD, Schiefelbein J (1998) A common position-dependent mechanism controls cell-type patterning and GLABRA2 regulation in the root, hypocotyl epidermis of Arabidopsis. *Plant Physiol* 117:73–84

Jones MA, Raymond MJ,Smirnoff N (2006) Analysis of the root-hair morphogenesis transcriptome reveals the molecular identity of six genes with roles in root-hair development in Arabidopsis. *Plant J* 45:83–100

Jones MA, Raymond MJ, Yang Z, Smirnoff N (2007) NADPH oxidase-dependent reactive oxygen species formation required for root hair growth depends on ROP GTPase. *J Exp Bot* 58:1261–1270

Jones MA, Shen JJ, Fu Y, Li H, Yang Z, Grierson CS (2002) The Arabidopsis Rop2 GTPase is a positive regulator of both root hair initiation and tip growth. *Plant Cell* 14:763–776

Ketelaar T, Faivre-Moskalenko C, Esseling JJ, de Ruijter NC, Grierson CS, Dogterom M, Emons AM (2002) Positioning of nuclei in Arabidopsis root hairs: an actin-regulated process of tip growth. *Plant Cell* 14:2941–2955

Ketelaar T, Allwood EG, Hussey PJ (2007) Actin organization and root hair development are disrupted by ethanol-induced overexpression of Arabidopsis actin interacting protein 1 (AIP1). *New Phytol* 174:57–62

Kim DW, Lee SH, Choi SB, Won SK, Heo YK, Cho M, Park YI, Cho HT (2006) Functional conservation of a root hair cell-specific cis-element in angiosperms with different root hair distribution patterns. *Plant Cell* 18:2958–2970

Kirik V, Simon M, Huelskamp M, Schiefelbein J (2004) The ENHANCER OF TRY AND CPC1 gene acts redundantly with TRIPTYCHON and CAPRICE in trichome and root hair cell patterning in Arabidopsis. *Dev Biol* 268:506–513

Knox K, Grierson CS, Leyser O (2003) AXR3 and SHY2 interact to regulate root hair development. *Development* 130:5769–5777

Koornneef M (1981) The complex syndrome of ttg mutants. *Arabidopsis Inf Serv* 18:45–51

Koornneef M, Dellaert SWM, van der Veen JH (1982) EMS- and radiation-induced mutation frequencies at individual loci in *Arabidopsis thaliana* (L.) Heynh. *Mutat Res* 93:109–123

Koshino-Kimura Y, Wada T, Tachibana T, Tsugeki R, Ishiguro S, Okada K (2005) Regulation of CAPRICE transcription by MYB proteins for root epidermis differentiation in Arabidopsis. *Plant Cell Physiol* 46:817–826

Kurata T, Ishida T, Kawabata-Awai C, Noguchi M, Hattori S, Sano R, Nagasaka R, Tominaga R, Koshino-Kimura Y, Kato T et al (2005) Cell-to-cell movement of the CAPRICE protein in Arabidopsis root epidermal cell differentiation. *Development* 132:5387–5398

Kwak SH, Schiefelbein J (2006) The role of the SCRAMBLED receptor-like kinase in patterning the Arabidopsis root epidermis. Dev Biol doi:10.1016/j.ydbio.2006.09.009

Larkin JC, Brown ML, Schiefelbein J (2003a) How do cells know what they want to be when they grow up? Lessons from epidermal patterning in Arabidopsis. *Annu Rev Plant Physiol Plant Mol Biol* 54:403–430

Larkin JC, Brown ML, Schiefelbein J (2003b) How do cells know what they want to be when they grow up? Lessons from epidermal patterning in Arabidopsis. *Annu Rev Plant Biol* 54:403–430

Larkin JC, Marks MD, Nadeau J, Sack F (1997) Epidermal cell fate and patterning in leaves. *Plant Cell* 9:1109–1120

Lavy M, Bloch D, Hazak O, Gutman I, Poraty L, Sorek N, Sternberg H, Yalovsky S (2007) A Novel ROP/RAC effector links cell polarity, root-meristem maintenance, and vesicle trafficking. *Curr Biol* 17:947–952

Leavitt RG (1904) Trichomes of the root in vascular cryptograms and angiosperms. *Proc Boston Soc Nat Hist* 31:273–313

Lee MM, Schiefelbein J (1999) WEREWOLF, a MYB-related protein in Arabidopsis, is a position-dependent regulator of epidermal cell patterning. *Cell* 99:473–483

Lee MM, Schiefelbein J (2002) Cell pattern in the Arabidopsis root epidermis determined by lateral inhibition with feedback. *Plant Cell* 14:611–618

Lee SH, Cho H.-T. (2008) Auxin and root hair morphogenesis. In: Emons AMC, Ketelaar T (eds) Root hairs: excellent tools for the study of plant molecular cell biology. Springer, Berlin Heidelberg New York. doi:10.1007/7089_2008_16

Leyser HM, Pickett FB, Dharmasiri S, Estelle M (1996) Mutations in the AXR3 gene of Arabidopsis result in altered auxin response including ectopic expression from the SAUR-AC1 promoter. *Plant J* 10:403–413

Limpens E, Bisseling T (2008) Nod factor signal transduction in the Rhizobium-Legume symbiosis. In: Emons AMC, Ketelaar T (eds) Root hairs: excellent tools for the study of plant molecular cell biology. Springer, Berlin Heidelberg New York. doi:10.1007/7089_2008_10

Lin Y, Schiefelbein J (2001) Embryonic control of epidermal cell patterning in the root and hypocotyl of Arabidopsis. *Development* 128:3697–3705

Lopez-Bucio J, Cruz-Ramirez A, Herrera-Estrella L (2003) The role of nutrient availability in regulating root architecture. *Curr Opin Plant Biol* 6:280–287

Lopez-Bucio J, Hernandez-Abreu E, Sanchez-Calderon L, Perez-Torres A, Rampey RA, Bartel B, Herrera-Estrella L (2005) An auxin transport independent pathway is involved in phosphate stress-induced root architectural alterations in Arabidopsis. Identification of BIG as a mediator of auxin in pericycle cell activation. *Plant Physiol* 137:681–691

Lopez-Bucio J, Campos-Cuevas JC, Hernandez-Calderon E, Velasquez-Becerra C, Farias-Rodriguez R, Macias-Rodriguez LI, Valencia-Cantero E (2007) Bacillus megaterium rhizobacteria promote growth and alter root-system architecture through an auxin- and ethylene-independent signaling mechanism in Arabidopsis thaliana. *Mol Plant Microbe Interact* 20:207–217

Lynch JP (2007) Roots of the second green revolution. *Aust J Bot* 55:493–512

Ma Z, Bielenberg DG, Brown KM, Lynch JP (2001) Regulation of root hair density by phosphorus availability in Arabidopsis thaliana. *Plant Cell Environ* 24:459–467

Masucci JD, Schiefelbein JW (1994) The rhd6 Mutation of Arabidopsis thaliana Alters Root-Hair Initiation through an Auxin- and Ethylene-Associated Process. *Plant Physiol* 106:1335–1346

Masucci JD, Schiefelbein JW (1996) Hormones act downstream of TTG and GL2 to promote root hair outgrowth during epidermis development in the Arabidopsis root. *Plant Cell* 8:1505–1517

Masucci JD, Rerie WG, Foreman DR, Zhang M, Galway ME, Marks MD, Schiefelbein JW (1996) The homeobox gene GLABRA2 is required for position-dependent cell differentiation in the root epidermis of Arabidopsis thaliana. *Development* 122:1253–1260

Menand B, Yi K, Jouannic S, Hoffmann L, Ryan E, Linstead P, Schaefer DG, Dolan L (2007) An ancient mechanism controls the development of cells with a rooting function in land plants. *Science* 316:1477–1480

Miura K, Rus A, Sharkhuu A, Yokoi S, Karthikeyan AS, Raghothama KG, Baek D, Koo YD, Jin JB, Bressan RA et al (2005) The Arabidopsis SUMO E3 ligase SIZ1 controls phosphate deficiency responses. *Proc Natl Acad Sci USA* 102:7760–7765

Molendijk AJ, Bischoff F, Rajendrakumar CS, Friml J, Braun M, Gilroy S, Palme K (2001) Arabidopsis thaliana Rop GTPases are localized to tips of root hairs and control polar growth. *Embo J* 20:2779–2788

Muller M, Schmidt W (2004) Environmentally induced plasticity of root hair development in Arabidopsis. *Plant Physiol* 134:409–419

Nagpal P, Walker LM, Young JC, Sonawala A, Timpte C, Estelle M, Reed JW (2000) AXR2 encodes a member of the Aux/IAA protein family. *Plant Physiol* 123:563–574

Nielsen E (2008) Plant cell wall biogenesis during tip growth in root hair cells. In: Emons AMC, Ketelaar T (eds) Root hairs: excellent tools for the study of plant molecular cell biology. Springer, Berlin Heidelberg New York. doi:10.1007/7089_2008_11

Ohashi Y, Oka A, Rodrigues-Pousada R, Possenti M, Ruberti I, Morelli G, Aoyama T (2003) Modulation of phospholipid signaling by GLABRA2 in root-hair pattern formation. *Science* 300:1427–1430

Ojangu EL, Jarve K, Paves H, Truve E (2007) Arabidopsis thaliana myosin XIK is involved in root hair as well as trichome morphogenesis on stems and leaves. *Protoplasma* 230:193–202

Oyama T, Shimura Y, Okada K (2002) The IRE gene encodes a protein kinase homologue and modulates root hair growth in Arabidopsis. *Plant J* 30:289–299

Parker JS, Cavell AC, Dolan L, Roberts K, Grierson CS (2000) Genetic interactions during root hair morphogenesis in Arabidopsis. *Plant Cell* 12:1961–1974

Perry P, Linke B, Schmidt W (2007) Reprogramming of root epidermal cells in response to nutrient deficiency. *Biochem Soc Trans* 35:161–163

Pesch M, Hulskamp M (2004) Creating a two-dimensional pattern de novo during Arabidopsis trichome and root hair initiation. *Curr Opin Genet Dev* 14:422–427

Pitts RJ, Cernac A, Estelle M (1998) Auxin and ethylene promote root hair elongation in Arabidopsis. *Plant J* 16:553–560

Preuss ML, Schmitz AJ, Thole JM, Bonner HK, Otegui MS, Nielsen E (2006) A role for the RabA4b effector protein PI-4Kbeta1 in polarized expansion of root hair cells in Arabidopsis thaliana. *J Cell Biol* 172:991–998

Preuss ML, Serna J, Falbel TG, Bednarek SY, Nielsen E (2004) The Arabidopsis Rab GTPase RabA4b localizes to the tips of growing root hair cells. *Plant Cell* 16:1589–1603

Procissi A, Guyon A, Pierson ES, Giritch A, Knuiman B, Grandjean O, Tonelli C, Derksen J, Pelletier G, Bonhomme S (2003) KINKY POLLEN encodes a SABRE-like protein required for tip growth in Arabidopsis and conserved among eukaryotes. *Plant J* 36:894–904

Ramachandran S, Christensen HE, Ishimaru Y, Dong CH, Chao-Ming W, Cleary AL, Chua NH (2000) Profilin plays a role in cell elongation, cell shape maintenance, and flowering in Arabidopsis. *Plant Physiol* 124:1637–1647

Rentel MC, Lecourieux D, Ouaked F, Usher SL, Petersen L, Okamoto H, Knight H, Peck SC, Grierson CS, Hirt H et al (2004) OXI1 kinase is necessary for oxidative burst-mediated signalling in Arabidopsis. *Nature* 427:858–861

Rerie WG, Feldmann KA, Marks MD (1994) The GLABRA2 gene encodes a homeo domain protein required for normal trichome development in Arabidopsis. *Genes Dev* 8:1388–1399
Rigas S, Debrosses G, Haralampidis K, Vicente-Agullo F, Feldmann KA, Grabov A, Dolan L, Hatzopoulos P (2001) TRH1 encodes a potassium transporter required for tip growth in Arabidopsis root hairs. *Plant Cell* 13:139–151
Ringli C, Baumberger N, Diet A, Frey B, Keller B (2002) ACTIN2 is essential for bulge site selection and tip growth during root hair development of Arabidopsis. *Plant Physiol* 129:1464–1472
Ringli C, Baumberger N, Keller B (2005) The Arabidopsis root hair mutants der2-der9 are affected at different stages of root hair development. *Plant Cell Physiol* 46:1046–1053
Rosti J, Barton CJ, Albrecht S, Dupree P, Pauly M, Findlay K, Roberts K, Seifert GJ (2007) UDP-glucose 4-epimerase isoforms UGE2 and UGE4 cooperate in providing UDP-galactose for cell wall biosynthesis and growth of Arabidopsis thaliana. *Plant Cell* 19:1565–1579
Ruzicka K, Ljung K, Vanneste S, Podhorska R, Beeckman T, Friml J, Benkova E (2007) Ethylene regulates root growth through effects on auxin biosynthesis and transport-dependent auxin distribution. *Plant Cell* 19:2197–2212
Ryu KH, Kang YH, Park YH, Hwang I, Schiefelbein J, Lee MM (2005) The WEREWOLF MYB protein directly regulates CAPRICE transcription during cell fate specification in the Arabidopsis root epidermis. *Development* 132:4765–4775
Samaj J, Ovecka M, Hlavacka A, Lecourieux F, Meskiene I, Lichtscheidl I et al (2002) Involvement of the mitogen-activated protein kinase SIMK in regulation of root hair tip-growth. *Embo J* 21:3296–3306
Sanchez-Calderon L, Lopez-Bucio J, Chacon-Lopez A, Gutierrez-Ortega A, Hernandez-Abreu E, Herrera-Estrella L (2006) Characterization of low phosphorus insensitive mutants reveals a cross-talk between low phosphorus-induced determinate root development and the activation of genes involved in the adaptation of Arabidopsis to phosphorus deficiency. *Plant Physiol* 140:879–889
Santelia D, Vincenzetti V, Azzarello E, Bovet L, Fukao Y, Duchtig P, Mancuso S, Martinoia E, Geisler M (2005) MDR-like ABC transporter AtPGP4 is involved in auxin-mediated lateral root and root hair development. *FEBS Lett* 579:5399–5406
Schellmann S, Hulskamp M (2005) Epidermal differentiation: trichomes in Arabidopsis as a model system. *Int J Dev Biol* 49:579–584
Schellmann S, Schnittger A, Kirik V, Wada T, Okada K, Beermann A, Thumfahrt J, Jurgens G, Hulskamp M (2002) TRIPTYCHON and CAPRICE mediate lateral inhibition during trichome and root hair patterning in Arabidopsis. *Embo J* 21:5036–5046
Scheres B, Wlkenfelt H, Willemsen V, Terlouw M, Lawson E, Dean C, Weisbeek P (1994) Embryonic origin of the Arabidopsis primary root and root meristem initials. *Development* 120:2475–2487
Schiefelbein J (2003a) Cell-fate specification in the epidermis: a common patterning mechanism in the root and shoot. *Curr Opin Plant Biol* 6:74–78
Schiefelbein J (2003b) Cell-fate specification in the epidermis: a common patterning mechanism in the root and shoot. *Curr Opin Plant Biol* 6:74–78
Schiefelbein JW, Somerville C (1990) Genetic Control of Root Hair Development in Arabidopsis thaliana. *Plant Cell* 2:235–243
Schiefelbein J, Lee MM (2006) A novel regulatory circuit specifies cell fate in the Arabidopsis root epidermis. *Physiol Plant* 126:503–510
Schmidt W, Schikora A (2001) Different pathways are involved in phosphate and iron stress-induced alterations of root epidermal cell development. *Plant Physiol* 125:2078–2084
Schmidt W, Tittel J, Schikora A (2000) Role of hormones in the induction of iron deficiency responses in Arabidopsis roots. *Plant Physiol* 122:1109–1118
Seifert GJ, Barber C, Wells B, Dolan L, Roberts K (2002) Galactose biosynthesis in Arabidopsis: genetic evidence for substrate channeling from UDP-D-galactose into cell wall polymers. *Curr Biol* 12:1840–1845
Shi H, Zhu JK (2002) SOS4, a pyridoxal kinase gene, is required for root hair development in Arabidopsis. *Plant Physiol* 129:585–593
Shin R, Berg RH, Schachtman DP (2005) Reactive oxygen species and root hairs in Arabidopsis root respond to nitrogen, phosphorus, and potassium deficiency. *Plant Cell Physiol* 46:1350–1359

Shiu SH, Bleecker AB (2001) Receptor-like kinases from Arabidopsis form a monophyletic gene family related to animal receptor kinases. *Proc Natl Acad Sci USA* 98:10763–10768

Shiu SH, Bleecker AB (2003) Expansion of the receptor-like kinase/Pelle gene family and receptor-like proteins in Arabidopsis. *Plant Physiol* 132:530–543

Sieberer BJ, Timmers ACJ (2008) Microtubules in Plant Root Hairs and Their Role in Cell Polarity and Tip Growth. In: Emons AMC, Ketelaar T (eds) Root hairs: excellent tools for the study of plant molecular cell biology. Springer, Berlin Heidelberg New York. doi:10.1007/7089_2008_13

Sieberer BJ, Ketelaar T, Esseling JJ, Emons AM (2005) Microtubules guide root hair tip growth. *New Phytol* 167:711-719

Simon M, Lee MM, Lin Y, Gish L, Schiefelbein J (2007) Distinct and overlapping roles of single-repeat MYB genes in root epidermal patterning. *Dev Biol* 311:566–578

Sinnot EW, Bloch R (1939) Cell polarity and the differentiation of root hairs. *Proc Natl Acad Sci USA* 25:248–252

Song XF, Yang CY, Liu J, Yang WC (2006) RPA, a class II ARFGAP protein, activates ARF1 and U5 and plays a role in root hair development in Arabidopsis. *Plant Physiol* 141:966–976

Stenzel I, Ischebeck T, Konig S, Holubowska A, Sporysz M, Hause B, Heilmann I (2008) The type B Phosphatidylinositol-4-Phosphate 5-Kinase 3 is essential for root hair formation in Arabidopsis thaliana. Plant Cell 20:124–141

Stepanova AN, Yun J, Likhacheva AV, Alonso JM (2007) Multilevel interactions between ethylene and auxin in Arabidopsis roots. *Plant Cell* 19:2169–2185

Sugimoto-Shirasu K, Stacey NJ, Corsar J, Roberts K, McCann MC (2002) DNA topoisomerase VI is essential for endoreduplication in Arabidopsis. *Curr Biol* 12:1782–1786

Sugimoto-Shirasu K, Roberts GR, Stacey NJ, McCann MC, Maxwell A, Roberts K (2005) RHL1 is an essential component of the plant DNA topoisomerase VI complex and is required for ploidy-dependent cell growth. *Proc Natl Acad Sci USA* 102:18736–18741

Swarup R, Perry P, Hagenbeek D, Van Der Straeten D, Beemster GT, Sandberg G, Bhalerao R, Ljung K, Bennett MJ (2007) Ethylene upregulates auxin biosynthesis in Arabidopsis seedlings to enhance inhibition of root cell elongation. *Plant Cell* 19:2186–2196

Synek L, Schlager N, Elias M, Quentin M, Hauser MT, Zarsky V (2006) AtEXO70A1, a member of a family of putative exocyst subunits specifically expanded in land plants, is important for polar growth and plant development. *Plant J* 48:54–72

Takeda S, Gapper C, Kaya H, Bell E, Kuchitsu K, Dolan L (2008) Local positive feedback regulation determines cell shape in root hair cells. *Science* 319:1241–1244

Tanimoto M, Jowett J, Stirnberg P, Rouse D, Leyser O (2007) pax1–1 partially suppresses gain-of-function mutations in Arabidopsis AXR3/IAA17. *BMC Plant Biol* 7:20

Torii KU (2004) Leucine-rich repeat receptor kinases in plants: structure, function, and signal transduction pathways. *Int Rev Cytol* 234:1–46

Ueda M, Koshino-Kimura Y, Okada K (2005) Stepwise understanding of root development. *Curr Opin Plant Biol* 8:71–76

van Hengel AJ, Barber C, Roberts K (2004) The expression patterns of arabinogalactan-protein AtAGP30 and GLABRA2 reveal a role for abscisic acid in the early stages of root epidermal patterning. *Plant J* 39:70–83

Vicente-Agullo F, Rigas S, Desbrosses G, Dolan L, Hatzopoulos P, Grabov A (2004) Potassium carrier TRH1 is required for auxin transport in Arabidopsis roots. *Plant J* 40:523–535

Vincent P, Chua M, Nogue F, Fairbrother A, Mekeel H, Xu Y, Allen N, Bibikova TN, Gilroy S, Bankaitis VA (2005) A Sec14p-nodulin domain phosphatidylinositol transfer protein polarizes membrane growth of Arabidopsis thaliana root hairs. *J Cell Biol* 168:801–812

Wada T, Tachibana T, Shimura Y, Okada K (1997) Epidermal cell differentiation in Arabidopsis determined by a Myb homolog, CPC. *Science* 277:1113–1116

Wada T, Kurata T, Tominaga R, Koshino-Kimura Y, Tachibana T, Goto K, Marks MD, Shimura Y, Okada K (2002) Role of a positive regulator of root hair development, CAPRICE, in Arabidopsis root epidermal cell differentiation. *Development* 129:5409–5419

Walker AR, Davison PA, Bolognesi-Winfield AC, James CM, Srinivasan N, Blundell TL, Esch JJ, Marks MD, Gray JC (1999) The TRANSPARENT TESTA GLABRA1 locus, which regulates

trichome differentiation and anthocyanin biosynthesis in Arabidopsis, encodes a WD40 repeat protein. *Plant Cell* 11:1337–1350
Wang H, Lockwood SK, Hoeltzel MF, Schiefelbein JW (1997) The ROOT HAIR DEFECTIVE3 gene encodes an evolutionarily conserved protein with GTP-binding motifs and is required for regulated cell enlargement in Arabidopsis. *Genes Dev* 11:799–811
Wei N, Kwok SF, von Arnim AG, Lee A, McNellis TW, Piekos B, Deng XW (1994) Arabidopsis COP8, COP10, and COP11 genes are involved in repression of photomorphogenic development in darkness. *Plant Cell* 6:629–643
Xu J, Scheres B (2005) Dissection of Arabidopsis ADP-RIBOSYLATION FACTOR 1 function in epidermal cell polarity. *Plant Cell* 17:525–536
Xu CR, Liu C, Wang YL, Li LC, Chen WQ, Xu ZH, Bai SN (2005a) Histone acetylation affects expression of cellular patterning genes in the Arabidopsis root epidermis. *Proc Natl Acad Sci USA* 102:14469–14474
Xu J, Brearley CA, Lin WH, Wang Y, Ye R, Mueller-Roeber B, Xu ZH, Xue HW (2005b) A role of Arabidopsis inositol polyphosphate kinase, AtIPK2alpha, in pollen germination and root growth. *Plant Physiol* 137:94–103
Zarsky V, Fowler J (2008) ROP (Rho-related protein from Plants) GTPases for spatial control of root hair morphogenesis. In: Emons AMC, Ketelaar T (eds) Root hairs: excellent tools for the study of plant molecular cell biology. Springer, Berlin Heidelberg New York. doi:10.1007/7089_2008_14
Zhang YJ, Lynch JP, Brown KM (2003) Ethylene and phosphorus availability have interacting yet distinct effects on root hair development. *J Exp Bot* 54:2351–2361
Zheng H, Kunst L, Hawes C, Moore I (2004) A GFP-based assay reveals a role for RHD3 in transport between the endoplasmic reticulum and Golgi apparatus. *Plant J* 37:398–414

Intracellular Organization: A Prerequisite for Root Hair Elongation and Cell Wall Deposition

A.M.C. Emons(✉) and T. Ketelaar

Abstract Cell growth requires not only production of matter, but in addition, the targeting, transport, and delivery of this matter to the site of cell expansion. Thus, a proper organization of cell structure, the cytoarchitecture, is a necessity for cell elongation. The actual process of cell growth in a cell under turgor pressure is Golgi vesicle membrane insertion into the plasma membrane and, at the same time, discharge of its contents into the existing cell wall at the site of wall expansion. If one of these prerequisites is missing, growth will not occur. Thus, the Golgi vesicle is the unit of cell growth. The tip-growing cell with robust cell expansion at a defined site is a model system "par excellence" to study this process. In this chapter, we discuss the so-called tip-growth unit, i.e., the assemblage of nucleus, endoplasmic reticulum, polysomes, Golgi bodies, Golgi vesicles, exocytosis machinery, clathrin-coated vesicles, endosomes, and mitochondria that specifically accumulate in the (sub)apical region of tip-growing root hairs, all working in concert to enable apical growth. The last paragraph of this chapter reviews methods used for the visualization of cellulose microfibrils.

1 Introduction: The "Tip-Growth Unit" Concept

Root hairs have become model cells for (plant) cell growth studies since cell growth occurs exclusively, rapidly, and robustly at one end of the cell, the tip. Virtually all plant cells exhibit polar growth, with elongation in one orientation favored over elongation in other orientations, but tip-growing cells show this phenomenon in extremis. Since plant cells do not move within the plant body, cell division and cell elongation determine the overall shape and structure of a plant. Knowledge about localized growth, such as that occurring at root hair tips, can provide insight into

A.M.C. Emons
Laboratory of Plant Cell Biology, Wageningen University, Arboretumlaan 4, 6703 BD Wageningen, The Netherlands
Theory of Biomolecular Matter, FOM Institute AMOLF, Kruislaan 407, Amsterdam, The Netherlands
annemie.emons@wur.nl

Plant Cell Monogr, doi:10.1007/7089_2008_04

the sophisticated mechanisms that multicellular organisms, plants, animals, and fungi, employ to control growth direction.

Light microscopically visible root hair development starts with the bulging of the outer wall of a predestined root epidermal cell, a trichoblast. *Arabidopsis thaliana* has cell files with trichoblasts and atrichoblasts (Dolan et al. 1993), but in legumes studied so far all root epidermal cells appear to have the potential to form root hairs, as observed for *Vicia sativa* (de Ruijter et al. 1998), *Medicago truncatula* (Sieberer and Emons 2000), and *Lotus japonicus* (Sieberer et al. 2005). Thus all root epidermal cells in these legumes are trichoblasts. The bulge emerges when root epidermal cell elongation stops (Ketelaar and Emons 2001), and occurs at a species-specific location, initially indicated by a movement of the nucleus towards this site of emergence (Miller et al. 2000). Although the precise mechanism for this process is currently unknown, the cytoskeleton is clearly involved. During bulge formation the cell wall structure changes, although these changes have been poorly studied (Miller et al. 2000). After bulge formation, exocytotic vesicles start accumulating in the bulge, and this coincides with the stage at which the tip-growth unit assembles and becomes functional (Miller et al. 1997). We define the "tip-growth unit" as the combined structures localized in the apex and subapex of growing root hairs and that function in the tip-growth process.

Polarization of the cytoplasm is essential for fast, local, tip-focused growth: a tip-focused cellular organization directs, delivers, targets, and docks cell wall production cargo to the hair tip, culminating in exocytosis. At the same time, lipids and proteins are being supplied to the elongating plasma membrane, and polysaccharides and glycoproteins to the new cell wall (see Nielsen 2008). If the cell is under turgor pressure and takes up water, the existing cell wall extends. Before exocytosis can take place, exocytotic vesicles need to be brought to, and maintained at, the right place in the cell. In both of these processes, actin filaments (see Ketelaar and Emons 2008) and microtubules (see Sieberer and Timmers 2008) play central roles. Some of the proteins embedded in the vesicle membrane are ion channels that play important roles in the regulation of the exocytotic process (see Zarsky and Fowler 2008). Calcium ions function in the insertion of the vesicular membrane into the existing plasma membrane (see Bibikova and Gilroy 2008) and probably also in targeting the actin cytoskeleton (see Ketelaar and Emons 2008) as well as keeping the cell wall flexible at the site of cell extension (see de Keijzer et al. 2008).

Plant cell growth is an example of a typical homeostatic process. The constant renewal of the plasma membrane locally produces a membrane domain with specific proteins both within it and attached to it, such as the calcium ion channels (see Bibikova and Gilroy 2008) needed for exocytosis to continue. This phenomenon results in a specific local environment for the (actin) cytoskeleton and its binding proteins that subsequently sustain the delivery of new vesicles (see Ketelaar and Emons 2008). At the same time, the constant addition of new unbound cell wall matrix polymers allows for an expandable wall structure at the cell tip, which subsequently hardens in the subapex, a requirement for producing the tubular shape of a root hair (see de Keijzer et al. 2008).

Table 1 Cell structures involved in (root hair) elongation

Structure	Function	Dimensions
Vacuole	Turgor pressure	10 μm × >1,000 μm
Nucleus	Movement with speed of root hair growth	5 μm × 200 μm
Polysomes	Production of cytoplasmic proteins	25 nm × 200 nm
Endoplasmic reticulum	Production of membrane and wall proteins, reorients	50 μm × 1,000 μm
Golgi bodies	Production of Golgi vesicles and wall polysaccharides	20 nm × 500 nm
Exocytotic (Golgi) vesicles	Delivery of wall components and cellulose synthases	100 nm diameter
Coated pits/vesicles	Internalization of plasma membrane	60–120 nm
Endosome	Recycling of membrane	Unknown
Mitochondria	Energy production	50 nm × 1 μm
Actin cytoskeleton	Transportation of vehicles	7 nm × 7 μm
Fusion molecules, including exocyst	Exocytosis	Up to 1,200 kD
Microtubules	Determination of growth orientation	25 nm × >25 μm
Plasma membrane	Enlargement	6 nm thick
Cell wall	Production	100 nm thick

The whole production machinery of a cell that is involved in tip growth is listed in Table 1. We will describe each part of this machinery, and will discuss the tip-growth unit, which represents the structured, dynamic subapical cytoplasm positioned between the distal part of the vacuole and the root hair tip. Is this tip-growth unit, i.e., the assemblage of nucleus, endoplasmic reticulum, polysomes, Golgi bodies, Golgi vesicles, exocytosis machinery, clathrin-coated vesicles, endosomes, and mitochondria, also needed in intercalary plant cell elongation in which only the longitudinal walls elongate? Obviously, all of the components of the tip-growth unit are indeed required for plant cell elongation irrespective of direction, but the relative positioning of these components in the case of other types of plant cell elongation has not yet been studied in the same detail.

2 The Vacuole

In a growing root hair, the vacuole occupies most of the root hair body, but is almost absent from the subapex. It is a versatile structure, which is constantly changing shape because of actin dynamics (van der Honing et al. 2007). The central vacuole has thin extensions into the dense subapical cytoplasm, the area located between the hair tip and the main part of the central vacuole (Sieberer and Emons 2000). These extensions constantly change position, more or less extending into the dense subapical cytoplasm (Fig. 1). In fact, this subapical cytoplasm does not have a constant length. For example, the distance between the hair tip and the central vacuole in a

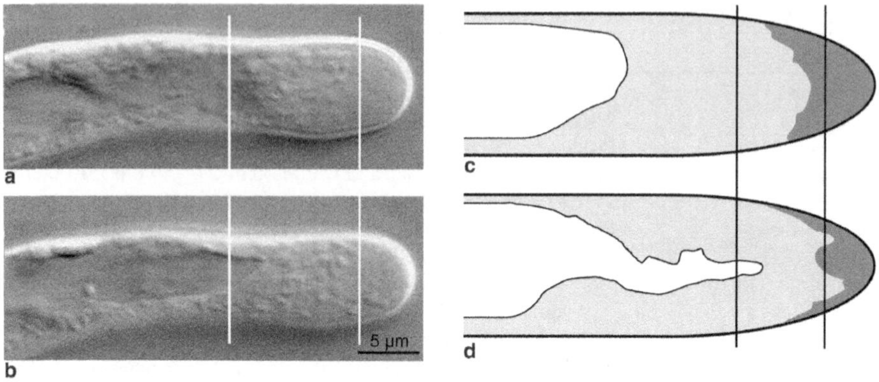

Fig. 1 The cytoarchitecture of a growing root hair is not static. The vesicle-rich area and the cytoplasmic dense area do not have a continuous straight border and this border undulates in time, mostly together with the undulation of the tonoplast, which is the border between the subapical and basal cytoplasm. (**a**, **b**) Two images from a movie of a growing Arabidopsis root hair under the differential interference contrast microscope. Image (**b**) taken 15 s after image (**a**). (**c**, **d**) Generalized images of the cytoarchitectures of growing root hairs. *White*, vacuole; *grey*, cytoplasmic dense area with organelles; *dark grey*, vesicle rich area in the hair tip

growing *Arabidopsis thaliana* root hair varies between a few micrometers and several tens of micrometers. Turgor pressure is the force that leads to plant cell elongation, but this pressure does not have a favored direction. Therefore, tube shape formation must derive from the localized addition of material, in combination with cell wall hardening behind the extending hemispherical tip.

3 The Nucleus Must Be at a Certain Distance from the Growing Root Hair Tip for Fast Tip Growth

We have studied nuclear positioning in *A. thaliana* root hairs, in which, as in all other growing root hairs, the nucleus follows the growing tip at a fixed distance from the apex (Ketelaar et al. 2002). The following observations showed that this positioning is essential for root hair growth:

- The nucleus migrates to a random position when cell elongation ceases.
- The nucleus moves from branch to branch in bifurcated root hairs of the A. thaliana cow mutant, and it is the branch containing the nucleus that starts or continues to grow. However, tip growth continues as long as vesicles are present in the branch tip from which the nucleus has moved away. Existing vesicles localized in that tip do not move away and cell elongation continues until the surplus of vesicles has been consumed.

- More direct evidence for the hypothesis that nuclear positioning is a key feature of root hair elongation was obtained by trapping the nucleus of growing root hairs in optical tweezers, and thus increasing the distance between the nucleus and the tip. This resulted in the arrest of cell elongation (Ketelaar et al. 2002).
- To determine the role of the actin and microtubular cytoskeleton, depolymerizing drugs were applied to growing root hairs. Actin visualization experiments showed that fine filamentous actin (fine F-actin), present in the cell's subapex, is the most important structure for positioning the nucleus.
- Injection of the cell with an antibody against plant villin, an actin-bundling protein, led to actin unbundling and an initial increase in individual cytoplasmic strands (Ketelaar et al. 2002), thus, of actin unbundling (van der Honing et al. 2007). Since this causes movement of the nucleus towards the hair tip, we conclude that bundled actin filaments at the tip side of the nucleus prevent the nucleus from approaching the apex.

The basipetal movement of the nucleus during root hair growth arrest was shown to require protein synthesis and a functional actin cytoskeleton in the root hair tube. In *A. thaliana* root hairs, microtubules do not seem to influence nuclear location (Ketelaar et al. 2003: 10), but experiments with root hairs of *Medicago truncatula* have shown that actin is involved in keeping the nucleus moving at constant distance from the hair tip, but that microtubules determine the distance, and that hair growth becomes slower when this distance increases (Sieberer et al. 2002). Jones et al. (2002: 334) showed simultaneous tip growth of multiple tips of a single root hair cell of plants overexpressing ROP2 GTPase. This is another example in which root hairs continue to grow even when there is a large distance between the nucleus and the tip.

Why should the nucleus not be closer to the growing tip, and at the same time not too far from it? In the following sections, we will discuss the requirement of the localization of this assembly of cellular components in the subapical part of a growing root hair. Clearly, if the nucleus was closer to the growing tip it would obstruct the transport of cargo to the tip where it is needed for growth. So why not further away? The bulk of the ground substance for cell wall production is the polysaccharide matrix contained in the exocytotic vesicles, and of course this material is needed for cell elongation to occur. Although the cellulose synthases that produce the cellulose microfibrils are embedded in the membrane of the exocytotic vesicles, the ground substance that makes them, UDP-glucose, comes from the cytoplasm (Delmer 1999). The cytoplasmic part of the cellulose synthase complex may also be recruited from the cytoplasm and presumably the origin of at least part of the machinery needed for the docking and insertion of the exocytotic vesicles is cytoplasmic (see de Zarsky and Fowler 2008). Thus, a nucleus located reasonably close to the site where these proteins are synthesized and required ensures that all of the mRNA coding for these proteins does not require targeted transport as well. Indeed, the subapex of root hairs, the area between the nucleus and the tip, is full of polysomes (transmission electron microscopy (TEM) data from *Equisetum hyemale* and *V. sativa*; Emons, A.M.C. unpublished), showing that many proteins are being produced in this area.

4 The Subapical Cytoplasmic Dense Area

4.1 Endoplasmic Reticulum

During bulge formation, which precedes actual tip growth, the endoplasmic reticulum (ER) consists of cisternae longitudinally aligning the plasma membrane of the root epidermal cell. When tip growth starts, cisternae are rearranged and new cisternae must be produced perpendicular to their previous orientation. These cisternae are abundant in the subapical part of the root hair in line with the actin cytoskeleton (Miller et al. 2000). The purpose of (R)ER location close to where cell wall production is needed is self-evident.

4.2 The Membrane System of Golgi Bodies, Vesicles, Coated Pits/Vesicles, and Endosomes

The subapical cytoplasm of growing root hairs has a high density of Golgi bodies (Miller et al. 2000). A closer look at electron microscopy images seems to indicate that only some of these are in close connection to the ER. This is coherent with the fact that the bulk of the material produced for the cell wall matrix is polysaccharide, and only a minor part is glycoprotein. However, all Golgi membranes possess membrane proteins as seen in freeze fracture images, indicating that all plant Golgi stacks originate from the ER.

In plant cells, all organelles move on (bundles) of actin filaments. Microtubules are not found in cytoplasmic strands (Holweg 2007). Endoplasmic (but not cortical) microtubules occur only at specific stages of plant cell development at specific locations. They can be found just before and after cell division (Vos et al. 2004), and in the cytoplasmic dense area of the subapical part of growing root hairs. For example, endoplasmic microtubules are abundant in growing *M. truncatula* root hairs (Sieberer et al. 2002) but less present in growing *A. thaliana* root hairs (Van Bruaene et al. 2004). In growing root hairs of the armadillo repeat-containing kinesin (ark1) mutant endoplasmic microtubules are abnormally abundant, suggesting that this kinesin acts by destabilizing endoplasmic, but not cortical, microtubules (Sakai et al. 2008). In the root hair tip subapex they are intermingled with the actin filaments, and their function is discussed in this volume (see Sieberer and Timmers 2008). In growing root hairs, as in all other plant cells, it is the actin cytoskeleton, and not the microtubules, that transports all organelles through the cell to the subapical area where they accumulate. It is not known whether Golgi bodies discharge their vesicles continuously during movement, or specifically when they arrive in the subapex, from where they go to the plasma membrane of the root hair. However, it is clear that the extreme tip of the hair contains only vesicles and not Golgi bodies. Indeed, an interesting topic in root hair research that has not yet been addressed is where, when, and how the vesicles are released from the bodies.

Vesicles present in the tip of tip-growing plant cells are usually assumed to be Golgi vesicles. A proportion of them must indeed be Golgi vesicles, while others are likely to be endocytotic coated pit-derived vesicles. Coated pits are abundant in root hairs (Emons and Traas 1986), particularly in the subapical shank of growing root hairs. Unfortunately, our knowledge about the exact content of exocytotic and endocytotic vesicles is extremely scarce. In TEM images of root hair tips, it is possible to observe vesicles with light and dark contents (Emons 1987; Miller et al. 2000). The dark vesicle contents resemble the color of the cell wall, when observed with the standard EM technique of glutaraldehyde/osmium tetroxide fixation and uranyl acetate/lead citrate staining of sections. However, it has not yet been determined whether the light/dark color relates to the exo/endocytotic nature of the vesicles.

Just as the boundary between the vacuole and the cytoplasmic dense area is dynamic and not a straight plane, the boundary between the cytoplasmic dense area and the vesicle-rich region is also undulating and dynamic (Fig. 1). Therefore, observation of cytoplasmic movement at a certain distance from the hair tip reveals two different functional movements, faster organelle movement in the cytoplasmic extensions functionally belonging to the subapical area and possibly slower vesicle movement in the vesicle area right at the tip. This is indeed what Wang and coworkers observed, using time-lapse evanescent wave imaging to visualize FM4–64-labeled structures in an optical slice proximal to the plasma membrane of *Picea meyeri* pollen tubes (Wang et al. 2006). The authors report that disruption of the actin cytoskeleton has a pronounced effect on vesicle mobility. This is in agreement with the fact that the observations were conducted not only on the functionally vesicle-rich region, but also on extensions of the cytoplasmic dense area. This can be seen when one uses differential interference contrast microscopy, a useful technique when working with fluorescently tagged proteins in living cells. In the work of Wang and coworkers, it is not clear whether Golgi bodies were also tracked in addition to vesicles. These authors refer to our unpublished work (a chapter in PhD thesis of Dr. T. Ketelaar), in which we are meant to have reported that Brownian motion is the source of vesicle movement in the vesicle-rich area of tip-growing cells. However, research addressing this question has not yet been reported. Using the best available methods, so that the cytoarchitecture of the labeled cell is closest to that of the living cell, no filamentous actin has been observed in the vesicle-rich area located at the extreme root hair tip (Miller et al. 1999). Unfortunately, no high-resolution TEM images showing actin filaments are available, and anti-actin immunogold electron microscopy has not yet been performed on root hairs. If the absence of actin filaments in the vesicle-rich area is real, we suggest that vesicle delivery at the base of this area, in combination with the consumption by exocytosis at the plasma membrane, is enough to account for vesicle movement over this submicrometer distance (based on EM images), with Brownian motion as the underlying mechanism of mobility.

The process of endocytosis retrieves vesicles from the plasma membrane together with the cargo contained in them. Components of the endocytotic machinery are present in the *A. thaliana* genome (Haas et al. 2007) and coated pits and vesicles are abundantly present in growing root hairs (Emons and Traas 1986). An obvious function for these structures in growing root hairs would be the retrieval of excess membrane brought to the plasma membrane by exocytosis. FM4–64 is an endocytotic dye that rapidly labels the whole membrane cascade from endocytotic vesicle to vacuole, but as expected, does not label the ER (van Gisbergen et al. 2008). What is internalized in these vesicles? Clathrin-mediated endocytosis appears to be the mechanism of constitutive internalization of the auxin efflux carriers (Dhonukshe et al. 2007). The endosomal compartment has also been called the prevacuolar system, and for a discussion on its many functions, see, for instance, Haas et al. (2007). Endosomal compartments transport material from one compartment to the other by budding and fusing, and clathrin is involved in this budding process. For a discussion on membrane trafficking, see Assaad (2008).

4.3 Mitochondria (and Plastids)

Golgi bodies remain in the subapical cytoplasm of root hairs, probably in order to deliver their vesicles with cell wall cargo to the vesicle-rich area for vesicle fusion with the plasma membrane. However, mitochondria and even plastids are also more abundant in this area than in the rest of the cell (Emons, A.M.C. and coworkers, unpublished). The organelles that arrive inside the subapical area are all excluded from entering the tip area where the vesicles accumulate. If these vesicles cannot accumulate because the specific actin configuration is no longer present, then growth stops. Thus, it is likely that Golgi bodies, mitochondria, and plastids are not targeted to this area specifically, but that the configuration of the actin cytoskeleton acts as a sieve preventing their passage.

5 The Cytoskeleton

5.1 The Actin Cytoskeleton is Necessary for Cell Elongation to Occur

The actin cytoskeleton has been well studied for root hairs of *A. thaliana* (Ketelaar et al. 2003). In legume root hairs the actin cytoskeleton has been studied during root hair development (Miller et al. 1999) and in response to rhizobacteria for *V. sativa* (de Ruijter et al. 1999), *Phaseolus* (Cardenas et al. 1998), and to a lesser extent in developing *M. truncatula* and *L. japonicus* root hairs (Sieberer et al. 2005b). Initial studies were earlier performed for *E. hyemale* (Emons and Derksen 1986). In the absence of GFP–actin or GFP–actin-binding fusion proteins capable of labeling all actin cytoskeleton configurations, the GFP:FABD2 fusion is the best currently available (Ketelaar et al. 2004). In such studies the confocal laser scanning microscope is a limiting factor in terms of sensitivity and speed of acquisition. Imaging of GFP:FABD2 fluorescence in growing *A. thaliana* root hairs with a spinning disc microscope, which is faster and more sensitive than a conventional point-scan confocal microscope, reveals a fine actin network that cannot be detected with conventional confocal microscope (see Ketelaar and Emons 2008). Even though the above-mentioned work has been carried out with conventional confocal microscopes, we can conclude that growing hairs possess bundled actin filaments in cytoplasmic strands in the nongrowing hair tube, just as in other mature plant cells. In addition, a finer actin configuration is present in the subapical tip, first shown in *V. sativa* (Miller et al. 1999) and later in *A. thaliana* (Ketelaar et al. 2003; Blancaflor et al. 2006) and *L. japonicus* root hairs (Weerasinghe et al. 2005).

In fact, there appear to be at least three different actin configurations in growing root hairs: bundles in the shank, fine bundles of actin filaments (fine F-actin) in the subapex, which are more dynamic, and even finer and more dynamic filaments closer to the tip. Not only the degree of bundling decreases, but also the dynamic activity increases towards the hair tip. This has been determined based on the sensitivity to the actin drugs cytochalasin and latrunculin. At low concentrations these drugs broaden the exocytosis area (Ketelaar et al. 2003), and at higher concentrations the fine F-actin of the subapical area depolymerizes, leading to an arrest of cell growth, but without inhibiting cytoplasmic streaming and depolymerizing the actin filaments in the cytoplasmic strands (Miller et al. 1999). These latter are only depolymerized at still higher drug concentrations, resulting in the cytoplasmic strands becoming thinner and eventually disappearing, but without killing the cell (Miller et al. 1999).

In the vesicle-rich apex of growing root hairs we have not observed filamentous actin with any of the methods used, including staining with fluorescent phalloidin, immunocytochemistry, and GFP:FABD2 labeling. Naturally, we cannot exclude the possibility that the binding sites for these probes of even finer actin bundles than those present in the subapex, including single filaments, are not accessible, or that the filaments are too dynamic for visualization with these approaches. For actin visualization methods and the role of actin-binding proteins on actin organization and dynamics, see Ketelaar and Emons (2008).

What is the function of the actin cytoskeleton in tip growth? A pushing activity comparable with the action of the actin cytoskeleton during the formation of membrane protrusions in fibroblasts (Giannone et al. 2007) is not a likely function for plant actin filaments since the force of actin filaments pushing against a cell wall would certainly be insufficient to deform it. Footer and coworkers used optical tweezers to measure forces required to stall small loose bundles of actin filaments, containing on average eight filaments per bundle. The stall force of such bundles of 1 pN is not more than that for a single filament (Footer et al. 2007). Although such actin bundles or networks, when connected in a framework, can push a plasma membrane forward (Mogilner and Rubinstein 2005), such forces by themselves are most unlikely to be enough to push a cell wall, although precise measurements still have to be performed. On the other hand, it is clear that actin bundles in cytoplasmic strands are involved in delivering the Golgi bodies or vesicles to the cytoplasmic dense subapical area which contains the fine F-actin. Whether organelles or vesicles move on this fine F-actin in a myosin motor-based process to the vesicle-rich area has not yet been demonstrated. The fine F-actin configuration in the subapex of tip-growing cells could be the structure on which molecular motor-driven transport of vesicles takes place, but it could also act as a sieve, as mentioned earlier. What is clear is that Golgi bodies and other organelles reverse their directional movement in this area, returning to the hair's shank, and that the vesicles proceed towards the hair tip. Since actual cell elongation is the result of exocytosis of the content of these vesicles into an area with a flexible cell wall under turgor pressure, the number of accumulated vesicles is not relevant to the speed of growth unless the stock of vesicles becomes limiting. It is obvious that the system is more robust and less dependent on environmental perturbations when a surplus of vesicles is constantly present behind the plasma membrane of the growing tip.

5.2 Microtubules Determine Root Hair Growth Orientation

Recovery of *A. thaliana* root hairs after a pulse with a low concentration of the actin-depolymerizing drugs cytochalasin or latrunculin leads to new growth in the original growth direction, whereas in the presence of oryzalin, a specific microtubule-depolymerizing drug (Anthony et al. 1999), this direction becomes random following a pulse of actin drugs. However, oryzalin alone, at the same concentration, has no influence on *A. thaliana* root hair elongation, but the hairs become wavy (Ketelaar et al. 2003). These results show that although actin filaments are required for growth per se, it is the microtubules that determine the orientation of cell elongation. The action of the microtubules seems to be mediated by mechanisms involving calcium ion distribution at the hair tip (Bibikova et al. 1999). These results provide an interesting indication that microtubules may be involved in setting up the new direction of legume root hair cell elongation after global application of rhizobial nodulation factors (Sieberer et al. 2005a,b). Application of nodulation factors in the Fåhreus slide assay causes root hairs that are at the developmental stage of growth arrest to swell and subsequently to re-initiate tip growth in a random orientation (Heidstra et al. 1997). More details about the microtubular cytoskeleton can be found in Sieberer and Timmers (2008), and their role in root hair curling around rhizobacteria in Limpens and Bisseling (2008).

6 Plasma Membrane, Cellulose Synthase Complex, Cellulose Microfibril

The ultimate purpose of the tip-growth unit is to produce cell wall material and bring it to the apical growth site of the cell in the form of vesicles containing biopolymers inside and callose and cellulose synthases in their membranes. The whole plant cell growth process is steered towards production of the cell wall, an assemblage of polymers (see Nielsen 2008), consisting of a matrix of hemicelluloses, pectins, and glycoproteins within which cellulose microfibrils are embedded, just as steel rods within reinforced concrete. All matrix molecules are delivered to the existing cell wall as the contents of exocytotic (Golgi) vesicles. The cellulose microfibrils are produced by cellulose synthase complexes which are initially embedded inside the (Golgi) vesicle membrane and then inserted into the plasma membrane by the process of exocytosis.

Cellulose microfibril production by these synthases is an active area of research since new tools have recently become available: XFP:cellulose synthase constructs now make it possible to directly visualize cellulose production in living cells under certain experimental conditions (Paredez et al. 2006; reviewed in Somerville 2006). Some of the questions that can now be answered using such and other methods, and for which root hairs could be a useful experimental system, are as follows:

- Are cellulose microfibrils in all plant species produced by a single cellulose synthase complex, or are there exceptions?
- How long are cellulose microfibrils?
- Do single cellulose microfibrils run within more than a single cell facet, moving from a longitudinal facet, across a transverse facet, and then again along a longitudinal facet?
- Are cellulose microfibrils bundled?
- Where in the plasma membrane does cellulose microfibril production start?
- What happens to cellulose microfibril production when root hair elongation has ceased?
- What is the role of cortical microtubules in the production of helicoidal cell walls in root hairs?
- Do the same enzymes, which normally make cellulose, switch to callose production if cells are wounded?

7 Cell Wall Texture

Cell wall texture, the architecture of cellulose microfibrils in a cell wall, has been studied in root hairs of a number of species in more depth than in other cell types. This is because, in addition to their accessibility for fixation, visualization, drug treatment, and signal molecule application, root hairs possess an interesting developmental feature. They grow at the cell tip where the primary cell wall is deposited, corresponding to cell expansion, and at the same time a secondary wall is deposited in the hair tube. Therefore, every stage of cell wall deposition can be observed in a longitudinal section along the plasma membrane, from the youngest wall at the tip to the fully mature wall at the root hair base (Fig. 2). In addition, because the secondary wall does not expand, the cellulose microfibrils remain in place and reveal the history of wall deposition in transverse sections. In this section, we review our knowledge about root hair cell wall texture, which has been instrumental in

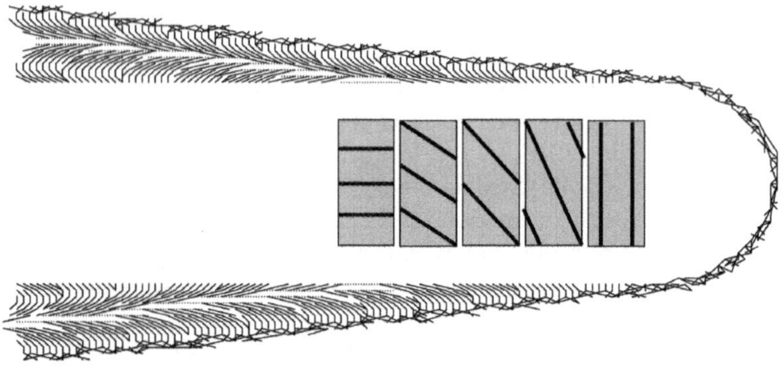

Fig. 2 A helicoidal cell wall texture explaining how, in tip-growing cells with secondary wall deposition, all developmental stages of wall deposition leading to different helicoidal orientations can be observed in the hair tube at all times. The history of wall texture deposition is seen in sections through the cell wall. *Outside:* drawing of a longitudinal section through a growing root hair with helicoidal wall texture showing the cellulose microfibrils. *Rectangles* inside indicate the direction of the cellulose microfibrils as seen in surface preparations of the cell wall at the plasma membrane

developing a more general model for texture formation (Emons 1994; Emons and Mulder 1998, 2000; Mulder and Emons 2001; Emons et al. 2002), and discuss methods for CMF visualization at the ultrastructural level.

7.1 Root Hair Cell Wall Texture

All root hairs studied until now have a so-called random primary cell wall which is deposited at the cell tip and remains random after moving over the tip's hemisphere to the outside of the root hair tube. Mathematical modelling has shown that this random wall texture could be helicoidal with widely spread cellulose microfibrils (Mulder et al. 2004). Inside this primary cell wall deposited at the cell tip, a secondary cell wall is deposited, which has an axial or helical texture, in which all the lamellae of cellulose microfibrils are either longitudinal to the cell axis or at the same angle to the cell axis. This wall texture has been observed in root hairs of terrestrial plants only (Emons 1987). All aquatic plants studied and all young growing root hairs of *Equisetum* species (Emons 1986) have a helicoidal cell wall texture, which is composed of one-microfibril-thick lamellae for which the microfibril angle with respect to a subsequent lamella is constant, and turns in the same direction. This secondary cell wall initiates behind the cell's hemispheric tip and continues to be deposited during cell tip growth. Thus, this secondary cell wall is thicker at the (older) base of the root hair than at the subapical region. When tip growth stops, the inside of the tip is then also covered with a secondary cell wall, for which EM data are available from *E. hyemale* (Emons and Wolters Arts 1983).

7.2 Comparison of Methods for Cell Wall Texture Evaluation

Artifact-free methods are rare, and one single method is not sufficient to reveal all the desired features. To measure the above-mentioned parameters relating to cell wall texture, a variety of methods need to be used. Here we discuss a number of published methods, bearing in mind that new methods will soon become available with electron tomography microscopy and liquid crystal polarization microscopy. We discuss electron microscopy methods for cell wall texture determination in this chapter because these are less well known than methods for cell structure and are becoming more important in combination with the GFP-fusion probes related to cell wall formation, such as GFP:cellulose synthase and GFP–tubulin. Elsewhere in this book, methods for analysis of cellular components are addressed.

7.2.1 Freeze Fixation Followed by Freeze Fracturing

Freeze fixation followed by freeze etching (with a knife) or fracturing (by breaking the material sandwiched between metal carriers) gives the most true-to-nature visualization of cellulose microfibrils, but is not useful for measuring the fibril width because a layer of platinum and carbon is sputtered onto the specimen. The biological material is dissolved from this platinum and carbon layer, and the platinum and carbon replica is analyzed in the TEM. Small-angle and wide-angle X-ray analyses are the methods of choice for determination of fibril width (Sturcova et al. 2004). The freeze etch or fracture procedure is also not helpful for determining the relation between cell content, for example, cortical microtubule and cellulose microfibril orientation, since the fracture direction cannot be controlled and, in addition, the last deposited lamella is rarely exposed directly. The last deposited cellulose microfibrils, but not the whole last deposited lamella, is seen when the extracytoplasmic fracture face or the plasmatic fracture face of the plasma membrane is exposed as protrusions in the extracytoplasmic fracture face or indentations in the plasmatic fracture face, respectively. For a description of this procedure see Emons (1985).

7.2.2 Dry Cleaving and Critical Point Drying

A method that uncovers the last deposited lamella is the dry-cleaving method, in which the material is fixed, cleaved, placed on an EM-grid, critical point dried, shadowed with platinum and carbon, and examined under TEM (Emons 1989). Generally, to observe individual cellulose microfibrils, the cell wall matrix must first be dissolved. However, to our surprise, the microfibrils in *V. sativa* root hairs can be seen after a simple water wash (Emons, unpublished data). A drawback of the dry-cleaving method is that the material shrinks during the procedure, which can be limited by using critical point drying. The dry-cleaving method has also been valuable for visualization of the microtubule cytoskeleton (Traas et al. 1985) and coated pits (Emons and Traas 1986).

7.2.3 Shadow Casting

The oldest method for the visualization of cellulose microfibrils is shadow casting (Frey-Wyssling 1949; Pluymakers 1982; Emons 1986). In this method, the cell content and cell wall matrix are extracted before or after breaking the cells open to expose the inside of the cell wall apposed to the plasma membrane, coating the specimen with platinum/carbon from one side or in a rotary shadowing procedure, prior to analysis in the transmission or scanning electron microscope. A variety of extraction methods have been used, including CDTA, Na_2CO_3, and KOH sequentially (McCann et al. 1990), and hydrogen peroxide glacial acetic acid from 30 to 120 min at 100°C. Using pyrolysis mass spectrometry, we have shown that this last extraction method leaves almost only crystalline cellulose (Wel et al. 1996). Critical in all these extraction procedures is the length of the extraction period: too short does not reveal the microfibrils, and too long causes bundling and wrinkling, as shown by comparison with freeze fracture images. In all these sputtering/shadowing procedures, the amount of platinum varies between 2 and 5 nm. Differential successive extraction steps can give insight into the relationship between the cellulose microfibrils and the rest of the polymers in the cell wall (McCann et al. 1990), especially when combined with Fourier transform infrared microscopy (McCann et al. 1993).

7.2.4 Thin Sectioning After Cell Wall Matrix Extraction

Generally, the regular TEM method of glutaraldehyde/osmium tetroxide fixation, embedding, thin sectioning, and uranyl acetate, lead citrate staining of the sections reveals stripes in the cell wall, but the stripes observed have never been proven to be cellulose microfibrils. On the other hand, it has been shown that these stripes cannot be cellulose microfibrils in *E. hyemale* root hairs (Emons 1988). Similarly, the staining methods with $KMnO_4$ or with PATAg are not specific. However, in these cases, the biological material is first extracted almost completely with hydrogen peroxide/glacial acetic acid at 100°C (Emons and Wolters Arts 1983), or partly with methylamine (Roland et al. 1975; Refrégier 2004). This latter procedure gives a less distinctive image. Subsequent staining of the sections with $KMnO_4$ or periodic acid thiocarbohydrazide–silver proteinate, respectively, reveals the cellulose microfibrils, not because these stains are specific, but because the other polysaccharides have been extracted. Thin sectioning after extraction and staining with $KMnO_4$ gives a cellulose microfibril of diameter 3.6 (SD 1.9) nm (Emons 1988), the same as measured with SAXS (Kennedy et al. 2007). This measure is more than the estimated 2.4 nm determined by WAXS and 2.6 nm determined by NMR analysis (Kennedy et al. 2007). In the literature, it is sometimes stated that cellulose microfibrils of higher plants are 10 nm in diameter, but this includes the platinum/carbon coat.

7.2.5 Atomic Force Microscopy

The method that theoretically gives the highest resolution is atomic force microscopy, and indeed subunits of cellulose microfibrils of algae have been observed that could be individual glucose units (Baker 1998). We have made an attempt to visualize cellulose microfibrils in root hairs of radish with this method (Wel et al. 1996), but the resolution is better when the material is analyzed under TEM. A problem with this scanning probe microscopy technique for biological application is convolution; unless the specimen is very flat so that the scanning tip can exactly follow the object, the measurement of the object dimension will always be inaccurate and exaggerated. Since algae have thick ribbons of cellulose microfibrils, it is possible to prepare a flat preparation, which until now has not been possible with the thin higher plant cellulose microfibrils. As so often with biological material, the limitation is not the microscope itself but the preparation methods that have to be carried out to reveal the required structures, and technical advances will allow more detailed observations in the future.

Acknowledgment We thank Dr. David Barker, Castanet Tolosan, France, for his useful comments on this chapter, and Dr. John Esseling for the drawing of Fig. 1. T.K. was supported by VENI fellowship 863.04.003 from the Dutch Science Foundation (NWO).

References

Anthony RG, Reichelt S, Hussey PJ (1999) Dinitroaniline herbicide-resistant transgenic tobacco plants generated by co-overexpression of a mutant alpha-tubulin and a beta-tubulin. Nature Biotechnology 17:712–716

Assaad FF (2008) The membrane dynamics of root hair morphogenesis. In: Emons AMC, Ketelaar T (eds) Root hairs: excellent tools for the study of plant molecular cell biology. Springer, Berlin Heidelberg New York. doi:10.1007/7089_2008_2

Baker AAMHW, Sugiyama J, Miles MJ (1998) Surface structure of native cellulose microcrystals by AFM. Appl Phys A 66:S559–S563

Bibikova TN, Blancaflor EB, Gilroy S (1999) Microtubules regulate tip growth and orientation in root hairs of Arabidopsis thaliana. Plant J 17:657–665

Bibikova T, Gilroy S (2008) Calcium in root hair growth. In: Emons AmC, Ketelaar T (eds) Root hairs: excellent tools for the study of Plant molecular cell biology. Springer, Berlin Heidelberg New York. doi:10.1007/7089_2008_3

Blancaflor EB, Wang YS, Motes CM (2006) Organization and function of the actin cytoskeleton in developing root cells. Int Rev Cytol 252:219–264

Cardenas L, Vidali L, Dominguez J, Perez H, Sanchez F, Hepler PK, Quinto C (1998) Rearrangement of actin microfilaments in plant root hairs responding to Rhizobium etli nodulation signals. Plant Physiol 116:871–877

de Keijzer MN, Emons AMC, Mulder BM (2008) Modeling tip growth: pushing ahead. In: Emons AMC, Ketelaar T (eds) Root hairs: excellent tools for the study of plant molecular cell biology. Springer, Berlin Heidelberg New York. doi:10.1007/7089_2008_7

de Ruijter NCA, Rook MB, Bisseling T, Emons AMC (1998) Lipochito-oligosaccharides re-initiate root hair tip growth in Vicia sativa with high calcium and spectrin-like antigen at the tip. Plant J 13:341–350

de Ruijter NCA, Bisseling T, Emons AMC (1999) Rhizobium Nod factors induce an increase in sub-apical fine bundles of actin filaments in Vicia sativa root hairs within minutes. Mol Plant Microbe Interact 12:829–832

Delmer DP (1999) Cellulose biosynthesis: exciting times for a difficult field of study. Annu Rev Plant Physiol Plant Mol Biol 50:245–276

Dhonukshe P, Aniento F, Hwang I, Robinson DG, Mravec J, Stierhof YD, Friml J (2007) Clathrin-mediated constitutive endocytosis of PIN auxin efflux carriers in Arabidopsis. Curr Biol 17:520–527

Dolan L, Janmaat K, Willemsen V, Linstead P, Poethig S, Roberts K, Scheres B (1993) Cellular-organization of the Arabidopsis-Thaliana root. Dev Biol 119:71–84
Emons AMC (1985) Plasma-membrane rosettes in root hairs of Equisetum hyemale. Planta 163:350–359
Emons AMC (1986) Cell-wall texture in root hairs of the genus Equisetum. Can J 64:2201–2206
Emons AMC (1987) The cytoskeleton and secretory vesicles in root hairs of equisetum and limnobium and cytoplasmic streaming in root hairs of Equisetum. Ann Bot 60:625–632
Emons AMC (1988) Methods for visualizing cell-wall texture. Acta Botanica Neerl 37:31–38
Emons AMC (1989) Helicoidal microfibril deposition in a tip-growing cell and microtubule alignment during tip morphogenesis – a dry-cleaving and freeze-substitution study. Can J Bot 67:2401–2408
Emons AMC (1994) Winding threads around plant-cells – a geometrical model for microfibril deposition. Plant Cell Environ 17:3–14
Emons AMC, Wolters Arts AMC (1983) Cortical microtubules and microfibril deposition in the cell-wall of root hairs of Equisetum hyemale. Protoplasma 117:68–81
Emons AMC, Derksen J (1986) Microfibrils, microtubules and microfilaments of the trichoblast of Equisetum hyemale. Acta Botanica Neerl 35:311–320
Emons AMC, Traas JA (1986) Coated pits and coated vesicles on the plasma-membrane of plant-cells. Eur J Cell Biol 41:57–64
Emons AMC, Mulder BM (1998) The making of the architecture of the plant cell wall: how cells exploit geometry. Proc Natl Acad Sci USA 95:7215–7219
Emons AM, Mulder BM (2000) How the deposition of cellulose microfibrils builds cell wall architecture. Trends Plant Sci 5:35–40
Emons AMC, Schel JHN, Mulder BM (2002) The geometrical model for microfibril deposition and the influence of the cell wall matrix. Plant Biol 4:22–26
Footer MJ, Kerssemakers JW, Theriot JA, Dogterom M (2007) Direct measurement of force generation by actin filament polymerization using an optical trap. Proc Natl Acad Sci USA 104:2181–2186
Frey-Wyssling A, Mühlethaler K (1949) Über den feinbau der zellwand von wurzelhaaren. Mikroskopie 4:257–266
Giannone G, Dubin-Thaler BJ, Rossier O, Cai Y, Chaga O, Jiang G, Beaver W, Dobereiner HG, Freund Y, Borisy G, Sheetz MP (2007) Lamellipodial actin mechanically links myosin activity with adhesion-site formation. Cell 128:561–575
Haas TJ, Sliwinski MK, Martinez DE, Preuss M, Ebine K, Ueda T, Nielsen E, Odorizzi G, Otegui MS (2007) The Arabidopsis AAA ATPase SKD1 is involved in multivesicular endosome function and interacts with its positive regulator LYST-INTERACTING PROTEIN5. Plant Cell 19:1295–1312
Heidstra R, Yang WC, Yalcin Y, Peck S, Emons AM, vanKammen A, Bisseling T (1997) Ethylene provides positional information on cortical cell division but is not involved in Nod factor-induced root hair tip growth in Rhizobium-legume interaction. Dev Biol 124: 1781–1787
Holweg CL (2007) Living markers for actin block myosin-dependent motility of plant organelles and auxin. Cell Motil Cytoskeleton 64:69–81
Jones MA, Shen JJ, Fu Y, Li H, Yang ZB, Grierson CS (2002) The Arabidopsis Rop2 GTPase is a positive regulator of both root hair initiation and tip growth. Plant Cell 14:763–776
Kennedy CJ, Cameron GJ, Sturcova A, Apperley DC, Altaner C, Wess TJ, Jarvis MC (2007) Microfibril diameter in celery collenchyma cellulose: x-ray scattering and NMR evidence. Cellulose 14:235–246
Ketelaar T, Emons AMC (2001) The cytoskeleton in plant cell growth: lessons from root hairs. New Phytol 152:409–418
Ketelaar T, Emons AMC (2008) The actin cytoskeleton in root hairs: a cell elongation device. In: Emons AMC, Ketelaar T (eds) Root hairs: excellent tools for the study of plant molecular cell biology. Springer, Berlin Heidelberg New York. doi:10.1007/7089_2008_8
Ketelaar T, de Ruijter NCA, Emons AMC (2003) Unstable F-actin specifies the area and microtubule direction of cell expansion in Arabidopsis root hairs. Plant Cell 15:285–292

Ketelaar T, Anthony RG, Hussey PJ (2004) Green fluorescent protein-mTalin causes defects in actin organization and cell expansion in Arabidopsis and inhibits actin depolymerizing factor's actin depolymerizing activity in vitro. Plant Physiol 136:3990–3998
Ketelaar T, Faivre-Moskalenko C, Esseling JJ, de Ruijter NCA, Grierson CS, Dogterom M, Emons AMC (2002) Positioning of nuclei in Arabidopsis root hairs: an actin-regulated process of tip growth. Plant Cell 14:2941–2955
Limpens E, Bisseling T (2008) Nod factor signal transduction in the Rhizobium-Legume symbiosis. In: Emons AMC, Ketelaar T (eds) Root hairs: excellent tools for the study of plant molecular cell biology. Springer, Berlin Heidelberg New York. doi:10.1007/7089_2008_10
McCann MC, Wells B, Roberts K (1990) Direct visualization of cross-links in the primary plant cell wall. J Cell Sci 96:323–334
McCann MC, Stacey NJ, Wilson R, Roberts K (1993) Orientation of macromolecules in the walls of elongating carrot cells. J Cell Sci 106(Pt 4):1347–1356
Miller DD, de Ruijter NCA, Emons AMC (1997) From signal to form: aspects of the cytoskeleton plasma membrane cell wall continuum in root hair tips. J Exp Bot 48:1881–1896
Miller DD, de Ruijter NCA, Bisseling T, Emons AMC (1999) The role of actin in root hair morphogenesis: studies with lipochito-oligosaccharide as a growth stimulator and cytochalasin as an actin perturbing drug. Plant J 17:141–154
Miller DD, Leferink-ten Klooster HB, Emons AM (2000) Lipochito-oligosaccharide nodulation factors stimulate cytoplasmic polarity with longitudinal endoplasmic reticulum and vesicles at the tip in vetch root hairs. Mol Plant Microbe Interact 13:1385–1390
Mogilner A, Rubinstein B (2005) The physics of filopodial protrusion. Biophys J 89:782–795
Mulder BM, Emons AMC (2001) A dynamical model for plant cell wall architecture formation. J Math Biol 42:261–289
Mulder BM, Schel JHN, Emons AMC (2004) How the geometrical model for plant cell wall formation enables the production of a random texture. Cellulose 11:395–401
Nielsen E (2008) Plant cell wall biogenesis during tip growth in root hair cells. In: Emons AMC, Ketelaar T (eds) Root hairs: excellent tools for the study of plant molecular cell biology. Springer, Berlin Heidelberg New York. doi:10.1007/7089_2008_11
Paredez AR, Somerville CR, Ehrhardt DW (2006) Visualization of cellulose synthase demonstrates functional association with microtubules. Science 312:1491–1495
Pluymakers HJ (1982) A helicoidal cell wall texture in root hairs of Limnobium stoloniferum. Protoplasma 112:107–116
Refrégier G, Pelletier S, Jaillard D, Höfte H (2004) Interaction between wall deposition and cell elongation in dark-grown hypocotyl cells in Arabidopsis. Plant Physiol 136:959–968
Roland JC, Vian B, Reis D (1975) Observations with cytochemistry and ultracryotomy on the fine structure of the expanding walls in actively elongating plant cells. J Cell Sci 19: 239–259
Sakai T, Honing H, Nishioka M, Uehara Y, Takahashi M, Fujisawa N, Saji K, Seki M, Shinozaki K, Jones MA, Smirnoff N, Okada K, Wasteneys GO (2008) Armadillo repeat-containing kinesins and a NIMA-related kinase are required for epidermal-cell morphogenesis in Arabidopsis. Plant J 53:157–171
Sieberer B, Emons AMC (2000) Cytoarchitecture and pattern of cytoplasmic streaming in root hairs of Medicago truncatula during development and deformation by nodulation factors. Protoplasma 214:118–127
Sieberer BJ, Timmers ACJ (2008) Microtubules in Plant Root Hairs and Their Role in Cell Polarity and Tip Growth. In: Emons AMC, Ketelaar T (eds) Root hairs: excellent tools for the study of plant molecular cell biology. Springer, Berlin Heidelberg New York. doi:10.1007/7089_2008_13
Sieberer BJ, Timmers ACJ, Lhuissier FGP, Emons AMC (2002) Endoplasmic Microtubules configure the subapical cytoplasm and are required for fast growth of Medicago truncatula root hairs. Plant Physiol 130:977–988
Sieberer BJ, Timmers AC, Emons AM (2005a) Nod factors alter the microtubule cytoskeleton in Medicago truncatula root hairs to allow root hair reorientation. Mol Plant Microbe Interact 18:1195–1204

Sieberer BJ, Ketelaar T, Esseling JJ, Emons AM (2005b) Microtubules guide root hair tip growth. New Phytol 167:711–719
Somerville C (2006) Cellulose synthesis in higher plants. Annu Rev Cell Dev Biol 22:53–78
Sturcova A, His I, Apperley DC, Sugiyama J, Jarvis MC (2004) Structural details of crystalline cellulose from higher plants. Biomacromolecules 5:1333–1339
Traas JA, Braat P, Emons AMC, Meekes H, Derksen J (1985) Microtubules in root hairs. Journal of Cell Science 76:303–320
Van Bruaene N, Joss G, Van Oostveldt P (2004) Reorganization and in vivo dynamics of microtubules during Arabidopsis root hair development. Plant Physiol 136:3905–3919
van der Honing HS, Emons AM, Ketelaar T (2007) Actin based processes that could determine the cytoplasmic architecture of plant cells. Biochim Biophys Acta 1773:604–614
van Gisbergen PAC, Esseling-Ozdoba A, Vos JW (2008) Microinjecting FM4-64 validates it as a marker of the endocytic pathway in plants. J. Microscopy
Vos JW, Dogterom M, Emons AMC (2004) Microtubules become more dynamic but not shorter during preprophase band formation: a possible "search-and-capture" mechanism for microtubule translocation. Cell Motil Cytoskeleton 57:246–258
Wang X, Teng Y, Wang Q, Li X, Sheng X, Zheng M, Samaj J, Baluska F, Lin J (2006) Imaging of dynamic secretory vesicles in living pollen tubes of Picea meyeri using evanescent wave microscopy. Plant Physiol 141:1591–1603
Weerasinghe RR, Bird DM, Allen NS (2005) Root-knot nematodes and bacterial Nod factors elicit common signal transduction events in Lotus japonicus. Proc Natl Acad Sci USA 102:3147–3152
Wel NNVD, Putman CAJ, Noort SJT, Grooth BGD, Emons AMC (1996) Atomic force microscopy of pollen grains, cellulose microfibrils, and protoplasts. Protoplasma 194:29–39
Zarsky V, Fowler J (2008) ROP (Rho-related protein from Plants) GTPases for spatial control of root hair morphogenesis. In: Emons AMC, Ketelaar T (eds) Root hairs: excellent tools for the study of plant molecular cell biology. Springer, Berlin Heidelberg New York. doi:10.1007/7089_2008_14

Auxin and Root Hair Morphogenesis

S.H. Lee and H.-T. Cho(✉)

Abstract Auxin is a potent hormonal effector of root hair development. A plethora of genetic and pharmacological studies have revealed that aberrations in auxin availability or signaling can cause defects in root hair growth and morphology. Recently identified components of auxin signaling and auxin transport have been implicated in root hair morphogenesis. The alteration of root hair morphogenesis by auxin also enables this single cell system to serve as an *in planta* biological marker through which the action mechanism of auxin can be examined.

1 Root Hairs Provide an Auxin-Responsive *In Planta* Single Cell System

Root hair development can be divided into two main stages: fate determination of hair or nonhair cells and hair morphogenesis (Grierson and Schiefelbein 2002). In *Arabidopsis*, genetic interactions mediating position-dependent hair or nonhair cell fate determination have been well characterized (for recent reviews, see Schiefelbein and Lee 2006; Grierson and Schiefelbein 2008). Downstream of fate determination, the hair-morphogenetic steps (hair initiation, bulge formation, and tip growth) are modulated by hormonal and environmental cues, as well as by developmental cues (Okada and Shimura 1994; Masucci and Schiefelbein 1994, 1996; Katsumi et al. 2000; Peterson and Stevens 2000; Schiefelbein 2000). In particular, auxin and ethylene are potent effectors of root hair morphogenesis. In this chapter, we focus on the role of auxin in root hair development.

Although the role of auxin during root hair development is likely to be universal across diverse angiosperm species (Katsumi et al. 2000), most studies have been conducted in *Arabidopsis*. Numerous studies with *Arabidopsis* roots have revealed

H.-T. Cho
Department of Biology, Chungnam National University, Daejeon, 305-764, South Korea
e-mail: htcho@cnu.ac.kr

Plant Cell Monogr, doi:10.1007/7089_2008_16

that both exogenous and endogenous auxin sources can affect root hair development. Exogenous auxin (2,4-dichlorophenoxyacetic acid) at concentrations as low as 5 nM could enhance root hair elongation in wild-type *Arabidopsis* (Pitts et al. 1998), and 10 nM indole-3-acetic acid (IAA) restored root hairs in the *root hair defective 6* (*rhd6*) mutant, which has an impairment in hair initiation (Masucci and Schiefelbein 1994). The phenotypes of two auxin-biosynthetic mutants also are consistent with the hypothesis that increased endogenous auxin levels stimulate root hair development. The auxin-overproducing *Arabidopsis* mutants *sur1 (superroot 1)* and *yucca* develop longer root hairs and more plentiful root hairs, because of reduced root epidermal cell length, than do wild-type plants (Boerjan et al. 1995; Zhao et al. 2001). Overexpression of the rice *YUCCA1* gene also greatly enhanced the root hair development in rice (Yamamoto et al. 2007).

Hormones such as auxin and ethylene do not appear to influence hair or nonhair cell fate determination in the root epidermis, as observed by Masucci and Schiefelbein (1996). Elevated auxin levels affect hair morphogenesis of epidermal cells only in the hair cell (H) position. The dominant auxin-signaling mutant *auxin resistant 2* (*axr2*), which is defective in root hair initiation and tip growth, maintains a normal atrichoblast-specific *GL2* expression pattern and a normal cytoplasmic staining pattern in which H-positioned cells are stained more densely by dyes than are non-hair-cell (N)-positioned cells. Additionally, the auxin-insensitive mutant *axr1* has a normal hair to nonhair cell ratio, although the root hairs are shorter (Pitts et al. 1998). Furthermore, although a loss of function in *CONSTITUTIVE TRIPLE RESPONSE 1* (*CTR1*), as well as treatment with 1-aminocyclopropane-1-carboxylic acid (the ethylene precursor), does elicit hair formation in the N position, the position-dependent cytoplasmic density of root epidermal cells is normal. Collectively, these data suggest that hair morphogenesis operates somewhat independently of the upstream fate-determining processes and that hormones are likely to act directly on hair morphogenesis rather than on fate determination.

Another question regarding the role of auxin in root hair development is whether auxin acts hair-cell autonomously. The hair or nonhair cell fate is determined in a noncell autonomous manner by lateral movement of the fate determinants (Schiefelbein and Lee 2006). In the whole plant system, auxin flows from young shoots through the stem and root down to the root tip and then upward to the zone of root epidermal elongation (Lomax et al. 1995). Considering the fundamental influences of auxin on diverse developmental processes and its mobile property, it is conceivable that auxin might act in a noncell autonomous manner in hair morphogenesis. In this regard, the biological properties of auxin have made it difficult to determine the nature of hair-cell (or non-hair-cell) autonomy in auxin-mediated hair morphogenesis, particularly when taking whole plant approaches using mutants or exogenous supplements of auxin or inhibitors. However, a study using a root-hair-cell-specific expression system has demonstrated that auxin mediates hair morphogenesis in a hair-cell autonomous manner (Lee and Cho 2006).

Recent discoveries of auxin receptors and transporters have greatly accelerated our understanding of the mechanism of auxin action (for a recent review, see Leyser 2006). Genetic studies have implicated many components for auxin signaling and transport in root hair development (summarized in Table 1). In this chapter, we

Table 1 Auxin-related *Arabidopsis* mutants and transgenic lines showing root hair phenotypes

Gene	Mutant/transgenic line	Gene status in mutant/transgenic line	Root hair phenotype	Reference
Auxin biosynthesis				
YUCCA	*yucca*	Gain-of-function mutation by *35S CAMV* enhancers	Longer and more plentiful root hairs	(1)
SUR1	*sur1-1*	Loss-of-function mutation	Increased root hair numbers	(2)
Auxin signaling				
TIR1, AFB1, AFB2, AFB3	P_{GVG}:*TIR1*	Overexpression under control of DEX-inducible promoter	Longer root hairs at the primary root tip	(3)
	tir1-1/afb2-1/afb3-1	Loss-of-function mutation	Class III seedlings with very few root hairs	(4)
	tir1-1/afb1-1/afb2-1/afb3-1	Loss-of-function mutation	Class III seedlings with very few root hairs	(4)
AXR1	*axr1-12*	Loss-of-function mutation	Shorter root hairs	(5–7)
	axr1-3	Loss-of-function mutation	Shorter root hairs	(5–7)
AXR2/IAA7	*axr2-1*	Gain-of-function mutation in domain II	Fewer root hairs, basal shift of root hair position, double hairs in a single cell	(8–14)
	P_{E7}:*axr2-1*	Root-hair-cell-specific overexpression	No root hair	(15)
AXR3/IAA17	*axr3-1*	Gain-of-function mutation in domain II	Very few root hairs	(16–18)
	HS:axr3-1	Overexpression under control of heat-shock promoter	Immediate inhibition of root hair development at the time of heat shock	(18)
	axr3-3	Gain-of-function mutation in domain II	No root hair except for a few at the root–hypocotyl junction	(16, 17)
	axr3-10	Loss-of-function mutation	Normal root hair number, slightly shorter root hairs than wild type	(18)
SHY2/IAA3	*shy2-2/shy2-1*	Gain-of-function mutation in domain II	Longer root hairs	(13, 18–20)
	shy2-3	Gain-of-function mutation in domain II	Longer root hairs	(13, 19)

(continued)

Table 1 (continued)

Gene	Mutant/transgenic line	Gene status in mutant/transgenic line	Root hair phenotype	Reference
	HS:shy2-6	Overexpression under control of heat-shock promoter	Gradual elongation of root hairs after heat shock	(18)
	shy2-31	Loss-of-function mutation	Fewer root hairs per root length, slightly shorter root hairs	(18)
SLR/IAA14	*slr-1*	Gain-of-function mutation in domain II	Fewer root hairs	(21)
	35S:mIAA14	Overexpression under control of *35S CAMV* promoter	Fewer root hairs	(21)
	P_{IAA14}*:mIAA14-GFP*	Overexpression under control of *IAA14* promoter	Reduced root hair formation	(21)
	P_{IAA14}*:mIAA14-GR*	Overexpression under control of DEX-inducible system	Almost no root hair formation after DEX treatment	(22)
IAA28	*iaa28-1*	Gain-of-function mutation in domain II	Fewer root hairs	(23)
Auxin transport				
AUX1	*aux1-7*	Loss-of-function mutation	Fewer and shorter root hairs, basal shift of hair position, (double hairs)	(7, 24, 25)
	aux1-21	Loss-of-function mutation	Basal shift of root hair position, double hairs	(24)
	aux1-22	Loss-of-function mutation	Fewer and shorter root hairs, basal shift of hair position, double hairs	(24–26)
PIN2	*pin2/eir1*	Loss-of-function mutation	Shorter root hairs	(27)
PIN3	P_{E7}*:PIN3-GFP*	Root-hair-cell-specific overexpression	Shorter root hairs	(15)
PGP4	*pgp4-1*	Loss-of-function mutation	Slightly longer root hairs and more variable hair lengths	(28)
	pgp4-2	Loss-of-function mutation	Slightly longer root hairs and more variable hair lengths	(28)
TRH1	*trh1*	Loss-of-function mutation	Poor development of root hairs, multiple hair initiations in a single cell	(29, 30)

PINOID	P^{E7}:*PID-GFP*	Root-hair-cell-specific overexpression	Shorter root hairs	(15)
GNOM	$gnom_{emb30\text{-}1/}$ $gnom_{B4049}$	Heterozygous loss-of-function mutation	Basal shift of root hair position	(31)
	$gnom_{van7}$	Partial loss-of-function mutation	Basal shift of root hair position	(31)

References: (1) Zhao et al. (2001); (2) Boerjan et al. (1995); (3) Gray et al. (1999); (4) Dharmasiri et al. (2005b); (5) Pitts et al. (1998); (6) Lincoln et al. (1990); (7) Cernac et al. (1997); (8) Wilson et al. (1990); (9) Timpte et al. (1992); (10) Masucci and Schiefelbein (1994); (11) Timpte et al. (1994); (12) Masucci and Schiefelbein (1996); (13) Nagpal et al. (2000); (14) Cho and Cosgrove (2002); (15) Lee and Cho (2006); (16) Leyser et al. (1996); (17) Rouse et al. (1998); (18) Knox et al. (2003); (19) Tian and Reed (1999); (20) Soh et al. (1999); (21) Fukaki et al. (2002); (22) Fukaki et al. (2005); (23) Rogg et al. (2001); (24) Grebe et al. (2002); (25) Rahman et al. (2002); (26) Marchant and Bennett (1998); (27) Cho et al. (2007); (28) Santelia et al. (2005); (29) Rigas et al. (2001); (30) Vicente-Agullo et al. (2004); (31) Fisher et al. (2006)

review the role of auxin signaling and transport in root hair morphogenesis and discuss key issues in auxin-mediated root hair development as well as the use of the root hair cell system to study the mechanism of auxin action.

2 Auxin Signaling and Root Hair Development

2.1 Auxin Receptors

Recently, two studies have demonstrated that TRANSPORT INHIBITOR RESPONSE 1 (TIR1), an F-box protein, is an auxin receptor that directly modulates the degradation of transcriptional repressors (Aux/IAAs, auxin/indole-3-acetic acids), which results in the expression of auxin-responsive genes (Dharmasiri et al. 2005a; Kepinski and Leyser 2005). Overexpression of *TIR1* under the control of the glucocorticoid-inducible promoter mimicked the effects of exogenous auxin treatment of wild-type *Arabidopsis*, including enhanced root hair growth (Gray et al. 1999). This result indicates that TIR1 serves as an auxin receptor mediating the auxin signal for root hair development. Since TIR1 is localized to the nucleus (Dharmasiri et al. 2005b), the enhanced hair elongation in TIR1-overexpressing transformants is attributable to an increased responsiveness to intracellular auxin. However, the loss-of-function *tir1-1* mutant shows only minor defects in root hair elongation (our unpublished observations), suggesting that TIR1 is not the only auxin receptor affecting root hair development.

The TIR1-related F-box proteins AFB1, AFB2, and AFB3 (AUXIN SIGNALLING F-BOX protein) also bind auxin and interact with Aux/IAAs in an auxin-dependent manner (Dharmasiri et al. 2005b). Genetic analyses have further indicated that TIR1 and the AFBs function redundantly in auxin-mediated responses so that a subset (the class III) of the quadruple *tir1 afb1 afb2 afb3* mutants develop few root hairs (Dharmasiri et al. 2005*b*). This study suggests that AFBs also mediate auxin signaling in root hair development.

AUXIN-BINDING PROTEIN 1 (ABP1) has long been a candidate for an auxin receptor and has been implicated in extracellular auxin-mediated responses such as rapid ion (including H^+) flux-linked cell elongation and protoplast swelling (for reviews, see Napier et al. 2002; Badescu and Napier 2006). Because root hair morphogenesis is accompanied by H^+ movement from cytosol to apoplast at the point of the hair bulge emergence (Bibikova et al. 1998), it is conceivable that the extracellular auxin-mediated ABP1 signaling pathway is involved in root hair formation. However, a recent report implies that extracellular auxin is unlikely to affect root hair elongation. Root-hair-cell-specific overexpression of an auxin efflux carrier (PIN-FORMED 3, PIN3), which facilitates auxin transport from cytosol to apoplast, greatly suppressed root hair elongation (Lee and Cho 2006). This result indicates that intracellular auxin levels are critical for root hair elongation.

2.2 Activators of Auxin Receptors

The auxin receptor TIR1 is a component of SCF^{TIR1}, a ubiquitin protein ligase complex which is required for auxin responses via Aux/IAA degradation (Ruegger et al. 1998; Gray et al. 2001; summarized in Fig. 1). Proper activation of SCF^{TIR1} requires several posttranslational modifications, including covalent modification by RUB1 (Related to Ubiquitin 1, also known as Nedd8 in some species) (del Pozo and Estelle 1999; Parry and Estelle 2004). The RUB1 protein is, in turn, activated by the AUXIN RESISTANT 1 – E1 C-TERMINAL RELATED 1 (AXR1–ECR1_16) enzyme complex (del Pozo et al. 1998). A loss-of-function *axr1* mutant shows diverse defects in auxin responses (Lincoln et al. 1990), including defects in root hair development (Cernac et al. 1997; Pitts et al. 1998). Thus, a series of activating steps, AXR1–ECR1→ RUB1 → SCF^{TIR1}, appears to be required for auxin-mediated root hair development. However, the observation that exogenous auxin could fully restore root hair elongation in *axr1* plants (Pitts et al. 1998) indicates that AXR1 function could be redundant, most likely with its close paralog AXL1 (Hellmann and Estelle 2002). Although the loss of *AXL1* reveals no apparent phenotypic changes, the double mutant for *AXR1* and *AXL1* is lethal (Hellmann and Estelle 2002). RUB1 function also appears to overlap with its paralog RUB2 (Bostick et al. 2004).

Although there is a functional redundancy due to paralogous genes, the known components of auxin signaling are clearly operative in auxin-mediated root hair development.

2.3 Auxin/Indole-3-Acetic Acids

Aux/IAAs are early-auxin-response genes that encode short-lived nuclear proteins with four conserved domains, referred to as I, II, III, and IV (Reed 2001). Aux/IAA proteins act as transcriptional repressors of auxin-responsive reporter genes, as demonstrated by protoplast transfection assays (Tiwari et al. 2001). ARFs (auxin response factors) and Aux/IAAs form homo- and heterodimers both within and between families (Kim et al. 1997). Auxin promotes the interaction of Aux/IAAs with TIR1, and with AFBs, ultimately resulting in the degradation of the ubiquitinated Aux/IAAs by the 26S proteasome (Gray et al. 2001; Dharmasiri et al. 2005a,b). Domain II of the Aux/IAAs carries the amino acid residues required for destabilization of the Aux/IAAs. Thus, mutations in these residues can stabilize Aux/IAA proteins and thereby suppress auxin responses (Gray et al. 2001; Ramos et al. 2001).

Several studies have indicated that Aux/IAAs are involved in root hair development (Table 1). For example, the *axr2-1* mutant, a gain-of-function mutant in domain II of *AXR2/IAA7*, develops fewer and shorter root hairs than does the wild type (Wilson et al. 1990; Masucci and Schiefelbein 1996). Among other gain-of-function *Aux/IAA* mutants in which root hair development is inhibited are the mutants for *AXR3/IAA17* (*axr3-1* and *axr3-3*; Leyser et al. 1996), *SLR/IAA14* (*slr-1*; Fukaki et al. 2002), and *IAA28* (*iaa28*; Rogg et al. 2001). All these dominant

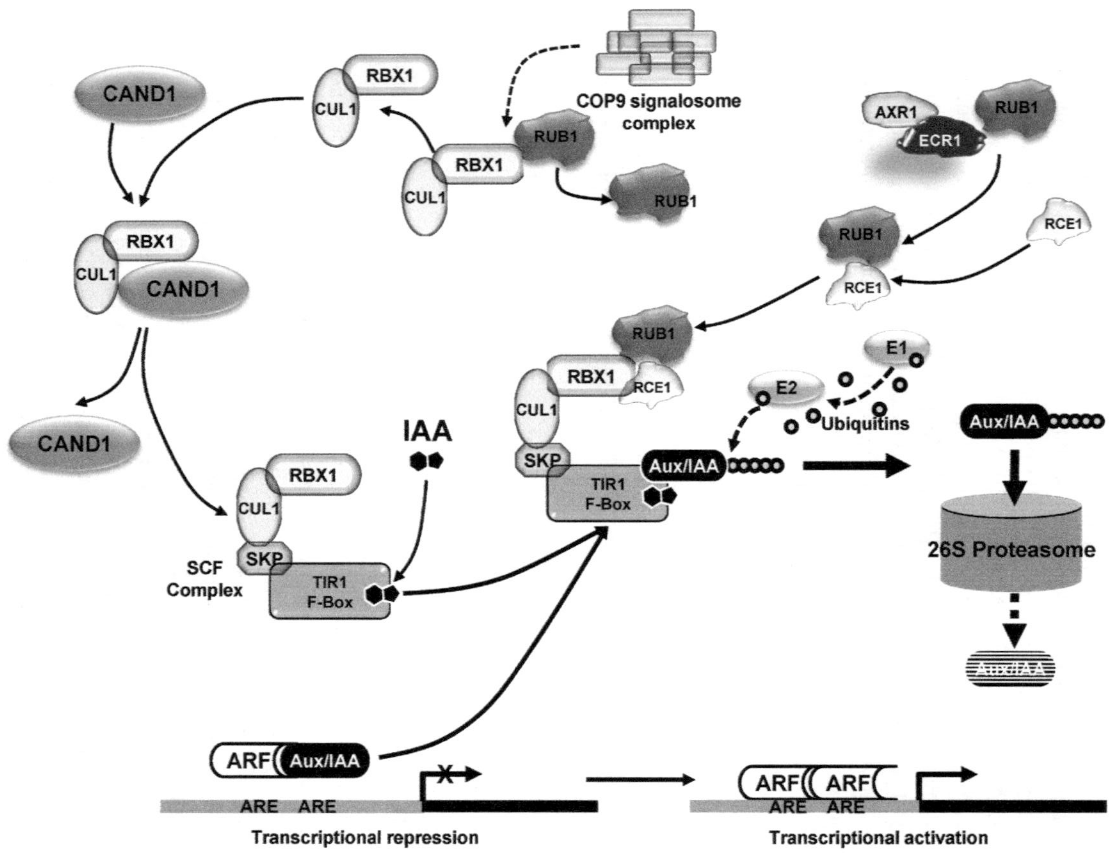

Fig. 1 Auxin signaling from auxin perception by TIR1 F-box to degradation of the transcriptional repressors Aux/IAAs, which takes place in the nucleus. For abbreviations and functions of the signaling components, see the text and Table 1

mutants develop fewer and shorter root hairs, indicating that they have overlapping functions in root hair development.

A study examining ectopic expression of the dominant *slr-1* gene supports, in part, the hair-cell autonomous nature of auxin-mediated root hair development. The expression of *slr-1* using stele (from the *SHORT ROOT* gene)- or endodermis (from the *SCARECROW* gene)-specific promoters could not suppress root hair formation, while the expression of the mutant gene in the root epidermis could (Fukaki et al. 2005). The result from a suppressor mutant of *slr-1* implicated the molecular structure of Aux/IAAs in the tissue specificity of Aux/IAA action (Fukaki et al. 2002). The suppressor mutant *slr-1R1* carries an additional amino acid substitution between domains I and II of the *slr-1* mutant gene. This suppressor mutant restored root hairs, but, intriguingly, not lateral roots, despite the presence of both defects in the *slr-1* mutant. This result suggests that cell-type-specific factors interact with different structural features of SLR/IAA14, or of Aux/IAAs in general.

Another gain-of-function Aux/IAA mutant showing root hair phenotypes is the *shy2* mutant which also carries mutations in domain II (Soh et al. 1999; Tian and Reed 1999). However, the dominant *shy2-2* and *shy2-3* mutants show normal, or even longer, root hair phenotypes than does wild-type *Arabidopsis* (Nagpal et al. 2000; Knox et al. 2003). The lack of inhibition of root hair development by these *shy2* mutations does not seem to be due to the lack of SHY2 expression in the root hair differentiation zone. The expression of SHY2 occurs mainly in the root tip but extends to the hair differentiation zone (Knox et al. 2003). Furthermore, when *shy2-6*, which includes a point mutation equivalent to that in *axr3-1* rather than to that in *shy2-2*, was expressed using a heat-shock-inducible promoter (*HS:shy2-6*), plants displayed more elongation of root hairs than did wild-type plants, whereas hair elongation was blocked by heat shock in the *HS:axr3-1* transformants (Knox et al. 2003). Assuming that *shy2* acts hair-cell autonomously, this opposing phenotypic effect of *shy2* mutations to that of other Aux/IAA mutations suggests that some molecular cues of Aux/IAAs are implicated in their action mechanism for root hair development.

Although the role of Aux/IAAs in root hair development has been shown by their dominant mutations, the common phenotypic effect of at least five dominant *Aux/IAA* mutants strongly supports the notion that Aux/IAA-mediated auxin signaling is involved in root hair development.

In contrast to Aux/IAAs, the role of ARFs in root hair development is not characterized. Molecular interactions between Aux/IAAs and ARFs have been reported in yeast (Kim et al. 1997), but evidence of their *in planta* interactions is rare (Hardtke et al. 2004). Several questions that need to be answered are, whether other Aux/IAAs also function to inhibit or enhance root hair development?; which ARFs are involved in root hair development?; and what interactions between Aux/IAAs and ARFs are operative in root hair development? Because it would be difficult to obtain *ARF* and additional *Aux/IAA* mutants which are defective in root hair morphogenesis, an alternative approach, such as root-hair-specific expression, needs to be considered. Ectopic expression of the dominant mutant *axr2-1* under a root-hair-specific promoter completely inhibited root hair formation in *Arabidopsis* (Lee and Cho 2006). Thus, root-hair-specific expression provides a reasonable bioassay system to assess the roles of those transcription regulators in root hair development and also to identify their molecular partners.

Although we have a large amount of evidence demonstrating that Aux/IAAs are involved in root hair development, their target genes functioning for hair growth and morphogenesis, as well as how auxin interacts with other hair-stimulating effectors, remain to be elucidated. On the other hand, the hierarchical relationship between auxin and ethylene in root hair development has been somewhat characterized. Several lines of evidence indicate that ethylene acts downstream of auxin. Auxin-induced restoration of root hairs in the *rhd6* mutant was almost completely blocked by 1-methylcyclopropene, a specific inhibitor of ethylene perception (Cho and Cosgrove 2002). The ethylene precursor 1-aminocyclopropane-1-carboxylic acid could restore root hairs in the auxin-resistant and root-hair-defective *slr-1* mutant (Fukaki et al. 2002). Auxin is known to stimulate ethylene-synthetic genes (McKeon et al. 1995), and it may therefore indirectly mediate root hair development through ethylene biosynthesis and action. In contrast, the fact that ethylene induces auxin synthesis in the *Arabidopsis* root (Stepanova et al. 2005) leaves the possibility that ethylene works upstream of auxin in root hair development.

3 Auxin Transport and Root Hair Development

Current experimental evidence indicates that root hair development requires both auxin movement from the source cells to the root hair cells and regulation of cellular auxin levels within the root hair cell.

The cell-to-cell movement of auxin is mediated by auxin-carrier (or -transporter) proteins that are localized in the plasma membrane (for reviews, see Morris et al. 2004; Kerr and Bennett 2007). The bacterial amino acid permease-like AUX1 (AUXIN RESISTANT 1) and AUX1 homologs, the LAXs (LIKE AUX1s), are auxin influx transporters, and PINs (PIN-FORMEDs) and p-glycoproteins (PGP1 and PGP19) are efflux transporters of auxin. In certain tissue types, the localization of auxin carrier proteins, particularly AUX1 and PINs, is asymmetrical in the plasma membrane, leading to a directional flow of auxin. Consistent with the asymmetrical subcellular localization of auxin carriers in the *Arabidopsis* root tissues, auxin flows acropetally (from the root–hypocotyl junction to the root tip) to the root tip columella through the root stele and stem cells, and then moves upward (basipetally; from the root tip to the root–hypocotyl junction) to the elongation zone via epidermal and lateral root cap cells (Blilou et al. 2005). AUX1 and PIN2 (/AGR1/EIR1/WAV6) are localized to the bottom and the top, respectively, of root epidermal and lateral root cap cell membranes, which entitles them to be the corresponding auxin carriers for the basipetal flow (Müller et al. 1998; Swarup et al. 2005).

The basipetal flow of auxin from the root tip to the epidermal elongation zone is likely to supply an amount of auxin sufficient for hair morphogenesis. The genetic evidence from loss-of-function *aux1* and *eir1/pin2* mutants is consistent with this hypothesis. The *aux1* mutant root develops fewer and shorter root hairs (Okada and Shimura 1994; Pitts et al. 1998; Rahman et al. 2002), but the wild-type phenotype can be restored by exposure to the diffusible auxin 1-naphthaleneacetic acid (Rahman et al. 2002). In addition to the probable role of AUX1 in the basipetal movement of auxin to the differentiating root hair cells, other scenarios can be implied. The lack of AUX1 could cause a depletion of auxin at the root tip because AUX1 is also responsible for acropetal auxin transport from the stem to the root tip through the protophloem cells. Alternatively, the differentiating root hair cell may require AUX1 function to take up apoplastic auxin.

The *eir1/pin2* mutant also develops shorter root hairs than does wild type (Cho et al. 2007). This defect in the *eir1* root is most likely due to the reduced auxin supply from the root tip to the root hair differentiation zone because PIN2 is the major basipetal auxin efflux transporter in this region (Müller et al. 1998).

A loss-of-function *pgp4* mutant grows longer root hairs than does wild type (Santelia et al. 2005; our observation). In the root epidermis, PIN2 is mainly expressed in the root tip and elongation regions, whereas PGP4 expression extends to the hair differentiation zone (Terasaka et al. 2005). This observation suggests that PGP4 may be one of the major auxin efflux carriers in root hair cells. The PGP4 defect may cause more retention of auxin in the hair cell, which in turn enhances hair elongation. Although auxin transport assays using heterologous systems such as HeLa and yeast cells found an influx activity of PGP4 (Santelia et al. 2005; Terasaka et al. 2005), the activity of PGP4 in plant cells remains to be demonstrated.

Another auxin-transporting protein, TRH1 (TINY ROOT HAIR 1, belonging to the K^+-transporter family), also influences root hair development. The loss of *TRH1* causes short and multiple root hair phenotypes (Rigas et al. 2001; Vicente-Agullo et al. 2004). TRH1 was shown to facilitate auxin efflux in yeast. Because the expression of *TRH1* is confined to the root columella and lateral root cap cells, TRH1 may play a role in the early basipetal movement of auxin from the root tip to the differentiating root hair cells, again suggesting that an auxin source in the root tip is required for root hair development.

Root-hair-cell-specific expression of an auxin efflux carrier has revealed that a particular level of intracellular auxin in the root hair cell is critical for root hair development (Lee and Cho 2006). When the auxin efflux carrier PIN3 was ectopically expressed in a root-hair-cell-specific manner using the *AtEXPA7* promoter (Cho and Cosgrove 2002; Kim et al. 2006), root hair elongation was greatly reduced without any noticeable phenotypic effects on other root tissues or other plant parts. PIN3 localized to the plasma membrane in both the *Arabidopsis* root hair cells and tobacco suspension cells and revealed an auxin efflux activity in tobacco suspension cells (Lee and Cho 2006). The shortened root hair phenotype resulting from root-hair-specific overexpression of PIN3 is likely to be the consequence of the high auxin efflux activity of PIN3, causing a reduction in the auxin level in the root hair cell and thereby inhibiting root hair elongation (Fig. 2). The hair elongation inhibited by PIN3 overexpression could be rescued not only by

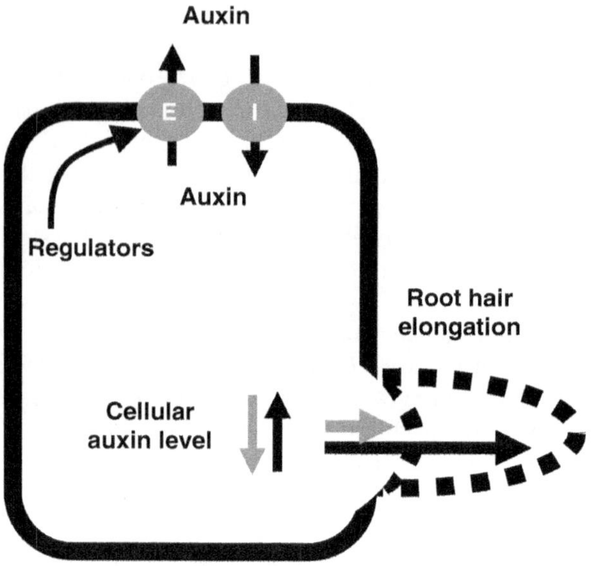

Fig. 2 The root hair cell system used to study auxin transport and its regulation. Auxin in the root hair cell ebbs and flows, depending on the activity of auxin efflux (*E*) and influx (*I*) transporters, or on the effect of regulatory molecules of auxin transport. Altered levels of auxin in the root hair cell then affect root hair elongation

naphthylphthalamic acid (NPA, an inhibitor of auxin efflux) or brefeldin A (BFA, an inhibitor of membrane trafficking of PINs) but also by staurosporine (a protein-kinase inhibitor), implying that protein phosphorylation positively regulates PIN-mediated auxin efflux activity.

4 Protein Phosphorylation, Auxin, and the Root Hair

Reversible phosphorylation of regulatory proteins is a key regulatory mechanism in cellular signaling and metabolism. The Ser/Thr protein kinase PINOID (PID) has been implicated in auxin transport (Benjamins et al. 2001; Friml et al. 2004; Lee and Cho 2006). Cellular levels of PID are important for the subcellular polarity of PIN localization and for PIN activity. High or low levels of PID, caused by overexpression or loss-of-function mutation, respectively, reversed the normal subcellular localization of PINs in the auxin-conducting cells of *Arabidopsis* (Friml et al. 2004). These data indicate that PID acts as a molecular switch controlling the subcellular polarity of PIN proteins and, thus, determines the directionality of auxin flow. In contrast, overexpression of PID using the root-hair-specific *AtEXPA7* promoter in *Arabidopsis* shortened root hair length; however, hair elongation could be restored by exogenous IAA, staurosporine, or inhibitors of auxin efflux activity such as NPA and BFA (Lee and Cho 2006). These results strongly suggest that the inhibition of root hair elongation by overexpression of PID was caused by PID activation of PIN activity, which results in reduced auxin levels in the hair cell and a shorter root hair (Fig. 2). The facilitation of auxin efflux activity by PID was further supported using a direct cellular auxin-transport assay with tobacco suspension cells, in which overexpression of PID increased auxin efflux from the cells. Although the role of PID in wild-type root hair cells remains to be elucidated, this study demonstrates that root hair cell development and the root-hair-cell-specific expression system can be used to study the role of regulators for auxin transport.

PID belongs to the AGC (cAMP-dependent protein kinase A, cGMP-dependent protein kinase G, and phospholipid-dependent protein kinase C) protein kinase family (Bögre et al. 2003). Although single loss-of-function mutants of *PID* and its closest homologs *WAG1* and *WAG2* in *Arabidopsis* do not show visible root hair phenotypes (Santner and Watson 2006; our unpublished observations), probably because of their overlapping functions, mutations of two other AGC kinase family members revealed root hair defects. Plants carrying a loss of function in *AGC2-1/OXI1*, which belongs to the AGCVIII subfamily together with PID, develop short root hairs (Anthony et al. 2004; Rentel et al. 2004). Plants with a knockout mutation of *INCOMPLETE ROOT HAIR ELONGATION* (*IRE*), which belongs to a minor AGC subfamily, also develop shorter root hairs than does wild type (Oyama et al. 2002). The regulation of AGC2-1/OXI1 by auxin suggests that this protein kinase probably mediates auxin-induced root hair development (Anthony et al. 2004).

5 Auxin and Polarity in Root Hair Morphogenesis

Another role of auxin in root hair morphogenesis is in determining the position of the root hair outgrowth. The root hair in *Arabidopsis* normally emerges near the apical (toward the root tip) end of a hair cell (Masucci and Schiefelbein 1994). Auxin and ethylene have been implicated in this polar positioning of root hairs, referred to as "planar polarity" (Grebe 2004; Fischer et al. 2007), and exogenous auxin induces an apical shift in the root-hair-initiating position (Masucci and Schiefelbein 1994; Grebe et al. 2002). In contrast, the auxin-insensitive *axr2-1* dominant mutation guides the hair position toward a more basal (toward the root base) location. In addition, the *axr2-1* mutant root develops multiple root hairs arising from multiple foci in the same hair cell (Masucci and Schiefelbein 1994). A similar phenotype is observed in mutant plants defective in auxin transport (*aux1* and *trh1*) or ethylene perception (*etr1*), as well as in wild-type plants treated with auxin transport inhibitors (NPA and BFA) (summarized by Fischer et al. 2007).

Recently, Fischer et al. (2006) reported that partially reduced GNOM activity in weak *gnom* mutants caused a basal shift in the root hair position. GNOM is an ADP-ribosylation factor (ARF)–GDP/GTP exchange factor (ARF-GEF) that plays a role in trafficking of auxin carrier proteins such as PINs to the plasma membrane (Geldner et al. 2001). BFA directly targets GNOM to inhibit PIN trafficking (Steinmann et al. 1999; Geldner et al. 2003), supporting the phenotypic consistency between effects by BFA treatment and weak *gnom* mutations. Moreover, double or triple mutants of *gnom*, *aux1*, and *ein2* (*ethylene insensitive 2*) resulted in more severe loss of the positional cue for hair initiation, suggesting that the combined activities of AUX1, GNOM, and EIN2 are necessary for proper hair positioning (Fischer et al. 2006). These results indicate that the auxin supply to the differentiating hair cell provides a polarity cue for root hair morphogenesis.

The current model for the planar polarity of hair positioning is that the auxin gradient, formed from the auxin maximum at the root tip, mediates the polarization process. This model is supported by some experimental evidence (Fischer et al. 2006). First, the root tip of a *gnom aux1 ein2* triple mutant displays a significant reduction in auxin concentration and DR5:GUS expression, indicating a weak auxin gradient toward the hair-initiating region. Second, when a local auxin maximum was reestablished by pinpointing the application of exogenous auxin either to the root tip or to the elongation zone of the triple mutant, the hair position shifted toward this newly formed maximum. The use of Rop2 (rho-of-plant 2) small GTPase as a marker for the root-hair-initiating site (Molendijk et al. 2001; Jones et al. 2002) revealed that the localization of this marker was affected by the auxin gradients, implying that auxin, ethylene, and GNOM affect the root hair planar polarity by acting upstream of Rop (Fischer et al. 2006).

However, the underlying connection between all these components and the polar morphogenesis of the root hair is yet to be elucidated. One hypothesis is that an appropriate auxin gradient from root tip to hair differentiation zone is required for hair planar polarity. To maintain the auxin maximum in the root tip and the normal basipetal redistribution of auxin to the hair differentiation zone, the function of not only AUX1 but also multiple PINs is required (Blilou et al. 2005). However, mutations in PINs do not cause significant changes in root hair positioning (Fischer et al. 2006). A simpler alternative hypothesis is that a proper auxin level within the differentiating root hair cell is important for hair positioning. AUX1 might be a major contributor to auxin uptake by the hair cell, but PINs could function redundantly in the hair cell, as indicated by an expression study with the root-hair-cell-specific RNA pool (Lee and Cho 2006). To assess this latter hypothesis, root-hair-cell-specific modulation of the molecular components mediating hair planar polarity would be useful.

6 The Root Hair Cell System to Study Auxin Transport

Root-hair-cell-specific gene manipulation in *Arabidopsis* can offer advantages, particularly in the study of auxin action, over whole-plant-level approaches, such as the use of mutants, gene overexpression by universal promoters, and pharmacological treatments. The effects of auxin on plant development are closely linked to movement of auxin in the plant tissues. Therefore, a specific developmental consequence after pharmacological or genetic alteration of auxin transport in a whole plant is often arguable because of the likelihood of pleiotropic or noncell autonomous effects. Such concerns can be alleviated by using a root-hair-cell-specific expression system combined with a single-cell-level developmental marker, namely, the root hair (Fig. 2).

Several single cell systems have profoundly contributed to the characterization of the components of auxin transport. These systems include plant cells such as *Arabidopsis* protoplasts, *Arabidopsis* cultured cells, and tobacco BY-2 (Bright Yellow 2) cells, as well as heterologous single cell systems such as yeast, HeLa cells, and *Xenopus* oocytes (Chen et al. 1998; Luschnig et al. 1998; Geisler et al. 2005; Santelia et al. 2005; Terasaka et al. 2005; Lee and Cho 2006; Petrášek et al. 2006; Yang et al. 2006). These systems have been used to measure auxin transport across the plasma membrane by auxin carrier proteins. In addition to these systems, *in planta* systems such as the root hair cell can be used to further our understanding of the role of auxin transporters and their regulators in the developmental context. Furthermore, measurement of root hair length can be used to quantify auxin-transporting activity, which is useful in comparing between transporters or in assessing a single transporter under various conditions.

Another useful property of the root hair cell is the robustness of hair growth in the presence of pharmacological agents. The membrane-trafficking inhibitor BFA and the protein-kinase inhibitor staurosporine are generally toxic to cell metabolism (Satiat-Jeunemaitre et al. 1996; Yamaki et al. 2002), but within a certain range of concentrations, these inhibitors restored root hair growth in PID- or PIN3-overexpressing transformant roots (Lee and Cho 2006). This result is surprising because the hair growth in wild-type roots continuously decreased with increasing concentrations of these chemicals. These results are indicative of the target-specific actions of these chemicals within a specific concentration range in root hair cells.

In spite of these advantages, when using the root hair system, caution should be taken to the plasticity of root hair development. Root hair development can be influenced by various environmental factors, including nutrients, pH, ions, light, and separation from the medium (Schiefelbein et al. 1992; Okada and Shimura 1994; Herrmann and Felle 1995; Bibikova et al. 1998; Peterson and Stevens 2000; Müller and Schmidt 2004). For example, water stress causes the development of aberrant root hairs in wild-type *Arabidopsis* which are reminiscent of root hair-defective mutants that display arrest in hair initiation, branching, waving, and bulging hair tips (Bibikova and Gilroy 2003). Maintenance of constant growing conditions and juxtaposition of control plants in the same agar plate are essential to avoid false results when observing root hair phenotypes.

In addition to the study of auxin transport, the root hair cell system also can be used to dissect the signaling mechanisms of auxin, ethylene, and other environmental stimuli that affect root hair development.

7 Measurement of Root Hair Length

Estimation of root hair length has been described in previous studies (Pitts et al. 1998; Santelia et al. 2005; Lee and Cho 2006). *Arabidopsis* seedlings are grown on the agarose medium containing 0.5× Murashige–Skoog nutrient mix, 1% sucrose, 0.5 g L^{-1} MES, pH 5.7 with KOH. Normally a 3-day cold treatment is conducted, but longer vernalization periods (for example, 5 days) enhance seed germination rate and synchronization of root growth. Root hair length can normally be measured from the third day after germination at 20–25°C with a 16-h light/8-h darkness photoperiod. Because we have observed that older roots sometimes tend to develop deformed root hairs, we suggest observation of root hairs on day 3 or 4. For treatment with inhibitors and effectors, 3-day-old seedlings are transferred to the chemical-containing medium, and root hairs are observed 24 h after the treatment. For large-scale (many roots) estimation, 12 cm × 12 cm square plates with 20–30 mL medium are used. However, in case of treatment with costly chemicals, 3-day-old seedlings are transferred onto the slide glass (7.5 cm × 2.5 cm) spread with 1–2 mL medium containing the chemicals. These slide glasses are sealed inside normal transparent plates to prevent the seedlings from water stresses. Plates are vertically located toward the gravity direction for the roots to grow along the medium surface. The final root hair length is measured with fully elongated root hairs in the mature root hair zone of the root. A low-magnification (20–100×) stereomicroscope can be used to take digital photographs of root hairs. Five consecutive hairs protruding perpendicularly from each side of the root, for a total ten hairs from both sides of the root, are measured from the photographs. Because root hair lengths of even the same genotype are often variable, a high number of root hairs are desirable to be observed; for example, more than a hundred root hairs.

Acknowledgments This work was supported by grants from the Korea Research Foundation (KRF-2004-041-C00366) and the Korea Science and Engineering Foundation (KOSEF) (R01-2007-000-10041-0).

References

Anthony RG, Henriques R, Helfer A, Mészáros T, Rios G, Testerink C, Munnik T, Deák M, Koncz C, Bögre L (2004) A protein kinase target of a PDK1 signalling pathway is involved in root hair growth in *Arabidopsis*. *EMBO J* 23:572–581

Badescu GO, Napier RM (2006) Receptors for auxin: will it all end in TIRs? *Trends Plant Sci* 11:217–223

Benjamins R, Quint A, Weijers D, Hooykaas P, Offringa R (2001) The PINOID protein kinase regulates organ development in *Arabidopsis* by enhancing polar auxin transport. *Development* 128:4057–4067

Bibikova T, Gilroy S (2003) Root hair development. *J Plant Growth Regul* 21:383–415

Bibikova TN, Jacob T, Dahse I, Gilroy S (1998) Localized changes in apoplastic and cytoplasmic pH are associated with root hair initiation in *Arabidopsis thaliana*. *Development* 125: 2925–2934

Blilou I, Xu J, Wildwater M, Willemsen V, Paponov I, Friml J, Heidstra R, Aida M, Palme K, Schere B (2005) The PIN auxin efflux facilitator network controls growth and patterning in *Arabidopsis* roots. *Nature* 433:39–44

Boerjan W, Cervera MT, Delarue M, Beeckman T, Dewitte W, Bellini C, Caboche M, Van Onckelen H, Van Montagu M, Inze D (1995) *Superroot*, a recessive mutation in *Arabidopsis*, confers auxin overproduction. *Plant Cell* 7:1405–1419

Bögre L, Okresz L, Henriques R, Anthony RG (2003) Growth signalling pathways in *Arabidopsis* and the AGC protein kinases. *Trends Plant Sci* 8:424–431

Bostick M, Lochhead SR, Honda A, Palmer S, Callis J (2004) Related to ubiquitin 1 and 2 are redundant and essential and regulate vegetative growth, auxin signalling, and ethylene production in *Arabidopsis*. *Plant Cell* 16:2418–2432
Cernac A, Lincoln C, Lammer D, Estelle M (1997) The SAR1 gene of *Arabidopsis* acts downstream of the AXR1 gene in auxin response. *Development* 124:1583–1591
Chen R, Hilson P, Sedbrook J, Rosen E, Caspar T, Masson PH (1998) The *Arabidopsis thaliana AGRAVITROPIC1* gene encodes a component of the polar-auxin-transport efflux carrier. *Proc Natl Acad Sci USA* 95:15112–15117
Cho H-T, Cosgrove DJ (2002) The regulation of *Arabidopsis* root hair initiation and expansin gene expression. *Plant Cell* 14:3237–3253
Cho M, Lee OR, Ganguly A, Cho H-T (2007) Auxin signaling: long and short. *J Plant Biol* 50:79–89
del Pozo JC, Estelle M (1999) The *Arabidopsis* cullin AtCUL1 is modified by the ubiquitin-related protein RUB1. *Proc Natl Acad Sci USA* 96:15342–15347
del Pozo JC, Timpte C, Tan S, Callis J, Estelle M (1998) The ubiquitin-related protein RUB1 and auxin response in *Arabidopsis*. *Science* 280:1760–1763
Dharmasiri N, Dharmasiri S, Estelle M (2005a) The F-box protein TIR1 is an auxin receptor. *Nature* 435:441–445
Dharmasiri N, Dharmasiri S, Weijers D, Lechner E, Yamada M, Hobbie L, Ehrismann JS, Jürgens G, Estelle M (2005b) Plant development is regulated by a family of auxin receptor F-box proteins. *Dev Cell* 9:109–119
Fischer U, Ikeda Y, Ljung K, Serralbo O, Singh M, Heidstra R, Palme K, Scheres B, Grebe M (2006) Vectorial information for *Arabidopsis* planar polarity is mediated by combined *AUX1*, *EIN2*, and *GNOM* activity. *Curr Biol* 16:2143–2149
Fischer U, Ikeda Y, Grebe M (2007) Planar polarity of root hair positioning in *Arabidopsis*. *Biochem Soc Trans* 35:149–151
Friml J, Yang X, Michniewicz M, Weijers D, Quint A, Tietz O, Benjamin R, Ouwerkerk PB, Ljung K, Sandberg G, Hooykaas PJ, Palme K, Offringa R (2004) A PINOID-dependent binary switch in apical-basal PIN polar targeting directs auxin efflux. *Science* 306:862–865
Fukaki H, Tameda S, Masuda H, Tasaka M (2002) Lateral root formation in blocked by a gain-of-function mutation in the *SOLITARY-ROOT/IAA14* gene of *Arabidopsis*. *Plant J* 29:153–168
Fukaki H, Nakao Y, Okushima Y, Theologis A, Tasaka M (2005) Tissue-specific expression of stabilized SOLITARY-ROOT/IAA14 alters lateral root development in *Arabidopsis*. *Plant J* 44:382–395
Geisler M, Blakeslee JJ, Bouchard R, Lee O, Vincenzetti V, Bandyopadhyay A, Peer WA, Bailly A, Richards EL, Edjendal KF, Smith AP, Baroux C, Grossniklaus U, Müller A, Hrycyna CA, Dudler R, Murphy AS, Martinoia E (2005) Cellular efflux of auxin catalyzed by the *Arabidopsis* MDR/PGP transporter AtPGP1. *Plant J* 44:179–194
Geldner N, Anders N, Wolters H, Keicher J, Kornberger W, Muller P, Delbarre A, Ueda T, Nakano A, Jürgens G (2003) The *Arabidopsis* GNOM ARF-GEF mediates endosomal recycling, auxin transport, and auxin-dependent plant growth. *Cell* 112:219–230
Geldner N, Friml J, Stierhof YD, Jürgens G, Palme K (2001) Auxin transport inhibitors block PIN1 cycling and vesicle trafficking. *Nature* 413:425–428
Gray WM, del Pozo JC, Walker L, Hobbie L, Risseeuw E, Banks T, Crosby WL, Yang M, Ma H, Estelle M (1999) Identification of an SCF ubiquitin-ligase complex required for auxin response in *Arabidopsis thaliana*. *Genes Dev* 13:1678–1691
Gray WM, Kepinski S, Rouse D, Leyser O, Estelle M (2001) Auxin regulates SCF^{TIR1}-dependent degradation of AUX/IAA proteins. *Nature* 414:271–276
Grebe M (2004) Ups and downs of tissue and planar polarity in plants. *Bioessays* 26:719–729
Grebe M, Friml J, Swarup R, Ljung K, Sandberg G, Terlou M, Palme K, Bennett MJ, Scheres B (2002) Cell polarity signaling in *Arabidopsis* involves a BFA-sensitive auxin influx pathway. *Curr Biol* 12:329–334
Grierson C, Schiefelbein J (2002) Root hairs. In: Somerville CR, Meyerowitz EM (eds) The Arabidopsis book. American Society of Plant Biologists, Rockville, doi:10.1199/tab.0060. http://www.aspb.org/publications/arabidopsis/

Grierson C, Schiefelbein J (2008) Genetics of root hair formation. In: Emons AMC, Ketelaar T (eds) Root hairs: excellent tools for the study of plant molecular cell biology. Springer, Berlin Heidelberg New York. doi:10.1007/7089_2008_15

Hardtke CS, Ckurshumova W, Vidaurre DP, Singh SA, Stamatiou G, Tiwari SB, Hagen G, Guilfoyle TJ, Berleth T (2004) Overlapping and non-redundant functions of the *Arabidopsis* auxin response factors *MONOPTEROS* and *NONPHOTOTROPIC HYPOCOTYL 4*. *Development* 131:1089–1100

Hellmann H, Estelle M (2002) Plant development: regulation by protein degradation. *Science* 297:793–797

Herrmann A, Felle HH (1995) Tip growth in root hair cells of *Sinapis alba* L.: significance of internal and external Ca^{2+} and pH. *New Phytol* 129:523–533

Jones MA, Shen JJ, Fu Y, Li H, Yang Z, Grierson CS (2002) The *Arabidopsis* Rop2 GTPase is a positive regulator of both root hair initiation and tip growth. *Plant Cell* 14:763–776

Katsumi M, Izumo M, Ridge RW (2000) Hormonal control of root hair growth and development. In: Ridge RW, Emons AMC (eds) Root hairs. Springer, Berlin Heidelberg New York, pp 101–114

Kepinski S, Leyser O (2005) The *Arabidopsis* F-box protein TIR1 is an auxin receptor. *Nature* 435:446–451

Kerr ID, Bennett MJ (2007) New insight into the biochemical mechanisms regulating auxin transport in plants. *Biochem J* 401:613–622

Kim J, Harter K, Theologis A (1997) Protein-protein interactions among the Aux/IAA proteins. *Proc Natl Acad Sci USA* 94:11786–11791

Kim DW, Lee SH, Choi S-B, Won S-K, Heo Y-K, Cho M, Park Y-I, Cho H-T (2006) Functional conservation of a root hair cell-specific *cis*-element in angiosperms with different root hair distribution patterns. *Plant Cell* 18:2958–2970

Knox K, Grierson CS, Leyser O (2003) AXR3 and SHY2 interact to regulate root hair development. *Development* 130:5769–5777

Lee SH, Cho H-T (2006) PINOID positively regulates auxin efflux in *Arabidopsis* root hair cells and tobacco cells. *Plant Cell* 18:1604–1616

Leyser O (2006) Dynamic integration of auxin transport and signaling. *Curr Biol* 16:R424–R433

Leyser HMO, Pickett FB, Dharmasiri S, Estelle M (1996) Mutations in the *AXR3* gene of *Arabidopsis* result in altered auxin response including ectopic expression from the *SAUR-AC1* promoter. *Plant J* 10:403–413

Lincoln C, Britton JH, Estelle M (1990) Growth and development of the *arx1* mutants of *Arabidopsis*. *Plant Cell* 2:1071–1080

Lomax TL, Muday GK, Rubery P (1995) Auxin transport. In: Davies PJ (ed) Plant hormones. Kluwer, Norwell, pp 509–530

Luschnig C, Gaxiola RA, Grisafi P, Fink GR (1998) EIR1, a root-specific protein involved in auxin transport, is required for gravitropism in *Arabidopsis thaliana*. *Genes Dev* 12:175–2187

Marchant A, Bennett MJ (1998) The *Arabidopsis AUX1* gene: a model system to study mRNA processing in plants. *Plant Mol Biol* 36:463–471

Masucci JD, Schiefelbein JW (1994) The *rhd6* mutation of *Arabidopsis thaliana* alters root hair initiation through an auxin- and ethylene associated process. *Plant Physiol* 106:1335–1346

Masucci JD, Schiefelbein JW (1996) Hormones act downstream of *TTG* and *GL2* to promote root hair outgrowth during epidermis development in the *Arabidopsis* root. *Plant Cell* 8:1505–1517

McKeon TA, Fernández-Maculet JC, Yang S-F (1995) Biosynthesis and metabolism of ethylene. In: Davies PJ (ed) Plant hormones. Kluwer, Dordrecht, pp 118–139

Molendijk AJ, Bischoff F, Rajendrakumar CS, Friml J, Braun M, Gilroy S, Palme K (2001) *Arabidopsis thaliana* Rop GTPases are localized to tips of root hairs and control polar growth. *EMBO J* 20:2779–2788

Morris DV, Friml J, Zažímalová E (2004) The transport of auxin. In: Davies PJ (ed) Plant hormones. Kluwer, Dordrecht, pp 437–470

Müller M, Schmidt W (2004) Environmentally induced plasticity of root hair development in *Arabidopsis*. *Plant Physiol* 134:409–419

Müller A, Guan C, Galweiler L, Tanzler P, Huijser P, Marchant A, Parry G, Bennett M, Wisman E, Palme K (1998) AtPIN2 defines a locus of *Arabidopsis* for root gravitropism control. *EMBO J* 17:6903–6911

Nagpal P, Walker LM, Young JC, Sonawala A, Timpte C, Estelle M, Reed JW (2000) *AXR2* encodes a member of the Aux/IAA protein family. *Plant Physiol* 123:563–574

Napier RM, David KM, Perrot-Rechenmann C (2002) A short history of auxin-binding proteins. *Plant Mol Biol* 49:339–348

Okada K, Shimura Y (1994) Modulation of root growth by physical stimuli. In: Meyerowitz EM, Somerville CR (eds) Arabidopsis. Cold Spring Harbor, New York, pp 665–684

Oyama T, Shimura Y, Okada K (2002) The IRE gene encodes a protein kinase homologue and modulates root hair growth in *Arabidopsis*. *Plant J* 30:289–299

Parry G, Estelle M (2004) Regulation of cullin-based ubiquitin ligases by the Nedd8/RUB ubiquitin-like proteins. *Semin Cell Dev Biol* 15:221–229

Peterson RL, Stevens KJ (2000) Evidence for the uptake of non-essential and essential nutrient ions by root hairs and their effect on root hair growth. In: Ridge RW, Emons AMC (eds) Root hairs. Springer, Berlin Heidelberg New York, pp 179–195

Petrášek J, Mravec J, Bouchard R, Blakeslee JJ, Abas M, Seifertová D, Wiśniewska J, Tadele Z, Kubeš M, ovanová M, Dhonukshe P, Sků pa P, Benková E, Perry L, K eček P, Lee OR, Fink GR, Geisler M, Murphy AS, Luschnig C, Zažímalová E, Friml J (2006) PIN proteins perform a rate-limiting function in cellular auxin efflux. *Science* 312:914–918

Pitts RJ, Cernac A, Estelle M (1998) Auxin and ethylene promote root hair elongation in *Arabidopsis*. *Plant J* 16:553–560

Rahman A, Hosokawa S, Oono Y, Amakawa T, Goto N, Tsurumi S (2002) Auxin and ethylene response interactions during *Arabidopsis* root hair development dissected by auxin influx modulators. *Plant Physiol* 130:1908–1917

Ramos JA, Zenser N, Leyser O, Callis J (2001) Rapid degradation of auxin/indoleacetic acid proteins requires conserved amino acids of domain II and is proteasome dependent. *Plant Cell* 13:2349–2360

Reed JW (2001) Roles and activities of Aux/IAA proteins in *Arabidopsis*. *Trends Plant Sci* 6:420–425

Rentel MC, Lecourieux D, Ouaked F, Usher SL, Petersen L, Okamoto H, Knight H, Peck SC, Grierson CS, Hirt H, Knight MR (2004) OXI1 kinase is necessary for oxidative burst-mediated signalling in *Arabidopsis*. *Nature* 427:858–861

Rigas S, Debrosses G, Haralampidis K, Vicente-Agullo F, Feldmann KA, Grabov A, Dolan L, Hatzopoulos P (2001) *TRH1* encodes a potassium transporter required for tip growth in *Arabidopsis* root hairs. *Plant Cell* 13:139–151

Rogg LE, Lasswell J, Bartel B (2001) A gain-of-function mutation in *iaa28* suppresses lateral root development. *Plant Cell* 13:465–480

Rouse D, Mackay P, Stirnberg P, Estelle M, Leyser O (1998) Changes in auxin response from mutations in an *AUX/IAA* gene. *Science* 279:1371–1373

Ruegger M, Dewey E, Gray WM, Hobbie L, Turner J, Estelle M (1998) The TIR1 protein of *Arabidopsis* functions in auxin response and is related to human SKP2 and yeast Grr1p. *Genes Dev* 12:198–207

Santelia D, Vincenzetti V, Azzarello E, Bovet L, Fukao Y, Düchtig P, Mancuso S, Martinoia E, Geisler M (2005) MDR-Like ABC transporter AtPGP4 is involved in auxin-mediated lateral root and root hair development. *FEBS Lett* 579:5399–5460

Santner AA, Watson JC (2006) The WAG1 and WAG2 protein kinases negatively regulate root waving in *Arabidopsis*. *Plant J* 45:752–764

Satiat-Jeunemaitre B, Cole L, Bourett T, Howard R, Hawes C (1996) Brefeldin A effects in plant and fungal cells: something new about vesicle trafficking? *J Microscopy* 181:162–177

Schiefelbein JW (2000) Constructing a plant cell: the genetic control of root hair development. *Plant Physiol* 124:1525–1531

Schiefelbein J, Lee MM (2006) A novel regulatory circuit specifies cell fate in the *Arabidopsis* root epidermis. *Physiol Plant* 126:503–510

Schiefelbein JW, Shipley A, Rowse P (1992) Calcium influx at the tip of growing root-hair cells of *Arabidopsis thaliana. Planta* 187:455–459

Soh MS, Hong S-H, Kim BC, Vizir I, Park DH, Choi G, Hong MY, Chung Y-Y, Furuya M, Nam HG (1999) Regulation of both light- and auxin-mediated development by the *Arabidopsis IAA3/SHY2* gene. *J Plant Biol* 42:239–246.

Steinmann T, Geldner N, Grebe M, Mangold S, Jackson CL, Paris S, Gälweiler L, Palme K, Jürgens G (1999) Coordinated polar localization of auxin efflux carrier PIN1 by GNOM ARF GEF. *Science* 286:316–318

Stepanova AN, Hoyt JM, Hamilton AA, Alonso JM (2005) A link between ethylene and auxin uncovered by the characterization of two root-specific ethylene-insensitive mutants in Arabidopsis. *Plant Cell* 17:2230–2242

Swarup R, Kramer EM, Knox K, Leyser O, Haseloff J, Bhalerao R, Bennett MJ (2005) Root gravitropism requires lateral root cap and epidermal cells for transport and response to a mobile auxin signal. *Nat Cell Biol* 7:1057–1065

Terasaka K, Blakeslee JJ, Titapiwatanakun B, Peer WA, Bandyopadhyay A, Makam SN, Lee OR, Richards EL, Murphy AS, Sato F, Yazaki K (2005) PGP4, an ATP-binding cassette P-glycoprotein, catalyzes auxin transport in *Arabidopsis thaliana* roots. *Plant Cell* 17:2922–2939

Tian Q, Reed JW (1999) Control of auxin-regulated root development by the *Arabidopsis thaliana SHY2/IAA3* gene. *Development* 126:711–721

Timpte CS, Wilson AK, Estelle M (1992) Effects of the *axr2* mutation of *Arabidopsis* on cell shape in hypocotyl and inflorescence. *Planta* 188:271–278

Timpte C, Wilson A, Estelle M (1994) The *axr2-1* mutation of *Arabidopsis thaliana* is a gain-of-function mutation that disrupts an early step in auxin response. *Genetics* 138:1239–1249

Tiwari SB, Wang X-J, Hagen G, Guilfoyle TJ (2001) Aux/IAA proteins are active repressors and their stability and activity are modulated by auxin. *Plant Cell* 13:2809–2822

Vicente-Agullo F, Rigas S, Desbrosses G, Dolan L, Hatzopoulos P, Grabov A (2004) Potassium carrier TRH1 is required for auxin transport in *Arabidopsis* roots. *Plant J* 40:523–535

Wilson AK, Pickett FB, Turner JC, Estelle M (1990) A dominant mutation in *Arabidopsis* confers resistance to auxin, ethylene and abscisic acid. *Mol Gen Genet* 222:377–383

Yamaki K, Hong J, Hiraizumi K, Ahn JW, Zee O, Ohuchi K (2002) Participation of various kinases in staurosporine induced apoptosis of RAW 264.7 cells. *J Pharm Pharmacol* 54:1535–1544

Yamamoto Y, Kamiya N, Morinaka Y, Matsuoka M, Sazuka T (2007) Auxin biosynthesis by the *YUCCA* genes in rice. *Plant Physiol* 143:1362–1371

Yang Y, Hammes UZ, Taylor CG, Schachtman DP, Nielsen E (2006) High-affinity auxin transport by the AUX1 influx carrier protein. *Curr Biol* 16:1123–1127

Zhao Y, Christensen SK, Fankhauser C, Cashman JR, Cohen JD, Weigel D, Chory J (2001) A role for flavin monooxygenase-like enzymes in auxin biosynthesis. *Science* 291:306–309

The Membrane Dynamics of Root Hair Morphogenesis

F.F. Assaad

Abstract Root hair elongation requires the delivery of cell wall materials and new membrane to the growing tip. This occurs via polarized secretion, a process mediated by proteins called SNARES (soluble *N*-ethylmaleimide-sensitive factor attachment protein receptors) on vesicle and target membranes. Although the *Arabidopsis* genome encodes an unprecedented number of SNARES, none have thus far been specifically implicated in root hair growth. Forward and reverse genetic approaches, however, have identified a Sec1 protein and a Rab GTPase implicated in root hair morphogenesis. Such proteins gate SNARE interactions and ensure the specificity and fidelity of membrane fusion. ARF GTPases, involved in the sorting of cargo upon vesicle formation, have also been implicated in root hair morphogenesis. In addition to a role in tip growth, polarized secretion may play a role in the establishment or maintenance of polarity, or both. Plasma membrane microdomains enriched in sterols or phosphatidylinositol phosphates may act as anchors for the polarization of the trafficking apparatus. In this chapter I discuss models whereby plant homologues of RHO GTPases (ROPs), master choreographers of cellular polarity, target plasma membrane domains for vesicle delivery and fusion, and how polarity, once established, is maintained.

1 Introduction

Root hair tip growth has been described as a process of polarized secretion. Rapid growth at the tip of the root hair requires the addition of new membranes and cell wall materials. Cellulose is synthesized at the plasma membrane, whereas the hemicelluloses (xyloglucan) and pectins, the bulk of the cell wall matrix, are synthesized in the Golgi apparatus. Membranes and cell wall materials are thought to be delivered to the root hair tip by Golgi-derived vesicles. I pose two questions in this chapter. First, how are Golgi vesicles targeted to the plasma membrane, as opposed to being

F.F. Assaad
Technische Universität München, Botanik, Am Hochanger 4, 85354, Freising, Germany
e-mail: Farhah.Assaad@wzw.tum.de

Plant Cell Monogr, doi:10.1007/7089_2008_2

targeted to other cellular membranes such as the vacuolar membrane? Second, how are these vesicles targeted to a specific, polarized site on the plasma membrane, corresponding to the root hair tip? While discussing the second question, I discuss models for the establishment and maintenance of polarity. To answer these two questions, I draw upon a vast literature on membrane trafficking in yeast and mammalian cells, and on genome analyses that suggest conserved mechanisms in plants. This is because there are too many gaps in our knowledge and too many missing players if the plant literature is taken on its own. Nonetheless, I place special emphasis on the players that have experimentally been shown to play a role in root hair morphogenesis in plants.

2 The Secretory Pathway

A plant cell contains billions of proteins, each of which needs to be targeted to its proper destination. There are a lot of destinations in a plant cell to which proteins can be targeted. These include peroxisomes, mitochondria, plastids, nuclei, the endoplasmic reticulum (ER), the Golgi apparatus, vacuoles, cell wall, and the plasma membrane. Proteins targeted towards the peroxisomes, mitochondria, plastids, and nucleus are synthesized on free ribosomes in the cytosol, and targeted by means of specific signal sequences that reside on the proteins themselves. Proteins destined for the vacuole, plasma membrane, cell wall, or apoplasm have a signal peptide at their N terminus that targets them to ER, and they are synthesized on ribosomes that are bound to the ER by virtue of this signal peptide during protein synthesis. ER studded with such ribosomes is called the rough endoplasmic reticulum or RER. At the RER, proteins are synthesized into the lumen of the ER or integrated into the ER membrane, in the case of integral membrane proteins. From the ER, the proteins enter *the secretory pathway*, an intracellular system of vesicles and cisternae that includes the ER, Golgi stacks, the tonoplast, and the plasma membrane (see Fig. 1). Anterograde transport through the secretory system moves newly synthesized proteins via vesicles from the ER to the Golgi apparatus, which consists of the complement of the Golgi stacks and the *trans*-Golgi network (TGN). From the *trans*-Golgi, proteins are packaged into vesicles that are transported to the cell surface or to the vacuole.

During the process of root hair morphogenesis, a number of lipids, proteins, and polysaccharides need to be delivered to the growing tip. This requires the coordinated function of the ER and Golgi apparatus. In the ER, proteins that enter the secretory pathway are synthesized, processed, and sorted. The synthesis of diverse lipid molecules and the addition of glycans to proteins also take place in the ER. In the Golgi apparatus, complex (branched) polysaccharides of the cell such as pectins and xyloglucans are assembled. In addition, the oligosaccharide side chains of glycoproteins in the secretory pathway are synthesized, as are glycolipids for the plasma membrane and tonoplast. An impairment of ER to Golgi traffic, as seen in the *Arabidopsis rhd3* mutant (Zheng et al. 2004), results in abnormal, crooked root hairs. The secretory pathway, probed using a secreted form of GFP, the actin cytoskeleton,

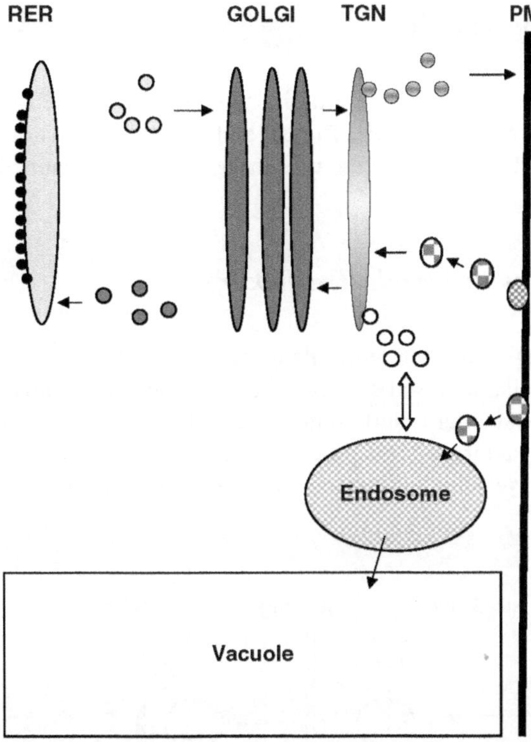

Fig. 1 The secretory pathway. This is an intracellular system of vesicles and cisternae that includes the endoplasmic reticulum (ER), Golgi stacks, the tonoplast (vacuolar membrane), and the plasma membrane (PM). Proteins that enter the secretory pathway are synthesized on ribosomes that are bound to the ER. ER studded with such ribosomes is called the rough endoplasmic reticulum (RER). At the RER, proteins are synthesized into the lumen of the ER or integrated into the ER membrane, in the case of integral membrane proteins. From the ER, the proteins enter the secretory pathway. Anterograde transport through the secretory system moves newly synthesized proteins via vesicles from the ER to the Golgi apparatus, which consists of the complement of the Golgi stacks and the *trans*-Golgi network (TGN). From the *trans*-Golgi, proteins are packaged into vesicles and transported to the cell surface or to the vacuole. The TGN is involved in early endocytosis and membrane recycling, such that while it is in principle generated by the Golgi stacks it can either function as or be mistaken for an early endosome

and endocytosis are impaired in *rhd3* mutants (Zheng et al. 2004), and it is not entirely clear which of these cellular defects are the primary consequences of mutation at the *RHD3* locus. In addition to ER and Golgi traffic, root hair morphogenesis entails a considerable expansion of the vacuolar membrane, which would also require the fusion of vesicles to the tonoplast. Because exocytosis adds more membrane than what is required for growth, an extensive amount of endocytosis or membrane recycling also occurs at the tip (Emons and Traas 1986; for a recent review, see Samaj et al. 2006). Membrane recycling could play a role in conserving resources by maintaining phospholipids and membrane-bound proteins at the growing tip.

3 Membrane Traffic

The question I address in this section is, how vesicles are delivered to their target membrane in a way that is both efficient and faithful. Unless otherwise specified, the mechanisms discussed here are true for both yeast and animal cells.

3.1 *Five Steps in Vesicle Trafficking*

The issue of bringing a given molecule to its proper destination within a cell can be thought of as a reliable postal service. Thus, the right cargo must be packaged into a given vesicle, as a letter is stuffed into an envelope, and the vesicle must be delivered to the right address. The process requires a number of steps, as depicted in Fig. 2. First, vesicles containing the right cargo must be formed, a process

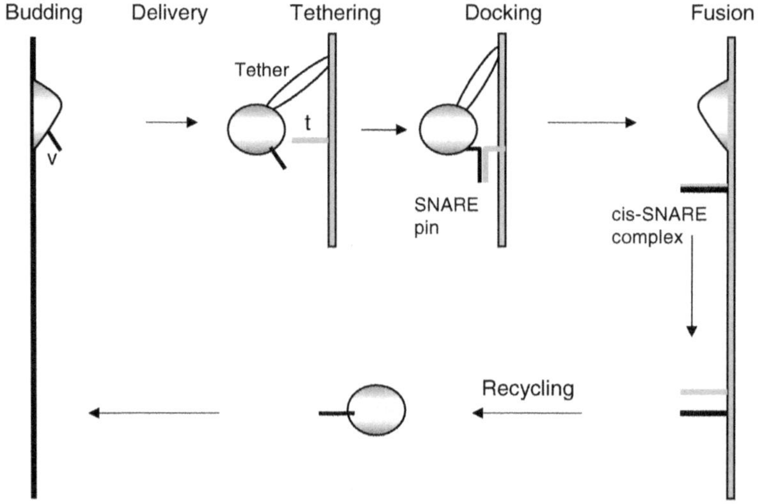

Fig. 2 Six major steps in vesicle traffic. *Budding:* A vesicle buds off a donor membrane. *Delivery:* The vesicle is delivered to a target membrane. *Tethering:* The vesicle is tethered to an acceptor or target membrane by virtue of a tethering molecule or complex. These are typically long molecules or multisubunit complexes that hold the vesicle in the vicinity of the target membrane. *Docking:* the tethered vesicle becomes tightly docked when the v- and t-SNAREs zipper to form a SNARE pin. This is referred to as a *trans*-SNARE complex, as the SNAREs are on different membranes. *Fusion:* The vesicle and target membranes fuse, thereby delivering vesicle content (cargo) to the target compartment. The SNARE complex is now called a *cis*-SNARE complex as the vesicle membrane is now contiguous with the target membrane. *Recycling*: The t- and v-SNAREs are now recycled. This means that the tight complex needs to be disrupted, and this is done by virtue of the NSF-ATPase/SNAP complex, where NSF stands for NEM-sensitive fusion protein and is an AAA ATPase. Adapted from Mellman and Warren 2000

described as budding from a donor membrane. Second, these vesicles must be transported and delivered to their final destination. Third, vesicles are tethered and subsequently docked at their target membrane. Fourth, the vesicle and target membranes are fused to each other, such that the cargo is delivered to its destination while the vesicle membranes become contiguous with the target membrane. Finally, the machinery for membrane traffic is recycled, enabling new rounds of membrane fusion (see Fig. 2; Mellman and Warren 2000).

3.1.1 Vesicle Formation

This entails the recruitment of cargo, and is intimately related to the recruitment of coat proteins. The COPII coat complex is implicated in ER to Golgi traffic, and COPI-coated vesicles mediate retrograde transport from the *cis*-Golgi to the ER as well as traffic between the Golgi stacks. Clathrin-coated vesicles transport cargo that exits the TGN. Key players in the formation of coated vesicles are ARF proteins. ARF proteins are small GTP-binding proteins that regulate both membrane traffic and actin remodeling (D'Souza-Schorey and Chavrier 2006; Randazzo and Hirsch 2004). Similar to other Ras-related GTP-binding proteins, ARF proteins cycle between their active GTP-bound and inactive GDP-bound states. The exchange of GDP for triphosphate nucleotide is mediated by guanine nucleotide exchange factors or GEFs. Conversely, the hydrolysis of bound GTP is mediated by GTPase-activating proteins or GAPs (Randazzo and Hirsch 2004). Current models based on the well characterized ARF1 protein postulate that ARF proteins in their active GTP-bound state recruit coat proteins and subsequently trap cargo. Coat proteins induce the local membrane deformations associated with vesicle budding; ARF GTPases thereby play an important role both in the sorting of cargo and in vesicle budding. They also help prevent the formation of vesicles that lack cargo (Randazzo and Hirsch 2004).

The *Arabidopsis* genome encodes 21 ARF GTPases, compared with 6 in mammals, and 6 of these are in the ARF1 family (Randazzo and Hirsch 2004; Vernoud et al. 2003). The study of ARF1 proteins and their interactors or inhibitors has highlighted the importance of vesicle trafficking for the establishment of polarity as of the very onset of root hair initiation. ARF1 and ARF-GEF are the targets of the vesicle transport inhibitor Brefeldin A, which inhibits the conformational changes in ARF1 required for the dislodging of GDP, thereby affecting the protein's rate of guanine nucleotide exchange. Brefeldin A treatment randomizes the root hair initiation site of *Arabidopsis* trichoblast cells and interferes with the localization of auxin influx and efflux carriers (Geldner et al. 2001, 2003). A number of ARF1 or ARF1-related mutants have root hair phenotypes: GTP- or GDP-locked mutants of ARF1, an ARF-GAP mutant called *rpa*, and the ARF-GEF mutant *gnom* (Fischer et al. 2006; Song et al. 2006; Xu and Scheres 2005). As described in more detail in Table 1, the root hair phenotypes in these mutants include an effect on the planar polarity of root hair positioning along the trichoblast's apical–basal axis, on root hair initiation and tip growth. ARF1 in *Arabidopsis* localizes to the Golgi

Table 1 Root hair mutants in which membrane polarity, traffic, or cycling are impaired

Cellular process	Mutant	Molecular function	Root hair phenotype	Mislocalization	Reference
Phosphatidylinositol signaling	PI-4Kβ1 PI-4Kβ2[a]	Phosphatidylinositol 4-OH kinase	Shorter, branched, bulged, jagged, wavy, or bulbous tip (in double mutant)	RabA4b-labelled compartments	Preuss et al. 2006
Sterol synthesis	*smt1*[orc]	Sterol methyl transferase	Random root hair position	PIN1 and PIN3	Willemsen et al. 2003
ER–Golgi traffic	*rhd3*		Short, wavy, sometimes branched	–	Zheng et al. 2004
Golgi function and endocytosis	*rpa*	ARF-GAP	Short, bulbous/isotropically expanded, branched	–	Song et al. 2006
	arf1(GTP- or GDP-locked)[b]	ARF1	Root hair initiation and/or tip growth significantly inhibited, and planar polarity in root hair positioning affected	ROP2, PIN2	Xu and Scheres 2005
	gnom[c]	ARF-GEF	Basal shift in root hair position	PIN1	Fisher et al. 2006
Post-Golgi traffic	*keule*	Sec1	Absent, stunted, or swollen	–	Assaad et al. 2001
Polarized secretion	*roothairless*	Sec3 subunit of exocyst	Root hairs are initiated but fail to elongate	–	Wen et al. 2005
	AtEXO70A1	Exo70 exocyst subunit	Polar growth disturbed	–	Synek et al. 2006
Endocytosis[d]	*AtRac10*[ca]	ROP GTPase	Balloon shaped, swollen	–	Bloch et al. 2005

[a]This is a double mutant between two paralogues

[b]These are mutants with an inducible cell-type-specific expression of gain-of-function or dominant-negative GTP or GDP-locked mutants of ARF1. Note that the *Arabidopsis* genome encodes six genes in the ARF1 family that are ubiquitously expressed and for which single loss-of-function mutants have no readily discernable phenotypes

[c]This is seen in partial loss-of-function *gnom* mutants (allele *van7*)

[d]This is a constitutively activated mutant in which FM4–64 internalization is impaired

apparatus and, surprisingly, to endocytic organelles (Xu and Scheres 2005), showing that Golgi function and/or endocytosis are required for all aspects of root hair morphogenesis.

3.1.2 Transport or Delivery

Vesicles are often actively transported through the cytoplasm towards their target. Although this has not been proven, it is thought that vesicles are transported by using either actin-dependent motors (myosins) or microtubule-dependent motors (kinesins or dyneins). For further details, please see the chapters on actin (Ketelaar and Emons 2008) and tubulin (Sieberer and Timmers 2008).

3.1.3 Tethering

Tethering is a process that brings and holds vesicles in close proximity to their target membranes. There are two classes of tethering factors. The first class consists of long coiled-coil proteins and includes p115, EEA1, and the Golgins (Grosshans et al. 2006). The second class consists of large multisubunit complexes and includes the exocyst, the TRAPP complexes, the Sec34/35p oligomeric Golgi complex, and the VPS complex (Grosshans et al. 2006). Each of these tethering factors resides on a specific cellular compartment and mediates a specific series of membrane fusion events in the secretory pathway. Root hair morphogenesis in principle requires the tethering of numerous types of vesicles to the entire complement of membrane compartments of the secretory pathway. First, ER vesicles should tether to the Golgi apparatus. Second, Golgi vesicles should tether to the different Golgi stacks. Third, secretory vesicles should tether to polarized domains on the plasma membrane. Fourth, a number of vesicles must fuse to the vacuolar membrane. Fifth, endocytotic vesicles from the plasma membrane must fuse with the TGN and early endosomal compartment. By analogy to animal systems, a number of tethering molecules and complexes would be required, and the *Arabidopsis* genome harbors possible homologues for the majority of such proteins (Latijnhouwers et al. 2005). The only tethering complex that has, to date, been shown to be required in root hair morphogenesis, however, is the exocyst. The exocyst is a large, multisubunit tethering complex that resides on the plasma membrane and is required for polarized exocytosis. Its role in root hair morphogenesis is discussed in Sect. 4.4.

To function as tethers, tethering factors need to be attached to donor and acceptor membranes. Rab GTPases play a critical role in the tethering reaction in that they capture tethering factors when they are in their GTP- (but not GDP-)bound forms (Fig. 3a). Rab proteins belong to a family of small GTP-binding proteins, which function as molecular switches that cycle between "active" and "inactive" states. This cycle is linked to the binding and hydrolysis of GTP. The *Arabidopsis* genome encodes 57 Rab GTPases, and among these the RabA family is the most expanded, with 26 members (Vernoud et al. 2003). RabAs are homologous to

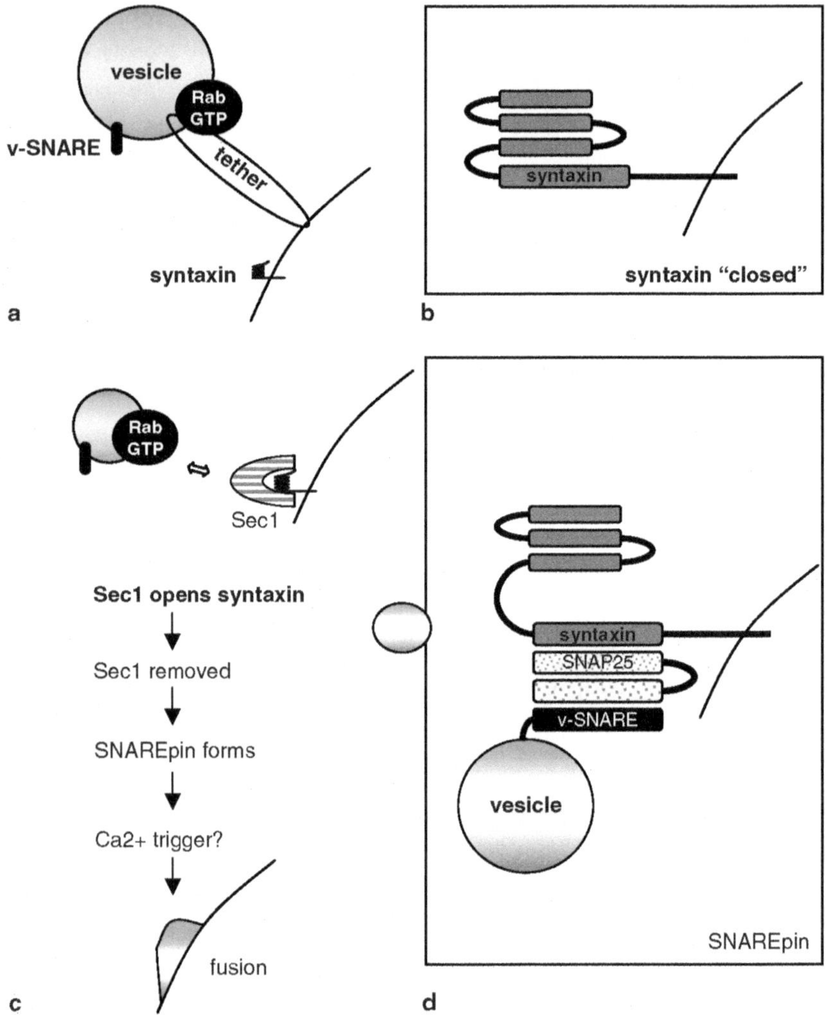

Fig. 3 From tethering to fusion. (**a**) *Tethering:* The vesicle is tethered to an acceptor or target membrane by virtue of a tethering molecule or complex. In their GTP- (but not GDP-)bound forms, Rab GTPases bound to the vesicle and target membranes capture the tethering molecules. (**b**) *Closed syntaxin on target membrane:* Target SNAREs (t-SNAREs) are localized on the target membrane and belong to two different families, the syntaxin-like family and the SNAP-25-like family. The syntaxins are transmembrane proteins that contain several regions with coiled-coil propensity in their cytosolic part, which form a bundle when the syntaxin is in its "closed" form. (**c**) *From docking to fusion:* A Sec1 or SM protein binds to the syntaxin in its closed form. The Rab GTPase and Sec1 can interact through a Rab effector. The Sec1 causes conformational changes in the syntaxin, causing it to open. The Sec1 is removed from the syntaxin and accelerates SNARE-pin formation, but only for cognate SNAREs. Subsequently, a calcium signal triggers membrane fusion. (**d**) *The SNARE-pin or core complex:* This is formed by four SNARE motifs (two from SNAP25 and one each from a v-SNARE and a syntaxin) that are unstructured in isolation but form a parallel four-helix bundle on assembly. SNAP-25 is a protein consisting of two coiled-coil regions, which is associated with the membrane by lipid anchors

mammalian Rab11 proteins and are thought to regulate post-Golgi traffic, occurring on TGN and post-Golgi vesicles. A member of this expanded family, AtRabA4b, has been implicated in root hair morphogenesis. RabA4b localizes to the tip of growing, but not mature, root hairs (Fig. 4a, b). This localization is disrupted in a

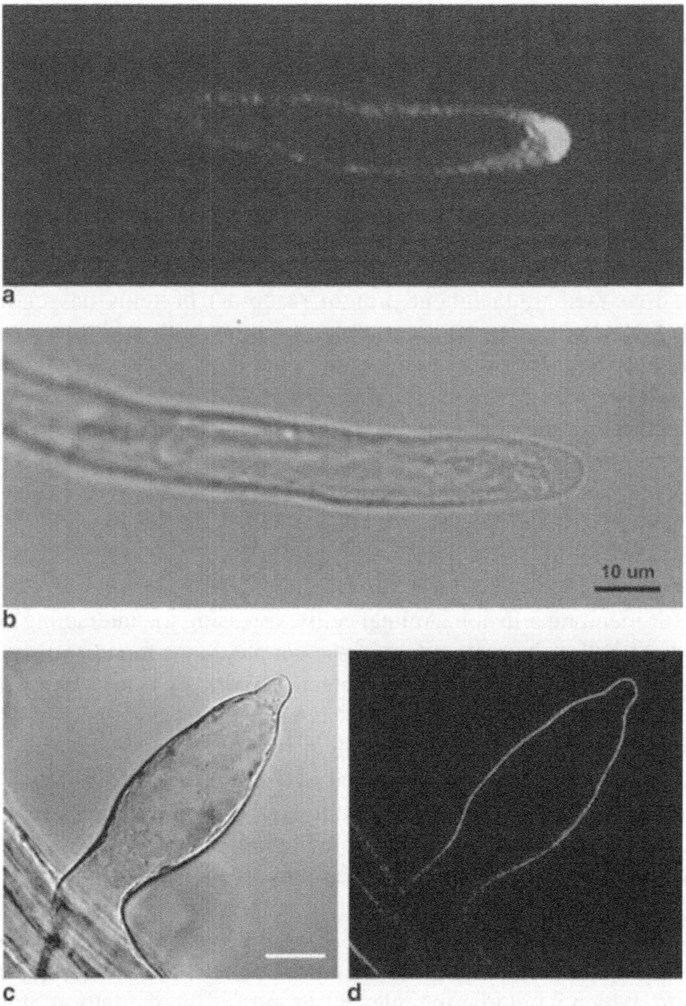

Fig. 4 Rab and ROP GTPases and membrane traffic or recycling in root hairs. EYFP-RabA4b is localized to the tips of growing root hairs. Images of *Arabidopsis thaliana* seedlings stably expressing EYFP-RabA4b were collected by confocal microscopy with a 40 × DIC water lens on a Zeiss LSM 510 microscope. (**a**) Medial section is shown. (**b**) Corresponding DIC image. Courtesy of Julie Thole and Erik Nielsen, University of Michigan. Root hair cells of GFP-*Atrac10*^{CA} (*ROP11*) seedlings that have been shown to have a defect in membrane cycling. Cells were visualized by DIC (**c**) or GFP (**d**). Bar in (**c**) for (**c**) and (**d**): 20 μm. Reproduced with permission from Bloch et al. 2005

number of root hair mutants by collapsing the tip-focused calcium gradient (Preuss et al. 2004, 2006). It is not entirely clear what membrane compartments RabA4b marks at the root hair tip, but most likely these correspond to secretory compartments budded off the TGN (Erik Nielsen, 2007, personal communication).

3.2 Membrane Fusion

Membranes are designed not to fuse but to remain separate, such that membrane-bound cellular compartments remain distinct from each other in a cell. To achieve membrane fusion, barriers to fusion must be overcome. Furthermore, vesicles and target membranes must be matched. Early models of membrane fusion postulated that a tag on the vesicle served as an address to be matched with the proper address on the target membrane. The discovery of SNARE (soluble *N*-ethylmaleimide-sensitive factor attachment protein receptor) proteins has endorsed this model. SNAREs are coiled-coil membrane proteins with a highly conserved 60 amino acid SNARE motif next to the membrane anchor. The SNARE or cognate pairing hypothesis suggests that pairs of SNAREs described as vesicle (v-) SNAREs and target membrane (t-) SNAREs interact specifically to control and mediate intracellular membrane fusion events, with a given v-SNARE being able to form a complex only with a specific set of t-SNAREs (Mellman and Warren 2000). v-SNAREs have also been called VAMPs and a subset of t-SNAREs syntaxins.

Very elegant experiments have shown that SNAREs are both necessary and sufficient for membrane fusion. Animal cells expressing the interacting domains of v- and t-SNAREs on the cell surface, for example, were found to fuse spontaneously (Hu et al. 2003). When SNARE molecules pair they form a four helix bundle in which one helix is derived from a v-SNARE, one from a t-SNARE, and two additional helices are provided by a protein referred to as SNAP25 (Fig. 3b, d). Current models postulate that as unfolded SNARE proteins fold into a four helix bundle, the free energy released is used to merge bilayers. In other words, protein folding is thermodynamically coupled to membrane fusion (Hu et al. 2003).

The *Arabidopsis* genome encodes an unprecedented number of SNAREs (Assaad 2001; Sanderfoot et al. 2000). It is therefore perhaps surprising that none have thus far been shown to play a role in root hair morphogenesis. This may be because the most expanded group of SNARES is in the plasma membrane group, which would be the relevant group for the mediation of exocytosis at the root hair tip. Phylogenetic analyses suggest that as a target membrane, the plasma membrane is defined by at least five syntaxins, namely, SYP121, SYP122, SYP123, SYP124, and SYP125, where SYP stands for *sy*ntaxin of *p*lants (Sanderfoot et al. 2000). SYP121 and SYP122 have been shown to reside on the plasma membrane (Assaad et al. 2004; Collins et al. 2003), and the others have as yet not been characterized. The *syp121 syp122* double mutant is severely dwarfed and necrotic (Assaad et al. 2004), showing that the two syntaxins are partially redundant. SYP122 plays a role in cell wall biogenesis, and the *syp121 syp122* double mutant shows a complex change in cell wall composition, possibly because of a defect in secretion, but a root hair defect has not been reported (Assaad et al. 2004). Either the other plasma membrane syntaxins mediate exocytosis at the root hair tip, or they are redundant with SYP121 and SYP122 in this regard. It is also not clear which v-SNARES are required for root hair morphogenesis, yet identifying such players would help us define the vesicles involved in this process.

3.3 The Gating of Membrane Fusion

Although SNAREs are major determinants of specificity in membrane fusion events, it is now recognized that SNARE interactions are promiscuous and cannot alone account for the specificity we observe in membrane fusion events. Furthermore, SNARE fusion is spontaneous, and one question is how fusion is clamped until it is desirable. Recent studies have therefore focused on the regulation and gating of SNARE interactions. Key regulators of membrane fusion include Sec1/Munc18, or SM proteins, discovered in both yeast and animal cells. X-ray crystallography has shown that the larger Sec1 protein binds to syntaxins or t-SNAREs and induces conformational changes in these proteins (Misura et al. 2000). In the case of mammalian cells, it is thought that the Sec1 binds to the syntaxin in its closed form and causes a conformational change to the open form. The closed form being inaccessible for an interaction with a v-SNARE, the Sec1 plays an important role in priming the SNARE interaction. Sec1 proteins are large, however, and in the tight complex they form with a syntaxin, they physically obstruct contact with other SNAREs, thereby precluding membrane fusion (Fig. 3c). More recent findings suggest that just as the Sec1 can bind the t-SNARE in its closed form, so it can bind the SNARE pin, as these are both four helix bundles (see Fig. 3b, d). Current models suggest that the SM protein in fact accelerates SNARE-pin assembly and membrane fusion but only for cognate SNAREs, thereby enhancing fusion specificity (Shen et al. 2007).

The *Arabidopsis* genome encodes six Sec1 proteins and one, *KEULE*, has been shown to play a role in both root hair morphogenesis and cytokinesis (Assaad et al. 2001). Root hairs in *keule* mutants are severely stunted and branched. Compared to root hair morphogenesis in other cytokinesis-defective seedlings, the phenotype of *keule* root hairs is extreme, suggesting that it is not an indirect consequence of the cytokinesis defect. By contrast, the reduced number and aberrant position of root hairs at the basal end of *keule* seedlings are likely to be an indirect consequence of the cytokinesis defect, as this defect impacts cell differentiation. KEULE has been shown to have all the properties of a Sec1 protein, including syntaxin binding (Assaad et al. 2001). The requirement of a Sec1 in root hair morphogenesis provides indirect evidence that SNAREs are involved, even though they have as yet not been identified by mutation.

3.4 The Role of Calcium in Membrane Fusion

As described by Bibikova and Gilroy (2008), calcium plays a critical role in root hair morphogenesis. As regards membrane traffic, calcium may play an important role in promoting membrane fusion at the root hair tip. Current models postulate that docked vesicles fuse with their target membranes only if they are actively triggered to do so. The trigger is widely accepted to be calcium (see Fig. 3c), and the protein that transduces the calcium signal is likely to be synaptotagmin, which can cause membrane curvature and thereby reduce the activation energy of membrane fusion when triggered by a pulse of calcium (Martens et al. 2007). Although synaptotagmin has homologues in *Arabidopsis* (Craxton 2001), their role in root hair morphogenesis remains to be demonstrated.

3.5 Fidelity, Speed, and Efficiency in Membrane Fusion

The issue of fidelity in membrane fusion is one of how the right cargo is packaged into the right vesicle and delivered to the right address. From vesicle budding to membrane fusion, specificity is tightly regulated at each step. Key regulators such as ARF and Rab GTPases and Sec1 proteins have all been shown to be required for root hair morphogenesis. The missing players are the SNAREs themselves. Known players in membrane traffic during root hair tip growth tend to be members of expanded families in plants, such as the ARF1 or RabA/Rab11 families, implicated in Golgi body function or post-Golgi traffic. The plasma membrane SNAREs being the most expanded family of SNARES, it is likely that functional redundancy has precluded their identification in the process of polarized exocytosis during root hair tip growth.

As root hairs of *Arabidopsis* grow at a rate of 100 $\mu m\ h^{-1}$, the issues of speed and efficiency in membrane fusion are as important as the issue of fidelity. Speed in trafficking is achieved by setting the process of vesicle fusion up until the docking stage, and then triggering the fusion event at the right time at the right place. The tip-focused calcium gradient may play an important role as a trigger for, and as a spatial regulator of, membrane fusion. As regards efficiency, membrane recycling may play an important role in conserving resources at the rapidly expanding tip.

4 The Establishment and Maintenance of Polarity

Having discussed the issues of fidelity and specificity in membrane fusion, I now turn to polarity in exocytosis. In well-characterized models such as the yeast mating cell, a hierarchy of steps is thought to be required for the establishment of polarity. First, a site on the cell surface is selected by extrinsic or intrinsic cues. Second, this site is marked by the recruitment of lipid microdomains and the deposition of landmark proteins. Third, cell polarity is established by the activation of small GTPases, with Rho proteins being the major choreographers. Fourth, actin cables are assembled in a polarized fashion, guiding post-Golgi vesicles to the site of cell growth, and maintaining polarity (Bagnat and Simons 2002; Proszynski et al. 2006). The description of plasma membrane domains and their role in the establishment of cellular polarity in plant systems is limited. One example that is well characterized at the cellular level, however, is the plant response to fungal infection. In this instance, the fungal penetration peg provides a clear extrinsic cue that gives rise to a cell wall domain, and subsequently to a plasma membrane domain characterized by the localized expression of a plasma membrane t-SNARE and reminiscent of lipid rafts (Assaad et al. 2004; Bhat et al. 2005; Bhat and Panstruga 2005). These positional cues at the cell surface and plasma membrane orchestrate the polarization of the actin cytoskeleton and of the secretory apparatus that is required for the plant's defense response. A tentative flow chart of the events that might establish and maintain polarity in root hair initiation and elongation is depicted in Fig. 5. I describe these steps here.

4.1 An Intrinsic Positional Cue

In contrast to straightforward examples of cellular polarity such as the yeast mating cell or the plant response to fungal infection, in which the mating partner or fungal pathogen provides a clear extrinsic cue at the cell surface, it is a priori not intuitively clear how the root hair cell knows exactly where to position a hair along its apical–basal axis. The hair cell's ability to do so is referred to as the planar polarity of root hair positioning. In the absence of an extrinsic cue, one needs to postulate the existence of an intrinsic one. The current model is that auxin levels in the root hair cell might provide a positional cue (Fischer et al. 2006). The supporting evidence, which is scant and indirect, is discussed by Lee and Cho (2008).

4.2 Lipid Rafts

Plasma membrane microdomains enriched in sterols or in phosphatidylinositol phosphates may form in response to a positional cue and act as anchors that would dock proteins to specific sites. In animal cells, such domains are referred to as lipid rafts and are enriched in cholesterol (Bagnat and Simons 2002). A number of sterols structurally related to cholesterol, the most abundant of which is sistosterol, may form lipid rafts in plants. Lipid rafts form platforms for polarized protein delivery and membrane compartmentalization. Different proteins such as GPI (glycosylphosphatidylinositol)-anchored proteins specifically associate with lipid rafts and are thus sorted or retained in a polarized fashion (Bagnat and Simons 2002;

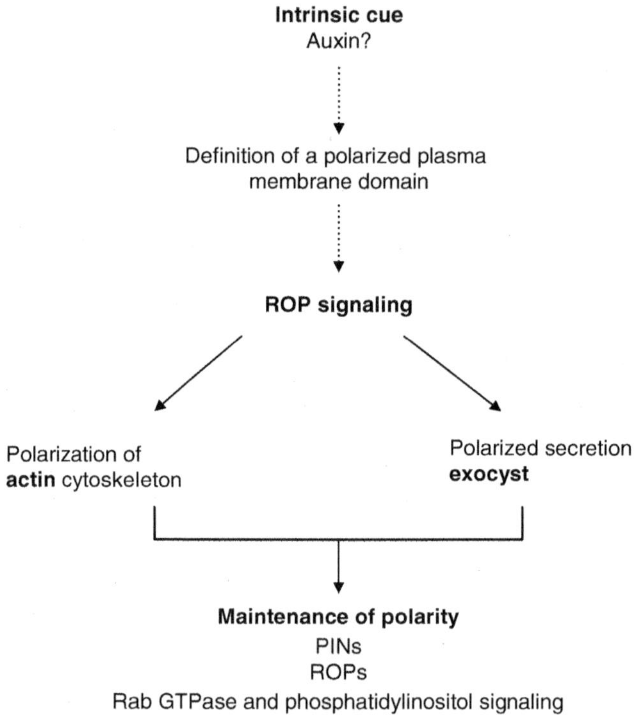

Fig. 5 Tentative flow chart of the hierarchy of processes that establish and maintain polarity in root hair initiation or tip growth. *Dotted lines* depict a process for which the evidence in the plant literature on root hair morphogenesis is scant. It has been suggested that auxin acts as a potential positional cue (Fischer et al. 2006). This would place auxin upstream of ROP signaling. How auxin gradients could be translated into positional information is unclear. What is clear is that positional information must be transmitted to a domain on the plasma membrane that recruits ROP proteins in a polarized fashion. Whether auxin is the intrinsic positional cue, and whether it binds unknown receptors that allow it to act at the plasma membrane is unclear. Evidence for a potential action of auxin upstream of lipid signaling is indirect and is limited to the analysis of a sterol mutant (Willemsen et al. 2003). For references and details see text

Wachtler and Balasubramanian 2006). An interesting mutant in this context is the *orc* mutant, also called *cephalopod* or *smt1*^{orc}. The *orc* mutation, identified as a single recessive allele in a large-scale screen for root development mutants, represents an allele of the *STEROL METHYLTRANSFERASE1* gene, which is required for appropriate sterol levels and catalyzes an alkylation step during sterol biosynthesis (Willemsen et al. 2003). The initiation of root hairs in *smt1*^{orc} mutants is randomized over the apical–basal axis, showing that sterols contribute to positional information during the determination of the root hair initiation site (Willemsen et al. 2003).

Application of exogenous auxin to *smt1*^{orc} mutants partially rescues the root hair phenotype, and applying similar concentrations of auxin to wild-type trichoblasts shifts root hairs to the basal end of the cell. SMT1 has been shown to be required for the polar localization of the putative auxin efflux carriers PIN1 and PIN3 at the plasma membrane (Willemsen et al. 2003). These data are reminiscent of a mutation in the yeast homolog ERG6, which enhances ergosterol levels and leads to the mislocalization of proteins that normally are polarized (Bagnat and Simons 2002). A careful analysis of *smt1*^{orc} mutants with the help of double mutant analysis and drug studies suggest that the primary defect in these mutants is one of cellular polarity. The developmental defects of *smt1*^{orc} mutants could be a secondary consequence of the mislocalization of PIN proteins (Willemsen et al. 2003). PIN proteins mark polarity changes but do not necessarily cause them, such that the available data are not inconsistent with a role for auxin as an intrinsic cue upstream of lipid rafts. Sterols are not the only components of lipid rafts, nor the only lipids that distribute asymmetrically within the lipid bilayer of the plasma membrane. As discussed by Ayoma (2008), phospholipid signaling is also thought to convey positional information in the context of root hair initiation and tip growth.

4.3 ROP Signaling

Plant Rho proteins are referred to as ROPs for RHO proteins of plants, and they are the master choreographers of polarized growth throughout root hair morphogenesis. In plants, ROP proteins are sometimes referred to as RAC proteins. An asymmetric distribution of ROP proteins at the sites of root hair initiation is one of the first visualized landmarks of polarity in root hair morphogenesis. The *Arabidopsis* genome encodes 11 ROP GTPases, of which four have been shown to affect root hair initiation and tip growth, as discussed by Zarsky and Fowler (2008). In the case of Rac10/Rop11 (Table 1, Fig. 4c, d), a role in membrane recycling at the root tip has been shown, and this may be independent of the well-documented role of ROP GTPases on the actin cytoskeleton. In mammalian cells, sterol-enriched microdomains target Rho GTPases to specific sites on the plasma membrane (Palazzo et al. 2004). While the randomized position of root hairs in *smt1*^{orc} mutants is consistent with a role for microdomains enriched in sterols upstream of ROP signaling, ROP localization has not been studied directly in these mutants, and the evidence for this in plants is scant.

In Fig. 5 we suggest that auxin, as a potential positional cue, acts upstream of ROP signaling. How this might happen, and how auxin gradients could be translated into positional information are unclear. What is clear is that positional information must be transmitted to a domain on the plasma membrane that recruits ROP proteins in a polarized fashion. Whether auxin is the intrinsic positional cue, and whether it binds unknown receptors that allow it to act at the plasma membrane are unclear. For details on the role of auxin, see Lee and Cho (2008). The role of ROP proteins on the actin cytoskeleton, tip-focused calcium gradient, reactive oxygen species, and phospholipid signaling is reviewed by Zarsky and Fowler (2008). In what follows, I describe how Rho/ROP proteins polarize the exocytotic apparatus.

4.4 The Exocyst

Root hair morphogenesis requires polarized exocytosis. This means that the machinery for exocytosis needs to assemble at the tip of the root hair, as opposed to being evenly distributed along the entire plasma membrane of the hair cell. How does this occur? And how is polarity established? Experiments in animals and yeast have shown that the exocyst complex mediates polarized exocytosis, in that it marks designated sites on the plasma membrane and tethers vesicles to these sites. The exocyst consists of eight components (Sec3, Sec5, Sec6, Sec8, Sec10, Sec15, Exo70, and Exo84). The Sec3 subunit has been proposed to represent a spatial landmark for polarized secretion in yeast (Guo et al. 2001). This is on the basis of the observation that its localization to sites of exocytosis is independent of actin, of membrane flux through the secretory pathway, and of the other exocyst subunits (Guo et al. 2001). These data point to the existence of spatial landmarks at the plasma membrane that are independent of, and that occur presumably prior to, the polarization of exocytosis and of the actin cytoskeleton. This poses the question as to what regulates the polarized localization of Sec3 at the plasma membrane. In yeast, a screen for mutants in which Sec3 was mislocalized identified certain *rho1* mutant alleles (Guo et al. 2001). Sec3 interacts directly with Rho1 in its GTP-bound form, and functional Rho1 is needed both to establish and to maintain the polarized localization of Sec3. Sec3 is not the only mediator of the effect of Rho1 on the exocyst, because some members of the complex are correctly targeted independently of the interaction between Rho1 and Sec3 (Guo et al. 2001). These results reveal the action of parallel pathways for the polarized localization of the exocytic machinery, and uncouple the role of Rho proteins on vesicle traffic from their role on the actin cytoskeleton.

The importance of the exocyst in root hair morphogenesis can be seen in the maize *roothairless 1* (*rth1*) mutant, in which root hairs fail to elongate (Wen et al. 2005). The *RTH1* gene encodes the exocyst subunit Sec3 (Wen et al. 2005). In plant and animal cells, Sec3 homologues lack the Rho-interaction domain, and recent experiments suggest that in plants Rho proteins recruit the Sec3 subunit through an effector protein that acts as a scaffold (Lavy et al. 2007). In animal cells, Exo70 rather than Sec3 is thought to be responsible for the assembly of the exocyst at the plasma membrane. It is interesting to note that there are at least 22 Exo70 subunits in *Arabidopsis*, compared with 3 in humans (Lavy et al. 2007) and one, *AtEXO70A1*, is required for root hair tip growth (Synek et al. 2006). For more details about the exocyst, see Zarsky and Fowler (2008).

4.5 The Maintenance of Polarity

At each step in the establishment of polarity, it is easy to see how polarity, once established, can perpetrate itself. Also, the trafficking apparatus is, at least as far as maintenance is concerned, implicated in each step of the polarization process.

4.5.1 Auxin

Trafficking mutants such as *arf1*, *rpa*, and *gnom*, all involved in vesicle budding and cargo selection, impact auxin transport and have root hair phenotypes (for references, see Table 1 and Sect. 3.1.1). This shows that membrane traffic is required to maintain auxin levels in the root hair cell and at the growing tip. Either as a secondary consequence of a defect in auxin flux, or because vesicle budding is directly required for the maintenance of polarity, ROP proteins are mislocalized in GTP- or GDP-locked arf1 mutants (Xu and Scheres 2005; see earlier text).

4.5.2 Lipid Rafts

In the example of lipid rafts, while I suggest that these act downstream of auxin, I also cite evidence that they are required for the polarized localization of PIN proteins, which would in turn help maintain auxin levels in the root hair. The compartmentalization of raft and nonraft membrane domains could start as early as the ER because raft association starts there and, after exit from the Golgi body, sorts in a polar fashion to the plasma membrane. In summary, whereas lipid rafts appear to be required for the correct positioning of Rho proteins, thereby orchestrating polarized exocytosis, their maintenance and dynamics might in turn depend on a certain degree of polarity in post-Golgi traffic to the plasma membrane.

4.5.3 Rab GTPases

In addition to their prominent role in tethering, Rab GTPases have been implicated in each step in membrane traffic, from vesicle formation to membrane fusion (Grosshans et al. 2006). Rab proteins act by recruiting a large number of cytosolic effector proteins. The exocyst, for example, is one of many Rab effectors in both yeast and mammalian cells. The current view is that Rabs, in coordination with Rho proteins (ROPs in plants), regulate vesicle tethering by promoting rearrangements within, or assembly of, the exocyst complex (Grosshans et al. 2006). An important example in the context of root hair morphogenesis is the recruitment of a phosphatidylinostol kinase by RabA4b (Preuss et al. 2006). A yeast two hybrid screen with activated RabA4b has identified a phosphatidylinostol kinase (PI-4Kβ1) and the interaction has been confirmed and corroborated by colocalization studies (Preuss et al. 2006). PI-4Kβ1 was also shown to interact with a calcium sensor by yeast two hybrid analysis. It has therefore been postulated that Rab GTPases could integrate membrane traffic, phosphatidylinostol signaling, and the perception of a tip-focused calcium (Preuss et al. 2006). Although no mutant phenotype has been reported for RabA4b insertion mutants, double mutants between PI-4Kβ1 and its close relative PI-4Kβ2 have both defective root hair morphology and aberrant RabA4b-labeled membrane compartments (Preuss et al. 2006). In the *pi-4Kβ1/β2* double mutants, root hairs are shorter than in the wild-type and often abnormal, but no reduction in root hair number has been reported (Preuss et al. 2006). This phenotype is consistent with a defect in tip growth rather than root hair initiation, and would point to a role for phosphatidylinostol signaling in the maintenance of rather than in the establishment of polarity. The delivery of phosphatidylinostol 4-OH kinase to the tip of the root hair would lead to the enrichment of PI-4P at the tip and thus play a role in the maintenance of polarity.

5 Open Questions and Conclusions

In conclusion, I have outlined in this chapter a series of steps in membrane traffic that are based on a vast and sophisticated yeast and mammalian literature. Two lines of evidence suggest that the fundamental mechanisms I describe are conserved across kingdoms. First, all the players mentioned are found in the plant genome. Second, molecular genetic, localization and protein interaction studies have shown that the key regulators are required for root hair morphogenesis.

An important question remains the establishment and maintenance of polarity, and the signaling upstream of ROP GTPases that determine how ROP action is confined to a polarized region at the plasma membrane at the very onset of and throughout root hair morphogenesis. What is the initial positional cue and how is it encoded, interpreted, and transduced? With respect to trafficking, we also need to define the potential role of calcium in promoting exocytosis at the root hair tip. The nature of the vesicles and of their cargo needs to be addressed by identifying the v-SNAREs and further Rab GTPases that ensure specificity in membrane traffic. We also need to identify the t-SNARES that define the plasma membrane as the target membrane in exocytosis. Although the predominant trafficking event in root hair morphogenesis is polarized exocytosis, the importance of ER to Golgi traffic, of endocytosis and of vacuolar enlargement should not be neglected. In spite of all the gaps in our knowledge, the root hair is evolving as one of the best systems for our understanding of polarized secretion in plants.

Acknowledgments I am very grateful to Erik Nielsen, Shaul Yalovsky, Hyung-Taeg Cho, Anne Mie Emons, Tijs Ketelaar, and J.E. Rothman for their stimulating discussions and images. Research in my laboratory is supported by grant AS110/4–3 from the DFG.

References

Aoyama T (2008) Phospholipid signaling in root hair development. In: Emons AMC, Ketelaar T (eds) Root hairs: excellent tools for the study of plant molecular cell biology. Springer, Berlin Heidelberg New York. doi:10.1007/7089_2008_1

Assaad FF (2001) Of weeds and men: what genomes teach us about plant cell biology. Curr Opin Plant Biol 4:478–487

Assaad FF, Huet Y, Mayer U, Jurgens G (2001) The cytokinesis gene KEULE encodes a Sec1 protein that binds the syntaxin KNOLLE. J Cell Biol 152:531–543

Assaad FF, Qiu JL, Youngs H, Ehrhardt D, Zimmerli L, Kalde M, Wanner G, Peck SC, Edwards H, Ramonell K et al. (2004) The PEN1 syntaxin defines a novel cellular compartment upon fungal attack and is required for the timely assembly of papillae. Mol Biol Cell 15:5118–5129

Bagnat M, Simons K (2002) Cell surface polarization during yeast mating. Proc Natl Acad Sci USA 99:14183–14188

Bhat RA, Panstruga R (2005) Lipid rafts in plants. Planta 223:5–19

Bhat RA, Miklis M, Schmelzer E, Schulze-Lefert P, Panstruga R (2005) Recruitment and interaction dynamics of plant penetration resistance components in a plasma membrane microdomain. Proc Natl Acad Sci USA 102:3135–3140

Bibikova T, Gilroy S (2008) Calcium in root hair growth. In: Emons AMC, Ketelaar T (eds) Root hairs: excellent tools for the study of plant molecular cell biology. Springer, Berlin Heidelberg New York. doi:10.1007/7089_2008_3

Collins NC, Thordal-Christensen H, Lipka V, Bau S, Kombrink E, Qiu JL, Huckelhoven R, Stein M, Freialdenhoven A, Somerville SC, Schulze-Lefert P (2003) SNARE-protein-mediated disease resistance at the plant cell wall. Nature 425:973–977

Craxton M (2001) Genomic analysis of synaptotagmin genes. Genomics 77:43–49

D'Souza-Schorey C, Chavrier P (2006) ARF proteins: roles in membrane traffic and beyond. Nat Rev Mol Cell Biol 7:347–358

Emons AM, Traas JA (1986) Coated pits and coated vesicles on the plasma membrane of plant cells. Eur J Cell Biol 41:57–64

Fischer U, Ikeda Y, Ljung K, Serralbo O, Singh M, Heidstra R, Palme K, Scheres B, Grebe M (2006) Vectorial information for Arabidopsis planar polarity is mediated by combined AUX1, EIN2, and GNOM activity. Curr Biol 16:2143–2149

Geldner N, Friml J, Stierhof YD, Jurgens G, Palme K (2001) Auxin transport inhibitors block PIN1 cycling and vesicle trafficking. Nature 413:425–428

Geldner N, Anders N, Wolters H, Keicher J, Kornberger W, Muller P, Delbarre A, Ueda T, Nakano A, Jurgens G (2003) The Arabidopsis GNOM ARF-GEF mediates endosomal recycling, auxin transport, and auxin-dependent plant growth. Cell 112:219–230

Grosshans BL, Ortiz D, Novick P (2006) Rabs and their effectors: achieving specificity in membrane traffic. Proc Natl Acad Sci USA 103:11821–11827

Guo W, Tamanoi F, Novick P (2001) Spatial regulation of the exocyst complex by Rho1 GTPase. Nat Cell Biol 3:353–360

Hu C, Ahmed M, Melia TJ, Sollner TH, Mayer T, Rothman JE (2003) Fusion of cells by flipped SNAREs. Science 300:1745–1749

Ketelaar T, Emons AMC (2008) The actin cytoskeleton in root hairs: a cell elongation device. In: Emons AMC, Ketelaar T (eds) Root hairs: excellent tools for the study of plant molecular cell biology. Springer, Berlin Heidelberg New York. doi:10.1007/7089_2008_8

Latijnhouwers M, Hawes C, Carvalho C (2005) Holding it all together? Candidate proteins for the plant Golgi matrix. Curr Opin Plant Biol 8:632–639

Lavy M, Bloch D, Hazak O, Gutman I, Poraty L, Sorek N, Sternberg H, Yalovsky S (2007) A novel ROP/RAC effector links cell polarity, root-meristem maintenance, and vesicle trafficking. Curr Biol 17:947–952

Lee SH, Cho H.-T. (2008) Auxin and root hair morphogenesis. In: Emons AMC, Ketelaar T (eds) Root hairs: excellent tools for the study of plant molecular cell biology. Springer, Berlin Heidelberg New York. doi:10.1007/7089_2008_16

Martens S, Kozlov MM, McMahon HT (2007) How synaptotagmin promotes membrane fusion. Science 316:1205–1208

Mellman I, Warren G (2000) The road taken: past and future foundations of membrane traffic. Cell 100:99–112

Misura KM, Scheller RH, Weis WI (2000) Three-dimensional structure of the neuronal-Sec1-syntaxin 1a complex. Nature 404:355–362

Palazzo AF, Eng CH, Schlaepfer DD, Marcantonio EE, Gundersen GG (2004) Localized stabilization of microtubules by integrin- and FAK-facilitated Rho signaling. Science 303:836–839

Preuss ML, Serna J, Falbel TG, Bednarek SY, Nielsen E (2004) The Arabidopsis Rab GTPase RabA4b localizes to the tips of growing root hair cells. Plant Cell 16:1589–1603
Preuss ML, Schmitz AJ, Thole JM, Bonner HK, Otegui MS, Nielsen E (2006) A role for the RabA4b effector protein PI-4Kbeta1 in polarized expansion of root hair cells in Arabidopsis thaliana. J Cell Biol 172:991–998
Proszynski TJ, Klemm R, Bagnat M, Gaus K, Simons K (2006) Plasma membrane polarization during mating in yeast cells. J Cell Biol 173:861–866
Randazzo PA, Hirsch DS (2004) Arf GAPs: multifunctional proteins that regulate membrane traffic and actin remodelling. Cell Signal 16:401–413
Samaj J, Muller J, Beck M, Bohm N, Menzel D (2006) Vesicular trafficking, cytoskeleton and signalling in root hairs and pollen tubes. Trends Plant Sci 11:594–600
Sanderfoot AA, Assaad FF, Raikhel NV (2000) The Arabidopsis genome. An abundance of soluble N-ethylmaleimide-sensitive factor adaptor protein receptors. Plant Physiol 124:1558–1569
Shen J, Tareste DC, Paumet F, Rothman JE, Melia TJ (2007) Selective activation of cognate SNAREpins by Sec1/Munc18 proteins. Cell 128:183–195
Sieberer BJ, Timmers ACJ (2008) Microtubules in Plant Root Hairs and Their Role in Cell Polarity and Tip Growth. In: Emons AMC, Ketelaar T (eds) Root hairs: excellent tools for the study of plant molecular cell biology. Springer, Berlin Heidelberg New York. doi:10.1007/7089_2008_13
Song XF, Yang CY, Liu J, Yang WC (2006) RPA, a class II ARFGAP protein, activates ARF1 and U5 and plays a role in root hair development in Arabidopsis. Plant Physiol 141:966–976
Synek L, Schlager N, Elias M, Quentin M, Hauser MT, Zarsky V (2006) AtEXO70A1, a member of a family of putative exocyst subunits specifically expanded in land plants, is important for polar growth and plant development. Plant J 48:54–72
Vernoud V, Horton AC, Yang Z, Nielsen E (2003) Analysis of the small GTPase gene superfamily of Arabidopsis. Plant Physiol 131:1191–1208
Wachtler V, Balasubramanian MK (2006) Yeast lipid rafts? – an emerging view. Trends Cell Biol 16:1–4
Wen TJ, Hochholdinger F, Sauer M, Bruce W, Schnable PS (2005) The roothairless1 gene of maize encodes a homolog of sec3, which is involved in polar exocytosis. Plant Physiol 138:1637–1643
Willemsen VFJ, Grebe M, van den Toorn A, Palme K, Scheres B (2003) Cell polarity and PIN Protein positioning in Arabidopsis require sterol methylase I function. Plant Cell 15: 612–625
Xu J, Scheres B (2005) Dissection of Arabidopsis ADP-RIBOSYLATION FACTOR 1 function in epidermal cell polarity. Plant Cell 17:525–536
Zarsky V, Fowler J (2008) ROP (Rho-related protein from Plants) GTPases for spatial control of root hair morphogenesis. In: Emons AMC, Ketelaar T (eds) Root hairs: excellent tools for the study of plant molecular cell biology. Springer, Berlin Heidelberg New York. doi:10.1007/7089_2008_14
Zheng H, Kunst L, Hawes C, Moore I (2004) A GFP-based assay reveals a role for RHD3 in transport between the endoplasmic reticulum and Golgi apparatus. Plant J 37:398–414

Plant Cell Wall Biogenesis During Tip Growth in Root Hair Cells

E. Nielsen

Abstract In plants, cells are surrounded by a rigid cell wall, which restricts changes in cell shape and size; therefore, polarized secretion and deposition of cell wall components take on a particular importance during plant growth and development. In recent years, significant advances have been made in discovering and characterizing enzymes and proteins involved in the synthesis of many of the main cell wall polysaccharides. However, despite these advances little is known of the membrane-trafficking pathways responsible for polarized secretion in plants, and how the specific delivery of cell wall components to regions of cell expansion is controlled. In root hair cells, the majority of new cell wall deposition occurs in a highly polarized manner at the expanding tips of these cells. Additionally, reinforcement of the cell wall and deposition of secondary cell wall components occur selectively in the more distal portions of the root hair cell. In this chapter we will discuss some of the major classes of polysaccharides and structural proteins that are found in plant cell walls and relevant evidence for their selective deposition and function during plant cell wall biogenesis during root hair tip growth.

1 Introduction: Plant Cell Walls and Tip Growth

In plants, cell shape is defined by the cell wall, and changes in cell shape and size are dictated by the ability to modify existing cell walls or deposit newly synthesized cell wall material (reviewed in Cosgrove 2005). In root hair cells this cellular expansion is highly unidirectional and restricted to the tip of the growing hair. This specific type of cellular expansion is termed tip growth, and this process is shared by root hairs, pollen tubes, fungal hyphae, and apical cells in filamentous algae and lower plants. Tip growth differs from diffuse, also called intercalary, growth, a form of cell expansion that occurs in most other plant tissues. In diffuse growth, cellular expansion occurs along the entire face of one or more sides of the cell, whereas in

E. Nielsen
Department of Molecular, Cellular, and Developmental Biology, University of Michigan,
830 North University Avenue, Ann Arbor, MI 48109, USA
e-mail: nielsene@umich.edu

Plant Cell Monogr, doi:10.1007/7089_2008_11

tip growth expansion is limited to a single region of the cell wall. Additionally, in diffuse growth expansion of the cell wall is often highly anisotropic, or directional, and the axis of expansion is linked transversely to the orientation of cellulose microfibrils deposited in the cell wall (reviewed in Baskin 2005). However, in tip-growing root hairs, expansion at the tip of the root hair is isotropic and there is no obvious directionality to the orientation of the cellulose microfibrils in this region of the root hair cell (Emons 1989). As a result, the longitudinal expansion that occurs at the tip of the root hair is a result of the shape of the dome and does not appear to be limited by cellulose microfibril orientation in the cell wall (Dumais et al. 2004; Shaw et al. 2000).

In *Arabidopsis*, once root-hair-forming epidermal cells cease expanding in the direction of root growth, they initiate a bulge at the end of the cell nearest the root meristem. The bulge transitions to tip-restricted expansion and a root hair of uniform diameter (~10 μm) grows out to lengths of up to a millimeter (reviewed in Carol and Dolan 2006). This transition to tip-restricted growth is accompanied by a localized accumulation of ROP GTPases in the plasma membrane of the bulge (Jones et al. 2002), and action of *RHD2*, an NADPH oxidase, results in a localized reactive oxygen species accumulation (Foreman et al. 2003). The localized accumulation of reactive oxygen species leads ultimately to activation of calcium channels and establishment of localized increase in cytosolic calcium concentration $[Ca^{2+}]_c$ (Schiefelbein et al. 1993; Wymer et al. 1997). During tip growth, localized cell wall deposition occurs through an accumulation of secretory vesicles containing cell wall components in a vesicle-rich zone (VRZ) directly beneath the expanding root hair tip (Dolan 2001). As the root hair grows, both the VRZ (Fig. 1) and the increase in $[Ca^{2+}]_c$ are maintained directly behind the growing tip (Dolan 2001). Behind the VRZ is a subapical region rich in Golgi bodies, plastids, mitochondria, and other organelles, which is maintained until growth ceases in the mature root hair (Fig. 1). Once expansion has ceased, the tip-focused $[Ca^{2+}]_c$ gradient dissipates, and the vacuole migrates into the tip, displacing the subapical cytoplasm-rich region (Miller et al. 1997).

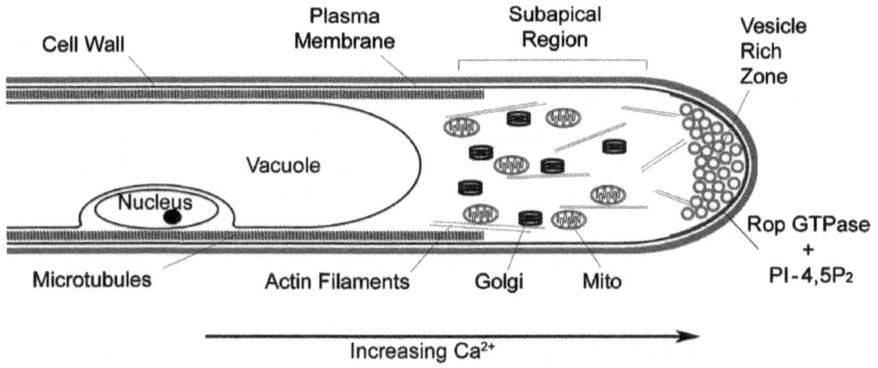

Fig. 1 A root hair cell

During tip-restricted growth, accurate delivery of membranes and new cell wall components is required for proper deposition of newly synthesized cellulose and delivery of hemicellulose polysaccharides and cell wall proteins. This is accomplished by polarized accumulation and fusion of secretory vesicles in the VRZ. While the nature of these secretory vesicles is still incompletely understood, they have previously been shown to contain hemicellulosic polysaccharides destined for the cell wall (Sherrier and VandenBosch 1994). More recently, *RABA4B*, a member of the RabA subfamily of Rab GTPases, has been shown to localize to vesicles that emerge from plant *trans*-Golgi network membranes and preferentially accumulate in the VRZ in growing root hairs (Preuss et al. 2004, 2006). Further, by immunoelectron microscopy, RabA4b-labeled vesicles from root meristem cells were labeled with monoclonal antibodies that recognize xyloglucan (Fig. 2), indicating that RabA4b-labeled vesicles represent at least one population of secretory vesicles involved in delivering newly synthesized cell wall materials to the growing plant cell walls.

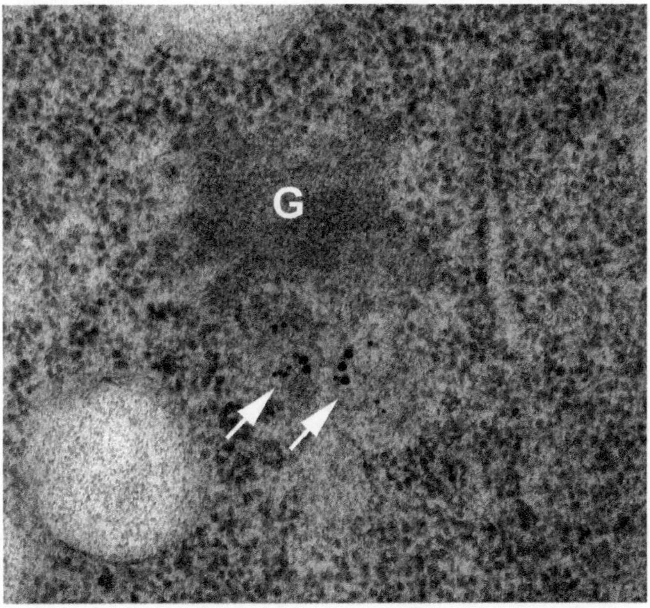

Fig. 2 Immunolocalization of RabA4b compartments xyloglucan polysaccharides. High-pressure frozen/freeze-substituted root tip cells from wild-type *A. thaliana* stably expressing EYFP-RabA4b were processed for immunoelectron microscopy analysis using affinity-purified anti-RabA4b antibodies (15 nm gold particles) and monoclonal anti-CCRC-M1 antibodies (5 nm gold particles). Anti-RabA4blabeled compartments specifically labeled tubular and vesicular elements (*arrows*) that were often in proximity to the *trans*-side of Golgi profiles (*G*). CCRC-M1 antibodies specifically label terminally fucosylated xyloglucan

2 Synthesis and Delivery of Cell Wall Components in Plants

Plant cell walls are composed of cellulose microfibrils embedded in a matrix of hemicelluloses and pectin (Cosgrove 1999). Synthesis of cellulose occurs at the plasma membrane of plant cells where microfibrils emerge from integral membrane complexes, termed rosette terminal complexes (Emons and Mulder 2000). These "rosette complexes" extrude the crystalline cellulose microfibril from the cell surface where they are incorporated into the expanding cell wall. In cells undergoing polarized growth, these "rosettes" were observed concentrated at the tips of *Adiantum* protonema filaments and in root hairs of *Equisetum hyemale*, and upon cessation of growth they rapidly disappeared from that location (Emons 1985; Schnepf et al. 1985; Wada and Staehelin 1981). These rosettes contain the catalytic subunit of cellulose synthase (Kimura et al. 1999), and have also been observed in Golgi cisternae and secretory vesicles (Haigler and Brown 1986). This distribution was interpreted as evidence that cellulose synthase complexes were delivered to the plasma membrane via membrane-trafficking intermediates (Emons and Mulder 2000). Interestingly, even though cellulose synthases localize to the plasma membrane, they do not have signal sequences characteristic of proteins entering the secretory pathway (Richmond and Somerville 2001). At present the membrane trafficking pathways responsible for the assembly and delivery of these complexes to the plasma membrane remain uncharacterized.

On the other hand, pectins and hemicellulosic polysaccharides are synthesized in the Golgi complex and then packaged into secretory vesicles for delivery to the plasma membrane via membrane trafficking. Incompletely characterized families of glycan synthases and glycosyltransferases synthesize noncellulosic polysaccharides (e.g., xyloglucan and pectin) and modify the sugar side chains of plant glycoproteins in plant Golgi complexes (Perrin et al. 2001). These cell wall components are then delivered via vesicle secretion to the plasma membrane for incorporation into the cell wall. Once released into the extracellular space, these components are incorporated into the cell wall by physical interactions or by cross-linking mechanisms of either chemical or enzymatic nature. Clearly, control and organization of plant membrane trafficking pathways must play critical roles during biogenesis and delivery of both cellulosic and noncellulosic components of plant cell walls. Additionally, cellulose synthases as well as several classes of cell wall proteins, such as arabinoglycan proteins, enter into the secretory pathway and are modified in the Golgi complex prior to delivery to the plasma membrane via membrane trafficking. So while the site of synthesis of these two classes of cell wall polysaccharides is distinct, both likely depend on proper sorting and targeting by membrane-trafficking events in order for proper cell wall deposition to occur.

3 Tip Growth: Cell Wall Deposition vs. Cell Wall Expansion

An important distinction during the consideration of cell wall deposition during tip-restricted growth in root hairs is that although cell wall deposition is clearly linked to cell expansion in the tip of the cell, continued cell wall deposition occurs in the nonexpanding tube of the root hair as well. Typically, cell wall deposition that occurs distal to the growing tip is considered secondary cell wall deposition (as these cell walls are no longer expanding), and it is likely that the components targeted to this region are distinct from those delivered to the expanding tip of the cell. Indeed, while cellulose microfibrils are deposited with random orientation at the tips of root hairs (Emons 1989), most studies indicated that microfibril deposition transitions from randomly oriented to axially-, helically-, or helicoidally oriented microfibril arrays in the more distal regions of the root hair tube (Emons and van Maaren 1987). Additionally, when cell walls of root hair cells were examined with monoclonal antibodies, CCRC-M1 antibodies, which recognized terminally fucosylated glycans (including xyloglucan and arabinogalactan), could be found throughout the cell wall, while CCRC-M2, which recognizes rhamnogalacturonan I (RGI), was restricted to the inner layers of the cell wall. CCRC-M7, which recognizes a arabinosyl (1 $\rightarrow$ 6)-linked β-galactan epitope, was observed only at the plasma membrane and the innermost layer of the cell wall in root hair cells (Freshour et al. 1996). This implies that components deposited in the primary cell wall differ from those incorporated at later periods during secondary cell wall deposition. Alternatively, these differences may simply be the result of masking of the reactive cell wall epitopes that these monoclonal antibodies recognize. However, this would still imply differences in the successive layers of the cell wall. Supporting evidence that cell wall components in tip-growing cells are altered over time comes from observed changes in the pectic wall components in *Physcomitrella patens*, in particular the monoclonal antibody LM6 specifically labeled walls at the apex of the protonemal cell, a position at which tip-restricted growth occurs (Lee et al. 2005). LM6 specifically recognizes arabinan side chains on pectin, suggesting that either root hair cell wall pectin is modified in more distal regions of the root hair tubes, or this epitope is masked by other cell wall components. Although similar studies have not yet been published in higher plants, treatment of tomato seedlings with β-Yariv's reagent, which binds to and interferes with pectic cell wall components and arabinoglycan proteins, results in reduced root hair initiation (Lu et al. 2001). Taken together, these observations highlight the likelihood that although cell wall deposition and the associated expansion in the root hair primarily occurs in the expanding tip, the additional cell wall deposition and modification of cell wall components occur at more distal regions of the root hair cell as well.

4 Cellulose Biosynthesis: Cellulose Synthases

In plants cellulose is the most abundant polymer in primary and secondary cell walls. In root hairs, the apex of the root hair cell contains only a thin primary cellulose cell wall for the first 20–30 μm of the root hair, at which point secondary cellulose layers are deposited (Galway et al. 1997; Newcomb and Bonnett 1965). In plants, cellulose is extruded from a plasma-membrane-localized complex consisting of at least three cellulose synthase subunits. Initially, cellulose synthase genes (CesA) were identified in bacteria (Aloni et al. 1982; Matthysse et al. 1995; Saxena et al. 1990), and then putative plant CesA genes were isolated from cotton fibers based on sequence similarity to the bacterial sequences and in vitro glycosyltransferase activity (Pear et al. 1996). In *A. thaliana* the mutant radial swollen 1 (RSW1) was identified to be a *CESA* protein (*CESA1*) by mutational analysis (Arioli et al. 1998). Subsequently, *A. thaliana* has been shown to contain ten *CESA* genes, of which at least three – *CESA1*, *CESA3*, and *CESA6*, are required for formation of a functional cellulose-synthesizing complex during primary cell wall formation (Scheible and Pauly 2004). Further secondary cellulose cell wall synthesis utilizes separate sets of genes, with *CESA4*, *CESA7*, and *CESA8* being required for secondary cell wall formation (Taylor et al. 2003). A single crystalline cellulose microfibril itself consists of a number of individual (1 → 4)-linked β-D-glucan polymers associated with one another by hydrogen bond linkages. Therefore, current models propose that the rosette complexes that produce cellulose microfibrils for primary and secondary cell wall synthesis are composed of six hexameric CesA complexes capable of synthesizing up to 36 individual (1 → 4)-linked β-D-glucan polymers.

The requirement for cellulose synthesis during tip growth can be shown by treating root hair cells with 2,6-dichlorobenzonitrile, a cellulose synthesis inhibitor, which causes elongating root hairs to burst rather than expand (Carol and Dolan 2002; Favery et al. 2001). Further, *rsw1* mutants containing a temperature-sensitive form of *CESA1* display radial swelling in all root tissues, including root hair epidermal cells, at nonpermissive temperatures (Williamson et al. 2001). Alternatively, in *prc1* mutants, which have defective *CESA6*, root hairs elongated properly but often displayed abnormal bulging at the base of the hair, suggesting that loss of this cellulose synthase subunit did not markedly affect tip growth (Desnos et al. 1996; Fagard et al. 2000). Together, these results indicate that the ability to deposit cellulose is required for root hair growth, but that mutations in different subunits of the cellulose synthase complex have different effects on tip expansion. Of particular interest is the observation that *CESA1* (*RSW1*) appears absolutely necessary for root hair tip growth, whereas *CESA6* (*PRC1*) is dispensable. Recent findings have shown that *CESA1* and *CESA3* subunits are required for proper formation of higher order cellulose synthase complexes, but that CesA6 is not and can be replaced with either CesA2 or CesA5 (Desprez et al. 2007). It is therefore possible that the lack of root hair phenotype in *prc1* mutants is simply due to redundancy with other CesA subunits, although the presence of cellulose synthase complexes with distinct subunit compositions also cannot be excluded. Clearly it will be interesting to determine whether cellulose synthase complexes containing these subunits are present in root hair cells and to what extent they participate in root hair tip growth vs. secondary cell wall deposition in the distal portions of root hair tubes.

5 Cellulose-Synthase-Like Proteins

The *CESA* cellulose synthases belong to a larger superfamily of proteins called cellulose-synthase-like (*CSL*) genes, which are in turn subdivided into eight subfamilies *CSLA-CSLH* (Richmond and Somerville 2000). Because many of the hemicellulose and pectin polysaccharides also contain β-D-linked glycan polymers, it is thought that members of the *CSL* gene families may provide the glycan synthase activities necessary for the production of the glycan backbones of a number of these cell wall polymers (e.g., xyloglucan, xylan, mannan, and glucomannan). Indeed, recent studies have shown that members of the *CSLA* gene family are responsible for synthesis of mannans and galactomannans (Dhugga et al. 2004; Liepman et al. 2005). Additionally, members of the *CSLF* gene family are only found in cereals, and these have been shown to be required for synthesis of the mixed linkage (1→3;1→4)-β-D-glucan polymers characteristic of the cell walls of grasses (Burton et al. 2006).

Intriguingly, in *A. thaliana* the *KJK* mutant was discovered to encode a member of the *CSLD* family (*CSLD3*; Favery et al. 2001; Wang et al. 2001). *OSCSLD1* was also shown to be necessary for proper root hair elongation in rice. It was expressed exclusively in root hairs, and in fact was the only member of four *OSCSLD* genes to be expressed in roots (Kim et al. 2007). Because the *CSLD* subfamily is most closely related to *CESA*, it was proposed that *CSLD* may participate in cellulose synthesis in the root hair cell, which would be consistent as *kjk* mutant root hairs often burst in a similar fashion as cells treated with the cellulose-synthase inhibitor 2,6-dichlorobenzonitrile (Carol and Dolan 2006). However, fluorescently tagged *CSLD3* was shown to reside on internal membranes, possibly the ER, a localization that would not match with a cellulose synthase activity in root hair plasma membranes. In addition, a recent study showed that *ATCSLD5* knock-out plants had lower accumulation of xylans in stems and reduced levels of xylan and homogalacturonan synthase activities (Bernal et al. 2007). Whether the CslD3 fluorescent fusions would localize properly has been called into question as chimeric CesA protein fusions have now been shown to only be functional when fusions are made at the amino-terminal end of the CesA protein (Wang et al. 2006). Indeed, AtCslD5 and AtCslD3 were both shown to mislocalize to the ER when fused with YFP at its carboxyl-terminal end. However, upon transferal of the fusion to the amino-terminal end, these two proteins localized to Golgi compartments when transiently expressed in *Nicotiana benthamiana* cells (Bernal et al. 2007). Clearly the question of *CSLD3* function in root hair cell wall biosynthesis cannot be answered until we know more about the function of this class of proteins.

6 Xyloglucan Synthesis and Remodeling

A major hemicellulose polysaccharide in primary cell walls is xyloglucan, which is thought to cross-link cellulose microfibrils either via direct linkage or indirectly (reviewed in Cosgrove 2005). This cross-linking role during cell enlargement has been confirmed by experiments in which plants are treated with auxin, which results in rapid breakdown of xyloglucan and associated cellular expansion. Further, when isolated plant cell walls were treated with an endoglucanase that degrades xyloglucan, the cell walls could be induced to expand (Yuan et al. 2001). As a result, regulation of xyloglucan deposition and control of its degradation or remodeling likely play important roles during the initial cell wall bulge formation and in regulating the shape of the root hair during tip-restricted expansion. Immunofluorescence labeling of *A. thaliana* seedlings with monoclonal CCRC-M1 antibodies that recognize terminally fucosylated xyloglucan normally labels both root hair cells and non-root-hair epidermal cells. In *mur1* mutants, initially isolated by their lack of fucose accumulation in aerial tissues, root tissues contain only ~40% of the levels of fucose in wild-type roots (Reiter et al. 1993). In *mur1* seedlings, only elongating root hair cells were strongly labeled with CCRC-M1, and not the cell walls in the body of the root hair cell (Freshour et al. 2003).

First evidence for a direct role of xyloglucan in root hair growth was uncovered by the cloning of the root hair defective mutant *rhd1*. *RHD1* was identified as a UDP-D-glucose 4-epimerase, which is responsible for conversion of UDP-D-glucose into UDP-D-galactose. Loss of *RHD1* activity resulted in impaired accumulation of galactosylated xyloglucan, as well as arabinosylated (1→6) β-D-galactan (Seifert et al. 2002). This in turn results in root epidermal cell bulging, including root hair cells. Because impaired production of UDP-D-galactose may affect biosynthesis of multiple polysaccharides, as well as arabinogalactan proteins, it was initially unclear which of these was ultimately responsible for root hair bulging (Andeme-Onzighi et al. 2002). More recent work on *REB1*, which is allelic to *RHD1*, has indicated that loss of *REB1/RHD1* function specifically alters the insoluble pool of xyloglucan and arabinoglycan proteins, but that the galactose-containing polysaccharides rhamnogalacturonan I and II remain unchanged (Nguema-Ona et al. 2006). The authors speculate that different forms of UDP-D-glucose 4-epimerase may function to provide distinct UDP-D-galactose pools which are incorporated into separate classes of polysaccharides.

Additional evidence that biosynthesis of xyloglucan is required for proper root hair expansion has recently been demonstrated in studies of two putative xyloglucan xylosyltransferase enzymes, *AtXT1* and *AtXT2* (Cavalier and Keegstra 2006; Faik et al. 2002). When expressed in insect cells, these xylosyltransferases have been shown to catalyze the addition of xylose to either cellopentaose or cellohexaose to produce xyloglucan-like oligosaccharides. Interestingly, *A. thaliana* plants in which both of these genes are knocked out lack detectable xyloglucan and display shortened and bulged root hairs (Cavalier, Lerouxel, and Keegstra, personal communication), further indicating the essential role of xyloglucan cell wall components in root hair growth.

Control of xyloglucan remodeling also appears to play important roles in root hair tip growth. Once newly synthesized xyloglucan polysaccharides are delivered to the plasma membrane via secretion, they must be incorporated into the existing network of cellulose cross-linked to xyloglucan. Integration into this network is thought to be mediated by a family of xyloglucan endotransglycosylases (XTH). These enzymes both hydrolyze xyloglucan β-glucan backbone chains and then also rejoin free ends of other xyloglucan chains (Nishitani and Tominaga 1992; Purugganan et al. 1997; Steele et al. 2001). By using fluorescently labeled xyloglucan oligosaccharides, it is possible to observe spatially where XTH activity is present in plant cell walls (Ito and Nishitani 1999). In root hairs, initial bulge formation was shown to be associated with a localized increase in XTH activity (Vissenberg et al. 2001). Interestingly, while bulge initiation required XTH activity, tip-restricted growth did not appear to require its maintenance. Furthermore, while XTH activity was present in more distal portions of the root hair tube, this did not appear to play a role in expansion but possibly in cell wall reinforcement (Vissenberg et al. 2001). These results highlight the necessity not only for the presence of xyloglucan but also for the proper control of xyloglucan remodeling as well during root hair growth and expansion.

7 Pectins

Pectins generally refer to a collection of heterogeneous polysaccharides that are easily extracted from cell wall preparations, and which are present in primary cell walls but absent or significantly reduced in secondary cell walls (Mohnen 1999). These complex polysaccharides may contain as many of 17 distinct monosaccharides, but generally can be defined by a high content of galacturonic acid (Willats et al. 2001). There are four main classes of pectic polysaccharides – homogalacturonan (HGA), rhamnogalacturonan I (RGI), rhamnogalacturonan II (RGII), and xylogalacturonan (reviewed in Cosgrove 2005; Willats et al. 2001). These pectic polysaccharides probably play distinct roles during primary plant cell wall deposition, and their modification likely alters cell wall characteristics. As a result deposition and modification of pectic polysaccharides likely also play important roles during tip growth in root hairs as well.

One of the major characteristics of HGA is its ability to form gels in complex with calcium ions. This calcium cross-linking and gel formation is probably essential for determining cell wall integrity and its physical characteristics (Jarvis 1984, 1992). During its synthesis in the Golgi, HGA is methyl esterified and is delivered to the plasma membrane in a highly (70–80%) methyl-esterified form. In this form it is unable to bind calcium ions and therefore unable to integrate into a pectin gel matrix. As a result, pectin deesterification is an important aspect determining cell wall characteristics. In tip-growing cells, continuous delivery of new cell wall components to the cell apex results in highly methyl-esterified HGA being localized to the tips of these cells. Pectin methyl esterases (PME) then remove the methyl groups, thus altering the physical characteristics of HGA. Although the actions of PMEs have not been extensively studied in root hairs, actions of these PMEs are known to be key regulators of pollen tip growth. Vanguard 1 (*VGD1*) is a PME-like protein expressed specifically in pollen, and in *vgd1* mutants pollen germinates normally but grows much more slowly and bursts much more readily than wild-type pollen (Jiang et al. 2005). Further, exogenous treatment of lily and tobacco pollen tubes with purified PME preparations resulted in inhibition of pollen tube elongation (Bosch et al. 2005). Interestingly, while overexpression of the PME domain alone of *NtPPME1* resulted in inhibition of tobacco pollen tube growth, expression of the full *NtPPME1* gene did not (Bosch et al. 2005), and antisense inhibition of *NtPPME1* accumulation also resulted in inhibition of tobacco pollen tube growth (Bosch and Hepler 2006). These results have been interpreted to indicate that PME activity is tightly regulated by both intramolecular domains, as well as being influenced by local pH and possibly by a class of small proteins termed pectin methylesterase inhibitors, or PMEI proteins (Di Matteo et al. 2005; reviewed in Bosch and Hepler 2005). Clearly, PMEs play important roles in pollen tube elongation; whether they play similar roles in root hair elongation remains to be determined.

Evidence for an important role of RGI and RGII in root hair expansion comes from an analysis of suppressor mutants that restore normal root hair morphology in a root hair developmental mutant *lrx1*. *LRX1* encodes an LRR (leucine-rich repeat)-extensin protein that is exclusively expressed in root hair epidermal cells, and *lrx1* mutants have short, irregular and sometimes branched root hair morphologies (Baumberger et al. 2001). Mutation of *RHM1*, which converts UDP-D-glucose to UDP-L-rhamnose (Oka et al. 2007; Reiter and Vanzin 2001), was sufficient to suppress the *lrx1* phenotype (Diet et al. 2006). In *rhm1* mutants, levels of 2-*O*-methyl-L-fucose and 2-*O*-methyl-D-xylose, which are specific markers for RGII, were reduced. Monoclonal antibodies specific to RGI showed reduced levels of accumulation in the *rhm1* mutant as well (Diet et al. 2006). These results suggest that the *lrx1* root hair defects are suppressed by alterations in the rhamnogalacturonan pectic polysaccharides.

8 Structural Cell Wall Proteins

Expansin proteins were initially identified in studies of "acid-growth" during cell wall expansion (Cosgrove and Li 1993; McQueen-Mason et al. 1992). Typically expanding cells maintain a cell wall pH between 4.5 and 6, which activates expansin activity. This rapid cell wall enlargement is mostly a result of the actions of wall-loosening enzymes, of which expansins are probably the dominant form (reviewed in Cosgrove et al. 2002). Expansins, which are subdivided into two subfamilies the α- and β-expansins, are 25–28-kDa proteins that contain two distinct conserved domains. The amino-terminal domain displays some similarity to endoglucanases, although no activity has been measured in vitro (McQueen-Mason and Cosgrove 1995). A second, carboxyl-terminal domain, which shows some similarity to pollen allergen proteins, has conserved aromatic and polar residues and may bind polysaccharides (Cosgrove 1997). The primary difference between the α- and β-expansin families is the absence of glycosylation sites in the a-expansin family, but the significance of this distinction remains unclear (Downes et al. 2001; Petersen et al. 1995; Pike et al. 1997). Expansins are thought to disrupt noncovalent binding of cell wall polysaccharides to one another (e.g., cellulose and xyloglucan), thereby allowing slippage and increase in cell size due to turgor pressure. It is likely that the disruption of polysaccharide interactions occurs between cellulose and other cell wall polysaccharides as artificial composites of cellulose and xyloglucan could be weakened by expansin treatment, but artificial composites of cellulose and other polysaccharides or of cellulose alone were not (Whitney et al. 2000).

In *A. thaliana*, initiation of root hair bulge formation is tightly associated with expression of two α-expansin genes, *AtEXP7* and *AtEXP18*, in root hair cells (Cho and Cosgrove 2002). Because specification of hair or non-hair cell fate occurs much earlier than bulge initiation during root hair development (reviewed in Schiefelbein 2000), the observation that expression of *AtEXP7* and *AtEXP18* occurs simultaneously with bulge initiation strongly suggests that the presence of these two α-expansins is functionally linked to the hair bulge formation. Interestingly, treatment of growing root hair cells with exogenously applied α-expansin protein resulted in either root hair bursting at high concentrations, or formation of root hair bulges at the growing tip of the cell at lower concentrations (Cosgrove et al. 2002). These results indicated that α-expansins likely modulate cell wall expansion in the tip region, but root hair cell walls are resistant to the cell wall loosening action of these proteins in more distal regions of the root hair tube. Taken together, these results indicate that the cell wall loosening properties of the α-expansin gene family play important roles during bulge initiation and in the isotropic expansion that occurs in the tip region of growing root hair cells. Whether these effects are required for root hair initiation and/or tip growth remains to be determined, as does the question of whether β-expansins play similar or distinct roles during tip growth.

Using reverse genetics techniques, a novel class of cell wall proteins that play important roles in root hair growth and development has been identified (Baumberger et al. 2001). These proteins, termed LRX proteins, consist of an amino-terminal signal peptide, followed by leucine-rich repeat domains, and a carboxyl terminal region that shares some degree of sequence similarity to extensin proteins (Baumberger et al. 2001). It should be noted, however, that the carboxyl-terminal extensin-like domain is somewhat distantly related to the α- and β-expansins and may perform distinct functions with regard to cell wall loosening or organization. In *A. thaliana*, *LRX1* was found to be exclusively expressed in root hair cells in the differentiation zone, with levels of expression reduced upon cessation of root hair elongation in more mature regions of the root, and in *lrx1* T-DNA insertional mutants root hairs failed to develop normally and often displayed branches or bulges along the length of the root hair (Baumberger et al. 2001). Double mutants of *LRX1* and a closely related *LRX2* showed synergistic effects, with root hairs often bursting early during tip growth of the cell (Baumberger et al. 2003). More recent analysis of suppressor mutants that alleviate the *lrx1* root hair phenotype has identified elements involved in RGI and RGII biosynthesis, indicating a potential role for this class of cell wall proteins in regulation or organization of pectic polysaccharides in the root hair cell wall (Diet et al. 2006).

Proline-rich proteins (PRP) were originally identified as proteins that accumulated in cell walls in response to wounding (Chen and Varner 1985; Tierney et al. 1988). More recently, this class of cell wall proteins has been shown to be expressed differentially during development of various tissues (Franssen et al. 1987; Hong et al. 1989; Sheng et al. 1991) and specific cell and tissue types (Suzuki et al. 1993; Wyatt et al. 1992). PRPs are thought to function in the cell wall by cross-linking to other cell wall components, and members of this class of proteins rapidly become insoluble in cell wall fractions upon initiation of defense responses (Bradley et al. 1992; Francisco and Tierney 1990). This has led to the notion that PRP proteins may be involved in strengthening and reinforcement of cell walls via their cross-linking characteristics. In *A. thaliana* two members of this class of proteins, *AtPRP1* and *AtPRP3*, are expressed in a root-hair-specific fashion (Bernhardt and Tierney 2000; Fowler et al. 1999). Promotion of root hair development by modulation of ethylene and auxin hormone levels resulted in higher levels of expression of these root-hair-specific PRP genes, whereas PRP expression levels were reduced in *A. thaliana* root hair mutants (Bernhardt and Tierney 2000). These results are consistent with *AtPRP1* and *AtPRP3* playing direct roles in root hair growth. Owing to the polarized nature of cellular expansion in tip-growing root hair cells, it would be interesting to determine whether subcellular distributions of these PRPs would be restricted to regions more distal to the expanding tips of the root hair cells where cell wall expansion is limited. Further experiments are clearly necessary to determine the possible roles of this class of proteins during root hair growth and development.

9 Conclusions and Directions

Over the last 10 years the integration of genetic techniques into the study of cell wall biogenesis has resulted in the identification of a number of the enzymes involved in cell wall biogenesis. This, combined with advanced microscopy and improved biochemical methods, has allowed for a rapid increase in our understanding of the biosynthesis of many of the main components of the cell walls of living cells. Although the machinery for synthesis of several cell wall polysaccharides has been identified (e.g., cellulose synthases), we still have a poor understanding of how others are made, and in general, we still have incomplete knowledge of how these cell wall polysaccharides are integrated into existing cell walls. Our understanding of the functions and roles of structural cell wall proteins is even less well developed. Clearly, there is a great deal to be learned about the basic mechanisms underlying plant cell wall synthesis and deposition. In addition, root hairs undergo a highly polarized type of expansion, which in a number of ways is quite distinct from that which occurs in most other plant cells. Understanding how synthesis and deposition of cell wall components in root hair cells differ from those taking place in cells undergoing the more common diffuse expansion will aid greatly in our knowledge of how these processes are integrated to allow for the changes in morphology that occur with cell differentiation.

10 Root Hair Immunolocalization Protocol

1. Need twice the number of slides as you have samples. Clean slides well with soap, rinse with DI water and then with ethanol, and air dry them.
2. Clean paintbrush with acetone, then make 0.5% PEI and brush it on slides (only half of the slides). let it air dry.

3. Make fixative – 4% paraformaldehyde, 0.1% glutaraldehyde, 0.5% Triton X-100, and 1× PME. Add 1mL to each slide, then add the seedling(s). Leave in humid chamber for 45 min.
4. After 45 min, dissolve cellulase/pectinase with liquid nitrogen. C/P is 1.5% cellulase, 0.15% pectinase, 1× PME (1 mL per slide) (0.06 and 0.006 for 4 mL).
5. Tip slides and remove most liquid; cut off shoots.
6. Place another slide on top at an angle (no PEI).
7. Smush, freeze in liquid nitrogen, and while still frozen, break slides apart.
8. Place PEI-coated slide (with roots) in humid chamber, thaw briefly, and cover with 1× PME (add to the side, not directly to the tissue).
9. Wash with 1× PME thrice, 5 min each. Definition of wash is to pick up the slide, dump off liquid, add wash liquid, dump off, then put in chamber and add 1× PME and sit for 5 min.
10. Incubate with cellulase/pectinase for 30 min.
11. Wash thrice with 1× PME.
12. Incubate in 0.5% Triton X-100, 1× PME for 10 min.
13. Wash thrice with 1× PME
14. Place slides in methanol at –20°C for 10 min. Do this in the freezer.
15. Transfer slides quickly to 1× PBS at RT for 10 min.
16. Place slides back in humid chamber and cover them with Ab dilution solution.
17. Incubate for 10 min at RT.
18. Remove all excess liquid with a paper towel. Incubate in primary Ab for 1 h at RT (used 1:100 dilution, made enough for 100 μL/slide).
19. Wash thrice in 1× PBS.
20. Remove all excess liquid with a paper towel. Incubate in secondary Ab for 45 min at RT under dark (used 1:200 dilution).
21. Wash thrice with 1× PBS – do these washes very well. Rinse twice in hand before standing them for 5 min with the wash. Keep lid on chamber during washes.
22. Remove all excess liquid and add 50–60 μL MOVIOL (at –20°C).
23. Add cover slip slowly using forceps.
24. Seal the four corners with nail polish. Once that is dry seal all the way around.
25. Observe the same day.

2 × PME

– 100 mM PIPES
– 2 mM $MgSO_4$
– 10 mM EGTA

Bring the pH to 6.9 with KOH.

Antibody dilution solution

– 3% BSA
– 0.05% Tween 20
– 0.02% NaN_3
– 1× PBS

Acknowledgments I thank David Cavalier, Oliver Lerouxel, and Ken Keegstra for sharing their unpublished results, and Byung-Ho Kang for providing the immunoelectron micrograph of RabA4b-labeled secretory vesicles containing cargo recognized by the CCCRC-M1 antibodies. This work was supported by a Department of Energy grant DE-FG02–03ER15412.

References

Aloni Y, Delmer DP, Benziman M (1982) Achievement of high rates of in vitro synthesis of 1, 4-beta-D-glucan: activation by cooperative interaction of the Acetobacter xylinum enzyme system with GTP, polyethylene glycol, and a protein factor. *Proc Natl Acad Sci USA* 79:6448–6452

Andeme-Onzighi C, Sivaguru M, Judy-March J, Baskin TI, Driouich A (2002) The reb1–1 mutation of Arabidopsis alters the morphology of trichoblasts, the expression of arabinogalactan-proteins and the organization of cortical microtubules. *Planta* 215:949–958

Arioli T, Peng L, Betzner AS, Burn J, Wittke W, Herth W, Camilleri C, Hofte H, Plazinski J, Birch R, Cork A, Glover J, Redmond J, Williamson RE (1998) Molecular analysis of cellulose biosynthesis in Arabidopsis. *Science* 279:717–720

Baskin TI (2005) Anisotropic expansion of the plant cell wall. *Annu Rev Cell Dev Biol* 21:203–222

Baumberger N, Ringli C, Keller B (2001) The chimeric leucine-rich repeat/extensin cell wall protein LRX1 is required for root hair morphogenesis in Arabidopsis thaliana. *Genes Dev* 15:1128–1139

Baumberger N, Steiner M, Ryser U, Keller B, Ringli C (2003) Synergistic interaction of the two paralogous Arabidopsis genes LRX1 and LRX2 in cell wall formation during root hair development. *Plant J* 35:71–81

Bernal AJ, Jensen JK, Harholt J, Sorensen S, Moller I, Blaukopf C, Johansen B, de Lotto R, Pauly M, Scheller HV, Willats WG (2007) Disruption of ATCSLD5 results in reduced growth, reduced xylan and homogalacturonan synthase activity and altered xylan occurrence in Arabidopsis. *Plant J* 52(5):791–802

Bernhardt C, Tierney ML (2000) Expression of AtPRP3, a proline-rich structural cell wall protein from Arabidopsis, is regulated by cell-type-specific developmental pathways involved in root hair formation. *Plant Physiol* 122:705–714

Bosch M, Hepler PK (2005) Pectin methylesterases and pectin dynamics in pollen tubes. *Plant Cell* 17:3219–3226

Bosch M, Hepler PK (2006) Silencing of the tobacco pollen pectin methylesterase NtPPME1 results in retarded in vivo pollen tube growth. *Planta* 223:736–745

Bosch M, Cheung AY, Hepler PK (2005) Pectin methylesterase, a regulator of pollen tube growth. *Plant Physiol* 138:1334–1346

Bradley DJ, Kjellbom P, Lamb CJ (1992) Elicitor- and wound-induced oxidative cross-linking of a proline-rich plant cell wall protein: a novel, rapid defense response. *Cell* 70:21–30

Burton RA, Wilson SM, Hrmova M, Harvey AJ, Shirley NJ, Medhurst A, Stone BA, Newbigin EJ, Bacic A, Fincher GB (2006) Cellulose synthase-like CslF genes mediate the synthesis of cell wall (1,3;1,4)-beta-D-glucans. *Science* 311:1940–1942

Carol RJ, Dolan L (2002) Building a hair: tip growth in Arabidopsis thaliana root hairs. *Philos Trans R Soc Lond B Biol Sci* 357:815–821

Carol RJ, Dolan L (2006) The role of reactive oxygen species in cell growth: lessons from root hairs. *J Exp Bot* 57:1829–1834

Cavalier DM, Keegstra K (2006) Two xyloglucan xylosyltransferases catalyze the addition of multiple xylosyl residues to cellohexaose. *J Biol Chem* 281:34197–34207

Chen J, Varner JE (1985) Isolation and characterization of cDNA clones for carrot extensin and a proline-rich 33-kDa protein. *Proc Natl Acad Sci USA* 82:4399–4403

Cho HT, Cosgrove DJ (2002) Regulation of root hair initiation and expansin gene expression in Arabidopsis. *Plant Cell* 14:3237–3253

Cosgrove DJ (1997) Relaxation in a high-stress environment: the molecular bases of extensible cell walls and cell enlargement. *Plant Cell* 9:1031–1041

Cosgrove DJ (1999) Enzymes and other agents that enhance cell wall extensibility. *Annu Rev Plant Physiol Plant Mol Biol* 50:391–417

Cosgrove DJ (2005) Growth of the plant cell wall. *Nat Rev Mol Cell Biol* 6:850–861

Cosgrove DJ, Li ZC (1993) Role of expansin in cell enlargement of oat coleoptiles (Analysis of developmental gradients and photocontrol). *Plant Physiol* 103:1321–1328

Cosgrove DJ, Li LC, Cho HT, Hoffmann-Benning S, Moore RC, Blecker D (2002) The growing world of expansins. *Plant Cell Physiol* 43:1436–1444

Desnos T, Orbovic V, Bellini C, Kronenberger J, Caboche M, Traas J, Hofte H (1996) Procuste1 mutants identify two distinct genetic pathways controlling hypocotyl cell elongation, respectively in dark- and light-grown Arabidopsis seedlings. *Development* 122:683–693

Desprez T, Juraniec M, Crowell EF, Jouy H, Pochylova Z, Parcy F, Hofte H, Gonneau M, Vernhettes S (2007) Organization of cellulose synthase complexes involved in primary cell wall synthesis in Arabidopsis thaliana. *Proc Natl Acad Sci USA* 104:15572–15577

Dhugga KS, Barreiro R, Whitten B, Stecca K, Hazebroek J, Randhawa GS, Dolan M, Kinney AJ, Tomes D, Nichols S, Anderson P (2004) Guar seed beta-mannan synthase is a member of the cellulose synthase super gene family. *Science* 303:363–366

Di Matteo A, Giovane A, Raiola A, Camardella L, Bonivento D, De Lorenzo G, Cervone F, Bellincampi D, Tsernoglou D (2005) Structural basis for the interaction between pectin methylesterase and a specific inhibitor protein. *Plant Cell* 17:849–858

Diet A, Link B, Seifert GJ, Schellenberg B, Wagner U, Pauly M, Reiter WD, Ringli C (2006) The Arabidopsis root hair cell wall formation mutant lrx1 is suppressed by mutations in the RHM1 gene encoding a UDP-L-rhamnose synthase. *Plant Cell* 18:1630–1641

Dolan L (2001) How and where to build a root hair. *Curr Opin Plant Biol* 4:550–554

Downes BP, Steinbaker CR, Crowell DN (2001) Expression and processing of a hormonally regulated beta-expansin from soybean. *Plant Physiol* 126:244–252

Dumais J, Long SR, Shaw SL (2004) The mechanics of surface expansion anisotropy in Medicago truncatula root hairs. *Plant Physiol* 136:3266–3275

Emons A (1989) Helicoidal microfibril deposition in a tip-growing cell and microtubule alignment during rip morphogenesis: a dry-cleaving and freeze -substitution study. *Can J Bot* 67:2401–2408

Emons AM (1985) Plasma membrane rosettes in root hairs of *Equisetum hyemale. Planta* 163:350–359

Emons AMC, van Maaren N (1987) Helicoidal cell wall texture in root hairs. Planta 170:145–151

Emons AM, Mulder BM (2000) How the deposition of cellulose microfibrils builds cell wall architecture. *Trends Plant Sci* 5:35–40

Fagard M, Desnos T, Desprez T, Goubet F, Refregier G, Mouille G, McCann M, Rayon C, Vernhettes S, Hofte H (2000) PROCUSTE1 encodes a cellulose synthase required for normal cell elongation specifically in roots and dark-grown hypocotyls of Arabidopsis. *Plant Cell* 12:2409–2424

Faik A, Price NJ, Raikhel NV, Keegstra K (2002) An Arabidopsis gene encoding an alpha-xylosyltransferase involved in xyloglucan biosynthesis. *Proc Natl Acad Sci USA* 99:7797–802

Favery B, Ryan E, Foreman J, Linstead P, Boudonck K, Steer M, Shaw P, Dolan L (2001) KOJAK encodes a cellulose synthase-like protein required for root hair cell morphogenesis in Arabidopsis. *Genes Dev* 15:79–89

Foreman J, Demidchik V, Bothwell JHF, Mylona P, Miedema H, Torres MA, Linstead P, Costa S, Brownlee C, Jones JDG, Davies JM, Dolan L (2003) Reactive oxygen species produced by NADPH oxidase regulate plant cell growth. *Nature* 422:442–446

Fowler TJ, Bernhardt C, Tierney ML (1999) Characterization and expression of four proline-rich cell wall protein genes in Arabidopsis encoding two distinct subsets of multiple domain proteins. *Plant Physiol* 121:1081–1092

Francisco SM, Tierney ML (1990) Isolation and characterization of a proline-rich cell wall protein from soybean seedlings. *Plant Physiol* 94:1897–1902

Franssen HJ, Nap JP, Gloudemans T, Stiekema W, Van Dam H, Govers F, Louwerse J, Van Kammen A, Bisseling T (1987) Characterization of cDNA for nodulin-75 of soybean: a gene product involved in early stages of root nodule development. *Proc Natl Acad Sci USA* 84:4495–4499

Freshour G, Clay RP, Fuller MS, Albersheim P, Darvill AG, Hahn MG (1996) Developmental and tissue-specific structural alterations of the cell-wall polysaccharides of arabidopsis thaliana roots. *Plant Physiol* 110:1413–1429

Freshour G, Bonin CP, Reiter WD, Albersheim P, Darvill AG, Hahn MG (2003) Distribution of fucose-containing xyloglucans in cell walls of the mur1 mutant of Arabidopsis. *Plant Physiol* 131:1602–1612

Galway ME, Heckman JW, Jr. Schiefelbein JW (1997) Growth and ultrastructure of Arabidopsis root hairs: the rhd3 mutation alters vacuole enlargement and tip growth. *Planta* 201: 209–218

Haigler CH Brown RM Jr (1986) Transport of rosettes from the golgi apparatus to the plasma membrane in isolated mesophyll cells of *Zinnia elegans.* during differentiation to tracheary elements in suspension culture. *Protoplasma* 134:111–120

Hong JC, Nagao RT, Key JL (1989) Developmentally regulated expression of soybean proline-rich cell wall protein genes. *Plant Cell* 1:937–943

Ito H, Nishitani K (1999) Visualization of EXGT-mediated molecular grafting activity by means of a fluorescent-labeled xyloglucan oligomer. *Plant Cell Physiol* 40:1172–1176

Jarvis MC (1984) Structure and properties of pectin gels in plant cell walls. *Plant Cell Environ* 7:153–164

Jarvis MC (1992) Control of thickness of collenchyma cell walls by pectins. *Planta* 187:218–220

Jiang L, Yang SL, Xie LF, Puah CS, Zhang XQ, Yang WC, Sundaresan V, Ye D (2005) VANGUARD1 encodes a pectin methylesterase that enhances pollen tube growth in the Arabidopsis style and transmitting tract. *Plant Cell* 17:584–596

Jones MA, Shen JJ, Fu Y, Li H, Yang Z, Grierson CS (2002) The Arabidopsis Rop2 GTPase is a positive regulator of both root hair initiation and tip growth. *Plant Cell* 14:763–776

Kim CM, Park SH, Je BI, Park SH, Park SJ, Piao HL, Eun MY, Dolan L, Han CD (2007) OsCSLD1, a cellulose synthase-like D1 gene, is required for root hair morphogenesis in rice. *Plant Physiol* 143:1220–1230

Kimura S, Laosinchai W, Itoh T, Cui X, Linder CR, Brown RM, Jr. (1999) Immunogold labeling of rosette terminal cellulose-synthesizing complexes in the vascular plant vigna angularis. *Plant Cell* 11:2075–2086

Lee KJ, Sakata Y, Mau SL, Pettolino F, Bacic A, Quatrano RS, Knight CD, Knox JP (2005) Arabinogalactan proteins are required for apical cell extension in the moss Physcomitrella patens. *Plant Cell* 17:3051–3065

Liepman AH, Wilkerson CG, Keegstra K (2005) Expression of cellulose synthase-like (Csl) genes in insect cells reveals that CslA family members encode mannan synthases. *Proc Natl Acad Sci USA* 102:2221–2226

Lu H, Chen M, Showalter AM (2001) Developmental expression and perturbation of arabinogalactan-proteins during seed germination and seedling growth in tomato. *Physiol Plant* 112:442–450

Matthysse AG, White S, Lightfoot R (1995) Genes required for cellulose synthesis in Agrobacterium tumefaciens. *J Bacteriol* 177:1069–1075

McQueen-Mason S, Durachko DM, Cosgrove DJ (1992) Two endogenous proteins that induce cell wall extension in plants. *Plant Cell* 4:1425–1433

McQueen-Mason SJ, Cosgrove DJ (1995) Expansin mode of action on cell walls. Analysis of wall hydrolysis, stress relaxation, and binding. *Plant Physiol* 107:87–100

Miller DD, Norbert CA et al (1997) From signal to form: aspects of the cytoskeleton-plasma membrane-cell wall continuum in root hair tips. *J Exp Bot* 48:1881–1896

Mohnen D (1999) Biosynthesis of pectins and galactomannans. In: Nakanishi K, Meth-Cohn O (eds) Comprehensive natural products chemistry, vol 3. Elsevier, *Amsterdam*, pp 497–527

Newcomb EH, Bonnett HT (1965) Cytoplasmic microtubules and cell wall microfibril orientation in root hairs of radish. *J Cell Biol* 27:575–589

Nguema-Ona E, Andeme-Onzighi C, Aboughe-Angone S, Bardor M, Ishii T, Lerouge P, Driouich A (2006) The reb1–1 mutation of Arabidopsis. Effect on the structure and localization of galactose-containing cell wall polysaccharides. *Plant Physiol* 140:1406–1417

Nishitani K, Tominaga R (1992) Endo-xyloglucan transferase, a novel class of glycosyltransferase that catalyzes transfer of a segment of xyloglucan molecule to another xyloglucan molecule. *J Biol Chem* 267:21058–21064

Oka T, Nemoto T, Jigami Y (2007) Functional analysis of Arabidopsis thaliana RHM2/MUM4, a multidomain protein involved in UDP-D-glucose to UDP-L-rhamnose conversion. *J Biol Chem* 282:5389–5403

Pear JR, Kawagoe Y, Schreckengost WE, Delmer DP, Stalker DM (1996) Higher plants contain homologs of the bacterial celA genes encoding the catalytic subunit of cellulose synthase. *Proc Natl Acad Sci USA* 93:12637–12642

Perrin R, Wilkerson C, Keegstra K (2001) Golgi enzymes that synthesize plant cell wall polysaccharides: finding and evaluating candidates in the genomic era. *Plant Mol Biol* 47:115–130

Petersen A, Becker WM, Moll H, Blumke M, Schlaak M (1995) Studies on the carbohydrate moieties of the timothy grass pollen allergen Phl p I. *Electrophoresis* 16:869–875

Pike RN, Bagarozzi D Jr, Travis J (1997) Immunological cross-reactivity of the major allergen from perennial ryegrass (Lolium perenne), Lol p I, and the cysteine proteinase, bromelain. *Int Arch Allergy Immunol* 112:412–414

Preuss ML, Santos-Serna J, Falbel TG, Bednarek SY, Nielsen E (2004) The Arabidopsis Rab GTPase RabA4b localizes to the tips of growing root hair cells. *Plant Cell* 16:1589–1603

Preuss ML, Schmitz AJ, Thole JM, Bonner HK, Otegui MS, Nielsen E (2006) A role for the RabA4b effector protein PI-4Kbeta1 in polarized expansion of root hair cells in Arabidopsis thaliana. *J Cell Biol* 172:991–998

Purugganan MM, Braam J, Fry SC (1997) The Arabidopsis TCH4 xyloglucan endotransglycosylase. Substrate specificity, pH optimum, and cold tolerance. *Plant Physiol* 115:181–190

Reiter WD, Vanzin GF (2001) Molecular genetics of nucleotide sugar interconversion pathways in plants. *Plant Mol Biol* 47:95–113

Reiter WD, Chapple CC, Somerville CR (1993) Altered growth and cell walls in a fucose-deficient mutant of Arabidopsis. *Science* 261:1032–1035

Richmond TA, Somerville CR (2000) The cellulose synthase superfamily. *Plant Physiol* 124:495–498

Richmond TA, Somerville CR (2001) Integrative approaches to determining Csl function. *Plant Mol Biol* 47:131–143

Saxena IM, Lin FC, Brown RM Jr (1990) Cloning and sequencing of the cellulose synthase catalytic subunit gene of Acetobacter xylinum. *Plant Mol Biol* 15:673–683

Scheible WR, Pauly M (2004) Glycosyltransferases and cell wall biosynthesis: novel players and insights. *Curr Opin Plant Biol* 7:285–295

Schiefelbein JW (2000) Constructing a plant cell. The genetic control of root hair development. *Plant Physiol* 124:1525–1531

Schiefelbein J, Galway M, Masucci J, Ford S (1993) Pollen tube and root-hair tip growth is disrupted in a mutant of Arabidopsis thaliana. *Plant Physiol* 103:979–985

Schnepf E, Witte O et al (1985) Tip cell growth and the frequency and distribution of particle rosettes in the plasmalemma: experimental studies in *Funaria* protonema cells. *Protoplasma* 127:222–229

Seifert GJ, Barber C, Wells B, Dolan L, Roberts K (2002) Galactose biosynthesis in Arabidopsis: genetic evidence for substrate channeling from UDP-D-Galactose into cell wall polymers. *Curr Biol* 12:1840–1845

Shaw SL, Dumais J, Long SR (2000) Cell surface expansion in polarly growing root hairs of Medicago truncatula. *Plant Physiol* 124:959–970

Sheng J, D'Ovidio R, Mehdy MC (1991) Negative and positive regulation of a novel proline-rich protein mRNA by fungal elicitor and wounding. *Plant J* 1:345–354
Sherrier DJ, VandenBosch KA (1994) Secretion of cell wall polysaccharides in *Vicia* root hairs. *Plant J* 5:185–195
Steele NM, Sulova Z, Campbell P, Braam J, Farkas V, Fry SC (2001) Ten isoenzymes of xyloglucan endotransglycosylase from plant cell walls select and cleave the donor substrate stochastically. *Biochem J* 355:671–679
Suzuki H, Wagner T, Tierney ML (1993) Differential expression of two soybean (Glycine max L.) proline-rich protein genes after wounding. *Plant Physiol* 101:1283–1287
Taylor NG, Howells RM, Huttly AK, Vickers K, Turner SR (2003) Interactions among three distinct CesA proteins essential for cellulose synthesis. *Proc Natl Acad Sci USA* 100:1450–1455
Tierney ML, Wiechert J, Pluymers D (1988) Analysis of the expression of extensin and p33-related cell wall proteins in carrot and soybean. *Mol Gen Genet* 211:393–399
Vissenberg K, Fry SC, Verbelen JP (2001) Root hair initiation is coupled to a highly localized increase of xyloglucan endotransglycosylase action in Arabidopsis roots. *Plant Physiol* 127: 1125–1135
Wada M Staehelin LA (1981) Freeze-fracture observations on the plasma membrane, the cell wall, and the cuticle of growing protonema of *Adiantum capillus-veneris. Planta* 151: 462–468
Wang X, Cnops G, Vanderhaeghen R, De Block S, Van Montagu M, Van Lijsebettens M (2001) AtCSLD3, a cellulose synthase-like gene important for root hair growth in arabidopsis. *Plant Physiol* 126:575–586
Wang J, Howles PA, Cork AH, Birch RJ, Williamson RE (2006) Chimeric proteins suggest that the catalytic and/or C-terminal domains give CesA1 and CesA3 access to their specific sites in the cellulose synthase of primary walls. *Plant Physiol* 142:685–695
Whitney SE, Gidley MJ, McQueen-Mason SJ (2000) Probing expansin action using cellulose/ hemicellulose composites. *Plant J* 22:327–334
Willats WG, McCartney L, Mackie W, Knox JP (2001) Pectin: cell biology and prospects for functional analysis. *Plant Mol Biol* 47:9–27
Williamson RE, Burn JE, Birch R, Baskin TI, Arioli T, Betzner AS, Cork A (2001) Morphology of rsw1, a cellulose-deficient mutant of Arabidopsis thaliana. *Protoplasma* 215:116–127
Wyatt RE, Nagao RT, Key JL (1992) Patterns of soybean proline-rich protein gene expression. *Plant Cell* 4:99–110
Wymer CL, Bibikova TN, Gilroy S (1997) Cytoplasmic free calcium distributions during the development of root hairs of Arabidopsis thaliana. *Plant J* 12:427–439
Yuan S, Wu Y, Cosgrove DJ (2001) A fungal endoglucanase with plant cell wall extension activity. *Plant Physiol* 127:324–333

Modeling Tip Growth: Pushing Ahead

M.N. de Keijzer(✉), A.M.C. Emons, and B.M. Mulder

Abstract Tip growth, the localized extension of a cell at one of its ends, is a beautiful example of morphogenesis. Because of the highly localized nature of the growth process, it is relatively amenable to analysis. Hence, it has attracted the attention of experimentalists and theorists alike, who over the years, have sought to elucidate the mechanisms underlying this form of development, the latter through explicit mathematical models. This review provides an overview of the modeling of tip-growing cells in general, and that of plant root hairs in particular, as it has developed during the last decades. Two main lines of modeling can be distinguished. In geometrical models, the focus is on the shape of the cells alone, while the aim of biomechanical models is to clarify the underlying physical mechanisms. So far, only a few attempts have been made to combine these two approaches. Yet, the incorporation of the mechanical properties of the nascent cell wall and the forces exerted on it is very likely needed to fully understand and ultimately control tip growth. This synthesis would pave the way to fully predictive models and hence could also guide new experiments to verify them. We provide an outlook on possible routes towards this goal.

1 Introduction

Tip growth, the localized extension of a cell at one of its ends, has captured the imagination of many scientists, experimentalists, and theoreticians alike. This is because it provides a beautiful example of cell growth. Although tip growth at first appears simple as it is highly localized, it is a complex phenomenon. Many different and intertwined processes within the cell are involved in the construction of the cell wall: the production of wall-building proteins and polymers at respectively the endoplasmic reticulum and the Golgi system, their processing into vesicles by the Golgi apparatus, the transport of these vesicles towards the plasma membrane, the concomitant exocytosis of the Golgi vesicle content, incorporation of building materials into the cell wall, and finally the expansion of the cell wall itself. However,

M.N. de Keijzer
FOM Institute AMOLF, Kruislaan 407, 1098SJ Amsterdam, The Netherlands
e-mail: dekeijzer@amolf.nl

Plant Cell Monogr, doi:10.1007/7089_2008_07

in tip growth the product of all these processes is targeted to one specific area of the cell, the tip. The question therefore arises as to how this mode of growth is controlled. Is tip growth a self-regulated homeostatic process? In other words, is the growth regulated by external controlling mechanisms such as morphogens or is the growth stability determined by the dynamics themselves? Arguably, it is hard to draw definite conclusions from experiments about possible controlling mechanisms without having a concrete conceptual framework in mind. This is exactly what a good model can provide. That is, it can reveal the explicit dependencies between the controlling parameters and the observed variables and can therefore make predictions amenable to experimental verification or, equally important, falsification. It has been argued that only fully quantitative models can both provide useful new information and be falsifiable (Popper 1935). They allow us to test our hypotheses and their consequences, and point to the design of effective new experiments. So far, only parts of the tip-growth process have been investigated, and often not in the same species or even the same developmental stage of a cell. The ultimate aim of modeling tip growth is the full elucidation of the underlying physical mechanisms and the discovery of probable self-regulating principles. However, since quite a few of the different processes at work during tip growth are yet not fully understood, building a theoretical framework for describing it remains a challenging problem.

Two main approaches to modeling tip growth can be distinguished. The first approach has its focus on the geometry of the growth process, which is described using purely geometrical concepts. The power of these geometrical models is their relative simplicity. In this framework, it is also easier to take into account the complications of nonaxisymmetric cells and time-dependent problems. Moreover, they can often be (at least partially) solved analytically. This often provides insights that are more difficult to obtain by a purely numerical way of tackling a problem. Especially, they can prove useful for pointing towards key ingredients of growth. By using such models, it is relatively easy to see to which parameters the cell shape is most sensitive. However, it is unlikely that these models can provide a true mechanistic explanation of the morphogenesis. For this, too many underlying physical or biological mechanisms are neglected.

The other approach focuses on the biomechanical processes involved in tip growth, in which specific mechanisms that contribute to cell growth are modeled in detail. The aim is to provide a better understanding of the underlying principles involved. This type of models is often called biomechanical for convenience. However, only a few of such models have yet been created for tip growth. On topics like cytoskeleton dynamics (e.g. Surrey et al. 2001; Kruse et al. 2005) and cell wall ultrastructure production (e.g. Emons and Mulder 1998; Mulder and Emons 2001) quite a few biomechanical models have been proposed, but in the context of morphogenesis, and in particular, that of tip growth, they are nearly nonexistent. In addition, no models have been proposed that really take into account the dependencies between different parts of the growth process. Although current biomechanical models do describe and explain some of the mechanics, they often do not yet provide predictions of the shape and properties of the cell wall. Because of the complexity involved, it is hard to tackle such models fully in three dimensions (3D) without imposing constraints such as steady-state growth or axisymmetry of the cell.

Moreover, even when making such assumptions, they pose significant technical challenges. However, the "Holy Grail" of modeling in this field consists of a geometrical description of the genesis of the shape of the cell, while simultaneously revealing the ways in which the underlying biomechanical processes regulate and constrain this geometry. In this way, a synthesis between the two modeling approaches is created. To date, such a model has not yet been constructed.

Tip growth occurs not just in plants, but also in fungi, algae, and cells from the animal kingdom, e.g. nerve cells. Accordingly, the modeling of tip growth has been done from different perspectives, and frequently research groups were not even aware of each other's work. Nevertheless, many shared elements can be found, and therefore the question arises whether a generic and unified theory for all kinds of tip-growing cells with a cell wall is possible. Quite a few modelers have already focused on fungal hyphae. It is probably easier to model the essential mechanisms underlying this type of morphogenesis by studying these organisms instead of plants, because the geometry of the mechanism of vesicles delivery to the growth zone is perhaps the simplest (we return to this topic in the next section). Therefore, both as a source of inspiration, because of the common feature of having a cell wall, and in view of the historical development, we also address models of fungal hyphae in this review.

It was as early as 1892 that Reinhardt proposed the first "model" of tip growth for walled cells (Reinhardt 1892). He came up with the hypothesis that tip-growing cells have a "soft spot." This means that the newly incorporated material at the tip can flow and therefore easily deform. On the basis of chemical measurements this hypothesis acquired an experimental basis (Sietsma and Wessels 1990), which provoked new discussions (Koch 1994; Harold 1997). Furthermore, Reinhardt already stated that the expansion of the cell wall has to be orthogonal to the tangent plane of the cell wall. This seems like a natural assumption, as the turgor pressure exerts a force perpendicular to the cell wall. Bartnicki-García et al. (2000) claimed to have corroborated this assumption for hyphae of Rhizoctonia solani by tracking marked particles in the expanding cell wall in time. They reach this conclusion by comparing the observed trajectories to three different growth scenarios, one of which (the one corresponding to the hyphoid; see later) could have been rejected a priori, and showing a good fit only with orthogonal expansion. However, it is unclear whether the resolution of the experiment is enough to rule out small but systematic deviations from true orthogonal expansion. Moreover, very close to the tip (the most sensitive area) there are no observations. Thus, although the data seem to indicate that the growth is to a large extent orthogonal, they may be insufficient to really prove the assumption of completely orthogonal growth. In fact, it is quite unlikely that fully orthogonal growth is the true expansion mode. As the mechanical properties of the cell wall, such as plasticity and rigidity, vary for different locations along the cell wall (Sietsma and Wessels 1990), one expects inhomogeneities in the wall stresses even upon loading by an orthogonal force. These inhomogeneities in turn would cause, apart from the dominant orthogonal component, a small but significant tangential component in the trajectories of material points. Dumais et al. (2004) attempted to map the expansion mode for Medicago root hairs. Their data seem to imply that fully orthogonal growth does not occur in plant root hairs.

Money (2001) suggested that turgor pressure is not the only driving force in the tip growth of the Oomycete Saprolegnia ferax. When during his experiments the pressure was reduced to 0.1–0.2 MPa, the growth rate was higher than at normal pressure. He suggested that the growth was induced by wall-loosening substances, creating a very fluid cell wall structure. Even a small pressure could inflate such a highly plastic tip (Money 1997). However, this observation could also be explained by a cell, which in a reaction to the lower turgor pressure, starts to overcompensate the expression of cell-wall-loosening proteins such as expansins. In any case, turgor pressure was still needed for the inflation of the Oomycetes cells. This is in line with earlier measurements of Money on the Oomycete Achlya bisexualis (Money and Harold 1992). During plant cell growth, new polymers not yet linked to each other are continually exocytosed at the site of growth, and proton efflux acidifies this cell wall. This acidification induces cell wall loosening (see, e.g., the reviews by Cosgrove, 1998, 2000). These data from non-tip-growing plant cells imply that the cell wall at the tip is softer, more expandable than the wall of the hair tube. For plants, Oomycetes, algae, and fungi the consensus is that the driving force for cell growth, including tip growth, is pressure, but that this pressure is only effective if the wall is, or is made, expandable. Proseus and Boyer (2006) concluded from measurements on the alga Chara corallina that the growth velocity and the rate of wall deposition actually do decrease when the turgor pressure is lowered from 0.5 to 0.1 MPa.

Before we are ready for our exploration of existing tip-growth models, we briefly summarize the many aspects involved in tip growth in root hairs. This is to get a feeling of the scope that a proper model should have. The protein building materials for the cell wall are produced at the endoplasmic reticulum. Subsequently, they are transported to the Golgi apparatus. In the Golgi apparatus, polysaccharides except cellulose and callose are produced. Furthermore, the cell-wall-building materials are further processed and packaged into vesicles of the Golgi bodies. The Golgi bodies are transported through the cytoplasm, including the subapical cytoplasm of root hairs, by motor proteins over bundles of actin filaments. Having reached the vicinity of the tip, the Golgi bodies reverse direction and the vesicles move towards the extreme tip of the cell. The exact whereabouts of the release of vesicles from the Golgi bodies has not yet been shown. Since actin filaments have not been observed in this extreme apex, the vesicles have been assumed to freely diffuse towards the plasma membrane (Emons and Ketelaar 2008). Here building materials are extruded from the vesicles by means of exocytosis while the vesicle membrane with its associated proteins is inserted into the plasma membrane. This process is regulated by calcium ions and hence by Ca^{2+} channels brought into the plasma membrane by the exocytosis process itself (Bibikova and Gilroy 2008). Once arrived in the extracellular space, the building polymers are incorporated into the cell wall. The wall has varying material properties depending, for example, on the amount of cross-links between the polymers (Nielsen 2008). Combined, these properties determine how a cell wall responds, by stretching and bending, to exerted pressure. See Fig. 1 for a schematic overview of all the processes involved.

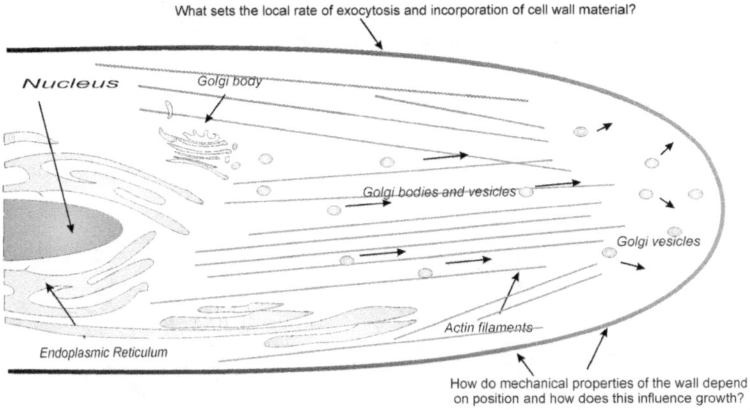

Fig. 1 An impression of important concepts and questions for the process of tip growth in plant cells. Motor proteins transport Golgi bodies over actin filaments. Once the released vesicles have entered the cell apex, they move by an as-yet-unresolved mechanism, which does not exclude diffusion (Tindemans et al. 2006) towards the cell membrane. By means of exocytosis, building polymers are incorporated in the existing cell wall. Since new material is added constantly after which polymers are linked up to each other, the material properties of the cell wall are not constant from tip to base of the hair

The ultimate model of tip growth has to integrate all these ingredients, explaining how they work together and how they are regulated.

2 The Delivery of the Vesicles

We now start our voyage through the root hair towards the tip, to try to follow the various processes involved in delivering the growth vesicles to their proper destination. We do not look in detail into genetic networks regulating the production of vesicles and building polymers. We therefore assume, as do all present models of tip growth, that a steady amount of material is produced per unit time.

In light of modeling tip growth of root hairs, the transport of Golgi bodies by motor proteins over actin filaments is an unexplored field. However, for fungal hyphae, Sugden et al. (Sugden et al. 2007) have proposed a model that could be useful as a source of inspiration. Within fungal hyphae motor proteins transport Golgi bodies over microtubules (instead of over the actin filaments) from the production site towards the Spitzenkörper. The Spitzenkörper is an accumulation of vesicles and actin filaments, and is found close to the tip in many fungal hyphae. It is assumed to release vesicles into the cytoplasm and to guide the direction of the tip growth. See Fig. 2 for an illustration of the vesicle release from the Spitzenkörper in a fungal hypha, contrasted to a more widely spread vesicle source

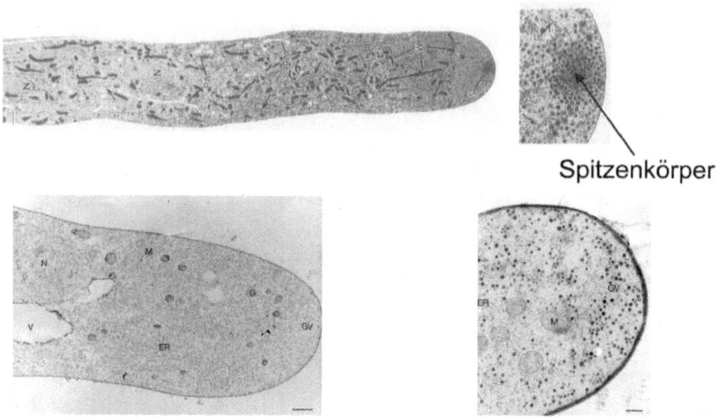

Fig. 2 The upper electron microscopy micrograph shows a longitudinal section through a hyphal tip of Sclerotium rolfsii (Roberson and Fuller 1988) with a Spitzenkörper. In the lower pictures, the more widely-spread apical accumulation of vesicles (the Golgi vesicle, GV, area) in the plant root hair cell of Vicia sativa can be seen [Miller et al. 2000 (lower left picture)]

in the subapex of a growing plant root hair. This wider vesicle "disc" in plants, root hairs (e.g. Miller et al. 1999) and pollen tubes, has been compared to the Spitzenkörper in fungi (de Ruijter et al. 2000).

Sugden et al. (2007) used a "modern" technique in statistical physics called the "totally asymmetric simple exclusion process" to model transport of particles over microtubules. They regard the microtubules as parallel lines extending from a "production area" to the tip. To mimic the essentially one-dimensional nature of microtubules, one-dimensional lattices were used, in which each site can be occupied by at most one particle at a certain time. Far away from the tip, the system is fed with vesicles seen as particles that enter the lattice at a certain rate α. On the tip end, two things can happen: the particles can either leave the microtubule at a certain rate β or a new lattice site can be formed at rate γ (growth). This assumption gives a growth velocity that is a function of delivery rate (α) and the hyphal-extension efficiency ($\gamma/(\beta + \gamma)$). Two ways of solving the problem have been attempted: a mean-field and a Monte Carlo approach. Although the results do not completely agree with each other and β was mostly neglected, it was shown that a regime of parameters exists where a high density of vesicles near the tip having a length-scale of the order of that of the Spitzenkörper is established.

A similar approach could be useful for modeling the delivery of exocytotic vesicles containing wall-building polymers in plant root hairs over actin filaments to the vesicles-rich area. As long as the filaments are aligned among each other, this approach is probably sufficient for mimicking the delivery of particles. However, when describing asymmetrical dynamical processes such as curling (e.g. Esseling et al. 2004; Sieberer et al. 2005) of root hairs, a 2D (two-dimensional) or even a full 3D approach is needed.

Once released, the vesicles have to find their way out to the plasma membrane. Since the extreme tip of growing root hairs does not seem to have actin filaments (see Ketelaar et al. 2002; Ketelaar and Emons 2008; Emons and Ketelaar 2008; though endoplasmic microtubules in Arabidopsis root hairs enter this region close to the plasma membrane (Timmers et al. 2007; see Sieberer and Timmers 2008), the mode of transport in this part of the process is still uncertain, but simple diffusion, combined with delivery at one side and consumption at the other, could be a plausible scenario. For fungi, several models have been developed with this concept in mind. However, before we start investigating them, we dedicate a few words to more straightforward completely geometrical models of tip growth. In such models the cell boundary (the plasma membrane plus the cell wall) is described as a mathematical surface, sometimes also called a shell. The expansion of the cell boundary is, in these approaches, taken to be completely determined by the intrinsic geometrical properties of the surface such as local curvature.

Obviously, cytoskeleton dynamics could also be a factor in a tip-growth model. However, although the cytoskeleton in general has attracted many scientists, not much modeling of it has been done in the light of tip growth. Regalado (1998) proposed a model for fungi in which he built a microscopic fundament for the model by Goodwin and Trainor (1985). He did this by describing the distribution of wall-building vesicles by a viscoelastic cytoskeleton. The rheology (deformation propensity) of the actin filaments is regulated by Ca^{2+} gradients on which they respond in a nonlinear way. The model also predicts an accumulation of vesicles near the tip. Moreover, it provides a mechanism for osmotic pressure control.

Free cytoplasmic Ca^{2+} gradients have been measured in root hair tips showing that a Ca^{2+} gradient is required for tip growth (de Ruijter et al. 1998; Wymer et al. 1997; Bibikova and Gilroy 2008). Apart from this gradient and the high vesicle concentration near the tip, not much experimental evidence has been given for Regalado's model except for several measurements by Yin and Gollnick (Yin 1980; Gollnick et al. 1991): consider, for example, Regalado's assumption of tip growth regulation by osmotic-pressure variations caused by cytoskeleton contractions, which is unsupported by measurements. The Ca^{2+} gradient is ad hoc in this model: it is not self-regulating. Furthermore, no connection has been made yet with the geometry of the cell.

Van Batenburg et al. (1986) performed tip-growth simulations of root hair cell boundaries pushed ahead by the turgor pressure (modeled by orthogonal growing shells with chosen growth-rate functions). They also described the curling of the root hair in this way. However, not only were intracellular processes not considered, but also the expansion rate at every distance from the tip was imposed in an ad hoc way. Therefore, this model does not address possible self-regulating mechanisms. Different points on the cell boundary are also not explicitly connected to each other in this model; any point on the boundary can move without dragging the rest of the boundary with it. However, as long as the growth proceeds smoothly, this defect will implicitly be corrected. Yet, the growth process proves to be quite unstable in this approach: even small fluctuations in the growth profile cause a completely different evolution of the shape. Nevertheless, it does make a dynamic non-steady-state (2D) description possible. Because of the many numerical and analytical problems encountered when attempting to model a more realistic nonstationary (dynamical) growth, modeling of tip-growth dynamics is still a relatively unexplored area.

In his lecture notes, Todd (1986) proposed an approach to modeling tip growth using differential geometry. Pelcé and Pocheau (1992) continued even further along this line by creating a model of the growth of (the mainly circular cross-sections of) algae inspired by a description of dendrites (see, for example, the book of Pelcé (2004) for an overview). The evolution of the cell boundary is a function of the local curvature, its second derivate and the tangential velocity of the wall "particles." When these particles are moving perpendicularly to the boundary, instabilities are found analogous to those causing microstructure formation in crystal growth. As with most geometrical models, only shapes are derived, without explaining the underlying mechanics. Nevertheless, insight is provided into the possible shapes, given several very general assumptions.

We now turn to a few attempts to combine modeling of the intracellular processes with the boundary. In the so-called vesicle supply center (VSC) models an autonomously moving VSC is assumed releasing vesicles containing the wall-building polymers. All existing VSC models have been developed for fungi: the VSC is a point-like representation of the Spitzenkörper. In 1989 Bartnicki-García et al. (1989) proposed the first VSC model for predicting the shape of fungal hyphae. Once released isotropically (equally in all directions) from the VSC, vesicles are assumed to be transported ballistically (motion over straight lines) towards the cell boundary. The incorporation rate into the cell wall and therefore its expansion rate are simply proportional to the number of arriving vesicles. For a schematic picture of this model, see Fig. 3.

Stationary, or steady-state, growth is assumed by Bartnicki-García et al. (1989). This means that for every moment in time the shape of the tip looks the same. Moreover, the tip moves with a constant velocity. An analytical description of the shape of fungal hyphae is then derived: a so-called shape equation. Its solution gives a shape, dubbed the hyphoid, which is given by the formula

$$z = r \cot \frac{Vr}{N},$$

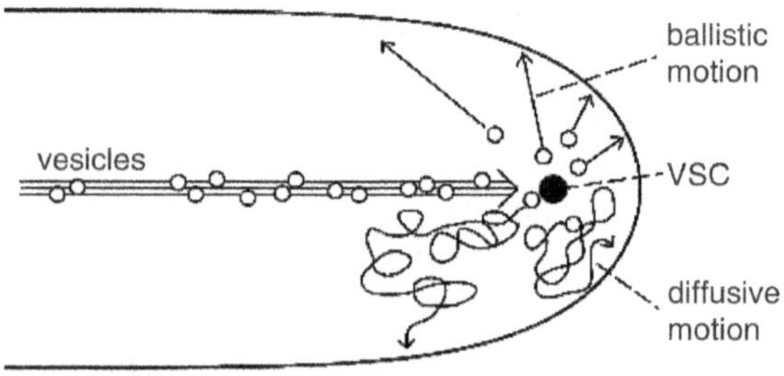

Fig. 3 A schematic picture of the vesicle supply center (VSC) model. Vesicles are delivered to a VSC by motor proteins, moving over microtubules, from where they are released into the cytoplasm and head towards the cell wall. Bartnicki-García et al. (Bartnicki-García et al. 1989; Gierz and Bartnicki-García 2001) assume a ballistic motion of the vesicles from the VSC to the plasma membrane, while Tindemans et al. (2006) assume a diffusive one

where V is the velocity of the VSC moving along axis z, r the distance between cell axis and cell boundary for a certain z perpendicular to the z-axis, and N the amount of particles per unit time released from the VSC. We refer to Fig. 4 for a clarification of the geometrical variables involved. Although the hyphoid shape has been assumed again by other researchers as a valid description of hyphae, it is in part artificial. This is because the actual growth calculated is that of thickening of the cell wall (volume growth) instead of an expansion of its surface (Koch 1994). Such an (artificial) expansion leads to the scenario that the inner side of the cell wall does not move in space while the wall becomes thicker and thicker because the outer boundary expands: in other words, massive cells (full of cell wall material) are obtained. Furthermore, an additional sine term is involved when going from a 2D model to a 3D one. We will come back to this transition later.

It is a valid question whether the proposed VSC model really describes a stable growth process. In this light, Heath and van Rensburg (1996) investigated the stability of tip growth. They showed that for models in which vesicles are ballistically transported from a VSC towards the cell boundary only the case with an isotropic release of vesicles is stable. This opposed to the many stable shapes that they obtained by just imposing a distribution of delivered vesicles at the cell boundary without taking attention to the actual delivery mechanism. Therefore, it was concluded that the VSC model was unlikely to be universally applicable. However, they made the same assumptions as those in the just mentioned article of Bartnicki-García (Bartnicki-García et al. 1989), and therefore implicitly only the case of volume growth of the cell wall was considered. Furthermore, the conclusion also pointed towards the need for a less-constrained 3D model.

The just mentioned models are all 2D. It was generally assumed that because tips are nearly axisymmetric this was not a problematic simplification. In 2001 Bartnicki-García's group finally broke with his tradition and assumed an axisymmetric 3D steady-state model. They derived a shape equation from geometrical considerations

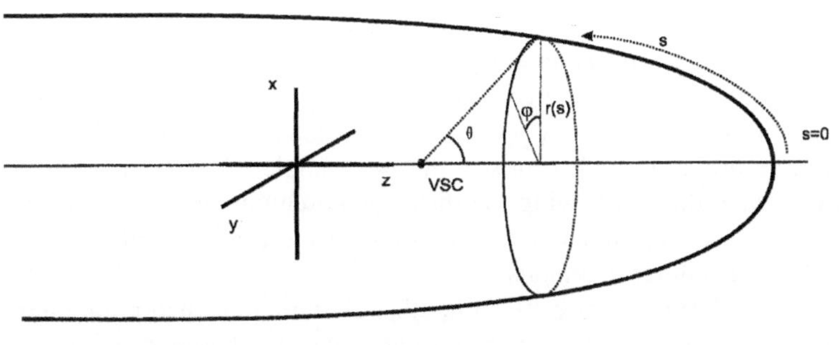

Fig. 4 A schematic overview of variables used in this review for describing tip growth, where x, y, and z are the Cartesian coordinates, with z lying along the cell axis; s is the arc length in the longitudinal direction; υ is the angle between the z-axis and the line from the VSC to the associated point on the cell wall; and φ is the angle in the meridional direction

and they showed that 3D tip shapes are different from 2D ones because of an additional sin θ term (see Fig. 4 for a clarification of the variables used in this review). They also showed that a VSC provides enough focusing towards the tip to let the tip growth occur. Furthermore, the experimentally observable quantity of the distance between VSC and apex (as a function of the ratio between the growth velocity and the cell radius) changed significantly compared to the original 2D model. This considerable change showed that a description in only 2D is inadequate for obtaining realistic outcomes (Gierz and Bartnicki-García 2001). The need for 3D models provided new challenges, because complexity levels are much higher when solving the 3D-shape equations. The requirement of 3D necessitates further research in which, at least, the more realistic 3D axisymmetry is assumed.

An important mechanism missing in the 3D model of Bartnicki-García's group (Gierz and Bartnicki-García 2001) is the concept of diffusion: the spontaneous spatial spreading of vesicles. The model assumes that particles are being shot in completely straight lines in the direction of the cell boundary. Although little is known about the precise delivery mechanism, it may be obvious that in a viscous, crowded environment like a cell, ballistic transport of submicrometer-sized vesicles is highly unrealistic. Even though one can argue about the magnitude of the actual error introduced upon making this assumption, this defect provoked a new model by Tindemans et al. (2006). They assume that vesicles diffuse freely from the VSC before being absorbed at the cell boundary. The additional effect of drag exerted by a possible flow in the cytoplasm (advection) is neglected. The local incorporation rate into the cell wall is proportional to the amount of vesicles available, although in reality saturation at high vesicle concentration occurs. The proportionality constant, called the absorption propensity at the cell wall, compared to the diffusion rate, is a free parameter in the model. The obtained absorption profile is put into the same shape equation for orthogonal growth as derived by Gierz and Bartnicki-García (2001). However, this time the more convenient cylindrical coordinates are used instead of Cartesian ones. For a clarification of this model, see Fig. 3.

A complicating factor for finding the ultimate tip theory for walled cells is that solving diffusion equations can prove quite difficult. For 2D tips, diffusion equations are still analytically solvable. However, as soon as 3Ds are considered, this is often not the case anymore. For more involved shapes than idealized geometrical objects such as a sphere, the diffusion equation can typically only be solved numerically. Currently, this is very computer-time-consuming, and many numerical problems arise as well. Therefore, many more or less advanced solving methods have been developed in different fields of research, for example, fluid dynamics and electrodynamics. Nevertheless, solving 3D diffusion equations remains very demanding. The method used by Tindemans et al. is called the fundamental solution method and is often used in electrostatics.

The model of Tindemans et al. shows that by tuning the incorporation rate of the material at the cell boundary, the cell can tune the sharpness of the tip shape. Furthermore, the different apex-VSC distances can also be obtained. However, the spectrum of different shapes that can be described this way cannot completely explain all the observed ones. In particular, the sharper and the blunter ones are not addressed. It should also be pointed out that the shape per se is perhaps not the most sensitive test for the success of a model, as potentially many models could lead to the same shape, and the shapes found under different conditions are fairly similar.

Exocytosis is regulated by calcium and, hence, calcium channels. Until now, the dependence of the exocytosis rate of vesicles at the plasma membrane on a Ca^{2+} gradient has not been modeled explicitly in light of tip-growth morphogenesis. In theory, such a gradient could regulate the incorporation rate of the vesicles. But, since exocytosis is a fast process, and thus not rate-limiting for the whole process, its rate is not deemed so important.

For all their success in unlocking parts of the tip-growth process to analysis, all these models neglect one very important concept: the material properties of the cell wall are not taken into account at all. A true model of the growth process has to capture the viscoplastic properties of the polymer network of the cell wall, if its ambition reaches farther than just geometrically describing the morphogenesis process. Moreover, this has to be done in an a priori way instead of an a posteriori way. Otherwise, it seems impossible to properly explain, let alone predict, how the loosening of the cell wall could induce growth, bulging (e.g. Cosgrove 2000; Foreman and Dolan 2001), and curling of root hairs.

3 The Biomechanics of the Cell Wall

The plant cell wall withstands the osmotic pressure of the cytoplasm, and therefore, it is a mechanically strong and fairly rigid structure. It is permanently under high tensile stress, and in order to let cells grow, it must also be able to expand. The plant cell wall consists of long cellulose microfibrils embedded in a polysaccharide matrix. The microscopic properties of these cell-wall polymers determine viscosity, plasticity, and elasticity of the cell wall. Collectively these determine how much stress it can bear, as well as how it deforms given a certain stress. A crucial role in this is played by the number of physical and chemical contacts between the polymers, which is the basis of the soft spot hypothesis. See, for example, the work of Sietsma and Wessels for measurements of varying chemical structure over the cell wall for fungal hyphae of Schizosaccharomyces pombe and Schizophyllum commune (Sietsma and Wessels 1990; Sietsma et al. 1995).

Fungal cells contain cell-wall molecules that are different from those of plant cells. The most striking difference is that fungal cell walls contain chitin molecules instead of cellulose. When considered in detail, this inevitably leads to different parameter values for cell-wall properties such as yield stress and extensibility. Nevertheless, it is likely that the same type of formalism could still be used to describe both kinds of walls. That is because macromolecular systems like this are often described by coarse-grained models in which whole groups of molecules are lumped together into effective "particles," because only the mechanical properties on a macroscopic scale matter. The differences between molecules then often only lead to an adjustment of the parameters describing the macroscopic mechanical properties.

A lot of modeling work has been done on systems similar to cell walls. In particular, a lot can be learned from the field of viscoelasticity. This is an active field in polymer physics, in which the relationships between imposed stress on a material and its subsequent expansion, often called strain, are being studied. The corresponding physics is described in many books about this topic. We refer the interested reader to the textbook by Landau and Lifschitz (1986).

In 1970, Green et al. stated a relationship between expansion of the cell surface and the stress that caused this deformation. They described this by assuming that the extensibility of the cell wall was reciprocal to its viscosity after subtracting a threshold value. This relationship is essentially what is now known as the Lockhart equation (Lockhart 1965). In fact, both equations are nothing but the Bingham equation originally proposed to describe materials such as paint (Bingham 1922). Bingham's law states that the plastic rate of deformation of this type of material is proportional to the excess stress above a certain threshold.

Veytsman and Cosgrove (1998) constructed a model in which the extensibility and the yield threshold in the Lockhart equation are derived from microscopic properties and thermodynamical considerations. The model is roughly as follows. The cell shape is modeled by a long cylinder length L and radius R. It is assumed to expand in the longitudinal direction. Although thus not a tip-growth model, it is still very illuminating for this more specialized case as well.

The internal wall stress s is defined as the derivative of the free energy of the wall G:

$$\sigma = \frac{1}{2\pi Rh}\frac{\partial G}{\partial L}$$

where h is the cell wall thickness. By considering the mechanical work that is performed by the turgor pressure P, the cell wall, and water absorption, a general form of the Lockhart equation is derived.

$$\frac{d\ln L}{dt} = \frac{K_1 K_2}{K_1 + K_2}\left(P - \frac{2h}{R}\sigma\right),$$

where K_1 is the hydraulic conductance (a measure of how easily water can flow through pores) and K_2 is a wall-yield coefficient. These parameters themselves do not follow from the formalism.

The formation of hydrogen bonds is the only interaction taken into account. Furthermore, only those among the cellulose and the glucan chains and those among the water molecules present are explicitly modeled. All the other interactions are implicitly taken into account. Nevertheless, using more realistic reaction potentials could change the outcome. The elastic free energy of the polymers is calculated using a formula given by Doi and Edwards (1986). The interaction energy is obtained by calculating the partition function of randomly oriented glucan- and cellulose segments each capable of forming one hydrogen bond. The sum of both gives the total free energy.

It turns out that in such a model the yield threshold is mostly determined by the cellulose and to a lesser extent by the concentration of glucan chains. Thus, a definite relationship is established between the microscopic properties involving constituent molecules and the macroscopic behavior. As the authors have already mentioned, this framework is far from finished. The assumption of orientational randomness of polymers is an approximation as well as the assumption that water does not contribute to wall stress. Moreover, the (small) flexibility of the cellulose fibrils is not considered. Finally, using only hydrogen-bonding interaction potentials is quite a simplification. Nevertheless, this model provides a first step in the right direction.

Tabor and Goriely then took up the gauntlet by producing a model of tip growth in which the cell wall is described as a biomechanical shell (Goriely and Tabor 2003a,b; Goriely et al. 2005). The thickness of the cell wall is here assumed to be very small, compared to the dimensions of the cell as a whole that it can be approximated by an infinitely thin shell. Indeed, the wall thickness of growing Arabidopsis root hairs is only 85 ± 16 nm, compared to a root hair diameter of around 7–8 µm, and its variation is only small (Ketelaar et al. 2008). So the approximation of assuming a constant wall thickness seems a reasonable one.

Goriely and Tabor employ a formalism for describing the dynamics of thin shells that was developed earlier for biomembranes by Skalak et al. (1973). This is actually the same formalism as is often used to describe the inflation of a balloon. It assumes that mechanical equilibrium is maintained throughout.

The main parameters in this model are the turgor pressure P, the principal local curvatures in the s and the φ direction κ_s and κ_φ, and the local tensions in these directions t_s and t_φ. See Fig. 4 for a clarification of all the parameterizations used.

The equilibrium equations obtained from the biomembrane model read as follows:

$$P = \kappa_s t_s + \kappa_\varphi t_\varphi$$

$$\frac{\partial(rt_s)}{\partial s} - t_\varphi \frac{\partial r}{\partial s} + r\tau_c = 0$$

where the shear stress τ_c is assumed to be zero.

The tip is described as a shell with certain given elastic and plastic properties that depend on the distance from the apex. These properties, together with the curvature, determine the local expansion rate of the cell wall. Actually, the idea of using curvature–stress relations for the cell wall is already quite old. Koch (1982), for example, already suggested the use of the Young–Laplace law for the surface tension in describing the evolution of the cell wall. The stretching of the wall, strain, and stress are linked to each other using the theory of Evans and Skalak (1980). The stress has to be in equilibrium with the turgor pressure. To model the extensibility rate, a variable effective pressure is used. This is a function of the elastic modulus and the turgor pressure. It is obtained by fitting the rate of incorporated building materials to experimental data.

Tabor and Goriely's model provides a simple yet powerful formalism. Stresses naturally follow from the deformation, thereby reducing the rate of deformation. It can easily be expanded towards a model describing cell wall mechanics more precisely. The formalism can describe a wide variety of possible cell shapes. For example, it can recreate the bubble shape at the end of a swollen hyphal Streptomyces coelicolor tip after treatment with lysozymes. Perhaps its only weakness is the fact that the important parameters of the model are not determined self-consistently by the theory itself. The effective pressure, for example, is ad hoc and can only be obtained by fitting to an experimentally determined shape. Nevertheless, this model can point out which mechanisms may be dominant in tip growth.

It is clear that any degree of alignment of the wall polymers would significantly influence the mechanical properties of the cell wall. However, in all the work discussed so far the cell wall material is assumed to have isotropic (the same in all directions) mechanical properties. This implies that the wall-building polymers are completely disordered, i.e., on average they are randomly orientated. Although the cellulose microfibrils deposited in the tip are known to be randomly oriented within the plane of the wall (Emons 1989), it is not clear whether the same holds for the less rigid matrix polymers that are probably subjected to anisotropic longitudinal stress during the growth process. In fact, one should distinguish different degrees of wall anisotropy, as illustrated in Fig. 5.

The influence of anisotropic viscoplasticity of the cell wall on the shape of the cell was addressed by Dumais et al. (2004, 2006), inspired by the work on (visco-) plasticity by Prager and Hill (Prager 1937; Hill 1950). Similar to the models by Tabor and Goriely (Goriely and Tabor 2003a; Goriely et al. 2005), the model of Dumais et al. (2006) consists of a thin shell representing the cell wall. Every point on the shell again has its local stress, curvature, and material properties that determine the deformation of the surface. However, in this case, emphasis is placed on the fact that the material properties of the cell wall are anisotropic. This anisotropy is needed to account for observed patterns of wall expansion.

Dumais et al. show that using shell theory without addressing anisotropy in the material properties is not sufficient to account for the wall expansion profiles observed in tip-growing cells such as the rhizoid of the alga Chara or Medicago truncatula root hairs. By simply assuming at least transverse isotropy of the polymers, the fit with the observed wall expansion already increases. So it seems that the final cell wall theory should also take into account anisotropic properties of the cell wall or at least the transverse part of it. One of the shortcomings of this model is that it is still in some sense ad hoc. Given an imposed extensibility profile of the cell wall or a flow-coupling profile the shape is determined. But this profile again does not follow from the theory itself, as the combined mechanism of material delivery and the subsequent

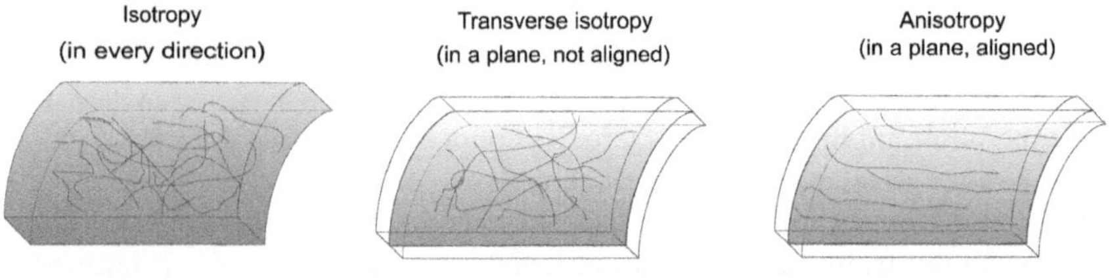

Fig. 5 An illustration of different ways cell-wall polymers can be aligned. The left one is the complete isotropic case, the middle one is the transversal one, and the right one is the completely aligned case

evolution of its mechanical properties is not included. As such, this approach is therefore still in the same category as that of Tabor and Goriely (Goriely et al. 2005).

Recently, Bernal et al. (2007) proposed a novel model, inspired by the work of Hejnowicz et al. (1977). The latter had already pointed out that the inflation process of cylindrical rubber balloons is very similar to the morphogenesis of tip-growing cells. On the basis of quantitative observations on expanding balloons, Bernal et al. constructed a “simple” mechanical model to describe tip growth. The expansion profile of the cell wall is determined by cell geometry alone and provides a good match to the experimental data which also show that the point of biggest extensibility is located slightly behind the tip. However, the local elastic compliance, i.e., the propensity of the cell wall to expand as a reaction to stress, is again determined by fitting to the observed shapes. As such, the role of the underlying dynamics of addition of new material through vesicle content absorption and the concomitant change in material properties through cross-linking of the wall-building polymers are not really addressed.

4 Outlook

We argue that a full understanding of steady-state tip growth as a self-organized phenomenon requires the integration of all the relevant component processes: vesicle delivery, incorporation of new wall material, evolution of the mechanical properties, and the resultant geometrical growth all dynamically coupled. Our group has taken a first step in the direction for a simple model of tip growth in fungal hyphae (Eggen et al., in preparation). The aim was to study the influence of the cross-linking rate of the polymers, and the corresponding hardening rate of the cell wall, on the shape of the tip. The hypothesis was that cell-wall hardening by cross-linking could be a self-regulating principle underlying the stability of tip growth by influencing the polymer incorporation rate. In the new model, we again start from a VSC from where the vesicles diffuse towards the plasma membrane. However, the incorporation rate of the particles into the cell wall is now not only a function of local density of available vesicles, but also a function of the local plasticity. This local plasticity is determined by the “age” distribution of the wall material, determined by keeping track of the time elapsed since the incorporation of the various polymer fractions of which it is composed. As the polymers “age,” they become increasingly cross-linked and this translates into an increased stiffness, or loss of plasticity of the wall material. By letting the degree of cross-linking of the wall polymers feed back into the incorporation rate, which determines the “rejuvenation” of the wall material, we obtain a self-regulating mechanism for tip growth, which in a sense implements the classical soft spot hypothesis (Koch 1982). The sharpness of the tip turns out to be strongly dependent on this cross-linking rate. In this way, a much wider family of observed tip shapes can be described than was possible in the earlier VSC models.

The formalism of this model can also be used to describe the regulation of the exocytosis rate by the presence of calcium ions. In this version of the model, slowly decaying calcium channels are assumed to be incorporated into the plasma membrane at the same rate as the vesicles. The only extra postulate made is that the concentration of calcium channels is to first approximation linearly proportional to the calcium density and hence to the exocytosis rate. So if new experimental data find evidence that indeed the calcium-facilitated exocytosis rate turns out to be the rate-limiting step in the whole growth process, the new formalism shows that this could be self-regulated as well. We refer the interested reader for an overview of recent experiments on this topic to the chapter of Bibikova and Gilroy (2008).

In spite of all the progress made towards explaining tip shapes, it should be pointed out that most of the existing models start out by assuming the existence of steady-state growth. To show that such a stable steady state exists requires some form of stability analysis. Here we should distinguish between the type of analysis as performed by Heath and van Rensburg (Heath and van Rensburg 1996), which probes the stability of the steady-state solutions of a model under variations of its parameters, and the much more stringent, dynamical stability analysis, which shows whether the assumed steady-state solutions are stable over time. To perform the latter type of analysis, one needs a fully solvable model for the dynamical evolution of the tip shape from a wide variety of initial conditions. So far, only the models proposed by Goriely and Tabor (2003b) and Dumais et al. (2006) meet this requirement, and they were able to show how the dynamics they assume leads by itself to steady state in which an invariantly shaped tip evolves out of an initially planar configuration, reminiscent of the way a root hair tip develops from an epidermis cell.

This brings us to more complicated growth phenomena such as bulging and curling of the root hair. To understand, for example, root hair curling, one would clearly need a 3D model that allows the description of tip growth whose direction of motion is no longer constrained to a straight line. In this way the implications of various mechanistic scenarios could be worked out to understand the response of a root hair upon exposure to nodulation factor excreted by Rhizobium bacteria leading up to the development of a "shepherd's crook." See, for example, Limpens and Bisseling (2008), Hirsch et al. (2008), and Sieberer and Timmers (2008), for more details. A number of questions come to mind. Do autolysins induced by the expression of early nodulation genes loosen the part of the cell wall opposite to the application point of the nodulation factor? Alternatively, could a rotation of the whole aligned subapical F-actin structure, which causes a nonsymmetrical release of vesicles steering the growth in another direction, explain this? Measurements of Esseling et al. (2003) seem to indicate this. These questions challenge us to surmount the limitations of the current generation of models.

At the very least a proper way of describing the shape of the source from which vesicles are released into the apical zone is needed to distinguish between plants and hyphae. In plants and Oomycetes, the vesicles are not released from a pointlike Spitzenkörper, as is the case for many fungi. Instead, they are released from the subapically located spatially extended fine-bundled actin. So as an alternative to the pointlike VSC a spatially extended source is needed, such as an annulus or a disc. The measurements of Dumais et al. (2004) indicate that in Medicago root hairs the rate of growth and the plasticity are at a maximum slightly behind the apex, an observation recently corroborated by Zonia and Munnik (2008) in pollen tubes. This finding is indeed consistent with vesicle release from a structure slightly behind the apex whose edges are closer to the cell boundary locations behind the tip. Our group is currently working on this problem.

Finally, we believe that it is important to derive a model for the cell-wall mechanics, starting from basic polymer physics. Plasticity and elasticity are then no longer externally specified quantities, but follow self-consistently from the evolution of the inserted material and the subsequent pattern of growth. Obviously there is as yet a scarcity of reliable information on the actual physicochemical processes taking place in the nascent cell wall. As already mentioned, classic measurements on the cell wall by, for example, Sietsma and Wessels (Sietsma and Wessels 1990; Sietsma et al. 1995) did show that the chemical structure of the fungal cell wall varies over the cell boundary because of cross-linking of (1–6) β-glucan. Near the tip the wall mainly contains chitin, (1–3) β-glucan, and no (1–6) β-glucan, and far from the tip the chitin is embedded in a β-glucan matrix containing (1–6) β-glucan cross-links. As a result, the cell wall is soft in the apex, but hardens towards the subapex. We have tried to capture this aspect in our upcoming model, but stress that the wall plasticity function we introduce is a heuristic device and not properly derived using polymer physics. This remains a challenge for the future.

References

Bartnicki-García S (2002) Hyphal tip growth: outstanding questions. In: Osiewacz HD (ed) Molecular biology of fungal development. Dekker, New York pp 29–58

Bartnicki-García S, Hergert F, Gierz G (1989) Computer simulation of fungal morphogenesis and the mathematical basis for hyphal (tip) growth. Protoplasma 153:46–57

Bartnicki-García S, Bracker CE, Gierz G, López-Franco R, Lu H (2000) Mapping the growth of fungal hyphae: orthogonal cell wall expansion during tip growth and the role of turgor. Biophys J 79:2382–2390

Bernal R, Rojas ER, Dumais J (2007) The mechanics of tip growth morphogenesis: what we have learned from rubber balloons. J Mech Mat Struct 2 (6):1157–1168

Bibikova T, Gilroy S (2008) Calcium in root hair growth. In: Emons AMC, Ketelaar T (eds) Root hairs: excellent tools for the study of plant molecular cell biology. Springer, Berlin Heidelberg New York. doi:10.1007/7089_2008_3

Bingham EC (1922) Fluidity and plasticity. Mc-Graw-Hill, New York

Cosgrove DJ (1998) Cell wall loosening by expansins. Plant Physiol 118:333–339

Cosgrove DJ (2000) Loosening of plant cell walls by expansins. Nature 407:321–326

de Ruijter NCA, Rook MB, Bisseling T, Emons AMC (1998) Lipochito-oligosaccharides re-initiate root hair tip growth in Vicia sativa with high calcium and spectrin-like antigen at the tip. Plant J 13 (3):341–350

de Ruijter NCA, Esseling JJ, Emons AMC (2000) "The roles of calcium and the actin cytoskeleton in regulation of root hair tip growth by rhizobial signal molecules". In: Geitmann A (ed) Cell biology of plant and fungal tip growth. IOS Press, Amsterdam, p 161

Doi M, Edwards SF (1986) The theory of polymer dynamics. Clarendon, Oxford

Dumais J, Long SR, Shaw SL (2004) The mechanics of surface expansion anisotropy in Medicago truncatula root hairs. Plant Physiol 136:3266–3275

Dumais J, Shaw SL, Steele CR, Long SR, Ray PM (2006) An anisotropic-viscoplastic model of plant cell morphogenesis by tip growth. Int J Dev Biol 50:209–222

Emons AMC (1989) Helicoidal microfibril deposition in a tip growing-cell and microtubule alignment during tip morphogenesis: a dry-cleaving and freeze-substitution study. Can J Bot 67:2401–2408

Emons AM, Ketelaar T (2008) Intracellular organization: a prerequisite for root hair elongation and cell wall deposition. In: Emons AMC, Ketelaar T (eds) Root hairs: excellent tools for the study of plant molecular cell biology. Springer, Berlin Heidelberg New York. doi:10.1007/7089_2008_4

Emons AMC, Mulder BM (1998) The making of architecture of the plant cell wall: How cells exploit geometry. Proc Natl Acad Sci USA 95 (12):7215–7219

Esseling JJ, Lhuissier FGP, Emons AMC (2003) Nod factor induced root hair curling: Continuous polar growth towards the point of nod factor application. Plant Phys 132:1982–1988

Esseling JJ, Lhuissier FGP, Emons AMC (2004) A nonsymbiotic root hair tip growth phenotype in NORK-mutated legumes: implications for nodulation factor-induced signaling and formation of a multifaceted root hair pocket for bacteria. Plant Cell 16 (4):933–944

Evans EA, Skalak R (1980) Mechanics and thermodynamics of biomembranes. CRC Press, Boca Raton

Foreman J, Dolan L (2001) Root hairs as a model system for studying plant cell growth. Ann Bot 88: 1–7

Gierz G, Bartnicki-García S (2001) A three-dimensional model of fungal morphogenesis based on the vesicle supply center concept. J Theor Biol 208:151–164

Gollnick F, Meyer R, Stockem W (1991) Visualization and measurement of calcium transients in Amoeba proteus by fura-2 fluorescence. Eur J Cell Biol 55:262–271

Goodwin BC, Trainor LEH (1985) Tip and whorl morphogenesis in Acetabularia by calcium-regulated strain fields. J Theor Biol 117:79–106

Goriely A, Tabor M (2003a) Biomechanical models of hyphal growth in Actinomycetes. J Theor Biol 222:211–218

Goriely A, Tabor M (2003b) "Self-similar tip growth in filamentary organisms." Phys Rev Lett 90 (10):108101:1–108101:4

Goriely A, Károlyi G, Tabor M (2005) Growth induced curve dynamics for filamentary micro-organisms. J Math Biol 51:355–366

Green PB, Erickson RO, Richmond PA (1970) On the physical basis of cell morphogenesis. Ann NY Acad Sci 175(Article 2):712–731

Harold FM (1997) How hyphae grow: morphogenesis explained? Protoplasma 197:137–147

Heath IB, van Rensburg EJJ (1996) Critical evaluation of the VSC model for tip growth. Mycoscience 37:71–80

Hejnowicz Z, Heinemann B, Sievers A (1977) Tip growth: patterns of growth rate and stress in the Chara rhizoid. J Plant Phys 81:409–424

Hill R (1950) The mathematical theory of plasticity. Oxford University Press, Oxford

Hirsch AM, Lum MR, Fujishige NA (2008) Microbial encounters of a symbiotic kind: attaching to roots and other surfaces. In: Emons AMC, Ketelaar T (eds) Root hairs: excellent tools for the study of plant molecular cell biology. Springer, Berlin Heidelberg New York. doi:10.1007/7089_2008_6

Katelaar T, Emons AM (2008) The actin cytoskeleton in root hairs: a cell elongation device. In: Emons AMC, Ketelaar (eds) Root hairs: excellent tools for the study of plant molecular cell biology. Springer, Berlin Heidelberg New York. doi:10.1007/7089_2008_8

Ketelaar T, Faivre-Moskalenko C, Esseling JJ, de Ruijter NCA, Grierson CS, Dogterom M, Emons AMC (2002) Positioning of nuclei in Arabidopsis root hairs: An actin-regulated process of tip growth. Plant Cell 14(11):2941–2955

Ketelaar T, Galway M, Mulder BM, Emons AMC (in press) A study into exocytosis and endocytosis rates in Arabidopsis root hairs and pollen tubes. J Microsc

Koch AL (1982) The shape of the hyphal tips of fungi. J Gen Microbiol 128: 947

Koch AL (1994) The problem of hyphal growth in streptomycetes and fungi. J Theor Biol 171:137–150

Kruse K, Joanny JF, Jülicher F, Prost J, Sekimoto K (2005) Generic theory of active polar gels: a paradigm for cytoskeletal dynamics Eur Phys J E Soft Matter 16(1):5–16
Landau LD, Lifschitz EM (1986) "heory of elasticity Pergamon, New York
Limpens E, Bisseling T (2008) Nod factor signal transduction in the Rhizobium-Legume symbiosis. In: Emons AMC, Ketelaar T (eds) Root hairs: excellent tools for the study of plant molecular cell biology. Springer, Berlin Heidelberg New York. doi:10.1007/7089_2008_10
Lockhart JA (1965) An analysis of irreversible plant cell elongation. Bot Rev 6:515–574
Miller DD, de Ruijter NCA, Bisseling T, Emons AMC (1999) The role of actin in root hair morphogenesis: studies with lipochito-oligosaccharide as a growth stimulator and cytochalasin as an actin perturbing drug. Plant J 17:141–154
Miller DJ, Leferink-ten Klooster HB, Emons AMC (2000) Lipochito-oligosaccharide nodulation factors stimulate cytoplasmic polarity with longidunal Endoplasmic Reticulum and vesicles at the tip in vetch root hairs. MPMI 13(12):1385–1390
Money NP (1997) Wishful thinking of turgor revisited: the mechanics of fungal growth. Fungal Genet Biol 21:173–187
Money NP (2001) Functions and evolutionary origin of hyphal turgor pressure. In: Geitmann A (ed) Cell biology of plant and fungal tip growth. IOS Press, Amsterdam, p 161
Money NP, Harold FM (1992)"Extension growth of the water mold Achlya: interplay of turgor and wall strength. Proc Natl Acad Sci USA 89:4245–4249
Mulder BM, Emons AMC (2001) A dynamic model for plant cell wall architecture formation. J Mat Biol 42:261–289
Nielsen E (2008) Plant cell wall biogenesis during tip growth in root hair cells. In: Emons AMC, Ketelaar T (eds) Root hairs: excellent tools for the study of plant molecular cell biology. Springer, Berlin Heidelberg New York. doi:10.1007/7089_2008_11
Pelcé P (2004) New visions of form and growth. Fingered, growth, dendrites and flames. Oxford University Press, Oxford
Pelcé P, Pocheau A (1992) Geometrical approach to the morphogenesis of unicellular algae. J Theor Biol 156:197–214
Popper KR (1935) Logik der Forschung. Springer, Berlin Heidelberg New York
Proseus TE, Boyer JS (2006) Periplasm turgor pressure controls wall deposition and assembly in growing Chara corallina cells. Ann Bot 98:93–105
Prager MW (1937) Mécanique des solides isotropes au delà du domaine élastique. Mem Sci Math 87:1–66
Regalado CM (1998) Roles of calcium gradients in hyphal tip growth: a mathematical model. Microbiology 144:2771–2782
Reinhardt MO (1892) Das wachstum der pilzhyphen. Jahrb Wissenschaft Bot 23:479–566
Roberson RW, Fuller MS (1988) Ultrastructural aspects of the hyphal tip of Sclerotium rolfsii preserved by freeze substitution. Protoplasma 146:143–149
Skalak R, Tozeren A, Zarda RP, Chien S (1973) Strain energy function of red blood cell membranes. Biophys J 13:245–264
Sieberer BJ, Timmers AC, Emons AMC (2005) Nod factors alter the microtubule cytoskeleton in Medicago truncatula root hairs to allow root hair reorientation. Mol Plant Microbe Interact 18(11):1195–1204
Sieberer BJ, Timmers ACJ (2008) Microtubules in Plant Root Hairs and Their Role in Cell Polarity and Tip Growth. In: Emons AMC, Ketelaar T (eds) Root hairs: excellent tools for the study of plant molecular cell biology. Springer, Berlin Heidelberg New York. doi:10.1007/7089_2008_13
Sietsma JH, Wessels JGH (1990) The occurrence of glucosaminoglycan in the wall of Schizosaccharomyces pombe. J Gen Microbiol 136 (11):2261–2265
Sietsma JH, Wösten HAB, Wessels JGH (1995) Cell-wall growth and protein secretion in fungi. Can J Bot 73:388–395
Sugden KEP, Evans MR, Poon WCK, Read ND (2007):Model of hyphal tip growth involving microtubule-based transport. Phys Rev E 75:031909:1–031909:5

Surrey T, Nedelec F, Leibler S, Karsenti E (2001) Physical properties determining self-organization of motors and microtubules. Science 292 (5519):1167–1171

Timmers ACJ, Vallotton P, Heym C, Menzel D (2007) Microtubule dynamics in root hairs of *Medicago truncatula*. Eur J Cell Biol 86: 69–83

Tindemans SH, Kern N, Mulder BM (2006) The diffusive vesicle supply center model for tip growth in fungal hyphae. J Theor Biol 238 (4):937–948

Todd PH (1986) Intrinsic geometry of biological surface growth. Springer Lecture Notes in Biomathematics 67

Van Batenburg FHD, Jonker R, Kijne JW (1986) *Rhizobium* induces marked root hair curling by redirection of tip growth: a computer simulation. Physiol Plant 66:476–480

Veytsman BA, Cosgrove DJ (1998) A model of cell wall expansion based on thermodynamics of polymer networks. Biophys J 75:2240–2250

Wymer C, Bibikova T, Gilroy S (1997) Calcium distributions in growing root hairs of Arabidopsis *thaliana*. Plant J 12:427–439

Yin HL, Zaner KS, Stossel TP (1980) Ca2+ control of actin gelation. Interaction of gelsolin with actin filaments and regulation of actin gelation. J Biol Chem 255: 9494–9500

Zonia L, Munnik T (2008) Vesicle trafficking dynamics and visualisation of zones of exocytosis and endocytosis in tobacco pollen tubes. J Exp Bot 59:861–873

Root Hair Electrophysiology

R.R. Lew

Abstract In this review, the value of root hairs as a model system for a broad range of electrophysiological experiments is highlighted by an introductory case study of their use for in situ measurements of vacuolar electrophysiology. Then, the basic tenets of electrophysiological measurements are presented: the electrical properties of the cell and how they are measured. Recent research on ion transporters (pumps, channels, and coupled transporters) is reviewed, followed by a discussion of the role of ion transport in root hair morphogenesis.

1 Introduction

When roots grow into a soil substrate, root hairs often develop behind the expansion zone of the root tip. They can serve a variety of roles. Structurally, they strengthen the anchoring of the root in the soil. They also serve to increase the surface area of the root, creating a three-dimensional network of tubular extensions that will ramify through the soil and increase the soil volume in near contact with the plant. Thus, they can increase the availability of ions and water for uptake by the growing root and the aboveground shoots and leaves (Peterson and Stevens 2000). Although root hairs are not an obligatory cell type, they are known to preferentially express ion transporters in the plasma membrane, corroborating their expected role in ion uptake by the plant (Lew 2000).

Root hairs are an excellent model cell for studies of ion transport in higher plant roots. They are accessible, extending out from the surface of the cell, and so they are easily impaled with the micropipette(s) necessary for electrophysiological characterization of ionic currents. Imaging, including fluorescence microscopy, is unimpeded by other cells, and external probes, such as ion-selective microelectrode and oxygen electrodes, can be brought directly to the surface of the cell. After

R.R. Lew
Department of Biology, York University, Toronto, Canada
e-mail: planters@yorku.ca

Plant Cell Monogr, doi:10.1007/7089_2008_9

impalement, the micropipette tip location can be confidentially assigned to the cytoplasm, especially in younger (10–100 μm long) cytoplasm-rich root hairs, based on injection of fluorescent molecules. Alternatively, the root hair can be impaled in the vacuole by impaling at the vacuolated base of the root hair, as indicated later. Root hair tip growth is relatively fast (in *Arabidopsis thaliana*, for instance, about 1 μm min^{-1}), so that ion transport measurements can be made in the context of a dynamic, physiologically competent, growing cell. Much of this chapter will describe the electrophysiological properties of root hairs, and review studies on ion transport in which root hairs played a crucial role. Much of this research emphasizes the role of root hairs in ion uptake by the plant, and the interplay of ion transport and cellular morphogenesis. However, to highlight the breadth of advantages that root hairs confer as a model cell, a case study (electrical properties of in situ vacuoles) will introduce the versatility of root hair research.

2 In Situ Vacuolar Electrophysiology Using Root Hairs

In plant cells, the vacuole often constitutes 80–90% of the cell volume. It must play an important role in ionic homeostasis. Yet, much of our understanding of the nature of ion fluxes across the vacuolar membrane relies on vacuoles isolated from their normal milieu, the only way to access them for patch clamp (Sakmann and Neher 1995) or other direct transport measurements. How transport properties change as a consequence of removal from the cell is unknown, but vacuolar properties may change markedly. Growing root hairs are an ideal system for measuring the electrical properties of the vacuole in situ, because of the accessibility of the cytoplasmic and vacuolar compartments in a dynamic, growing cell.

The electrical network of the root hair shown in Fig. 1 indicates the complexity of plasma membrane and vacuole electrical properties in series, and the technique that allows the electrical properties of the vacuole to be measured in situ – voltage clamping while the cytoplasm is maintained at a virtual ground to isolate the electrical responses of the vacuole from those of the cytoplasm. An example of the impalements is shown in Fig. 2. The experiments uncovered an increased ionic conductance in response to hyperosmotic stress in root hairs, presumably due to ion channel activation (Lew 2004). The results highlight the close coordination between plasma membrane (Shabala and Lew 2002) and vacuolar membrane responses (Lew 2004) to osmotic conditions in plant cells. Such in situ electrical measurements complement analyses of ion tracer efflux, such as those performed by MacRobbie (2006) in turgor-active guard cells responding to external changes in osmotic conditions.

In situ measurements could be taken in other cells, from other organisms. That root hairs can be used is provident, since they are often an integral part of the root system, and play a role in ion and water uptake during the growth and development of the plant. Research on root hairs enlightens our understanding of diverse aspects of plant physiology, and is relevant to the physiology of crop yield. They are accessible to a broad array of techniques in support of scientific enquiry.

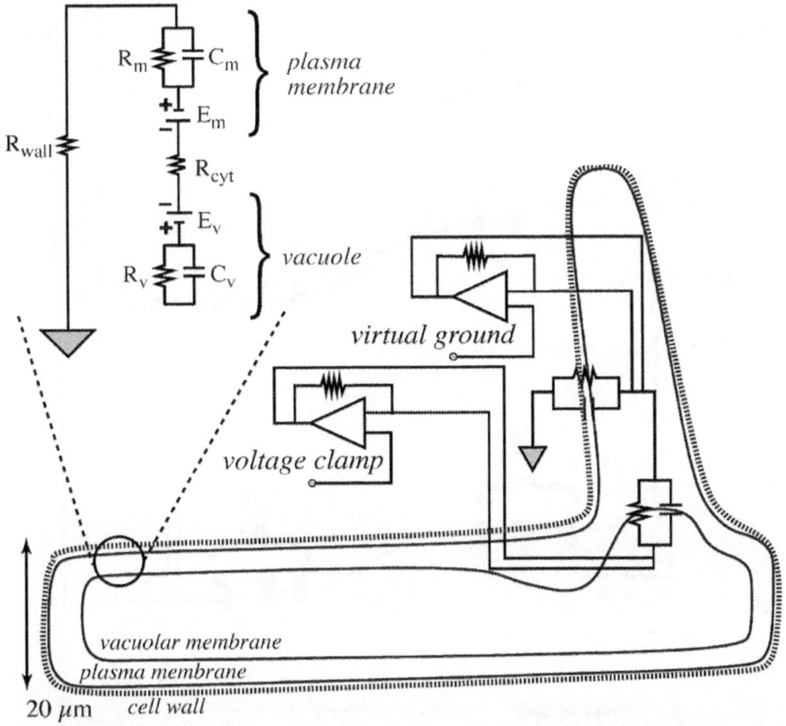

Fig. 1 The electrical networks of the root hair. The plasma membrane and vacuolar membrane network of resistances (R_m and R_v, respectively), capacitance (C_m and C_v), and potentials (E_m and E_v) are shown to the *left*. The electrical circuits required to measure the electrical properties of the vacuolar membrane in situ are shown to the *right*, overlaid on a diagram of the root hair

To undertake electrophysiological research on root hairs, understanding the electrical properties of the root hair is important. These properties are directly related to the net charge movements that can occur because of ion transport across the membranes.

2.1 Electrical Model of the Root Hair

The electrical properties of cells, including root hairs, are normally described in the same terms as those used in electronics: voltage, current, resistance, capacitance, and impedance (which is related to both resistance and capacitance). The first three electrical properties are related through Ohm's law, which describes the voltage created by a current passing through a resistance: *E* (voltage) = *I* (current) × *R* (resistance). An equivalent electrical circuit for the root hair is shown in Fig. 1. The voltage across the plasma membrane (E_m) of *A. thaliana* root hairs is normally

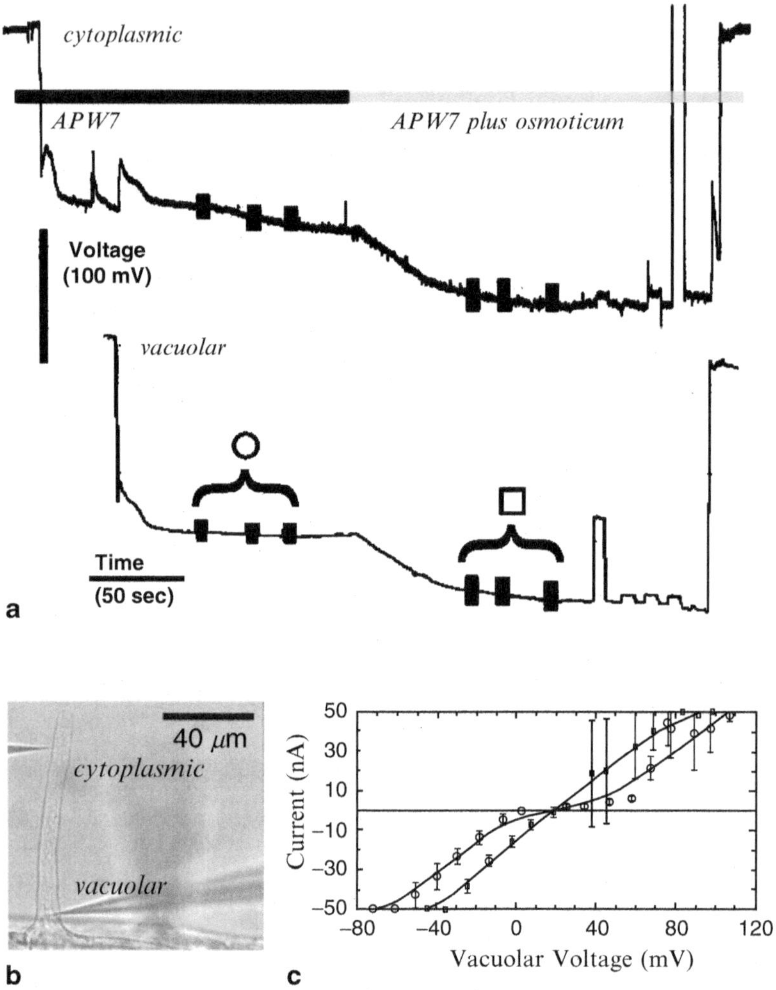

Fig. 2 An example of in situ measurement of the electrical properties of the vacuolar membrane. (**a**) Impalements into the cytoplasm and vacuole, as marked. (**b**) The dual impalements. (**c**) The current–voltage relations for the vacuolar membrane before (*circles*) and after (*squares*) a hyperosmotic treatment. These were measured while the cytoplasm was maintained at a "virtual ground" by the electrical circuitry shown in Fig. 1. Note that hyperosmotic treatment causes a hyperpolarization of the plasma membrane, recorded by both the cytoplasmic and vacuolar microelectrodes (since the vacuolar membrane is in electrical series with the plasma membrane). Using the virtual ground to isolate the vacuole from the plasma membrane, the vacuole membrane response to hyperosmotic treatment is increased conductance without a change in the positive electrical potential of the vacuole (**c**)

about −180 mV (negative inside), and the resistance (R_m) is normally about 20–40 M . The actual magnitude of the voltage varies depending on the state of the cell and the composition of the extracellular medium. Some of the voltage difference is caused by the asymmetric distribution of ions on either side of the plasma membrane.

These ions will contribute to the voltage based on their relative permeability through the plasma membrane. Permeability will be affected by ion transport through the membrane. There is a relation between resistance (R) and ionic permeability (P), but not a simple one, since resistance depends upon both the permeability and concentrations of each of the contributing ions (Schultz 1980). In root hairs, K^+ permeability dominates. If K^+ were the sole permeant ion (which it is not), the voltage at *net* zero current would be described by the Nernst potential: $E_{Nernst} = (RT/F)\ln(c_o/c_i)$, where R is the gas constant, T the temperature, F the Faraday constant, and c_o and c_i are the ion concentrations outside and inside, respectively. At room temperature, the Nernst potential (in millivolts) can be simplified to $E_{Nernst} = 55 \log_{10}(c_o/c_i)$. In addition to asymmetric distributions of ions, the plasma membrane H^+-ATPase generates an outward H^+ current which also contributes to the voltage (Felle 1982; Lew 1991). The relative contributions in *Arabidopsis* root hairs can be estimated from the effect of metabolic inhibitors which deplete cellular ATP (Lew and Spanswick 1984): The H^+-ATPase contributes about −135 mV, the Nernst potential about −45 mV (Lew 1991).

Capacitance, the fourth electrical property, is the ability of the root hair to hold charge, just as a battery is capable of holding charge. Capacitance determines the net imbalance of positive vs. negative ions in the cell required to generate a voltage difference (cf. Noble 1991). The net charge, Q (in coulombs), is related to capacitance (C, in farads, for which the fundamental units are coulombs per volt) by $Q = C\Delta E$, where ΔE is the voltage difference. The charge, Q, depends upon the volume of the cell: $Q = V_{roothair} c_{ion} F$, where $V_{roothair}$ is the root hair volume, c_{ion} is the concentration of ions resulting in the net charge imbalance, and F is the Faraday constant (96,480 C mol^{-1}, to convert concentration to charge). Assuming typical newly developing *Arabidopsis* root hair dimensions (cylindrical) of 15 by 87.5 μm for the epidermal part and 10 by 75 μm for the hair, the root hair area is 6.57×10^{-5} cm^2, and the volume is 21.35×10^{-12} L. The root hair capacitance is about 66 pF (based on the specific capacitance for biological membranes, about 1 μF cm^{-2}, multiplied by the root hair area). The net charge imbalance required to create a voltage difference of −180 mV would be $c_{ion} = (66 \times 10^{-12}$ C V$^{-1} \times (-0.18$ V$))/(21.35 \times 10^{-12}$ L $\times$ 96,480 C mol^{-1}), or 5.8 μM, a very small ion concentration difference. This calculation does not account for the larger *effective* volume (contributed by the volume of adjacent cells due to plasmodesmatal connections of the interconnected cells; Sect. 2.3), but even if the *effective* volume were tenfold larger, the net ion concentration difference would remain low compared to typical ion concentrations in a root cell. Thus, very small *net* charge movements across the membrane will have large effects on the voltage difference: A single ion channel with a flux of 1×10^6 molecules s^{-1} (1 pA current) would have to remain open only 12 s to cause this large a voltage difference in the root hair. In a "real" root hair, there is an ensemble of transporters functioning simultaneously; yet, the potential is remarkably constant.

Capacitance has another impact on the electrophysiological measurements. It affects the time dependence of voltage changes in response to changes in ionic current flow across the membrane. Formally, we first take the derivative with respect to time of the basic equation, $Q = C\Delta E$: $dQ/dt = C(dE/dt)$. Since current (I, coulombs per second) is equal to dQ/dt, then $I = C(dE/dt)$. Substituting Ohm's law, $E = IR$ (or $I = E/R$), we obtain $E/R = C(dE/dt)$. This last equation can be solved to yield $E = A$

$e^{-t/RC}$, where A is the voltage at time zero. The solution describes how a change in voltage will occur, as an exponential change over time. The rapidity of the change will depend upon the product of the resistance and capacitance of the cell (RC, with units of time, often abbreviated as $\tau = RC$, the time constant). For the root hair, the time constant, RC, is about 2–40 ms (Lew, unpublished). This becomes important in some measurement techniques, such as the discontinuous voltage clamp, which rely upon differences in time responses to separate the resistance of the cell from the resistance of the microelectrode impaled into the cell. It can also be important in instrumentation, because it places limits on the measurability of fast events. The fifth electrical property is impedance, a measurement of resistance that accounts for the effect of capacitance on time-varying changes. It is normally used in the context of the frequency dependence of changes in voltage or current.

Ion transport across the membrane, as long as net charge movements are occurring, is the fundamental act that can be explored in electrophysiological measurements.

2.2 *Electrophysiological Measurements*

Voltage differences and resistances are measured by inserting microelectrodes into the cell. Microelectrode construction and use are described by Thomas (1978), Purves (1981), and Blatt (1991). Ion-selective microelectrodes are described in detail by Ammann (1986). The *Plymouth Workshop Handbook* (Ogden 1994) includes contributed chapters describing a variety of experimental techniques and analyses. Volkov (2006) is a recent compilation of microelectrode (and other) techniques.

Impalements with a single microelectrode can be used to measure the root hair voltage, but measurements of resistance and capacitance are complicated by the fact that current injection occurs through the same micropipette as the measurement of voltage. Since micropipette tips have high resistances (often 20–40 MΩ), the tip resistance obscures any measurement of the resistance across the root hair plasma membrane. Therefore, the "standard" for measurements of voltage *and* resistance is impalements with two microelectrodes (Etherton et al. 1977). One microelectrode is used to inject current, while the second one monitors voltage changes. This is a feasible approach with root hairs (Lew 1994, 1996, 2004), but technically challenging. With a single microelectrode, it is possible to use a technique called discontinuous voltage clamp to obtain estimates of both voltage and resistance (Finkel and Redman 1984), a technique used successfully in a characterization of Nod factor effects on multiple ion transporters (H^+-ATPase, K^+, and anion channels) of clover root hairs (Kurkdjian et al. 2000), part of the signaling pathway prior to root nodulation (Kurkdjian 1995), which also involves intracellular Ca^{2+} spiking (Kanamori et al. 2006; see Limpens and Bisseling 2008). The discontinuous voltage-clamp technique relies upon the differences in the response times of the microelectrode and cell, which in turn depend upon their resistance and capacitance (the time constant RC described earlier). In general, it is difficult to assure that resistance and capacitance of the microelectrode would not be changed during the impalement; this is not a problem with multiple impalements.

Double-barrel micropipettes offer the ability to impale only once, yet be able to inject current and monitor voltage in separate microelectrodes. Their fabrication and technical aspects of their use are described by Lew (2006). Current and voltage measurements are normally taken using voltage clamp, to characterize the voltage dependence of ion transport.

2.3 *Cable Properties and Electrical Coupling*

When current is injected into an *Arabidopsis* root hair, and the voltage deflection measured in that root hair (root hair I) and the adjacent root hair (root hair II), the voltage deflections in root hair II are quite large (ca. 50% of the deflection in root hair I) (Lew 1994). This is direct evidence for electrical coupling between adjacent root hairs, presumably due to ion flux through the plasmodesmata. The cells act like an electrical cable, through which ionic currents flow freely from cell to cell. Electrical signals will readily propagate cell-to-cell, but ionic compositions would likely be heterogeneous, because of the time required for ion diffusion through the relatively long distances of an interconnected network of multiple cells.

Technically, cell-to-cell electrical coupling causes problems with quantitation of voltage clamping. Voltage clamping requires clamp fidelity, in which the voltage is clamped to the same value throughout all regions of the cellular space. Because ionic currents flow into adjacent cells, a complete space clamp cannot be achieved. It is a technical drawback that must be acknowledged, and would not be solved until a quantitative map of cable properties (including sub-epidermal cells) is created. Such a quantitative map would enlighten our understanding of symplastic ion transport through root cells.

2.4 *Beyond Potential and Resistance*

There are a number of electrophysiological techniques available beside intracellular microelectrodes. These include electrophysiological techniques for the measurement of cytoplasmic ion activities, for the measurement of diffusive ion gradients extracellularly, which can be converted into ion fluxes, for direct measurements of turgor with a pressure probe, and even for measurements of other molecules, such as oxygen and metabolites such as auxin. These techniques will be described in the following sections.

2.4.1 Ion-Selective Microelectrodes

Ion-selective microelectrodes are normally constructed by placing an ion-sensitive liquid ion-exchanger (Ammann 1986; Thomas 1978) in the tip of the microelectrode.

To avoid displacement due to turgor when the microelectrode is impaled into the cell, poly(vinyl chloride) is often included with the ion-selective cocktail so that the ion-exchanger is held firmly in place (Felle 1993). The resistance of the resulting membrane is very high, and so a high input impedance electrometer (>10^{15} Ω) must be used for measurements. The ion-selective microelectrode usually exhibits a close to Nernstian response to the ion it is specific for. When the ion-selective microelectrode is impaled into a cell, it measures both the cell potential and the Nernst potential for the ion. To separate the two voltages, the cell must be impaled with another microelectrode to measure the cell potential. This may be a separate impalement, or a double-barrel microelectrode can be used (Miller and Wells 2006). Root hairs are very suitable for these types of measurements, since the cytoplasmic compartment can be impaled with ease, obviating the need for any indirect assignment of tip location, based, for example, on the measured activity of the selected ion.

2.4.2 Extracellular Measurements of Net Ion Fluxes

Not only can ion-selective electrodes be used to measure intracellular (and extracellular) free ion concentrations, but with modifications to the measuring technique, they can be used to measure ion fluxes. This is done by measuring the ion concentration at two distances from the cell or tissue being examined. The fundamental basis for inferring flux from two measurements of ion concentration (c_1 and c_2 measured Δx distance apart) is that a diffusive gradient will be created, either inward or outward depending on whether influx or efflux is occurring. This technique was one of the methods used by Felle and Hepler (1997) to demonstrate a tip-localized Ca^{2+} influx in growing root hairs. The method has been described in detail by Volkov (2006). The general procedure of measuring diffusive gradients created by net fluxes across the plasma membrane has been extended to include a wide variety of substances that can be sampled using microelectrode technology, including oxygen.

2.4.3 Fluorescent Probes

An emerging technique is the use of fluorescent fusion proteins to explore gene expression, and cytological localization and trafficking dynamics. The use of GFP (green fluorescent protein) constructs has been described by Goedhart and Gadella (2000). With the development of relatively simple techniques to study the expression of fluorescent protein constructs in root epidermis (Campanoni et al. 2007), researchers will gain insight into cytological features of root hairs, to provide a better understanding of expression and trafficking of transport proteins (Campanoni et al. 2007).

2.4.4 Measurements of Turgor

Measurements of the cell hydrostatic pressure (turgor) would not normally be considered an *electro*physiological technique. However, the required micromanipulation methods used to impale the cell are the same, and turgor is closely allied with transport of both water and ions across cellular membranes (both the plasma membrane and vacuolar membrane). In this technique, micropipettes are pulled to a coarser tip, to minimize the obstruction to mass flow through the tip. The micropipette is filled with low-viscosity silicone oil, which improves the ease of hydraulic flow and does not mix with the cell cytoplasm. Upon impalement, the internal pressure of the cell pushes the oil/cytoplasm meniscus into the micropipette. Pressure is applied to the micropipette via a piston: the pressure required to "push" the meniscus back to the tip is the initial turgor of the cell. The pressure is measured with a transducer in contact with the silicone oil. Typical turgor of *Arabidopsis* root hairs is about 680 ± 200 kPa (Lew 1996). Turgor is regulated rapidly: within 40–50 min of a hyperosmotic treatment that causes a rapid decrease in turgor, the turgor recovers close to its original level. At least in *Arabidopsis*, turgor recovery occurs in concert with activation of the plasma membrane H^+-ATPase and uptake of osmotically active ions (mostly K^+ and Cl^-) (Shabala and Lew 2002).

3 Ion Transport in Root Hairs

Having summarized some of the cell biophysical techniques used to measure the physiological dynamics of root hairs, what roles would ion transport play in the growth and differentiation of the root hair?

Maintaining osmotic balance of the root hair during cellular expansion is essential to maintain a constant turgor. With growth rates of 1 μm min^{-1}, a root hair of 8 μm diameter will increase its volume about 50 fL min^{-1}. If osmotic balance is to be maintained by ion influx, the root hair must take up about 25 fmol min^{-1}. Much of this uptake will be into the vacuole, which increases in size as the cell expands. Root hairs may also be a pathway for ion uptake to support the osmotic balance of other cells within the plant body. Thus ionic balance is crucial to the root hair. In addition, it is common for plant cells to accumulate ions selectively. That is, K^+ accumulation is preferred to Na^+; Ca^{2+} is actively excluded from the cytoplasm; Cl^- is normally taken up as the counterion.

Ions may also play a role in signaling. Ca^{2+} influx to maintain a tip-high Ca^{2+} gradient during tip growth of the hair has already been noted (Sect. 2.4.2) (see Bibikova and Gilroy 2008). Ca^{2+} may act as a mediator of the vesicle fusion required for continued cell expansion, and is known to activate both kinases and phosphatases in the cytoplasm. Protons are crucial to the life of the cell, not only in the context of cytoplasmic pH regulation, but modulation of extracellular pH, and as a signaling molecule analogous to Ca^{2+} during the onset of nodulation (Felle and Herrmann 2000). The ubiquitous role of protons is further supported by the central role of the H^+-ATPase in nutrient transport (Palmgren 1998, 2001; Sondergaard et al. 2004).

3.1 The Plasma Membrane H^+-ATPase

The plasma membrane H^+-ATPase is a P-type active ion pump (Lauger 1991) common in walled cells of the plant and fungal kingdoms, and many of the algae. Owing to the electrogenic nature of the ATP-dependent H^+ extrusion, the H^+-ATPase contributes significantly to the negative-inside potential of the cell, a driving force for uptake of a number of nutrients. In *Arabidopsis* root hairs, direct evidence for the role of the H^+-ATPase was based upon the depolarizing effects of an inhibitor of P-type active pumps, vanadate, and cytoplasmic ATP depletion by cyanide (Lew 1991). The magnitude of the depolarization, about 135 mV from an initial value of about –180 mV, is evidence for its dominant role generating the potential (Lew 1991). The gene(s) for the H^+-ATPase have been cloned from a number of species, and it is known to exist in various isoforms (Michelet and Boutry 1995). A number of gene subfamilies (Arango et al. 2003) exhibit differential expression, which suggests different functional roles (Moriau et al. 1999; Lefebvre et al. 2005). In *Nicotiana plumbaginifolia*, one of the genes (*PMA4*) of a subfamily that appears to be expressed in tissues that have a role in ion uptake is strongly expressed in root hairs of seedlings, implicating root hairs as a major site of ion uptake for young roots (Moriau et al. 1999). The voltage-dependent conductance of the membrane before and after vanadate inhibition of the H^+-ATPase is relatively unchanged at voltages from –200 to 0 mV (Lew 1991). Thus, the pump exhibits a voltage-independent "constant current" activity. Estimates of the current density are problematic because of the problems discussed regarding the root hair cable properties and cell-to-cell coupling, but it appears to be quite high, indicating a high level of H^+-ATPase activity in root hairs (Lew 1991).

It is unclear whether root hair H^+-ATPase activity is regulated by intracellular ions, such as calcium (known to activate the *Neurospora crassa* H^+-ATPase [Lew 1989]) or protons, one of the substrates for the enzyme (although the H^+-ATPase does not appear to regulate cytosolic pH [Felle 1996]). Experiments ionophoresing either Ca^{2+} or H^+ ions directly into the root hair revealed no compelling evidence for regulation by either ion. Increasing cytoplasmic $[Ca^{2+}]$ rapidly inhibits cytoplasmic streaming; only at higher levels does Ca^{2+} depolarize the potential and inhibit cell-to-cell coupling (Lew 1994). Increasing cytoplasmic $[H^+]$ causes a rapid increase in vacuolar area (Lew, unpublished) but has no effect on the membrane potential and minimal effects on cell-to-cell coupling (Lew 1994).

The plant hormones auxin (Tretyn et al. 1991; Ayling et al. 1994) and cytokinin (Silverman et al. 1998) affect the membrane potential of root hairs. Exogenous, but not intracellular injection of, cytokinin hyperpolarizes the potential, consistent with H^+-ATPase activation, and stimulates root hair growth (Silverman et al. 1998). Auxin either depolarizes (Ayling et al. 1994; Tretyn et al. 1991) or hyperpolarizes the potential, depending on the auxin concentration (Tretyn et al. 1991). In addition to effects on the electrical properties, auxin is implicated in root hair initiation (Masucci and Schiefelbein 1994). For information about the effects of auxin on root hair cells, see Lee and Cho (2008).

Because of the central role of the H^+-ATPase in nutrient uptake, its regulation is of considerable interest (Palmgren 2001). The fungal toxin fusicoccin is known to activate the H^+-ATPase and does so by activating an endogenous transduction pathway of H^+-ATPase phosphorylation to cause binding of a 14-3-3 protein (cf. Kanczewska et al. 2005). Root hairs presumably contain the appropriate signaling components, since fuscoccin hyperpolarizes the potential of *Limnobium stolonifera* root hairs (Ullrich and Novacky 1990). There is a recent report on fusicoccin activation of an MAP kinase cascade in tomato leaves (Higgins et al. 2006). An MAP kinase cascade directly activates the H^+-ATPase in the fungus *N. crassa* in a pathway separate from activation of gene expression (Lew et al. 2006). It is part of a turgor-regulating system in the fungus; a similar system may exist in higher plants (Shabala and Lew 2002). In plant root hairs, it is likely that continuing research will reveal a web of pathways controlling H^+-ATPase activity, either directly or via regulation of H^+-ATPase gene(s) expression.

As noted earlier, on the basis of inhibitor studies, the H^+-ATPase is a major contributor to the membrane potential of the cell. The potential is one component of the proton motive force (pmf), which is the sum of the potential and pH differences between the inside of the cell and the extracellular environment: pmf (mV) = ΔVoltage + 55ΔpH. The potential or the pH difference, or both, can be used to drive the uptake of other nutrients and ions into the cell. Of these ions, K^+ and Cl^- are accumulated in large amounts inside the cell.

3.2 K^+ Transport

K^+ is an essential nutrient, actively accumulated from the soil solution, which normally contains low concentrations (0.3–5 mM) compared to 100–200 mM in the cell. Much of the identification of K^+ transporters has relied on functional complementation of yeast K^+ transport mutants using cDNA library screens (Fox and Guerinot 1998). There are at least five K^+ transporter families that include K^+ channels, cotransporters, and antiporters (Maser et al. 2001; Ashley et al. 2006).

Uptake through channels would be energized by the negative-inside potential of the cell (Gassmann and Schroeder 1994; Maathuis and Sanders 1993, 1995). If the potential is about –170 mV and internal [K^+] is 100 mM, then active accumulation through the channel can occur at soil solution [K^+] greater than about 0.1 mM. One example of an inward channel is AKT1, known to be expressed in *Arabidopsis* roots (Legarde et al. 1996). The *akt1* mutant exhibits reduced K^+ uptake (using ^{86}Rb as the radioactive tracer) and apparently lacks the inward K^+ channel (Hirsch et al. 1998; Ivashikina et al. 2001), as measured by patch clamp. Direct evidence for the K^+ channel's role in K^+ uptake in intact root hairs was obtained by using extracellular ion-selective microelectrodes to monitor K^+ sequestration after oligochitin-elicitor treatment. The elicitor first induced K^+ release, followed by reuptake in wild-type root hairs, but reuptake was delayed in the *akt1* mutant (Ivashikina et al. 2001). The kinetics of the inward K^+ channel (AKT1) are modulated by AtKC1, a regulatory subunit (Reintanz et al. 2002); it is regulated by a Ca^{2+}-activated protein kinase pathway (Xu et al. 2006; Li et al. 2006).

Outward K^+ channels (GORK) are also expressed in root hairs (Ivashikina et al. 2001). Because the voltage-sensitive gating of this outward K^+ channel is affected by extracellular [K^+] (activation shifts to more positive voltages at higher [K^+]), it is possible that the channel functions in K^+-"sensing" and may have a role in the polar growth of the cell (Ivashikina et al. 2001). Alternatively, the outward channel may maintain the potential within well-defined limits: at low external [K^+], the potential will be more negative, hence the channel would activate at more negative potentials should the potential depolarize; at high external [K^+], when the potential is more positive, activation at a more positive potential would avoid excessive K^+ loss from the cell.

Another K^+ permease family plays a role in K^+ uptake: *KT/KUP*. In *Arabidopsis*, 10 of the 13 *AtKT/KUP* genes are expressed in root hairs (Ahn et al. 2004), which is an indirect evidence for the role of root hairs in K^+ uptake. Functional complementation of the genes in an *E. coli* mutant deficient in K^+ uptake demonstrated the protein products function in K^+ transport (Ahn et al. 2004). A knock-out mutant of the *AtKT3/KUP4* member (Ahn et al. 2004) of the *KT/KUP* family causes a "tiny root hair" (*trh1*) phenotype, specifically the absence of root hair elongation after initiation (Rigas et al. 2001). Whether the phenotype is caused by a disruption of K^+ transport is unclear, since elevating extracellular [K^+] did not rescue the phenotype. However, even in wild type, elevated [KCl] causes a decrease in root hair length, a response also observed in the already short root hairs of the *trh1* mutant (Desbrosses et al. 2003).

Much of the above-mentioned research has relied upon characterization of the ion channels with the patch clamp technique. Root hairs are special as a model system for examining issues related to transport of K^+ and other osmotically active ions because uptake can be examined in the context of cellular growth, in situ. Time-dependent inward and outward currents, inhibitable by quaternary ammoniums (tetraethylammonium and tetrapentylammonium) and Cs^+ ("standard" K^+ channel inhibitors, Hille 1984), have been measured in intact root hairs by using a discontinuous single-electrode voltage clamp technique (Bouteau et al. 1999), vindicating in vitro characterizations. The inward K^+ current does appear to have a role in cellular expansion. To maintain turgor during growth, the root hair must accumulate osmotically active substances. K^+ uptake plays a role: inhibition of the inward K^+ channel with tetraethylammonium inhibits an inward K^+ current and *Arabidopsis* root hair growth (Lew 1991). The inhibition of growth is transient (growth resumes after about 4 min), and so K^+ uptake is not an obligatory mechanism for maintaining the internal osmolarity at a level sufficient to "drive" cellular expansion.

Net K^+ influx was measured in growing *Limnobium stoloniferum* root hairs using the vibrating ion-selective probe technique (Jones et al. 1995). The root hairs have TEA-sensitive inward (and outward) K^+ currents (Grabov and Bottger 1994). Root hair growth is inhibited by Al^{3+} (half-maximally at about 7 μM), but K^+ influx continued after inhibition of growth by 20 μM Al^{3+} (Jones et al. 1995). Aluminum does inhibit inward K^+ channels of wheat root hairs (half-maximally at about 8 μM;

Gassmann and Schroeder 1994). So, Al^{3+} levels sufficient to inhibit an inward K^+ channel measured on root hair protoplasts with the patch clamp technique do not affect K^+ influx in intact root hairs, even though growth inhibition is observed. The conclusion is that there is no one-to-one correspondence between K^+ uptake and growth. This may be unfortunate for scientists, but certainly beneficial for plants subjected to diverse stresses during their growth and survival, such that multiple mechanisms for maintaining growth are crucial.

Because of their large size, the electrical properties of *Limnobium* root hairs are easily measured (cf. Ullrich and Novacky 1990) with a technique uncommon in higher plant electrophysiology: the sucrose gap (Purves 1981). The sucrose gap electrically isolates two regions of the root hair, so that the voltage dependence of ionic currents can be measured by voltage clamping. With this technique, Grabov and Bottger (1994) identified TEA-sensitive inward and outward K^+ channels that were active at potentials more negative and more positive than the normal membrane potential, respectively. Modifying the redox potential of the root hairs by adding the electron acceptor hexacyanoferrate(III) to the external solution activated the inward K^+ channels and inhibited the outward K^+ channels. The physiological role of redox modulation is not known, but it may be related to a redox system located on the plasma membrane that mediates iron uptake (Moog et al. 1995; Robinson et al. 1999). Redox modification (with reductive agents) is known to control root hair morphogenesis (Sanchez-Fernandez et al. 1997).

Other factors beside redox poise may regulate K^+ transport. Many K^+ channels are regulated by ATP. It would not be surprising to uncover a linkage between "energy charge" of the cell and ion transport, both direct effects (for example, the ATP dependence of the H^+-ATPase activity) and indirect effects (regulation of channel activity). Such an energy charge linkage exists between the Na^+/K^+-ATPase and K^+ channel activity in renal proximal tubules (Tsuchiya et al. 1992). In yeast, Ramirez et al. (1989) reported altered K^+ channel activation by cytoplasmic ATP in a H^+-ATPase mutant (*pma1–105*). In higher plants, there is no direct biochemical linkage between K^+ uptake and the H^+-ATPase (Briskin and Gawienowski 1996), but indirect linkages may eventually be discovered. Certainly, nutritional status regulates K^+ transport (Ashley et al. 2006). Root hairs are a relatively "simple" system in which to explore such regulation.

K^+ uptake alone cannot account for osmotic balance in growing cells; it must be accompanied by a counterion to assure the maintenance of electrical balance. Cl^- is the major anion actively accumulated in plant cells.

3.3 Cl^- Transport

In growing *Arabidopsis* root hairs, Cl^- influx was measured directly with the ion-selective vibrating probe (Lew 1998b). Cl^- uptake cannot be passive through a Cl^- channel, because the negative inside voltage of the cell would electrophoretically "expel" the Cl^- anion from the cell. Instead, uptake relies upon an nH^+/Cl^- symport, so that Cl^- uptake is coupled to the proton motive force generated by the H^+-ATPase. Evidence for a symport mechanism is based upon measurements of the electrical potential and pH, both cytoplasmic and extracellular, in *Limnobium* root hairs (Ullrich and Novacky 1990). The extracellular addition of Cl^- causes a large depolarization of the potential (about 60–100 mV) along with acidification of the cytoplasm and alkalinization of the extracellular medium. This is consistent with cotransport of more than one H^+ with each Cl^- ion: the net positive charge influx would cause the depolarization. The pH changes are consistent with the fact that HCl is being transported. More extensive experiments supporting the presence of a H^+/Cl^- symport were performed by Felle (1994), who measured Cl^- accumulation directly with an intracellular Cl^--selective microelectrode in *Sinapis alba* root hairs. Felle (1994) also measured pH with a H^+-selective microelectrode, as well as the potential and resistance of the root hair. A shift to an acidic extracellular pH (from 9.5 to 4.5) was sufficient to increase intracellular $[Cl^-]$ about 2.5-fold, even though the membrane potential depolarizes under these conditions. This indicates a kinetic dependence of H^+/Cl^- symport activity on the ΔpH component of the proton motive force, separate from the ΔVoltage component.

Chloride channels would function in Cl^- efflux from the cell. Nod factor signaling includes a transient depolarization of the potential which occurs concomitant with the appearance of Cl^- in the extracellular medium and a decrease in cytoplasmic $[Cl^-]$. The Cl^- efflux is believed to be caused by a Ca^{2+}-activated Cl^- channel in the root hair (Felle et al. 1998). Anion channels from root epidermal cells have been characterized using patch clamp (Diatloff et al. 2004). The channels are permeable to a wide range of anions (including citrate). Organic anion selective channels are regulated by extracellular Al^{3+} and the phosphate nutritional status of the plant; these and other physiological roles of anion channels in roots (e.g., osmoregulation) have been recently reviewed by Roberts (2006).

3.4 *Nitrogen (NH_4^+ and NO_3^-), Sulphate, and Phosphate Transport*

Although they are not normally accumulated in large amounts in the plant, nitrogen species (usually NH_4^+ and NO_3^-), sulphate, and phosphate are essential ionic nutrients and must be actively imported by the cell.

Ammonium uptake must involve multiple mechanisms that would depend upon pH, and thus the relative contributions of the unprotonated ammonia (NH_3) and protonated NO_3^-. The two forms are interconvertible by the following reactions: $NH_3(aq)+H_2O \leftrightarrow NH_4^+ + OH^-(pK_b, 4.75)$ and $NH_4^+ \leftrightarrow NH_3 + H^+(pK_a, 9.25)$. At neutral pH, the dominant form would be the ammonium ion (NH_4^+), but NH_3 may still play a role as a membrane-permeable form. However, the addition of NH_4^+ extracellularly causes depolarization of the potential in barley and tomato root hairs (Ayling 1993). This is consistent with positive charge entry into the cell and implies that NH_4^+ is the transported form. Because K^+ channels are often significantly permeable to NH_4^+, the ammonium ion could be transported by an inward K^+ channel, but ammonium transporters have been identified in root hairs (Lauter et al. 1996). The *LeAMT1* was discovered by screening a root-hair-specific cDNA library of tomato; its functional identification was based on rescue of a yeast mutant deficient in ammonium transport (Lauter et al. 1996). Four *AMT* genes that are expressed in roots have been identified in *Arabidopsis* (Loque et al. 2006), some are expressed in response to nitrogen deficiency, and have been assigned a role in high-affinity uptake by analyzing T-DNA insertion mutants. When the tomato genes *LeAMT1;1* and *LeAMT1;2* are expressed in oocytes, the addition of NH_4^+ caused inward currents, consistent with uptake of the positively charged ammonium ion (Ludewig et al. 2003).

In general, NH_4^+ is a preferred nitrogen species, because NO_3^- must be reduced to ammonium for production of amino acids, but NO_3^- is more common in the soil solution. As an anion, it is expected that active accumulation would rely on an nH^+/NO_3^- symport. Using *Limnobium* root hairs, Ullrich and Novacky (1990) examined the effect of NO_3^- on the membrane potential and cytoplasmic pH. Nitrate addition depolarized the potential, but, unexpectedly, it caused an alkalinization of both the cytoplasm and extracellular solution. This is not consistent with an nH^+/NO_3^- symporter, which should acidify the cytoplasm, but could be due to rapid reduction of NO_3^- in the cytoplasm, consuming the imported H^+. Using *Arabidopsis* root hairs, Meharg and Blatt (1995) voltage clamped to assess the voltage dependence of the ionic currents induced by NO_3^- transport. They observed larger NO_3^--induced currents at voltages more negative than the normal membrane potential and at acidic extracellular pH, supporting the presence of a voltage-dependent nH^+/NO_3^- symporter. Nitrate (NO_3^-) transporters have been cloned (cf. Wang and Crawford 1996; Wang et al. 1998; Lauter et al. 1996), and some are preferentially expressed in root hairs (Lauter et al. 1996). Heterologous expression of a nitrate transporter (CHL1) in *Xenopus laevis* oocytes caused pH-dependent inward currents consistent with an nH^+/NO_3^- symporter (Tsay et al. 1993). Mutations in both the *Arabidopsis* CHL1 (NRT1) and NRT2 nitrate transporters minimize the membrane potential changes observed after additions of low levels of NO_3^- (Wang and Crawford 1996; Wang et al. 1998).

The importance of urea uptake has been recently demonstrated. Urea is normally broken down to ammonium relatively quickly. However, Kojima et al. (2007) identified a gene for high-affinity urea uptake, *AtDUR3*, and documented its expression in the plasma membrane of *Arabidopsis* root hairs by immunocytochemistry, and its expression under nitrogen-deficient conditions with a GFP (green fluorescent protein) fusion construct.

As is true for other anions, phosphate appears to be taken up via an nH^+/Pi symporter (Ullrich and Novacky 1990). Phosphate transporters have been cloned, and do appear to be expressed in root epidermal cells, including root hairs (Daram et al. 1998; Liu et al. 1998). The promoter for the phosphate transporter gene has been proposed as a biotechnological tool since it causes induction of heterologous genes in roots, including root hairs (Schunmann et al. 2004).

Sulfate is normally reduced in photosynthetic tissues, but must first be transported into the plant. A number of genes encoding sulfate transporters have been cloned from a variety of plant species and their function has been established by heterologous expression in yeast (cf. Yoshimoto et al. 2002). The transporters are expressed in root hairs, and are induced by sulfate deficiency (Maruyama-Nakashita et al. 2004). Sulfate deficiency also induces the expression of enzymes responsible for the reduction of SO_4^-, but since most reduction and assimilation occur in the leaves, regulatory mechanisms are complex (Hopkins et al. 2005), and include sulfate transport into vacuoles as a mechanism controlling cytoplasmic concentrations (Kataoka et al. 2004; the regulation of sulfate assimilation was recently reviewed by Kopriva 2006). The transport mechanism is as yet uncharacterized in situ, but since sulfate is taken up as the anion, a nH^+/SO_4^- symport is very likely, and supported by the observation that uptake is stimulated by acidic pH when plant sulfate transporters (SHST1 and SHST2 from *Stylosanthes hamata*) are expressed in yeast (Smith et al. 1995).

4 Ion Transport and Root Hair Morphogenesis

As a tip-growing cell, a root hair offers the ability to explore potential roles that ion transport may have during root hair initiation and elongation. As osmotically active agents accumulated in large amounts, K^+ and Cl^- may function to maintain intracellular osmolarity during growth. In addition, transported ions may act as a second messenger during signal transduction. Hormones and other intracellular components of signal transduction may regulate ion transport as part of their regulation of morphogenesis.

4.1 Ca^{2+} Transport and Root Hair Morphogenesis

Ca^{2+} is very important to the physiological function of plant cells, primarily because of its role as a second messenger. The importance of Ca^{2+} in signal transduction will be detailed elsewhere in this book (see Bibikova and Gilroy 2008).

4.2 Control of Root Hair Morphogenesis by Ionic Nutrients

Besides the signaling role of Ca^{2+}, other ions also affect morphogenesis, or more accurately, root hair initiation and growth appear to be induced under conditions of nutrient deficiency. The general role of root hairs is believed to be an increase in the soil volume in near proximity to the root, increasing the effective zone of diffusive supply of ions to the root (Jungk 2001). Root hair length and number are increased by nutrient deficiencies (Peterson and Stevens 2000): for example, phosphorus (Bates and Lynch 1996), nitrate (Jungk 2001), potassium (Desbrosses et al. 2003), iron (Moog et al. 1995), and manganese (Konno et al. 2006). The regulation of root hair initiation and growth under nutrient deficiencies is unclear, and the "sensing" of nutrient deficiency is unlikely to be localized specifically to root hairs. However, Shin et al. (2005) suggest that reactive oxygen species may play a role in response to nutrient deficiencies. H_2O_2 is induced within 30 h of nutrient deprivation in wild type, but not in *Arabidopsis* root hair mutants (for potassium and nitrogen, but not for phosphorus deprivation); so it is possible that root hairs have a role in sensing potassium and nitrogen deficiencies. The hormone auxin may control root hair formation in response to iron or phosphorus deficiency, since root hair and transfer cell formation is inhibited in auxin mutants (*axr2*), completely for iron deficiency and partially for phosphorus deficiency (Schikora and Schmidt 2001).

The concept that root hairs can play a role in nutrient uptake is supported by their induction under deficient conditions, but they may not be *obligatory* for nutrient uptake. Comparisons of rice mutants lacking root hairs or lateral roots revealed that lateral roots are more important for silicon uptake (Ma et al. 2001). The density of the root network ramifying through the soil would be expected to compensate for the absence of root hairs.

5 Concluding Remarks

Even if root hairs are not crucial for the survival of the plant, they can play important roles in nutrient uptake. They are a remarkable model system for intracellular manipulations, electrophysiology, and cell imaging. It is probably fair to say that no other cell type in the plant can be used with such remarkable experimental sophistication in situ, intact and growing. They are an excellent single cell test tube; in conjunction with molecular biological manipulations, they can be used to unravel molecular mechanisms of ion transport that underpin the growth and development of a plant cell. In fact, one could envision experiments modifying ion concentrations inside the cell by ionophoresis, using ion-selective fluorescent probes to map ion activity in the cell, while measuring ion transport across the cell with voltage clamp, and, the concordant ion fluxes with an extracellular ion-selective electrode. Such experiments could reveal not only the basic transport mechanisms (such as stoichiometries and energetics), but how ion transporters act cooperatively in a growing cell.

References

Ahn SJ, Shin R, Schachtman DP (2004) Expression of *KT/KUP.* genes in *Arabidopsis* and the role of root hairs in K^+ uptake. *Plant Physiol* 134:1135–1145

Ammann D (1986) Ion-selective microelectrodes. Principles, design and application. Springer, Berlin Heidelberg New York, pp 1–346

Arango M, Gevaudant F, Oufattole M, Boutry M (2003) The plasma membrane proton pump ATPase: The significance of gene subfamilies. *Planta* 216:355–365

Ashley MK, Grant M, Grabov A (2006) Plant responses to potassium deficiencies: a role for potassium transport proteins. *J Exp Bot* 57:425–436

Ayling SM (1993) The effect of ammonium ions on membrane potential and anion flux in roots of barley and tomato. *Plant Cell Environ* 16:297–303

Ayling SM, Brownlee C, Clarkson DT (1994) The cytoplasmic streaming response of tomato root hairs to auxin: observations of cytosolic calcium levels. *J Plant Physiol* 143:184–188

Bates TR, Lynch JP (1996) Stimulation of root hair elongation in *Arabidopsis thaliana* by low phosphorus availability. *Plant Cell Environ* 19:529–538

Bibikova T, Gilroy S (2008) Calcium in root hair growth. In: Emons AMC, Ketelaar T (eds) Root hairs: excellent tools for the study of plant molecular cell biology. Springer, Berlin Heidelberg New York. doi:10.1007/7089_2008_3

Blatt MR (1991) A primer in plant electro hysiological methods. In: Hostettmann K (ed) Methods in plant biochemistry, vol 6. Assays for bioactivity. Academic, London pp 281–321 (ISBN: 0124610161)(xi and 360 pp), pp 281–321 (ISBN: 012461061)

Bouteau F, Pennarun A-M, Kurkdjian A, Convert M, Cornel D, Monestiez M, Rona J-P, Bousquet U (1999) Ion channels of intact young root hairs from *Medicago sativa. Plant Physiol Biochem* 37:889–898

Briskin DP, Gawienowski MC (1996) Role of the plasma membrane H^+ -ATPase in K^+ transport. *Plant Physiol* 111:1199–1207

Campanoni P, Sutter JU, Davis CS, Littlejohn GR, Blatt MR (2007) A generalized method for transfecting root epidermis uncovers endosomal dynamics in *Arabidopsis* root hairs. *Plant J* 51:322–330

Daram P, Brunner S, Persson BL, Amrhein N, Bucher M (1998) Functional analysis and cell-specific expression of a phosphate transporter from tomato. *Planta* 206:225–233

Desbrosses G, Josefsson C, Rigas S, Hatzopoulos O, Dolan L (2003) *AKT1* and *TRH1* are required during root hair elongation in *Arabidopsis. J Exp Bot* 54:781–788

Diatloff E, Roberts M, Sanders D, Roberts SK (2004) Characterization of anion channels in the plasma membrane of *Arabidopsis* epidermal root cells and the identification of a citrate-permeable channel induced by phosphate starvation. *Plant Physiol* 136:4136–4149

Etherton B, Keifer DW, Spanswick RM (1977) Comparison of three methods for measuring electrical resistances of plant cell membranes. *Plant Physiol* 60:684–688

Felle H (1982) Effects of fusicoccin upon membrane potential, resistance and current-voltage characteristics in root hairs of *Sinapis alba. Plant Sci Lett* 25:219–225

Felle HH (1993) Ion-selective microelectrodes: their use and importance in modern cell biology. *Bot Acta* 106:5–12

Felle HH (1994) The H^+/Cl^- symporter in root hairs of *Sinapis alba*: an electrophysiolgoical study using ion-selective microelectrodes. *Plant Physiol* 106:1131–1136

Felle HH (1996) Control of cytoplasmic pH under anoxic conditions and its implication for plasma membrane proton transport in *Medicago sativa* root hairs. *J Exp Bot* 47:967–973

Felle H, Hepler PK (1997) The cytosolic Ca^{2+} concentration gradient of *Sinapsis alba* root hairs as revealed by Ca^{2+} -selective microelectrode tests and fura-dextran ratio imaging. *Plant Physiol* 114:39–45

Felle HH, Herrmann A (2000) pH regulation in and by root hairs. In: Ridge RW, Emons AMC (eds) Root hairs. Cell and molecular biology. Springer, Berlin Heidelberg New York, pp 165–178

Felle HH, Kondorosi É, Kondorosi Á, Schultze M (1998) The role of ion fluxes in Nod factor signalling in *Medicago sativa. Plant J* 13:455–463

Finkel AS, Redman S (1984) Theory and operation of a single microelectrode voltage clamp. *J Neurosci Meth* 11:101–127

Fox TC, Guerinot ML (1998) Molecular biology of cation transport in plants. *Annu Rev Plant Physiol Plant Mol Biol* 49:669–696

Goedhart J, Gadella TWJ Jr (2000) Fluorescence microspectroscopic methods. In: Ridge RW, Emons AMC (eds) Root hairs. Cell and molecular biology. Springer, Berlin Heidelberg New York, pp 65–94

Gassmann W, Schroeder JI (1994) Inward-rectifying K^+ channels in root hairs of wheat. A mechanism for aluminum-sensitive low-affinity K^+ uptake and membrane potential control. *Plant Physiol*105:1399–1408

Grabov A, Bottger M (1994) Are redox reactions involved in regulation of K^+ channels in the plasma membrane of *Limnobium stoloniferum* root hairs? *Plant Physiol* 105:927–935

Higgins R, Lockwood T, Holley S, Yalamanchili R, Stratmann JW (2006) Changes in extracellular pH are neither required nor sufficient for activation of mitogen-activated protein kinases (MAPKs) in response to systemin and fusicoccin in tomato. Planta doi 10.1007/s00425–006–0440–8

Hille B (1984) Ionic channels of excitable membranes. Sinauer Associates, Sunderland, ix and 426 pp

Hirsch RE, Lewis BD, Spalding EP, Sussman MR (1998) A role for the AKT1 potassium channel in plant nutrition. *Science* 280:918–921

Hopkins L, Parmar S, Blaszczyk A, Hesse H, Hoefgen R, Hawkesford MJ (2005) O-Acetylserine and the regulation of expression of genes encoding components for sulfate uptake and assimilation in potato. *Plant Physiol* 138:433–440

Ivashikina N, Becker D, Ache P, Meyerhoff O, Felle HH, Hedrich R (2001) K^+ channel profile and electrical properties of *Arabidopsis* root hairs. *FEBS Lett* 508:463–469

Jones DL, Shaff JE, Kochian LV (1995) Role of calcium and other ions in directing root hair tip growth in *Limnobium stoloniferum.* I. Inhibition of tip growth by aluminum. *Planta* 197:672–680

Jungk A (2001) Root hairs and the acquisition of plant nutrients from soil. *J Plant Nutr Soil Sci* 164:121–129

Kanamori N, Madsen LH, Radutoiu S, Frantescu M, Quistgaard EM, Miwa H, Downie JA, James EK, Felle HH, Haaning LL, Jensen TH, Sato S, Nakamura Y, Tabata S, Sandal N, Stougaard J (2006) A nucleoporin is required for induction of Ca^{2+} spiking in legume nodule development and essential for rhizobial and fungal symbiosis. *Proc Natl Acad Sci USA* 103:359–364

Kanczewska J, Marco S, Vandermeeren C, Maudoux O, Rigaud J-L, Boutry M (2005) Activation of the plant plasma membrane H^+ -ATPase by phosphorylation and binding of 14–3–3 proteins converts a dimer into a hexamer. *Proc Natl Acad Sci USA* 102:11675–11680

Kataoka T, Watanabe-Takahashi A, Hayashi N, Ohnishi M, Mimura T, Buchner P, Hawkesford MJ, Yamaya T, Takahashi H (2004) Vacuolar sulfate transporters are essential determinants controlling internal distribution of sulfate in Arabidopsis. *Plant Cell* 16:2693–2704

Kojima S, Bohner A, Gassert B, Yuan L, von Wirén N (2007) AtDUR3 represents the major transporter for high-affinity urea transport across the plasma membrane of nitrogen-deficient *Arabidopsis* roots. *Plant J* 52:30–40

Konno M, Ooishi M, Inoue Y (2006) Temporal and positional relationships between Mn uptake and low-pH-induced root hair formation in *Lactuca sativa* cv. Grand Rapids seedlings. *J Plant Res* 119:439–447

Kopriva S (2006) Regulation of sulfate assimilation in *Arabidopsis* and beyond. *Ann Bot* 97:479–495

Kurkdjian AC (1995) Role of the differentiation of root epidermal cells in Nod factor (from *Rhizobium meliloti*)-induced root-hair depolarization of *Medicago sativa. Plant Physiol* 107:783–790

Kurkdjian A, Bouteau F, Pennarun A-M, Convert M, Cornel D, Rona J-P, Bousquet U (2000) Ion currents involved in early Nod factor response in *Medicago sativum* root hairs: a discontinuous single-electrode voltage-clamp study. *Plant J* 22:9–17

Lauger P (1991) Electrogenic Ion Pumps. Sinauer Associates, Sunderland, xi + 313 pp.

Lauter F-R, Ninnemann O, Bucher M, Riesmeier JW, Frommer WB (1996) Preferential expression of an ammonium transporter and of two putative nitrate transporters in root hairs of tomato. *Proc Natl Acad Sci USA* 93:8139–8144

Lee SH, Cho H.-T. (2008) Auxin and root hair morphogenesis. In: Emons AMC, Ketelaar T (eds) Root hairs: excellent tools for the study of plant molecular cell biology. Springer, Berlin Heidelberg New York. doi:10.1007/7089_2008_16

Lefebvre B, Arango M, Oufattole M, Crouzet J, Purnelle B, Boutry M (2005) Identification of a *Nicrotiana plumbaginifolia* plasma membrane H^+ -ATPase gene expressed in the pollen tube. *Plant Mol Biol* 58:775–787

Legarde D, Basset M, Lepetit M, Conejero G, Gaymard F, Astruc S, Grignon C (1996) Tissue-specific expression of *Arabidopsis* AKT1 gene is consistent with a role in K^+ nutrition. *Plant J* 9:195–203
Lew RR (1989) Calcium activates an electrogenic proton pump in *Neurospora* plasma membrane. *Plant Physiol* 91:213–216
Lew RR (1991) Electrogenic transport properties of growing *Arabidopsis thaliana* root hairs The plasma membrane proton pump and potassium channels. *Plant Physiol* 97:1527–1534
Lew RR (1994) Regulation of electrical coupling between *Arabidopsis* root hairs. *Planta* 193:67–73
Lew RR (1996) Pressure regulation of the electrical properties of growing *Arabidopsis thaliana* L. root hairs. *Plant Physiol* 112:1089–1100
Lew RR (1998a) Mapping fungal ion channel locations. *Fung Genet Biol* 24:69–76
Lew RR (1998b) Immediate and steady state extracellular ionic fluxes of growing Arabidopsis thaliana root hairs under hyperosmotic and hypoosmotic conditions. *Physiol Plant* 104:397–404
Lew RR (2000) Root hair electrobiology. In: Ridge RW, Emons AMC (eds) Root Hairs. Cell and molecular biology. Springer, Berlin Heidelberg New York, pp 115–139
Lew RR (2004) Osmotic effects on the electrical properties of *Arabidopsis thaliana* L. root hair vacuoles in situ. *Plant Physiol* 134:352–360
Lew RR (2006) Use of double barrel micropipettes to voltage-clamp plant and fungal cells. In Volkov AG (ed) Plant electrophysiology. Theory and methods. Springer, Berlin Heidelberg New York, pp 139–154
Lew RR, Spanswick RM (1984) Characterization of the elctrogenicity of soybean (Glycine max L.) roots: ATP dependence and the effect of ATPase inhibitors. *Plant Physiol* 75:1–6
Lew RR, Levina NN, Shabala L, Anderca MI, Shabala SN (2006) Role of a MAP kinase cascade in ion flux-mediated turgor regulation in fungi. *Eukaryotic Cell* 5:480–487
Li L, Kim B-G, Cheong YH, Pandey GK, Luan S (2006) A Ca^{2+} signaling pathway regulates a K^+ channel for low-K response in *Arabidopsis. Proc Natl Acad Sci USA* 103:12625–12630
Limpens E, Bisseling T (2008) Nod factor signal transduction in the Rhizobium-Legume symbiosis. In: Emons AMC, Ketelaar T (eds) Root hairs: excellent tools for the study of plant molecular cell biology. Springer, Berlin Heidelberg New York. doi:10.1007/7089_2008_10
Liu C, Muchhal US, Uthappa M, Kononowicz AK, Raghothama KG (1998) Tomato phosphate transporter genes are differentially regulated in plant tissues by phosphorus. *Plant Physiol* 116:91–99
Loque D, Yuan L, Kojima S, Gojon A, Wirth J, Gazzarrini S, Ishiyama K, Takahashi H, Wiren N von (2006) Additive contribution of AMT1;1 and AMT1:3 to high-affinity ammonium uptake across the plasma membrane of nitrogen-deficient *Arabidopsis* roots. *Plant J* 48:522–534
Ludewig U, Wilken S, Wu B, Jost W, Obrdlik P, El Bakkoury M, Marini A-M, Andre B, Hamacher T, Boles E, von Wiren N, Frommer WB (2003) Homo- and hetero-oligomerization of ammonium transporter-1 NH_4^+ uniporters. *J Biol Chem* 278:45603–45610
Ma JF, Goto S, Tamai K, Ichii M (2001) Role of root hairs and lateral roots in silicon uptake by rice. *Plant Physiol* 127:1773–1780
Maathuis FJM, Sanders D (1993) Energization of potassium uptake in *Arabidopsis thaliana. Planta* 191:302–307.
Maathuis FJM, Sanders D (1995) Contrasting roles of two K^+ -channel types in root cells of *Arabidopsis thaliana.Planta* 197:456–464
MacRobbie EAC (2006) Osmotic effects on vacuolar ion release in guard cells. *Proc Natl Acad Sci USA* 103:1135–1140
Maruyama-Nakashita A, Nakamura Y, Yamaya T, Takahashi H (2004) Regulation of high-affinity sulphate transporters in plants: towards a systematic analysis of sulphur signalling and regulation. *J Exp Bot* 55:1843–1849
Maser P, Thomine S, Schroeder JI, Ward JM, Hirschi K, Sze H, Talke IN, Amtmann A, Maathuis FJM, Sanders D, Harper JF, Tchieu J, Gribskov M, Persans MW, Salt DE, Kim SA, Guerinot

ML (2001) Phylogenetic relationships within cation transporter families of *Arabidopsis. Plant Physiol* 126:1646–1667

Masucci JD, Schiefelbein JW (1994) The rhd6 mutation of *Arabidopsis thaliana* alters root-hair initiation through an auxin- and ethylene-associated process. *Plant Physiol* 106:1335–1346

Meharg AA, Blatt MR (1995) NO_3^- transport across the plasma membrane of *Arabidopsis thaliana* root hairs: Kinetic control by pH and membrane voltage. *J Memb Biol* 145:49–66

Michelet B, Boutry M (1995) The plasma membrane H^+ -ATPase A highly regulated enzyme with multiple physiological functions. *Plant Physiol* 108:1–6

Miller AJ, Wells DM (2006) Electrochemical methods and measuring transmembrane ion gradients. In Volkov AG (ed) Plant electrophysiology. Theory and methods. Springer, Berlin Heidelberg New York, pp 15–34

Moog PR, der Kooij TAW, van Bruggemann W, Schielfelbein JW, Kuiper PJC (1995) Responses to iron deficiency in *Arabidopsis thaliana*: The Turbo iron reductase does not depend on the for-mation of root hairs and transfer cells. *Planta* 195:505–513

Moriau L, Michelet B, Bogaerts P, Lambert L, Michel A, Oufattole M, Boutry M (1999) Expression analysis of two gene sub-families encoding the plasma membrane H^+ -ATPase in *Nicotiana plumbaginifolia* revelas the major transport functions of this enzyme. *Plant J* 19:31–41

Noble PS (1991) Physicochemical and environmental plant physiology. Academic, San Diego, xx and 635 pp

Ogden D (1994) Microelectrode techniques. The Plymouth workshop handbook, 2nd edn. The Company of Biologists Ltd, Cambridge, x and 448 pp

Palmgren MG (1998) Proton gradients and plant growth: role of the plasma membrane H^+-ATPase. *Adv Bot Res* 28:1–70

Palmgren MG (2001) Plant plasma membrane H^+ -ATPases: powerhouses for nutrient uptake. *Annu Rev Plant Physiol Plant Mol Biol* 52:817–845

Peterson RL, Stevens KJ (2000) Evidence for the uptake of non-essential ions and essential nutrient ions by root hairs and their effect on root hair development In Ridge RW, Emons AMC (eds) Root hairs. Cell and molecular biology. Springer, Berlin Heidelberg New York, pp 179–195

Purves RD (1981) Microelectrode methods of intracellular recording and ionophoresis. Academic, London, x and 146 pp

Ramirez JA, Vacata V, McCusker JH, Haber JE, Mortimer RK, Owen WG, Lecar H (1989) ATP-sensitive K^+ channels in a plasma membrane H^+ -ATPase mutant of the yeast *Saccharomyces cerevisiae.Proc Natl Acad Sci USA* 86:7866–7870

Reintanz B, Szyroki A, Ivashikina N, Ache P, Godde M, Becker D, Palme K, Hedrich R (2002) AtKC1, a silent *Arabidopsis* potassium channel alpha-subunit modulates root hair K^+ influx. *Proc Natl Acad Sci USA* 99:4079–4084

Rigas S, Debrosses G, Haralampidis K, Vicente-Agullo F, Feldmann KA, Grabov A, Dolan L, Hatzopoulos P (2001) *TRH1* encodes a potassium transporter required for tip growth in *Arabidopsis* root hairs. *Plant Cell* 13:139–151

Roberts SK (2006) Plasma membrane anion channels in higher plants and their putative functions in roots. *New Phytol* 169:647–666

Robinson NJ, Procter CM, Connolly EL, Guerinot ML (1999) A ferric-chelate reductase for iron uptake from soils. *Nature* 397:694–697

Sakmann B, Neher E (1995) Single channel recording, 2nd edn. Plenum, New York, xxii and 700pp

Sanchez-Fernandez R, Fricker M, Corben LB, White NS, Sheard N, Leaver CJ, Montagu M, Van Inze D, May MJ (1997) Cell proliferation and hair tip growth in the *Arabidopsis* root are under mechanistically different forms of redox control. *Proc Natl Acad Sci USA* 94:2745–2750

Schultz SG (1980) Basic principles of membrane transport. Cambridge University Press, Cambridge, xii and 144 pp

Schunmann PH, Richardson AE, Smith FW, Delhaize E (2004) Characterization of promoter expression patterns derived from the Pht1 phosphate transporter genes of barley (*Hordeum vulgare* L.) *J Exp Bot* 55:855–865

Schikora A, Schmidt W (2001) Acclimative changes in root epidermal cell fate in response to Fe and P deficiency: a specific role for auxin? *Protoplasma* 218:67–75
Shabala S, Lew RR (2002) Turgor regulation in osmotically stressed *Arabidopsis thaliana* epidermal root cells: Direct support for the role of inorganic ion uptake as revealed by concurrent flux and cell turgor measurements. *Plant Physiol* 129:290–299
Shin R, Berg RH, Schachtman DP (2005) Reactive oxygen species and root hairs in Arabidopsis root response to nitrogen, phosphorus and potassium deficiency. *Plant Cell Physiol* 46:1350–1357
Silverman FP, Assiamah AA, Bush DS (1998) Membrane transport and cytokinin action in root hairs of Medicago sativa. *Planta* 205:23–31
Smith FW, Ealing PM, Hawkesford MJ, Clarkson DT (1995) Plant members of a family of sulfate transporters reveal functional subtypes. *Proc Natl Acad Sci USA* 92:9373–9377
Sondergaard TE, Schulz A, Palmgren MG (2004) Energization of transport processes in plants. Roles of the plasma membrane H+ -ATPase. *Plant Physiol* 136:2475–2482
Thomas RC (1978) Ion-sensitive intracellular microelectrodes. How to make and use them. Academic, London, xiii and 110 pp
Tretyn A, Wagner G, Felle HH (1991) Signal transduction in *Sinapis alba* root hairs: Auxins as external messengers. *J Plant Physiol* 139:187–193
Tsay Y-F, Schroeder JI, Feldmann KA, Crawford NM (1993) The herbicide sensitivity gene CHL1 of *Arabidopsis* encodes a nitrate-inducible nitrate transporter. *Cell* 72: 705–713
Tsuchiya K, Wang W, Giebisch G, Welling PA (1992) ATP is a coupling modulator of parallel Na,K-ATPase-K-channel activity in the renal proximal tubule. *Proc Natl Acad Sci USA* 89:6418–6422
Ullrich CI, Novacky AJ (1990) Extra- and intracellular pH and membrane potential changes induced by K^+, Cl^-, $H_2PO_4^-$, and NO_3^- uptake and fusicoccin in root hairs of *Limnobium stoloniferum. Plant Physiol* 94:1561–1567
Volkov AG (2006) Plant Electrophysiology. Theory and Methods. Springer, Berlin Heidelberg New York , xxi and 508 pp
Wang R, Crawford NM (1996) Genetic identification of a gene involved in constitutive, high-affinity nitrate transport in higher plants. *Proc Natl Acad Sci USA* 93:9297–9301
Wang R, Liu D, Crawford NM (1998) The *Arabidopsis* CHL1 protein plays a major role in high-affinity nitrate uptake. *Proc Natl Acad Sci USA* 95:15134–15139
Xu J, Li H-D, Chen L-Q, Wang Y, Liu L-L, He L, Wu W-H (2006) A protein kinase, interacting with two calcineurin B-like proteins, regulates K^+ transporter AKT1 in *Arabidopsis. Cell* 125:1347–1360
Yoshimoto N, Takahashi H, Smith FW, Yamaya T, Saito K (2002) Two distinct high-affinity sulfate transporters with different inducibilities mediate uptake of sulfate in *Arabidopsis* roots. *Plant J* 29:465–473

Calcium in Root Hair Growth

T. Bibikova and S. Gilroy(✉)

Abstract The growth of cells as diverse as fungal hyphae, pollen tubes, algal rhizoids, and root hairs is characterized by a highly localized control of cell expansion confined to the growing tip. The cellular regulators that have been shown to maintain this spatial localization of growth range from monomeric G-proteins and the actin cytoskeleton to protein kinases and phospholipid-modulating enzymes. A central theme in the proposed mode of action of most of these factors is either regulation of, or response to, the concentration of cytoplasmic Ca^{2+}. For example, a tip-focused Ca^{2+} gradient is associated with the growing point of the root hair and is thought to mediate spatial control of membrane trafficking and the cytoskeleton. Key advances in our understanding of how Ca^{2+} acts in this system have been the recent identification of some of the Ca^{2+} channels and transporters likely responsible for modulating this Ca^{2+} gradient and the likely central role for reactive oxygen species (ROS) in regulating these events. Similarly, molecular identification of Ca^{2+}-responsive elements, such as Ca^{2+}-dependent protein kinases, likely to interact with the tip-focused gradient has begun. This Ca^{2+}-signaling system appears to interact with many of the other components of the tip growth system, including monomeric G-proteins, phospholipases, and the cytoskeleton, helping integrate these activities to facilitate the localization of growth.

1 Introduction

Root hairs are tubular extensions from the surface of root epidermal cells. This hairlike morphology is essential for their role in nutrient uptake and anchoring the plant in the soil and is the result of the integration of two growth control processes acting in concert. Initially, the cellular growth machinery is centered around one

S. Gilroy
Department of Botany, University of Wisconsin, Birge Hall, 430 Lincoln Drive, Madison, USA
e-mail: sgilroy@wisc.edu

Plant Cell Monogr, doi:10.1007/7089_2008_3

point within the cell, leading to the secretion of new cell wall and insertion of plasma membrane material being limited to the growing apex. Following initiation, the young and flexible root hair cell wall yields to turgor pressure only in this tip region, leading to expansion being limited to the extreme apex of the elongating cell (see Emons and Ketelaar 2008). This spatial patterning of growth has led to root hairs emerging as a model system for understanding how plant cell growth is localized and controlled and for the identification of the molecular components that regulate this process. Myriad regulators have so far shown to be involved in driving tip growth (for example, see Aoyoma 2008; Assaad 2008; Zarsky and Fowler 2008), and of these, cytoplasmic Ca^{2+} has emerged as a key component controlling the spatial dynamics of growth.

Ca^{2+} is recognized as a ubiquitous regulator of plant growth and development, with elevation of cytoplasmic Ca^{2+} concentration being involved in developmental and physiological pathways ranging from hormone signaling and pollination, to leaf cell morphogenesis (Hetherington and Brownlee 2004). The calcium ion has been shown to affect the activity of many plant proteins, from kinases and phosphatases, to cytoskeletal-associated proteins and ion transporters, providing a wealth of potential cellular targets to control diverse plant processes. Elevated cytoplasmic Ca^{2+} is also thought to play a critical role in maintaining and directing the root hair growth machinery. For example, root hair elongation is always associated with a gradient in cytoplasmic Ca^{2+} focused to the growing tip, and dissipation of this gradient inhibits such polarized growth (see later).

In this chapter we will therefore review the role of Ca^{2+} in maintaining and directing the apical growth machinery of the root hair. We will also discuss the role of Ca^{2+} transporters in maintaining the tip-focused Ca^{2+} gradient required for apical growth and the cellular mechanisms that are likely involved in regulating these transporters. Finally we will review the current evidence for the possible targets of the elevated Ca^{2+} concentration at the root hair tip. It is important to note that while we will focus on the role of Ca^{2+} in growth, this ion has key signaling roles in the root hair, e.g., in the Nod factor response during the development of rhizobial symbiosis. For discussions on such signaling activities, see Limpens and Bisseling (2008).

2 Calcium Gradients in Root Hair Development

2.1 Root Hair Initiation

Root hair initiation is the process in which a localized bulge is formed on the outer cell wall of the trichoblast, the epidermal cell destined to form a root hair. The relative position of this initiation site is highly regulated, ensuring a consistent distribution of root hairs on the root surface. For example, in *Arabidopis thaliana* the root hair is always formed towards the apical end of the trichoblast, i.e., the end of the cell closest to the root apex (Dolan et al. 1994). In the main elongation zone of the root, the trichoblast cell undergoes rapid but diffuse growth along its entire length associated with elongation of the main root axis; however, upon entering a phase of root hair initiation this cell switches its growth habit to highly polarized expansion from one very localized site on the outer wall of the cell. This process is accompanied with the polarization of the cytoplasm, preferential localization of the growth machinery to the predefined future initiation site, and localized changes in the properties of the lateral cell wall to support hair outgrowth (reviewed in Bibikova and Gilroy 2002).

Intracellular Ca^{2+} gradients are recognized as ubiquitous determinants of polarity establishment and maintenance in many different eukaryotic organisms, including animals, fungi, protists, and plants (e.g., Schwab et al. 1997; Hyde and Heath 1995; Hurst and Kropf 1991; Watson and Barcroft 2001; Love et al. 1997; Malho 1998; Campanoni and Blatt 2007), and so provide an obvious candidate for localizing the cellular machinery required to initiate root hair formation. For example, the polarization of the mammalian epithelium is initiated by Ca^{2+}-dependent adhesion receptors that orient the actin cytoskeleton and induce subsequent localized changes in intracellular Ca^{2+} and H^{+} concentrations (Drubin and Nelson 1996; Alattia et al. 1997). Calcium also plays an important role in the process of polarity establishment in tip-growing fungi. Thus, an external gradient of Ca^{2+} ionophore can determine the site of hyphal emergence from protoplasts and cysts of *Saprolegnia ferax* and stimulate formation of a new hyphal tip in *Neurospora crassa* and *Achlia bisexualis* (Reissig and Kinney 1983; Hyde and Heath 1995). There is also evidence suggesting a role for Ca^{2+} in the process of axis specification and initial polarization during germination of fucoid algal zygotes. In this case, an influx of Ca^{2+} localized to the site of future rhizoid formation can be detected prior to visible polarization of the cytoplasm in the zygote (reviewed in Brownlee and Bouget 1998). Extracellular gradients of calcium ionophore also induce formation of the rhizoid at the site of higher Ca^{2+} concentration, and putative Ca^{2+} channels have been shown to localize to the site of future rhizoid emergence (Pu and Robinson 1998; Robinson and Jaffe 1973; Shaw and Quatrano 1996). In addition, the Ca^{2+}-dependent regulatory protein calmodulin was seen to localize to the future site of rhizoid emergence (Love et al. 1997). In the caulonemal cells of the moss *Funaria*, reorganization of growth is also accompanied by localized increase in cytoplasmic Ca^{2+} (Hahm and Saunders 1991; Demkiv et al. 1994; Hepler and Waayne 1985).

Despite this wealth of evidence supporting a role of localized cytoplasmic Ca^{2+} gradients in signaling or initiating polarization of different eukaryotic cells, the initial polarization events in root hairs appear to be Ca^{2+}-independent. Thus, measurement of cytoplasmic Ca^{2+} levels in the trichoblast prior to and during root hair initiation did not reveal elevation of Ca^{2+} levels associated with this process (Fig. 1). In addition, Ca^{2+} channel blockers, such as verapamil or $LaCl_3$, have no effect on the process of initiation. These data suggest that the signaling leading to the earliest events of trichoblast polarization and root hair initiation is likely not directed by Ca^{2+} (Wymer et al. 1997). Such ideas are consistent with evidence from some other organisms exhibiting polarized growth that Ca^{2+} might not always be the primary

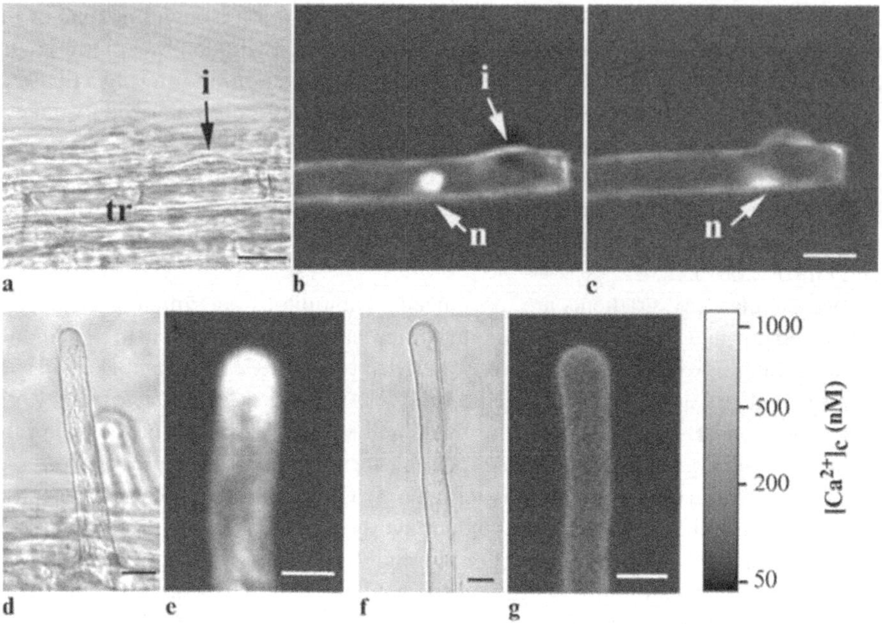

Fig. 1 (**a–g**) Cytoplasmic $[Ca^{2+}]$ in initiating and tip-growing root hairs of *Arabidopsis*. The root hairs were microinjected with the fluorescent Ca^{2+}-indicating dye Indo-1, and $[Ca^{2+}]$ was monitored by confocal ratio imaging. (**a**) Bright field and $[Ca^{2+}]$ in an initiating root hair. $[Ca^{2+}]$ (**b**) 5 min and (**c**) 15 min after the start of wall bulging. (**d, f**) Bright field and (**e, g**) $[Ca^{2+}]$ in a tip-growing and nongrowing root hair. Note the development of a tip-focused gradient in $[Ca^{2+}]$ when tip growth commences and its loss as the hair stops growing. *i* initiation site, *n* nucleus, *tr* trichoblast. Scale bar = 10 μm (Reproduced with permission from Bibikova and Gilroy 2000)

determinant of the cell polarization process. For example, the growth of the lobes of the alga *Microsterias* has been shown to be likewise Ca^{2+}-independent (Holzinger et al. 1995).

2.2 Root Hair Tip Growth

Following initiation, the root hair developmental program switches to apical growth. This process is thought to be similar in all tip-growing cells, such as algal rhizoids, fungal hyphae, pollen tubes, and root hairs. In these cells, deposition of new plasma membrane and primary wall materials is limited to the expanding tip, leading to an elongated, hairlike morphology. Such tip growth is driven by turgor and maintained by directed exocytosis of secretory vesicles at the growing tip.

Ultrastructural analysis of root hairs and other tip-growing cells reveals that their cellular structure is highly polarized (see Emons and Ketelaar 2008). The tips of these cells predominantly contain nonstreaming cytoplasm devoid of large organelles and enriched with endocytotic and exocytotic vesicles (Emons 1987; Sieberer and Emons 2003). Cytoplasm further from the tip contains large organelles, including Golgi bodies, mitochondria, endoplasmic reticulum, and plastids (Rosen et al. 1964; Sievers and Schnepf 1981; Lancelle and Hepler 1992; Galway et al. 1997; Miller et al. 2000). A tip-focused gradient of cytoplasmic Ca^{2+} is associated with this structural zonation. Both the polarized distribution of Ca^{2+} and zonation of the apical cytoplasm disappear with the cessation of apical growth and are thought to be essential parts of the tip growth machinery (reviewed in Carol and Dolan 2002).

Thus, apical growth is widely recognized as a Ca^{2+}-dependent process, with a tip-directed gradient in cytoplasmic Ca^{2+} supporting elongation of fungal hyphae, algal rhizoids, pollen tubes, and root hairs (reviewed in Brownlee et al. 1999; Bibikova and Gilroy 2000). Pollen tubes represent perhaps the most extensively studied plant system exhibiting apical growth. While it is clear that the maintenance of a tip-focused gradient in Ca^{2+} is essential for tube elongation (e.g. Holdaway-Clarke and Hepler 2003; Gu et al. 2005), interestingly, Ca^{2+} appears to be playing a more complex role than simply facilitating localized exocytosis. Detailed measurements of the kinetics of Ca^{2+} influx and the intracellular Ca^{2+} gradient during pulsatile growth of lily pollen tubes revealed that peaks of Ca^{2+} influx actually lag behind the peak of growth. However, the amplitudes of Ca^{2+} fluxes were proportional to the magnitude of the preceding growth pulses. Such a mismatch in timing between extracellular Ca^{2+} influx and growth might reflect refilling of an intracellular Ca^{2+} store or binding to Ca^{2+}-binding sites in the newly synthesized cell wall (Holdaway-Clarke et al. 1997; Messerli et al. 1999). When cytoplasmic Ca^{2+} was directly measured, the peaks of the increased tip-focused gradient were also seen to lag growth by about 4 s, leading to a model where a growth increase may actually give rise to subsequent Ca^{2+}-dependent secretion that is essential to prime the system for the next growth oscillation (Messerli et al. 2000).

As with pollen tubes, root hair apical growth is also Ca^{2+}-dependent. Studies with a self-referencing, Ca^{2+}-selective microelectrode revealed that Ca^{2+} influx is higher at the tip than at the base of the growing root hair, suggesting that a cytoplasmic Ca^{2+} gradient is maintained by the influx of Ca^{2+} through channels inserted into the apical plasma membrane (Herrmann and Felle 1995; Jones et al. 1995; Schiefelbein et al. 1992). This model of a tip-focused Ca^{2+} gradient was confirmed by direct measurements of Ca^{2+} levels at the root hair tip, showing they are elevated to several micromolars, compared to ~100 nM elsewhere in the cell (Fig. 1). For example, using either intracellular fluorescent probes or Ca^{2+}-selective microelectrodes, a tip-focused gradient of cytoplasmic Ca^{2+} has been reported in root hairs of *A. thaliana* (peak value of 1,500 nM; e.g. Wymer et al. 1997), *Medicago sativa* (peak value of 967 nM; Felle et al. 1999), *Sinapsis alba* (peak value of 700 nM; Herrmann and Felle 1995; Felle and Hepler 1997), and *Vicia sativa* (peak value of 2,000 nM; de Ruijter et al. 1998). In *Arabidopsis* root hairs, the steepness of the Ca^{2+} gradient correlates well with the rate of apical growth, being most pronounced in rapidly growing root hairs, whereas nongrowing root hairs show a uniform Ca^{2+} distribution of around 100 nM (Wymer et al. 1997). Further evidence for the important role Ca^{2+} plays in root hair growth comes from numerous observations that altering the gradient by buffering cytoplasmic Ca^{2+} to low levels, applying Ca^{2+} channel antagonists, or treating with Ca^{2+}-ionophore inhibits root hair growth (Wymer et al. 1997; Clarkson et al 1988; Herrmann and Felle 1995; Miller et al. 1992). Furthermore, when the direction of root hair growth is altered, for example, through disrupting the microtubule cytoskeleton (Bibikova et al. 1999), application of a touch stimulus (Bibikova et al. 1997), or addition of lipochito-oligosaccharide elicitor (Esseling et al. 2003), the new growth direction is always centered around the highest cytoplasmic Ca^{2+} concentration. Imposing an asymmetrical Ca^{2+} gradient onto the apex of a growing root hair (Bibikova et al. 1997) or pollen tube (Malho and Trewavas 1996) reorients growth in the direction of the new high Ca^{2+}, suggesting that in each case the elevated Ca^{2+} concentration at the tip does indeed somehow direct the cellular growth machinery. In addition, when root hairs are forced to produce multiple tips, for example, by genetic or pharmacological disruption, elevated Ca^{2+} is always associated with the new growth point (Bibikova et al. 1999; Molendijk et al. 2001). Taken together, these observations suggest a central role for the tip-focused Ca^{2+} gradient in regulating apical growth in the root hair.

3 Establishment of Tip-Focused Calcium Gradient

3.1 Calcium Channels and Their Regulation

Influx of Ca^{2+} into the tips of growing root hairs implies clustering of Ca^{2+} channels or localized activation of these channels at the root hair apical plasma membrane. Figure 2 summarizes the multitude of signals that have been proposed to regulate such channel activity, and each of these is discussed in greater detail in the following sections.

Of the Ca^{2+}-permeable channels that have been identified in plants, the hyperpolarization-, depolarization-, and stretch-activated channels (Foreman et al. 2003; Very and Davies 2000; Dutta and Robinson 2004; reviewed in White 1998) are all potential candidates for generating the localized Ca^{2+} influx supporting tip growth. For example, the tip-focused Ca^{2+} gradient in apically growing cells has been suggested to be mediated by Ca^{2+}-permeable channels that are sensitive to mechanical stress (Derksen 1996; Feijo et al. 2001; Dutta and Robinson 2004). The turgor-driven expansion at the apex of the tip-growing cell would lead to tension in the membrane and channel gating. Indeed, such stretch-sensitive Ca^{2+} channels have been identified in the apical plasma membrane of Fucus rhizoids where their activation caused a cytoplasmic Ca^{2+} increase (Taylor et al. 1996). Mechanosensitive Ca^{2+} channel activity was also detected at the apex of lily pollen tubes (Dutta and Robinson 2004) and *S. ferax* hyphae (Garril et al. 1993). Taken together, these observations suggest that in many tip-growing systems such as pollen tubes, algal rhizoids, and fungal hyphae, mechanical stress on the plasma membrane of the growing apex may open mechanically sensitive Ca^{2+} channels and thus determine

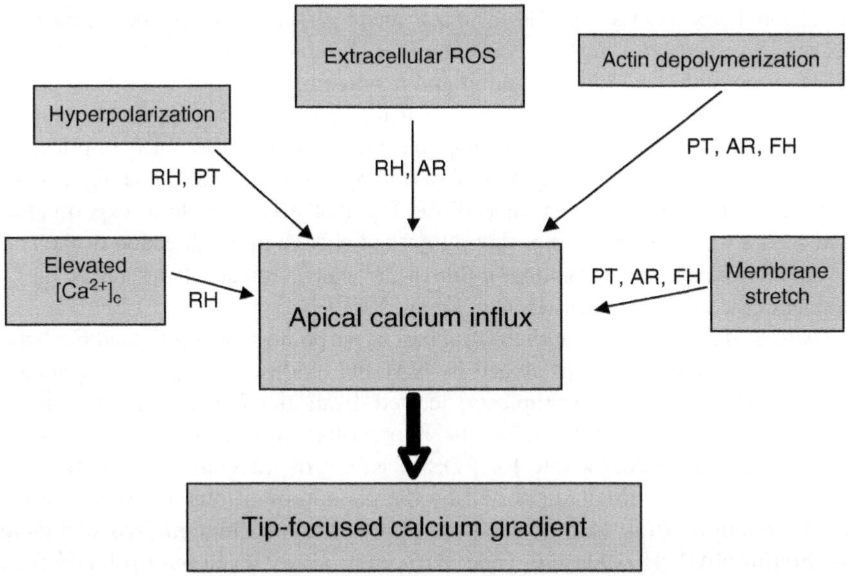

Fig. 2 Factors that can actuate influx of Ca^{2+} into the apex of different tip-growing cells. *RH* root hairs, *PT* pollen tubes, *FH* fungal hyphae, *AR* algal rhizoids. See text for details

the position and magnitude of the Ca^{2+} gradient. Mechanically sensitive Ca^{2+} channels have not been identified in root hairs; however, by analogy, it is likely that they may be a part of the mechanism establishing a Ca^{2+} gradient.

Algal rhizoids are also characterized by a tip-focused gradient of reactive oxygen species (ROS) production, which is required for rhizoid growth. These ROS provide another potential element regulating Ca^{2+} influx through channels in the apical plasma membrane (Speksnijder et al. 1989; Coelho et al. 2002). For example, a nonselective cation channel is thought to contribute to the establishment and maintenance of the tip-focused Ca^{2+} gradient in rhizoids. This conductance is activated by exogenous application of ROS (1 mM H_2O_2; Coelho et al. 2002). Inhibition of ROS production using diphenyleneiodonium causes immediate cessation of growth and dissipation of the Ca^{2+} gradient. Although an equivalent nonselective cation channel activity has not been identified in root hairs, evidence does suggest a role for ROS in root hair apical growth and channel gating (see later).

Using patch-clamp techniques, hyperpolarization-activated inward Ca^{2+} currents were identified in spheroplasts derived from the apical parts of *Arabidopsis* pollen tubes. The activity of the channels mediating this conductance was dependent on the dynamics of actin cytoskeleton (Wang et al. 2004), providing yet another possible regulator maintaining the Ca^{2+} gradient in tip-growing plant cells. Indeed, a likely candidate for the channel maintaining Ca^{2+} influx into the tips of growing root hairs is the hyperpolarization-activated Ca^{2+}-permeable channel identified in apical membranes from *Arabidopsis* root hairs (Very and Davies 2000). This conductance was inhibited by the

Ca^{2+} channel blockers La^{3+} and Gd^{3+} and activated by hyperpolarizing membrane potentials and increased cytosolic Ca^{2+} concentration. The characteristics of this conductance indicate a possible key role in creating and maintaining the tip-focused Ca^{2+} gradient (Very and Davies 2000). Activation of the channel by elevated cytosolic Ca^{2+} suggests a positive feedback mechanism ensuring apical Ca^{2+} influx. Thus, once a tip-localized Ca^{2+} influx is established, it may be self-sustaining, with high Ca^{2+} at the tip activating more Ca^{2+}-permeable channels. Indeed, the fact that it is possible to experimentally redirect root hair growth by imposing a transient change in the direction of the apical Ca^{2+} gradient suggests that localized influx of Ca^{2+} can recruit and then stabilize the Ca^{2+} gradient in the intact root hair (Bibikova et al. 1999).

There is also evidence that these root hair hyperpolarization-activated Ca^{2+} channels are regulated by ROS produced by NADPH oxidases. Exogenous application of hydroxyl radicals to protoplasts derived from the apical parts of growing *Arabidopsis* root hairs potentiated the hyperpolarization-activated Ca^{2+}-selective conductance, suggesting a role for ROS in gating of the channel (Foreman et al. 2003). This idea is further supported by the phenotype of mutants with disrupted ROS production. Thus, hairs formed on roots of the *rhd2* mutant (loss-of-function mutation in NADPH oxidase C gene, *AtrbohC*) fail to develop a tip-focused Ca^{2+} gradient and are unable to successfully undergo the transition from root hair initiation to tip growth. Exogenous application of ROS does not rescue the mutant phenotype, but induces formation of spherical root hair bulges (Foreman et al. 2003), consistent with a role for the ROS produced by the NADPH oxidase in regulating the localization of tip growth in these cells. The Rho GTPase GDP dissociation inhibitor (RhoGDI) supercentipede (SCN1) appears important for regulating the localization of the ROS produced by the NADPH oxidase C, hinting at a possible G-protein-mediated system imposing order on the tip growth machinery in these cells (Carol et al. 2005; see also Zarsky and Fowler 2008).

Indeed, the monomeric G-proteins have emerged as important regulators of targeted secretion, tip growth, and maintenance of the tip-focused Ca^{2+} gradient. For example, injection of anti-Rop antibodies disrupts the Ca^{2+} gradient and inhibits apical growth in pollen tubes (Li et al. 1999). On their own, these data are difficult to interpret since growth inhibition always correlates with the dissipation of Ca^{2+} gradient (Lin and Yang 1997; Kost et al. 1999; Li et al. 1999; Fu et al. 2001). However, the effects of manipulating ROP1 partners such as RIC3 (ROP-interactive CRIB-containing protein; Wu et al. 2001) reinforce the link to Ca^{2+}-dependent growth. Thus, overexpression of *RIC3* causes swelling of the pollen tube tip, protrusion of actin cables to the extreme apex, and delocalization of the Ca^{2+} gradient, all effects alleviated by the treatment with 100 μM of the Ca^{2+} channel blocker La^{3+} (higher La^{3+} levels cause growth arrest). Furthermore, the cellular consequences of *RIC3* overexpression are dependent on the extracellular Ca^{2+} concentration. Growth of wild-type *Arabidopsis* pollen is optimal at 5 mM extracellular Ca^{2+}, whereas the optimal concentration for apical growth of pollen tubes overexpressing RIC3 was much lower (0.5 mM). Based on these data, it may be that *RIC3* is acting to maintain an apical Ca^{2+} gradient by regulating influx of Ca^{2+} via channels localized to the apical membrane (Gu et al. 2005).

As with pollen tubes, the monomeric G-proteins also seem to play important roles in mediating tip growth in root hairs. Thus, genetic manipulation of the expression of *ROP6*, *ROP4*, and *ROP2* in *Arabidopsis* induces formation of branching, swollen root hairs. Expression of constitutively active *AtRop4* and *AtRop6* in root hairs of transgenic *Arabidopsis* plants abolished polarized growth and delocalized the tip-focused Ca^{2+} gradient. Interestingly, areas of high concentration of ROP-GFP protein do not correlate with cellular regions with higher Ca^{2+} concentration, suggesting that ROP-dependent generation of high Ca^{2+} concentration requires some other, as yet unknown, cytoplasmic factor(s) (Molendijk et al. 2001).

3.2 *Membrane Trafficking and the Tip-Focused Ca^{2+} Gradient*

Sec14p, the major yeast phosphatidylinositol transfer protein is involved in regulation of protein transport from Golgi membranes to the cell surface (Bankaitis et al. 1990). Recently, a family of Sec14p–nodulin proteins was described in*Arabidopsis* and evidence provided for the role of these proteins in the regulation of membrane trafficking in plants. Genetic deletion of AtSfh1p, a phosphatidylcholine- and phosphatidylinositol-binding/transfer protein, compromised polarized membrane trafficking and disrupted apical growth, leading to short, branched root hairs (Vincent et al. 2005). AtSfh1p acts to focus vesicle delivery to the root hair apical plasma membrane. Defects in polarized membrane trafficking in the mutant root hairs caused a uniform increase in cytoplasmic Ca^{2+} concentration throughout the mutant root hairs. Therefore *AtSfh1p* may act as an important factor in maintaining the tip-focused Ca^{2+} gradient by restricting the distribution of functional Ca^{2+} channels to the root hair apex, perhaps by their insertion via the exocytotic machinery. Interestingly, root hair initiation is not affected in the mutant, indicating that AtSfh1p is not required for epidermal cell polarization and selection of the initiation site, consistent with the Ca^{2+}-independent nature of the initiation process.

3.3 *Role of Intracellular Stores*

In addition to influx across the plasma membrane, many Ca^{2+}-dependent responses in plant cells involve release of Ca^{2+} from internal stores. TPC1 has recently been identified as the plant slow vacuolar, Ca^{2+}-permaeble conductance suggested to mediate Ca^{2+}-induced Ca^{2+} release from the vacuole (Peiter et al. 2005). Considering the self-reinforcing nature of the tip-focused Ca^{2+} gradient of the root hair noted earlier, Ca^{2+}-induced Ca^{2+} release would provide an obvious method to amplify the locally elevated Ca^{2+} levels through triggering Ca^{2+} release from the vacuole. TPC1 is encoded by a single copy gene in the *Arabidposis* genome, suggesting that a knockout mutant should likely reveal a functional phenotype. However, although *Arabidopsis* mutants with lesions in TPC1 show stomatal and germination phenotypes, they do not have obviously altered root hair formation (Peiter et al. 2005). In addition, the vacuole does not protrude into the cytoplasm-rich tip of growing root hairs, making it less likely to be contributing to maintenance of the tip-focused Ca^{2+} gradient.

Another attractive candidate for a site of internal Ca^{2+} release in tip-growing cells is an inositol-1,4,5-triphosphate (IP_3) mobilizable pool. PI-phospholipase C hydrolyzes phosphatidylinositol-4,5-bisphosphate (PIP_2) into IP_3 and diacylglycerol, two second-messengers that can modulate the activity of downstream proteins during cellular signaling. In animal cells, IP_3 activates a Ca^{2+} release channel associated with the endoplasmatic reticulum (Choe and Ehrlich 2006). Artificially increasing the cytoplasmic IP_3 concentration does cause elevation of (subapical) Ca^{2+} and alteration of pollen tube growth (Franklin-Tong et al. 1996; Monteiro et al. 2005), consistent with the presence of an IP_3-responsive Ca^{2+}-release system in the tube. Ca^{2+}-release channels have also been characterized in the endoplasmatic reticulum (Klusener et al. 1995; Muir and Sanders 1997) and it is feasible that release of Ca^{2+} from such internal stores is involved in the control of apical Ca^{2+} gradient in growing root hairs.

Thus, it has been proposed that PLC could regulate pollen tube growth by regulating intracellular Ca^{2+} dynamics (Trewavas and Malho 1997; Holdaway-Clarke and Hepler 2003). Consistent with this idea, overexpression of wild-type (*NtPLC3*; Helling et al. 2006) or a catalytically inactivated (*PetPLC1*; Dowd et al. 2006) PLC disrupted pollen tube growth, and in the latter case caused elevated Ca^{2+} levels in the swollen apex of the tube. However, both PetPLC1 and NtPLC3 are localized to plasma membrane that is not undergoing rapid extension, whereas their substrate, PIP_2, is enriched in regions of the apical plasma membrane undergoing rapid expansion. One model for the role of PLC is therefore that these enzymes are regulating tip growth by ensuring polarized distribution of PIP_2 in the apical plasma membrane rather than by generating IP_3 to directly alter Ca^{2+} levels. It is possible that disruption of actin dynamics caused by the inhibition of polarized distribution of PIP_2 regulates plasma membrane Ca^{2+} channels and causes changes in cytoplasmic Ca^{2+} concentration (Wang et al. 2004; Dowd et al. 2006). Monteiro et al. (2005) have reported that artificially increasing PIP_2 levels in pollen tubes causes transient growth arrest and Ca^{2+} increases by producing phosphatidic acid via coordinated PLC and diacylglycerol kinase action. Phosphatidic acid can also be produced via phospholipase D, an element that may be critical for root hair growth (Ohashi et al. 2003; but see also Li et al. 2006b). Such complex regulatory networks based on the activity of phospholipases and their phospholipid products are emerging as an important theme in plant responses and are discussed in detail in Aoyoma (2008).

At present, the hyperpolarization/ROS-sensitive channel found in *Arabidopsis* remains the best characterized candidate for a Ca^{2+} conductance mediating formation and maintenance of the apical Ca^{2+} gradient in the root hair. However, we are still lacking molecular information as to its identity. In addition, as described earlier, various treatments can redirect the Ca^{2+} gradient and apical growth, indicating that there are mechanisms that act to quickly inactivate or relocate the channels responsible for the gradient. Unfortunately, with the exception of a possible role for SCN1 in NADPH oxidase-C-dependent ROS production, we still lack molecular evidence as to the possible mechanisms of maintaining channel localization and down-regulating channel activity away from the apex.

3.4 Ca^{2+} Pumps and Tip Growth

To generate a tip-focused Ca^{2+} gradient, localized channel gating must be coupled to the activity of a Ca^{2+} efflux/sequestration system to limit the extent of the elevated Ca^{2+} to the extreme apex of the cell. Plants use a combination of pumps and cotransporters to control the cytoplasmic Ca^{2+} concentration. Thus, Ca^{2+}-ATPases and Ca^{2+}/proton exchangers work to maintain low resting Ca^{2+} in the unstimulated cell and are likely to be an important part of cellular machinery shaping the apical Ca^{2+} gradient in tip-growing cells. Indeed, evidence from gene disruption studies indicates that the plasma-membrane-localized Ca^{2+}-ATPase ACA9 is required for normal growth and orientation of *Arabidopsis* pollen tubes (Schiott et al. 2004). Disruption of this gene results in a semisterile phenotype. Mutant pollen is characterized by significantly reduced growth rate and a high frequency of aborted fertilizations. These results provide some of the first genetic evidence for the role of a defined Ca^{2+} transporter in apical growth. *aca9* mutants do not have an obvious root hair phenotype and so at present the pumps required for setting up the apical Ca^{2+} gradient remain to be defined.

In addition to Ca^{2+}-transporting ATPases, high capacity Ca^{2+}/proton antiporters use the energy of the transmembrane H^+ gradient to sequester Ca^{2+} and locally decrease cytoplasmic Ca^{2+} concentration. In *Arabidopsis* the CaX (for *Ca*tion e*X*changer) family of calcium/cation antiporters are well-characterized as transporting Ca^{2+} across the tonoplast (Shigaki and Hirschi 2006). However, as for the ACA pumps, a root hair phenotype for their *Arabidopsis* knockout mutants is yet to be reported. In the growing root hair the vacuole is excluded from the area where the Ca^{2+} gradient is maintained and therefore a role for tonoplast transporters in shaping the apical Ca^{2+} gradient may be less likely.

4 Ca^{2+}-Dependent Regulation of Apical Growth

If the tip-directed Ca^{2+} gradient is required to maintain and direct root hair elongation, Ca^{2+}-dependent processes and proteins must interpret the Ca^{2+} levels to mediate apical growth. Evidence points to the cytoskeleton, phospholipids, and Ca^{2+}-regulated proteins, such as calmodulin and calmodulin-like domain protein kinases (CDPKs), as primary candidates for such Ca^{2+}-responsive elements.

4.1 Ca^{2+}-Dependent Regulation of the Cytoskeleton

The structure of the cytoskeleton at the growing apex appears crucial for normal root hair development and is intimately linked to the Ca^{2+} gradient. For example, artificial dissipation of the tip-focused Ca^{2+} gradient causes disruption of normal actin dynamics and cessation of apical growth (Kohno and Shimmen 1987; Miller et al. 1999). In 1987, Kohno and Shimmen predicted the existence of Ca^{2+}-regulated actin filament severing proteins based on the observation that treatment with a Ca^{2+}-ionophore caused fragmentation of actin filaments in pollen tubes. In the last decade, biochemical evidence, such as isolation and characterization of Ca^{2+}-regulated actin-binding proteins in plants, has validated this hypothesis. These proteins include the villins, gelsolins, and actin-depolymerizing factors (ADFs). These actin-binding proteins are likely candidates for severing and capping activities that, upon stimulation by Ca^{2+}, alter the dynamics of the actin filaments at the root hair tip (Yokota et al. 2005; see also Ketelaar and Emons 2008).

The villins are actin-binding proteins that exhibit Ca^{2+}/calmodulin-activated actin bundling, severing, and capping activities (Tominaga et al. 2000; Yokota et al. 2005). Injection of antibodies raised against a villin homologue into growing root hairs or pollen tubes causes disruption of the normal actin cytoskeleton (Fan et al. 2004; Ketelaar et al. 2003; Tominaga et al. 2000). A villin/gelsolin-related protein was also isolated from lily and poppy pollen and shown to sever or cap actin filaments in a Ca^{2+}-dependent manner (Yokota et al. 2005; Huang et al. 2005). We must await identification of root-hair-specific villins, coupled with their functional analysis, to understand the role of these proteins in Ca^{2+}-dependent regulation of the root hair apical actin cytoskeleton.

Actin-depolymerizing factors increase actin dynamics by severing actin filaments and increasing depolymerization from the pointed end (Carlier et al. 1997; Gungabissoon et al. 1998; Blanchoin and Pollard 1999). The activity of the maize ADF, ZmADF3, is regulated by phosphorylation by a CDPK (Smertenko et al. 1998; Allwood et al. 2001), providing one obvious mechanism to translate the Ca^{2+} gradient into regulation of cytoskeletal dynamics. In contrast, a pollen-specific ADF, LIADF1, is not regulated by phosphorylation, but is sensitive to pH and phosphoinositides (Allwood et al. 2002), suggesting regulatory mechanisms in parallel to the Ca^{2+}-dependent system. Consistent with a general role of these proteins in tip growth, altering ADF expression levels affects both the organization of actin and morphology of *Arabidopsis* root hairs (Dong et al. 2001). Thus, root hairs of AtADF1-overexpressing plants are shorter and have irregular F-actin organization. Root hairs of AtADF1-underexpressing plants, on the contrary, are longer than wild type and contain an increased number of longitudinally oriented actin cables. However, the effect of genetic manipulations of AtADF1 on the relative size of other cell types was similar to its effect on the length of root hairs, indicating that the effect of genetic manipulation with the expression levels of AtADF1 on root hair growth might be nonspecific (Dong et al. 2001).

In contrast to actin, an intact microtubule cytoskeleton is not required for the tip growth machinery to operate, but it is involved in determining the direction of such growth (Bibikova et al. 1999). Thus, in plants treated with microtubule-disrupting drugs, although the directionality of elongation is altered, an apical Ca^{2+} gradient was seen to form at each region of the hair undergoing tip growth (Bibikova et al. 1999). Calcium and calmodulin have been shown to affect the dynamics of the microtubule cytoskeleton (Fisher et al. 1996), and microtubule-associated proteins are known to induce Ca^{2+}-dependent changes in microtubule structure (Cyr 1994). Similarly, in the *Arabidopsis ton2* mutant, where microtubule orientation is disrupted, calcium channel activity is also affected, with extended channel opening (Thion et al. 1998). It seems likely that control of root hair growth direction involves both stabilization of the Ca^{2+} gradient by the microtubule cytoskeleton and feedback control of this gradient on the microtubule dynamics themselves.

4.2 Ca^{2+}-Dependent Regulation of Secretion

In animal cells, Ca^{2+} is thought to be a key regulator of targeted exocytosis where it facilitates membrane fusion (Oheim et al. 2006; Burgoyne and Clague 2003). There is also a wealth of experimental evidence linking increases in cytoplasmic Ca^{2+} to secretion in plant cells (reviewed in Battey et al. 1999). For example, elevation in cytoplasmic Ca^{2+} concentration significantly stimulated exocytosis in maize root cap protoplasts (Carroll et al. 1998), maize coleoptile protoplasts (Thiel et al. 1994), and aleurone protoplasts (Zorec and Tester 1992, 1993; Homann and Tester 1997). This process was shown to be, at least partially, mediated by annexins (Carroll et al. 1998). Annexins are Ca^{2+}-dependent phospholipids-binding proteins that are involved in mediating secretion in animals (Gruenberg and Emans 1993) and likely plants (Clark and Roux 1995; Delmer and Potikha 1997; Lee et al. 2004). Accumulation of annexins in the apical cytoplasm of pollen tubes and fern rhizoids has been reported (Blackbourn et al. 1992; Clark et al. 1992), tentatively placing them in the apical sectretory machinery. The Ca^{2+} affinity of plant annexins ranges from 60 to several hundred micromolars (Delmer and Potikha 1997), which may be attained in the very highest region of the tip-focused Ca^{2+} gradient. Interestingly, animal annexins exhibit voltage-gated Ca^{2+} channel activity (Gerke and Moss 2002). The similarity between animal and plant annexins suggests that the plant proteins might also exhibit such an activity and so play a role in mediating exocytosis-coupled Ca^{2+} flux at the growing apex. Annexins also show Ca^{2+}-dependent actin binding (Ikebuchi and Wiseman 1990; Calvert et al. 1996), providing yet another site where they could play a role in Ca^{2+}-dependent regulation of the tip growth machinery. However, Ca^{2+}-dependent activation of plant annexins in the growing root hair apex and/or a Ca^{2+} transporting function of plant annexins has yet to be demonstrated.

Monomeric G-proteins have emerged as important regulators of polarized secretion in plant cells and provide another possible mechanistic link between Ca^{2+} and the promotion of secretory vesicle fusion to the apical plasma membrane. Thus, the Rab GTPase, RabA4b, is enriched in the vesicle-rich region of the *Arabidopsis* root hair apical cytoplasm, and RabA4b-labeled compartments are thought to deliver new cell wall and plasma membrane material to the expanding root hair tip (Preuss et al. 2004, 2006). RabA4b activates phosphtidylinositol 4-OH kinase (PI-4Kβ1), which acts to localize polarized vesicle trafficking required for normal root hair growth. In the yeast two-hybrid system, PI-4Kβ1 interacts with Ca^{2+} sensor CBL1. Based on analogy with animal systems, it is thought that the enzymatic activity of PI-4Kβ1 is stimulated upon binding to AtCBL1. Thus, G-protein-dependent and Ca^{2+}-dependent regulation of localized secretion may well become integrated through coregulation of the phosphoinositide-derived signaling molecules required for the organization of the secretory machinery at the apex (Preuss et al. 2006).

It is important to note that the timing of growth and Ca^{2+} fluxes suggests a complex role for Ca^{2+} in modulating exocytotic activity. Thus, the pulsatile growth of pollen tubes is associated with oscillations in the tip-focused Ca^{2+} gradient, and with oscillations in the rate of membrane trafficking (Holdaway-Clarke et al. 1997; Messerli and Robinson 1997; Feijo et al. 2001; Parton et al. 2001). The peak of the Ca^{2+} increase actually lags behind the peak of growth rate by about 4 s, rather than being coincident with it as would be the case if Ca^{2+} were simply driving exocytosis. Indeed, Messerli et al. (2000) have argued that the Ca^{2+} gradient and secretion are, in fact, uncoupled in pollen tubes. The relatively slower growth of root hairs compared with pollen tubes has complicated such a detailed temporal analysis of growth rate vs. Ca^{2+} gradient dynamics in root hairs. However, preliminary analyses suggest that a similar lag between increasing growth rate and elevation of apical Ca^{2+} exists in root hairs as well (Monshausen, Bibikova, Messerli, and Gilroy, unpublished).

4.3 Ca^{2+}-Dependent Regulation of Calmodulin and Protein Kinases in Tip Growth

Calmodulin accumulates at the growing apex of pollen tubes, fungal hyphae, and algal rhizoids (Hausser et al. 1984; Love et al. 1997; Wang et al. 2006) and shows localized activation towards the tip in pollen tubes (Rato et al. 2004). Application of calmodulin inhibitors also arrests tip growth (Hausser et al. 1984). Taken together, these data suggest that calmodulin is likely to be a part of the Ca^{2+}-dependent machinery mediating tip growth, although whether this role extends to the root hair, and what proteins it in turn regulates to promote growth (Zielinski 1998) remain to be determined.

In pollen tubes, putative calcium-dependent protein kinase activity has also been reported to be localized to the tip of the growing tube (Moutinho et al. 1998). Consistent with a functional role for such localized kinase activity, two CDPKs, *PiCDPK1* and *PiCDPK2*, have been identified in *Petunia* pollen tubes and demonstrated to have distinct and important roles in maintaining and directing apical growth (Yoon et al. 2006). Overexpression of *PiCDPK1* disrupts pollen tube growth polarity and causes an elevation of cytoplasmic Ca^{2+} concentration throughout the bulging tip, suggesting that PiCDPK1 participates in maintaining and directing the Ca^{2+} gradient, although its substrates remain to be determined. PiCDPK1 is localized to the plasma membrane, most likely by acylation, and this localization is essential for the biological activity of this kinase.

Evidence for a similar role of Ca^{2+}-dependent protein kinase activity in root hair growth is also emerging. An RNA-interference-based screen for gene function in *Medicago truncatula* identified a gene encoding Ca^{2+}-dependent protein kinase (*MrCDPK1*) as being required for normal root hair development (Ivashuta et al. 2005). In this study, silencing *MrCDPK1* caused a reduction in root hair length; however, these transgenic plants also exhibited significant changes in root cell length and increased cell wall lignification. We must await further analysis to determine whether MrCDPK1 is part of root hair tip growth machinery, or whether the RNA-interference-mediated root hair phenotype may be due to secondary effects of suppressing MrCDPK1 expression on general root growth (Ivashuta et al. 2005).

4.4 Calcium-Regulated Transporters

In addition to the localized fluxes of Ca^{2+} associated with generation of the tip-focused Ca^{2+} gradient, root hair apical growth is also associated with localized fluxes in a range of other ions such as H^+ and K^+ (reviewed in Gilroy and Jones 2000). The Ca^{2+} gradient may well play a role in the spatial regulation of the transporters responsible for these other fluxes. Thus, root hair growth is associated with influx of K^+ (Jones et al. 1998). On the basis of quantitative RT-PCR analysis of the expression of shakerlike K^+ channels in *Arabidopsis* root hairs, the inward rectifier AKT1 was shown to dominate this K^+ conductance. For example, root hair protoplasts isolated from plants showing *akt1* loss of function lacked inwardly rectifying K^+ currents (Reintanz et al. 2002). Activation of AKT1 channels in *Arabidopsis* root hairs in response to K^+ deprivation is Ca^{2+}-dependent. Following an increase in Ca^{2+}, the cytoplasmic Ca^{2+} sensors CBL1 and CBL9 are thought to activate the protein kinase CIPK23. CIPK23 in turn phosphorylates and activates AKT1, enhancing K^+ uptake (Li et al. 2006a). It would follow logically that the tip-focused Ca^{2+} gradient would therefore impose spatial restrictions on the activation of tip-localized AKT1. However, electrophysiological studies indicate that K^+ current is not focused around the tip but is symmetrical around the whole root hair surface (Jones et al. 1995), implying that Ca^{2+} may be just one of a range of regulators of these channels.

There is also an influx of protons into the apices of growing root hairs and pollen tubes (Herrmann and Felle 1995; Messerli et al. 1999; Jones et al. 1995). The activity of the plasma membrane H^+-ATPase has been shown to be inhibited by increasing cytoplasmic Ca^{2+} in stomatal guard cells (Kinoshita et al. 1995), suggesting a possible link between H^+ flux and the Ca^{2+} gradient. However, the transporters that determine the H^+ fluxes in root hairs have not been identified, and so at present, we can only speculate as to their possible link to Ca^{2+} and the tip growth machinery.

4.5 Calcium-Dependent Regulation of Apical Cell Wall Extensibility

In addition to the integral role of the secretory machinery delivering new membrane and wall material to the growing tip, apical growth is also driven by turgor. Tip-growing cells must maintain a careful balance between loosening of the cell wall at the tip to allow cell growth and maintaining sufficient rigidity of the wall to withstand turgor pressure and prevent bursting. Calcium has been linked to the regulation of the dynamic properties of the wall, primarily via regulation of the degree of pectin cross-linking. Pectin is a major component of pollen tube walls and is also present in the root hair cell wall. It is a polymer consisting of polygalacturonic acids, and Ca^{2+} binds to the anionic moieties of unesterified pectins, creating cross-links between pectin polymers and thus increasing wall rigidity. Pectins are secreted as methylesters and subsequently deesterified by the cell wall enzyme pectin methylesterase. Although the activity of pectin methylesterases is crucial for the apical growth of pollen tubes (e.g. Jiang et al. 2005), it is not clear to what degree the process of pectin cross-linking is regulated by Ca^{2+} availability in the root hair apoplast. Both root hair and pollen tube apical growth depends on the presence of extracellular Ca^{2+} (Picton and Steer 1983; Wymer et al. 1997). For example, increasing the Ca^{2+} concentration in the media reduces the rate of growth of oscillating pollen tubes and increases periods of growth oscillation (Holdaway-Clarke et al. 1997). An attractive model is that Ca^{2+} binding affects the mechanical properties of the wall, with optimal concentrations being high enough to rigidify the wall and prevent bursting, but low enough to maintain growth.

5 Conclusions and Perspectives

In this chapter we have focused on the role of Ca^{2+} in the growth of root hairs. Calcium has emerged as a central regulator of cell expansion in all tip-growing cells, where it is clearly playing a role in coordinating the activity of the tip growth machinery. The pronounced spatial and temporal patterning of Ca^{2+} changes such as the oscillating tip-focused gradient associated with growth hint at the complexity of the regulatory roles and information encoding likely invested in Ca^{2+} dynamics within tip-growing cells. Many of the recent insights into these diverse regulatory roles have come from molecular identification of the Ca^{2+}-responsive elements such as kinases and Ca^{2+} sensors that decode/respond to the information in the Ca^{2+} change. We can anticipate that similar insight will appear as the channels and transporters generating the regulatory Ca^{2+} fluxes are identified. The cloning of these channels and pumps remains a major challenge to our understanding of root hair formation and function. It is also worth noting that although there is a wealth of data linking the Ca^{2+} gradient to root hair elongation, observations such as the difference in timing between Ca^{2+} fluxes and growth maxima hint at a far more complex role for Ca^{2+} than the Ca^{2+} gradient simply "driving growth." The challenge for the future will undoubtedly be to place Ca^{2+}-dependent regulation within the context of the other control and response elements. It is becoming increasingly clear that it is the interactions between these systems that are required to robustly coordinate the growth machinery at the root hair tip.

6 Methods: Ca^{2+} Imaging in Root Hairs

To visualize the dynamics of Ca^{2+} in root hairs, a Ca^{2+}-responsive fluorescent probe is introduced into the cytoplasm of the trichoblast. Probes successfully used in root hairs include the dyes Indo-1 (e.g., Wymer et al. 1997; de Ruijter et al. 1998), Fura-2 (e.g., Ehrhardt et al. 1996), Oregon Green-BAPTA dextran (e.g., Kanamori et al. 2006), and the Ca^{2+}-responsive GFP-based sensors based on the cameleon construct (Miwa et al. 2006; Sun et al. 2007). The use of these probes requires (1) growing the plant in such a way that the root hairs are optically accessible on the microscope stage, (2) introducing the probe into the cytoplasm of the cell, and (3) choosing the appropriate imaging approach to visualize the Ca^{2+} dynamics. Each of these areas is outlined in what follows.

6.1 Plant Cultivation for Root Hair Imaging

In general, seedlings have been used to monitor root hair Ca^{2+} levels. The small size of the seedling facilitates placing the intact plant on the microscope stage, circumventing the need to dissect away the root. To image *Arabidopsis* seedling roots, place a sterile 40 × 22-mm^2 cover slip in a Petri dish and cover with 1 mL of 0.7% (w/v) phytagel (Sigma) in either half-strength Murashige and Skoog medium or half-strength Epstein's growth medium (Wymer et al. 1997). Growing *Arabidopsis* in full-strength Murashige and Skoog medium leads to severe disruption of root hair development. Pipette the molten phytagel onto the surface of the cover slip and ensure it flows to the edges. Once solidified, the 1–2-mm-thick layer is optically clear, allowing for high-resolution imaging. Plant individual, surface-sterilized (2 min in 70% (v/v) ethanol, rinse 3× in sterile distilled water) seeds at one end of the gel, ensuring the seed is pushed into the phytagel (otherwise the root system will not penetrate the gel). Place the Petri dish flat and wait for the seeds to just germinate and then place the dish at 80° with the seedling at the top of the gel. The root will now grow down onto, and then along, the cover slip under the gel. This places the root or root hairs next to the cover slip and allows high-resolution imaging of the root hairs using an inverted microscope.

6.2 Introducing the Probe into the Cytoplasm

Three major approaches have been taken to loading the Ca^{2+}-sensitive probes into the cytoplasm of the root hair: microinjection, acid-loading, and expression of transgenic reporters. Pressure microinjection has been successfully used to introduce the fluorescent reporters Indo-1, Fura-2, and a Ca^{2+}-sensitive version of Oregon Green into root hairs of several genera such as *Arabidopsis*, *Medicago*, and *Lotus* (Ehrhardt et al. 1996; Bibikova et al. 1997; Kanamori et al. 2006). The advantage of the microinjection approach is that it delivers the indicators in a highly controlled manner to the cytoplasm and it is compatible with use of dextran-coupled reporters. Conjugating the Ca^{2+}-sensing dye to a large dextran helps maintain the cytoplasmic locale of the probe, circumventing possible problems of compartmentalization of these small molecules that might otherwise occur. The disadvantage of the technique is that it is highly technically demanding to impale the root hair with the microinjection pipette, deliver the dye, and then remove the pipette while maintaining cell viability and normal growth. Empirical observations suggest that most damage to the root hair actually occurs during the removal of the pipette.

An alternative to microinjection that has been successfully applied to loading *Arabidopsis* roots with Ca^{2+} indicator is acid-loading (Wymer et al. 1997; de Ruijter et al. 1998). This approach takes advantage of the weak acid nature of the fluorescent Ca^{2+} indicators such as Indo-1 by lowering the medium pH and so protonating the negatively charged groups on the molecule. Once these charges are masked by the proton, the dye is nonpolar, becomes more lipid permeable, and so more likely to diffuse into the cell. The medium pH is buffered to 4.5 using 50 mM dimethyl glutaric acid, and the roots are incubated for 30 min to several hours with 25 μM dye. The root is then washed with normal growth medium. Careful monitoring of the buffer pH is required as these indicators precipitate at pH $\leq$ 4.0. This approach has the advantage of loading the dye into the cytoplasm without the requirement for specialized equipment or expertise. Furthermore, a whole population of root hairs is prepared for analysis, making statistical analysis possible. However, the extended incubation at low pH can affect normal root growth and the efficiency of loading is highly variable. In addition, the approach is not applicable to dextran-conjugated dyes.

The development of genetically encoded Ca^{2+} sensors such as the cameleons (Miyawaki et al. 1997) has opened the possibility of using transgenic approaches to monitoring root hair Ca^{2+} dynamics. Stable expression using the CaMV35S promoter drives expression in the root hair to levels suitable for imaging. Although the YC2.1 cameleon sensor has been used successfully to monitor root hair Ca^{2+}, e.g., revealing the nuclear Ca^{2+} spiking associated with Nod factor action in *Medicago* root hairs (Miwa et al. 2006; Sun et al. 2007), newer generations of these sensors, such as YC3.6 (Nagai et al. 2004), offer significantly increased dynamic range of signal and should be highly suited to noninvasive measurement of cytoplasmic Ca^{2+} during root hair growth. The large size of these proteins also helps reduce the likelihood of compartmentalization. The disadvantage of these genetically encoded sensors is that they operate through monitoring changes in their fluorescence resonance energy transfer (FRET) signal. FRET analysis requires very careful measurements of fluorescence intensity to avoid superimposing imaging artifacts on the Ca^{2+} measurements (Dixit et al. 2006).

6.3 *Imaging Technologies for Ca^{2+} Determination*

All the Ca^{2+} measurement approaches described here require ratio analysis to correct for optical artifacts associated with measuring the fluorescence signals form the Ca^{2+}-sensitive probes (Gilroy 1997). Ratio analysis involves capturing an image of the reporter at a non-Ca^{2+}-sensitive wavelength (a reference image) and a Ca^{2+}-dependent signal from the sensor and dividing the two images pixel by pixel using image analysis software to generate a ratio image. Such a ratio approach normalizes the signal for differences in probe concentration and distribution within the sample. Monitoring just the Ca^{2+}-dependent wavelength from these probes is highly sensitive to artifacts such as dye bleaching or dye accumulation during the measurement, which can easily be mistaken for changes in Ca^{2+}. The raw images for ratio analysis can be collected using a standard epifluorescence microscope coupled to a filter wheel system that changes between collecting the reference and Ca^{2+}-sensitive image. Alternatively, confocal microscopes with dual detector systems or two-photon microscopy can be used to obtain a clearer optical section of the root hair (Blancaflor and Gilroy 2000).

6.4 *Common Problems Associated with Ca^{2+} Imaging*

There are several very common problems that occur when imaging Ca^{2+} dynamics in root hairs. First, it is important to define whether the root hair being studied is growing. This determination is easily made by performing a time-lapse analysis of the hair. The development of root hairs is remarkably sensitive to the growth environment (Bibikova and Gilroy 2002) and so it is critical to know that the hair being imaged is in fact developing as expected. The disruption can be as subtle as a slight slowing of the growth rate when the microscope is collecting fluorescence images, an effect not easily noted without time-lapse analysis but that may have a large impact on the Ca^{2+} dynamics being observed. Indeed, one component of the microscope environment that is often overlooked but that can have profound effects on hair growth is the illumination itself. This is particularly noticeable with two-photon microscopy, where even brief laser illumination used to generate the fluorescence image can arrest hair elongation. However, growth inhibition triggered by illumination can also occur in both confocal and standard epifluorescence imaging. It is probable that intense illumination of the sample generates ROS, which then disrupt the growth. The approaches to avoid these problems are outlined in Dixit et al. (2006). One other extremely common artifact arises from imaging root hairs that are growing at an angle to the plane of focus of the microscope. When observing the tip, this orientation of the hair leads to the subapical regions being less in focus and so generates the appearance of an apical gradient in signal when a true mid-plane image would reveal none. Thus, it is possible to generate an artifactual apparent tip-focused Ca^{2+} gradient purely from choosing an inappropriate root hair for imaging. For a more in-depth discussion of Ca^{2+} imaging and potential artifacts, see Fricker et al. (2001).

Acknowledgments The authors thank Drs. Gabriele Monshausen and Sarah Swanson for their critical reading of the manuscript. This work was supported by grants from the National Science Foundation.

References

Alattia JR, Ames JB, Porumb T, Tong KI, Heng YM, Ottensmeyer P, Kay CM, Ikura M (1997) Lateral self-assembly of E-cadherin directed by cooperative calcium binding. FEBS Lett 417:405–408

Allwood EG, Anthony RG, Smertenko AP, Reichelt S, Drobak BK, Doonan JH, Weeds AG, Hussey PJ (2002) Regulation of the pollen-specific actin-depolymerizing factor LlADF1. Plant Cell 14:2915–2927

Allwood EG, Smertenko AP, Hussey PJ (2001) Phosphorylation of plant actin depolymerising factor by calmodulin-like domain protein kinase. FEBS Lett 499:97–100

Aoyama T (2008) Phospholipid signaling in root hair development. In: Emons AMC, Ketelaar T (eds) Root hairs: excellent tools for the study of plant molecular cell biology. Springer, Berlin Heidelberg New York. doi:10.1007/7089_2008_1

Assaad FF (2008) The membrane dynamics of root hair morphogenesis. In: Emons AMC, Ketelaar T (eds) Root hairs: excellent tools for the study of plant molecular cell biology. Springer, Berlin Heidelberg New York. doi:10.1007/7089_2008_2

Bankaitis VA, Aitken JR, Cleves AE, Dowhan W (1990) An essential role for a phospholipid transfer protein in yeast Golgi function. Nature 347:561–562

Battey NH, James NC, Greenland AJ, Brownlee C (1999) Exocytosis and endocytosis. Plant Cell 11:643–659

Bibikova TN, Blancaflor EB, Gilroy S (1999) Microtubules regulate tip growth and orientation in root hairs of *Arabidopsis thaliana*. Plant J 17:657–665

Bibikova TN, Gilroy S (2000) Calcium in root hair growth and development. In: Ridge R, Emons AM, (eds) The cellular and molecular biology of root hairs. Springer, Berlin Heidelberg New York

Bibikova TN, Gilroy S (2002) Root hair development. J Plant Growth Regul 21:383–415

Bibikova TN, Zhigilei A, Gilroy S (1997) Root hair growth in *Arabidopsis thaliana* is directed by calcium and an endogenous polarity Planta 203:495–505

Blackbourn HD, Barker PJ, Huskisson NS, Battey NH (1992) Properties and partial protein-sequence of plant annexins. Plant Physiol 99:864–871

Blanchoin L, Pollard TD (1999) Mechanism of interaction of *Acanthamoeba actophorin* (ADF/cofilin) with actin filaments. J Biol Chem 274:15538–15546

Blancaflor EB, Gilroy S (2000) Plant cell biology in the new millennium: new tools and new insights. Am J Bot 87:1547–1560

Brownlee C, Bouget FY (1998) Polarity determination in Fucus: from zygote to multicellular embryo. Semin Cell Dev Biol 9:179–185

Brownlee C, Goddard H, Hetherington AM, Peake L (1999) Specificity and integration of responses: Ca^{2+} as a signal in polarity and osmotic regulation. J Exp Bot 50:1001–1011

Burgoyne RD, Clague MJ (2003) Calcium and calmodulin in membrane fusion. Biochim Biophys Acta 1641:137–143
Calvert CM, Gant SJ, Bowles DJ (1996) Tomato annexins p34 and p35 bind to F-actin and exhibit nucleotide phosphodiesterase activity inhibited by phospholipid binding. Plant Cell 8:333–342
Campanoni P, Blatt MR (2007) Membrane trafficking and polar growth in root hairs and pollen tubes. J Exp Bot 58(1):65–74
Carlier MF, Laurent V, Santolini J, Melki R, Didry D, Xia GX, Hong Y, Chua NH, Pantaloni D (1997) Actin depolymerizing factor (ADF/cofilin) enhances the rate of filament turnover: implication in actin-based motility. J Cell Biol 136:1307–1322
Carol RJ, Dolan L (2002) Building a hair: tip growth in *Arabidopsis thaliana* root hairs. Phil Trans R Soc Lond B 357:815–821
Carol RJ, Takeda S, Linstead P, Durrant MC, Kakesova H, Derbyshire P, Drea S, Zarsky V, Dolan L (2005) A RhoGDP dissociation inhibitors patially regulates growth in root hair cells. Nature 438:1013–1016
Carroll AD, Moyen C, Van Kesteren P, Tooke F, Battey NH, Brownlee C (1998) Ca^{2+}, annexins, and GTP modulate exocytosis from maize root cap protoplasts. Plant Cell 10:1267–1276
Choe CU, Ehrlich BE (2006) The inositol 1,4,5-trisphosphate receptor (IP3R) and its regulators: sometimes good and sometimes bad teamwork. Sci STKE 363:15
Clark GB, Dauwalder M, Roux SJ (1992) Purification and immunolocalization of an annexin-like protein in pea-seedlings. Planta 187:1–9
Clark GB, Roux SJ (1995) Annexins of plant cells. Plant Physiol 109:1133–1139
Clarkson DT, Brownlee C, Ayling SM (1988) Cytoplasmic calcium measurements in intact higher plant cells results from fluorescence ratio imaging of fura-2. J Cell Sci 91:71–80
Coelho SM, Taylor AR, Ryan KP, Sousa-Pinto I, Brown MT, Brownlee C (2002) Spatiotemporal patterning of reactive oxygen production and Ca^{2+} wave propagation in Fucus rhizoid cells. Plant Cell 14:2369–2381
Cyr RJ (1994) Microtubules in plant morphogenesis - role of the cortical array. Annu Rev Cell Biol 10:153–180
Delmer DP, Potikha TS (1997) Structures and functions of annexins in plants. Cell Mol Life Sci 53:546–553
Demkiv OT, Khorkavtsiv YD, Kardash AR (1994) Intracellular pH during growth and differentiation of the gametophyte *Funaria hygrometrica* cells. Russ J Plant Physiol 41:84–87
Derksen J (1996) Pollen tubes: a model system for plant cell growth. Bot Acta 109:341–345
Dixit R, Cyr R, Gilroy S (2006) Using intrinsically fluorescent proteins for plant cell imaging. Plant J 45:599–615
Dolan L, Duckett CM, Grierson C, Linstead P, Schneider K, Lawson E, Dean C, Poethig S, Roberts K (1994) Clonal relationships and cell patterning in the root epidermis of *Arabidopsis*. Development 120:2465–2474
Dong CH, Xia GX, Honga Y, Ramachandrana S, Kosta B, Chua NH (2001) ADF proteins are involved in the control of flowering and regulate F-actin organization, cell expansion, and organ growth in *Arabidopsis*. Plant Cell 13:1333–1346
Dowd PE, Coursol S, Skirpan AL, Kao TH, Gilroy S (2006) Petunia Phospholipase C is involved in pollen tube growth. Plant Cell 18:1438–1453
Drubin DG, Nelson WJ (1996) Origins of cell polarity. Cell 84:335–344
Dutta R, Robinson KR (2004) Identification and characterization of stretch-activated ion channels in pollen protoplasts. Plant Physiol 135:1398–1406
Emons AMC (1987) The cytoskeleton and secretory vesicles in root hairs of Equisetum and Limnobium and cytoplasmic streaming in root hairs of Equisetum. Ann Bot 60:625–632
Emons AMC, Ketelaar T (2008) Intracellular organization: a prerequisite for root hair elongation and cell wall deposition. In: Emons AMC, Ketelaar T (eds) Root hairs: excellent tools for the study of plant molecular cell biology. Springer, Berlin Heidelberg New York. doi:10.1007/7089_2008_4
Ehrhardt DW, Wais R, Long SR (1996) Calcium spiking in plant root hairs responding to Rhizobium nodulation signals. Cell 85:673–681

Esseling JJ, Lhuissier FG, Emons AM (2003) Nod factor-induced root hair curling: continuous polar growth towards the point of nod factor application. Plant Physiol 132:1982–1988

Fan X, Hou J, Chen X, Chaudhry F, Staiger CJ, Ren H (2004) Identification and characterization of a Ca^{2+} - dependent actin-filament severing protein from lily pollen. Plant Physiol 136:3979–3989

Feijo JA, Sainhas J, Holdaway-Clarke T, Cordeiro MS, Kunkel JG, Hepler PK (2001) Cellular oscillations and the regulation of growth: the pollen tube paradigm. Bioessays 23:86–94

Felle HH, Hepler PK (1997) The cytosolic Ca^{2+} concentration gradient of *Sinapis alba* root hairs as revealed by Ca^{2+}- selective microelectrode tests and Fura-Dextran ratio imaging. Plant Physiol 114:39–45

Felle HH, Kondorosi E, Kondorosi A, Schultze M (1999) Nod factors modulate the concentration of cytosolic free calcium differently in growing and non-growing root hairs of *Medicago sativa* L. Planta 209:207–212

Fisher DD, Gilroy S, Cyr RJ (1996) Evidence for opposing effects of calmodulin on cortical microtubules. Plant Phys 112:1079–1087

Foreman J, Demidchik V, Bothwell JHF, Mylona P, Miedema H, Torres MA, Linstead P, Costa S, Brownlee C, Jones JD, Davies JM, Dolan L (2003) Reactive oxygen species produced by NADPH oxidase regulate plant cell growth. Nature 422:442–446

Franklin-Tong VE, Drøbak BK, Allan AC, Watkins PAC, Trewavas AJ (1996) Growth of pollen tubes of *Papaver rhoeas* is regulated by a slow moving calcium wave propagated by inositol triphosphate. Plant Cell 8:1305–1321

Fricker MD, Blancaflor EB, Meyer A, Parsons A, Plieth C, Tlaka M, Gilroy S (2001) Fluorescent probes for living cells. In: C Hawes and B Satiat-Jeunemaitre (eds) Cell Biology, A Practical Approach. Oxford University Press, Oxford, pp 35–85

Fu Y, Wu G, Yang Z (2001) Rop GTPase-dependent dynamics of tip-localized F-actin controls tip growth in pollen tubes. J Cell Biol 152:1019–1032

Galway ME, Heckman JW, Schiefelbein JW (1997) Growth and ultrastructure of *Arabidopsis* root hairs: the *rhd3* mutation alters vacuole enlargement and tip growth. Planta 201:209–218

Garril A, Jackson SL, Lew RR, Heath IB (1993) Ion channel activity and tip growth – tip-localized stretch-activated channels generate an essential Ca^{2+} gradient in the oomycete *Saprolegnia Ferax.* Eur J Cell Biol 60:358–365

Gerke V, Moss SE (2002) Annexins: From structure to function. Physiol Rev 82:331–371

Gilroy S, Jones DL (2000) Through form to function: root hair development and nutrient uptake. Trends Plant Sci 5:56–60

Gilroy S (1997) Fluorescence microscopy of living plant cells. Annu Rev Plant Physiol Plant Mol Biol 48:165–190

Gruenberg J, Emans N (1993) Annexins in membrane traffic. Trends Cell Biol 3:224–227

Gu Y, Fu Y, Dowd P, Li S, Vernoud V, Gilroy S, Yang Z (2005) A Rho family GTPase controls actin dynamics and tip growth via two counteracting downstream pathways in pollen tubes. J Cell Biol 169:127–138

Gungabissoon RA, Jiang CJ, Drobak BK (1998) Interaction of maize actin depolymerising factor with actin and phosphoinositides and its inhibition of plant phospholipase C. Plant J 16:689–696

Hahm SH, Saunders MJ (1991) Cytokinin increases intracellular calcium in *Funaria*: detection with Indo-1. Cell Calcium 12:675–681

Hausser I, Herth W, Reiss HD (1984) Calmodulin in tip-growing plant cells visualized by fluorescing calmodulin-binding phenothiazines. Planta 162:33–39

Helling D, Possart A, Cottier S, Klahre U, Kost B (2006) Pollen tube tip growth depends on plasma membrane polarization mediated by tobacco PLC3 activity and endocytic membrane recycling. Plant Cell 18:3519–3534

Hepler PK, Waayne RO (1985) Calcium and plant development. Annu Rev Plant Physiol 36:397–439

Herrmann A, Felle HH (1995) Tip growth in root hair-cells of *Sinapis-alba* L - significance of internal and external Ca^{2+} and pH. New Phytol 129:523–533

Hetherington AM, Brownlee C (2004) The generation of Ca^{2+} signals in plants. Annu Rev Plant Biol 55:401–427

Holdaway-Clarke TL, Feijo JA, Hackett GR, Kunkel JG, Hepler PK (1997) Pollen tube growth and the intracellular cytosolic calcium gradient oscillate in phase while extracellular calcium influx is delayed. Plant Cell 9:1999–2010

Holdaway-Clarke TL, Hepler PK (2003) Control of pollen tube growth: role of ion gradients and fluxes. New Phytol 159:539–563

Holzinger A, Callaham OA, Hepler PK, Meindl U (1995) Free calcium in *Mirasterias* – local gradients are not detected in growing lobes. Eur J Cell Biol 67:363–371

Homann U, Tester M (1997) Ca^{2+} - independent and Ca^{2+}/GTP-binding protein-controlled exocytosis in a plant cell. Proc Natl Acad Sci USA 94:6565–6570

Huang S, Robinson RC, Gao LY, Matsumoto T, Brunet A, Blanchoin L, Staiger CJ (2005) *Arabidopsis* VILLIN1 generates actin filament cables that are resistant to depolymerization. Plant Cell 17:486–501

Hurst SR, Kropf DL (1991) Ionic requirements for establishment of an embryonic axis in *Pelvetia* zygotes. Planta 185:27–33

Hyde GJ, Heath JB (1995) Ca^{2+}-dependent polarization axis establishment in the tip-growing organism, *Saprolegnia-ferax*, by gradients of the ionophore A23187. Eur J Cell Biol 67:356–362

Ikebuchi NW, Wiseman DM (1990) Calcium-dependent regulation of actin filament bundling by lipocortin-85. J Biol Chem 265:3392–342

Ivashuta S, Liu JY, Liu JQ, Lohar DP, Haridas S, Bucciarelli B, VandenBosch KA, Vance CP, Harrison MJ, Gantt JS (2005) RNA interference identifies a calcium-dependent protein kinase involved in *Medicago truncatula* root development. Plant Cell 17:2911–2921

Jiang L, Yang SL, Xie LF, Puah CS, Zhang XQ, Yang WC, Sundaresan V, Ye D (2005) VANGUARD1 encodes a pectin methylesterase that enhances pollen tube growth in the *Arabidopsis* style and transmitting tract. Plant Cell 17:584–596

Jones DL, Gilroy S, Larsen PB, Howell SH, Kochian LV (1998) Effect of aluminum on cytoplasmic Ca^{2+} homeostasis in root hairs of *Arabidopsis thaliana* (L.). Planta 206:378–387

Jones DL, Shaff JE, Kochian LV (1995) Role of calcium and other ions in directing root hair tip growth in *Limnobium stoloniferum*: I. Inhibition of tip growth by aluminum. Planta 197:672–680

Kanamori N, Madsen LH, Radutoiu S, Frantescu M, Quistgaard EM, Miwa H, Downie JA, James EK, Felle HH, Haaning LL, Jensen TH, Sato S, Nakamura Y, Tabata S, Sandal N, Stougaard J (2006) A nucleoporin is required for induction of Ca^{2+} spiking in legume nodule development and essential for rhizobial and fungal symbiosis. Proc Natl Acad Sci USA 103:359–364

Ketelaar T, Emons AMC (2008) The actin cytoskeleton in root hairs: a cell elongation device. In: Emons AMC, Ketelaar T (eds) Root hairs: excellent tools for the study of plant molecular cell biology. Springer, Berlin Heidelberg New York. doi:10.1007/7089_2008_8

Ketelaar T, de Ruijter NC, Emons AM (2003) Unstable F-actin specifies the area and microtubule direction of cell expansion in *Arabidopsis* root hairs. Plant Cell 15:285–292

Kinoshita T, Nishimura M, Shimazaki K (1995) Cytosolic concentration of Ca^{2+} regulates the plasma membrane H^+-ATPase in guard cells of Fava bean. Plant Cell 7:1333–1342

Klusener B, Boheim G, Liss H, Engelberth J, Weiler EW (1995) Gadolinium-sensitive, voltage-dependent calcium-release channels in the endoplasmic-reticulum of a higher-plant mechano-receptor organ. EMBO J 14:2708–2714

Kohno T, Shimmen T (1987) Ca^{2+} induced fragmentation of actin-filaments in pollen tubes. Protoplasma 141:177–179

Kost B, Lemichez E, Spielhofer P, Hong Y, Tolias K, Carpenter C, Chua NH (1999) Rac homologues and compartmentalized phosphatidylinositol 4, 5-bisphosphate act in a common pathway to regulate polar pollen tube growth. J Cell Biol 145:317–330

Lancelle SA, Hepler PK (1992) Ultrastructure of freeze substituted pollen tubes of *Lilium longiflorum*. Protoplasma 167:215–230

Lee S, Lee EJ, Yang EJ, Lee JE, Park AR, Song WH, Park OK (2004) Proteomic identification of annexins, calcium-dependent membrane binding proteins that mediate osmotic stress and abscisic acid signal transduction in *Arabidopsis*. Plant Cell 16:1378–1391

Li L, Kim BG, Cheong YH, Pandey GK, Luan S (2006a) A Ca^{2+} signaling pathway regulates a K^+ channel for low-K response in *Arabidopsis*. Proc Natl Acad Sci USA 103:12625–12630

Li M, Qin C, Welti R, Wang X (2006b) Double knockouts of phospholipases Dzeta1 and Dzeta2 in *Arabidopsis* affect root elongation during phosphate-limited growth but do not affect root hair patterning. Plant Physiol 140:761–770

Li H, Lin YK, Heath RM, Zhu MX, Yang Z (1999) Control of pollen tube tip growth by a Rop GTPase-dependent pathway that leads to tip-localized calcium influx. Plant Cell 11:1731–1742

Limpens E, Bisseling T (2008) Nod factor signal transduction in the Rhizobium-Legume symbiosis. In: Emons AMC, Ketelaar T (eds) Root hairs: excellent tools for the study of plant molecular cell biology. Springer, Berlin Heidelberg New York. doi:10.1007/7089_2008_10

Lin YK, Yang ZB (1997) Inhibition of pollen tube elongation by microinjected anti-Rop1Ps antibodies suggests a crucial role for Rho-type GTPases in the control of tip growth. Plant Cell 9:1647–1659

Love J, Brownlee C, Trewavas AJ (1997) Ca^{2+} and calmodulin dynamics during photopolarization in *Fucus serratus* zygotes. Plant Physiol 115:249–261

Malho R (1998) Expanding tip-growth theory. Trends Plant Sci 3:40–43

Malho R, Trewavas AJ (1996) Localized apical increases of cytosolic free calcium control pollen tube orientation. Plant Cell 8:1935–1949

Messerli MA, Creton R, Jaffe LF, Robinson KR (2000) Periodic increases in elongation precede increases in cytosolic Ca^{2+} during pollen tube growth. Dev Biol 222:84–98

Messerli MA, Danuser G, Robinson KP (1999) Pulsatile influxes of H^+, K^+ and Ca^{2+} tag growth pulses of *Lilium longiflorum* pollen tubes. J Cell Sci 112:1497–1509

Messerli M, Robinson KR (1997) Tip localized Ca^{2+} pulses are coincident with peak pulsatile growth rates in pollen tubes of *Lilium longiflorum*. J Cell Sci 110:1269–1278

Miller DD, Callaham DA, Gross DJ, Hepler PK (1992) Free Ca^{2+} gradient in growing pollen tubes of *Lilium longiflorum*. J Cell Sci 101:7–12

Miller DD, Leferink-ten Klooster HB, Emons AMC (2000) Lipochito-oligosaccharide nodulation factors stimulate cytoplasmic polarity with longitudinal endoplasmic reticulum and vesicles at the tip in vetch root hairs. Mol Plant Microbe Interact 13:1385–1390

Miller DD, de Ruijter NCA, Bisseling T, Emons AMC (1999) The role of actin in root hair morphogenesis: studies with lipochito-oligosaccharide as a growth stimulator and cytochalasin as an actin perturbing drug. Plant J 17:141–154

Miwa H, Sun J, Oldroyd GE, Downie JA (2006) Analysis of Nod-factor-induced calcium signaling in root hairs of symbiotically defective mutants of Lotus japonicus. Mol Plant Microbe Interact 19:914–923

Miyawaki A, Llopis J, Heim R, McCaffery JM, Adams JA, Ikura M, Tsien RY (1997) Fluorescent indicators for Ca^{2+} based on green fluorescent proteins and calmodulin. Nature 388:882–887

Molendijk AJ, Bischoff F, Rajendrakumar CS, Friml J, Braun M, Gilroy S, Palme K (2001) *Arabidopsis thaliana* Rop GTPases are localized to tips of root hairs and control polar growth. EMBO J 20:2779–2788

Monteiro D, Liu QL, Lisboa S, Scherer GEF, Quader H, Malho R (2005) Phosphoinositides and phosphatidic acid regulate pollen tube growth and reorientation through modulation of $[Ca^{2+}]_c$ and membrane secretion. J Exp Bot 56:1665–1674

Moutinho A, Trewavas AJ, Malho R (1998) Relocation of a Ca^{2+} - dependent protein kinase activity during pollen tube reorientation. Plant Cell 10:1499–1509

Muir SR, Sanders D (1997) Inositol 1,4,5-trisphosphate-sensitive Ca^{2+} release across nonvacuolar membranes in cauliflower. Plant Phys 114:1511–1521

Nagai T, Yamada S, Tominaga T, Ichikawa M, Miyawaki A (2004) Expanded dynamic range of fluorescent indicators for Ca(2+) by circularly permuted yellow fluorescent proteins. Proc Natl Acad Sci USA 101:10554–10559

Ohashi Y, Oka A, Rodrigues-Pousada R, Possenti M, Ruberti I, Morelli G, Aoyama T (2003) Modulation of phospholipid signaling by GLABRA2 in root-hair pattern formation. Science 300:1427–1430

Oheim M, Kirchhoff F, Stuhmer W (2006) Calcium microdomains in regulated exocytosis. Cell Calcium 40:423–439

Parton RM, Fischer-Parton S, Watahiki MK, Trewavas AJ (2001) Dynamics of the apical vesicle accumulation and the rate of growth are related in individual pollen tubes. J Cell Sci 114:2685–2695

Peiter E, Maathuis FJ, Mills LN, Knight H, Pelloux J, Hetherington AM, Sanders D (2005) The vacuolar Ca^{2+}-activated channel TPC1 regulates germination and stomatal movement. Nature 434(7031):404–408

Picton JM, Steer MW (1983) Evidence for the role of Ca^{2+} ions in tip extension in pollen tubes Protoplasma 115:1–17

Preuss ML, Schmitz AJ, Thole JM, Bonner HK, Otegui MS, Nielsen E (2006) A role for the RabA4b effector protein PI-4Kβ1 in polarized expansion of root hair cells in *Arabidopsis thaliana.* J Cell Biol 172:991–998

Preuss ML, Serna J, Falbel TG, Bednarek SY, Nielsen E (2004) The *Arabidopsis* Rab GTPase RabA4b localizes to the tips of growing root hair cells. Plant Cell 16(6):1589–603

Pu R, Robinson KR (1998) Cytoplasmic calcium gradients and calmodulin in the early development of the fucoid alga *Pelvetia compressa.* J Cell Sci 111:3197–3207

Rato C, Monteiro D, Hepler PK, Malho R (2004) Calmodulin activity and cAMP signalling modulate growth and apical secretion in pollen tubes. Plant J 38:887–897

Reintanz B, Szyroki A, Ivashikina N, Ache P, Godde M, Becker D, Palme K, Hedrich R (2002) AtKC1, a silent *Arabidopsis* potassium channel alpha-subunit modulates root hair K^{+} influx. Proc Natl Acad Sci USA 99:4079–4084

Reissig JL, Kinney SG (1983) Calcium as a branching signal in *Neurospora crassa.* J Bacteriol 154:1397–1402

Robinson KR, Jaffe LF (1973) Ion movements in a developing fucoid egg. Dev Biol 35:349–361

Rosen WG, Gawlik SR, Dashek WV, Siegesmund KA (1964) Fine structure and cytochemistry of *Lilium* pollen tubes. Am J Bot 51:61–71

de Ruijter NCA, Rook MB, Bisseling T, Emons AMC (1998) Lipochito-oligosaccharides re-initiate root hair tip growth in *Vicia sativa* with high calcium and spectrin-like antigen at the tip. Plant J 13:341–350

Schiefelbein JW, Shipley A, Rowse P (1992) Calcium influx at the tip of growing root-hair cells of *Arabidopsis thaliana.* Planta 187:455–459

Schiott M, Romanowsky SM, Baekgaard L, Jakobsen MK (2004) A plant plasma membrane Ca^{2+} pump is required for normal pollen tube growth and fertilization. Proc Natl Acad Sci USA 101:9502–9507

Schwab A, Finsterwalder F, Kersting U, Danker T, Oberleithner H (1997) Intracellular Ca^{2+} distribution in migrating transformed epithelial cells. Pflugers Arch 434:70–76

Shigaki T, Hirschi KD (2006) Diverse functions and molecular properties emerging for CAX cation/H^{+} exchangers in plants. Plant Biol (Stuttg) 8:419–429

Sieberer BJ, Emons AMC (2003) Cytoarchitecture and pattern of cytoplasmic streaming in root hairs of Medicago truncatula during development and deformation by nodulation factors. Protoplasma 214:118–127

Sievers A, Schnepf E (1981) Morphogenesis and polarity in tubular cells with tip growth. In: Kiermayer O (ed) Cytomorphogenesis in Plants. Springer, Berlin Heidelberg New York, pp 265–299

Shaw SL, Quatrano RS (1996) The role of targeted secretion in the establishment of cell polarity and the orientation of the division plane in Fucus zygotes. Development 122:2623–2630

Smertenko AP, Jiang CJ, Simmons NJ,Weeds AG, Davies DR, Hussey PJ (1998) Ser6 in the maize actin-depolymerizing factor, ZmADF3, is phosphorylated by a calcium stimulated protein kinase and is essential for the control of functional activity. Plant J 14:187–193

Speksnijder JE, Miller AL, Weisenseel MH, Chen TH, Jaffe LF (1989) Calcium buffer injections block fucoid egg development by facilitating calcium diffusion. Proc Natl Acad Sci USA 86:6607–6661

Sun J, Miwa H, Downie JA, Oldroyd GE (2007) Mastoparan activates calcium spiking analogous to Nod factor-induced responses in Medicago truncatula root hair cells. Plant Physiol 144:695–702

Taylor AR, Manison NFH, Fernandez C, Wood J (1996) Spatial organization of calcium signaling involved in cell volume control in the *Fucus* rhizoid. Plant Cell 8:2015–2031

Thiel G, Rupnik M, Zorec R (1994) Raising the cytosolic Ca^{2+} concentration increases the membrane capacitance of maize coleoptile protoplasts: Evidence for Ca^{2+}-stimulated exocytosis. Planta 195:305–308

Thion L, Mazars C, Nacry P, Bouchez D, Moreau M, Ranjeva R, Thuleau P (1998) Plasma membrane depolarization-activated calcium channels, stimulated by microtubule-depolymerizing drugs in wild-type *Arabidopsis thaliana* protoplasts, display constitutively large activities and a longer half-life in ton 2 mutant cells affected in the organization of cortical microtubules. Plant J 13:603–610

Tominaga M, Yokota E, Vidali L, Sonobe S, Hepler PK, Shimmen T (2000) The role of plant villin in the organization of the actin cytoskeleton, cytoplasmic streaming and the architecture of the transvacuolar strand in root hair cells of *Hydrocharis*. Planta 210:836–843

Trewavas AJ, Malho R (1997) Signal perception and transduction: The origin of the phenotype. Plant Cell 9:1181–1195

Very AA, Davies JM (2000) Hyperpolarization-activated calcium channels at the tip of *Arabidopsis* root hairs. Proc Natl Acad Sci USA 97:9801–9806

Vincent P, Chua M, Nogue F, Fairbrother A, Mekeel H, Xu Y, Allen N, Bibikova TN, Gilroy S, Bankaitis VA (2005) A Sec14p-nodulin domain phosphatidylinositol transfer protein polarizes membrane growth of *Arabidopsis thaliana* root hairs. J Cell Biol 168:801–812

Wang YF, Fan LM, Zhang WZ, Zhang W, Wu WH (2004) Ca^{2+}-permeable channels in the plasma membrane of *Arabidopsis* pollen are regulated by actin microfilaments. Plant Physiol 136:3892–3904

Wang G, Lu L, Zhang CY, Singapuri A, Yuan S (2006) Calmodulin concentrates at the apex of growing hyphae and localizes to the Spitzenkorper in *Aspergillus nidulans*. Protoplasma 228:159–166

Watson AJ, Barcroft LC (2001) Regulation of blastocyst formation. Front Biosci 6:D708–D730

White PJ (1998) Calcium channels in the plasma membrane of root cells. Ann Bot 81:173–183

Wu G, Gu Y, Li S, Yang Z (2001) A genome-wide analysis of *Arabidopsis* Rop-interactive CRIB motif-containing proteins that act as Rop GTPase targets. Plant Cell 13:2841–2856

Wymer CL, Bibikova TN, Gilroy S (1997) Cytoplasmic free calcium distributions during the development of root hairs of *Arabidopsis thaliana*. Plant J 12:427–439

Yokota E, Tominaga M, Mabuchi I, Tsuji Y, Staiger CJ, Oiwa K, Shimmen T (2005) Plant villin, lily P-135-ABP, possesses G-actin binding activity and accelerates the polymerization and depolymerization of actin in a Ca^{2+}-sensitive manner. Plant Cell Physiol 46:1690–1703

Yoon GM, Dowd PE, Gilroy S, McCubbin AG (2006) Calcium-dependent protein kinase isoforms in Petunia have distinct functions in pollen tube growth, including regulating polarity. Plant Cell 18:867–878

Zarsky V, Fowler J (2008) ROP (Rho-related protein from Plants) GTPases for spatial control of root hair morphogenesis. In: Emons AMC, Ketelaar T (eds) Root hairs: excellent tools for the study of plant molecular cell biology. Springer, Berlin Heidelberg New York. doi:10.1007/7089_2008_14

Zielinski RE (1998) Calmodulin and calmodulin-binding proteins in plants. Annu Rev Plant Physiol Plant Mol Biol 49:697–725

Zorec R, Tester M (1992) Cytoplasmic Ca^{2+} stimulates exocytosis in a plant secretory cell. Biophys J 63:864–867

Zorec R, Tester M (1993) Rapid pressure driven exocytosis–endocytosis cycle in a single plant cell. FEBS Lett 333:283–286

Phospholipid Signaling in Root Hair Development

T. Aoyama

Abstract Phospholipids, which are major components of the eukaryotic plasma membrane, play crucial roles in signal transduction, leading to not only total cellular responses via transcriptional regulation but also localized intracellular events such as membrane traffic and cytoskeletal reorganization, both of which underlie polarized cell morphogenesis. Although studies of phospholipid signaling have focused mainly on animals and fungi, evidence for its involvement in plant cell morphogenesis has also been accumulating. Because phospholipids function as site-specific signals on membranes, they likely play pivotal roles in localizing exocytosis and the fine F-actin configuration to regions of cell expansion, such as the tips of growing root hairs. In this chapter, evidence for the involvement of phospholipids in the regulation of root hair tip growth is described, with an emphasis on major signaling phospholipids, phosphoinositides and phosphatidic acid; in addition, a model signal transduction network for root hair tip growth, involving phospholipids, their metabolic enzymes, and their effector proteins is proposed.

1 Introduction

Root hair morphogenesis has been intensively studied as a model system for the molecular processes involved in shaping plant cells, because of the dispensability under laboratory conditions and accessibility for experimental observation of root hairs (Peterson and Farquhar 1996; see Grierson and Schiefelbein 2008). Root hairs are cellular protuberances resulting from the polar outgrowth of specific root epidermal cells called trichoblasts. During root hair morphogenesis in *Arabidopsis*, a bulge is initially formed at the distal end on the outer trichoblast surface; subsequently, the bulge protrudes further, perpendicular to the root surface, by highly polarized cell expansion, resulting in a thin cylindrical structure (Gilroy and Jones 2000; Ryan et al. 2001). This type of cell expansion is called tip growth, because all of the events involved in the growth, including cell wall deposition and plasma

T. Aoyama
Institute for Chemical Research, Kyoto University, Uji, Kyoto 611-0011, Japan
e-mail: aoyama@scl.kyoto-u.ac.jp

Plant Cell Monogr, doi:10.1007/7089_2008_1

membrane expansion, are limited to the tip (see Emons and Ketelaar 2008; de Keijzer et al. 2008; Nielsen 2008).

In growing root hairs, cytoskeletal reorganization and membrane traffic, both of which are involved in the deposition of materials for new plasma membranes and cell walls, are restricted to the apical region (see Assaad 2008; Ketelaar and Emons 2008), and a tip-focused cytoplasmic calcium ion gradient is known to ensure their proper localization (Hepler et al. 2001; Smith and Oppenheimer 2005; Cole and Fowler 2006; see Bibikova and Gilroy 2008). Recently, forward and reverse genetic studies of *Arabidopsis thaliana* have helped elucidate the molecular basis of sustained polar tip growth. Establishment of the calcium gradient has been shown to require reactive oxygen species (ROS) generated by the NADPH oxidase RHD2 (Foreman et al. 2003), and the RhoGDP dissociation inhibitor (GDI) SCN1 acts as an upstream regulatory component of the ROS accumulation (Carol et al. 2005). Moreover, Rho-related GTPases of plants (ROPs) have been shown to localize to the apical region of growing root hair tips, and the expression of their constitutively active forms disrupts polar growth (Molendijk et al. 2001; Jones et al. 2002). Such data suggest that ROP signaling leads to a calcium ion gradient via ROS accumulation (Carol and Dolan 2006; see Zarsky and Fowler 2008).

Phospholipids that function as site-specific signals on membranes are also important regulators of root hair tip growth. In fact, accumulating evidence suggests the involvement of phosphoinositides and phosphatidic acid (PA), major signaling phospholipids known in animal and fungal systems, in plant cell tip growth (Fischer et al. 2004). The *Arabidopsis* phosphatidylinositol (PI) transfer protein COW1/AtSFH1 and phosphatidylinositol 4-kinases (PI4Ks) PI-4Kβ1 and PI-4Kβ2 have been shown to regulate root hair morphogenesis (Bohme et al. 2004; Vincent et al. 2005; Preuss et al. 2006). It has also been found in *Arabidopsis* that the phospholipase D (PLD) PLDζ1 acts as a positive regulator of root hair development (Ohashi et al. 2003), and that the protein kinase AGC2-1 functions downstream of PA in root hair tip growth (Anthony et al. 2004). In the following sections, an overview of recent information regarding the function of phospholipids as site-specific signals in animals and fungi, which is mostly applicable to plants, will be presented. Then, the involvement of phosphoinositides, PA, and other phospholipid-related signals in root hair development will be described. Finally, a model signal transduction network for root hair tip growth, including phospholipid signaling, will be proposed.

2 Overview of Phospholipids as Site-Specific Signals

Phospholipids are major components of the plasma membrane in eukaryotic cells. In addition to their structural role, some phospholipids act as signals to recruit and regulate proteins with varied molecular functions. Phospholipid signals, like those of other signaling substances, induce total cellular responses (e.g., transcriptional responses to extracellular stimuli). However, in contrast to such soluble signaling substances as cAMP and calcium ions, phospholipids can act as site-specific signals on membranes. This distinctive character allows them to regulate spatially restricted cellular events, including localized cytoskeletal reorganization and membrane traffic, both of which are essential for polarized cell expansion (Martin 1998; Liscovitch et al. 2000).

Over the past decade, phosphoinositides have been most intensively studied as local intracellular signals (Di Paolo and De Camilli 2006; Gamper and Shapiro 2007; Krauss and Haucke 2007). PI, the precursor of phosphoinositides, is synthesized primarily in the endoplasmic reticulum and then delivered to other membrane compartments by vesicular transport or PI transfer proteins. The inositol head group of PI can be phosphorylated by phosphoinositide kinases at positions 3, 4, and 5, resulting in seven phosphoinositide species (Fig. 1; Fruman et al. 1998; Anderson et al. 1999; Rameh and Cantley 1999). Among them, phosphatidylinositol 4,5-bisphosphate (PI(4,5)P_2) has been historically associated with the generation of two second messengers, diacylglycerol (DAG) and inositol 1,4,5-triphosphate (IP_3), via phospholipase C (PLC)-mediated hydrolysis (Berridge and Irvine 1984). However, PI(4,5)P_2, as well as other phosphoinositides, is now known to function by itself in various types of intracellular signaling, especially those involved in cell polarization (Niggli 2005; Santarius et al. 2006). PI(4,5)P_2 is thought to be a key regulator of actin cytoskeletal reorganization, because it directly binds to and moderates the function of many actin regulatory proteins, including gelsolin, profilin, ADF/cofilin, and WASP (Yin and Janmey 2003; Logan and Mandato 2006).

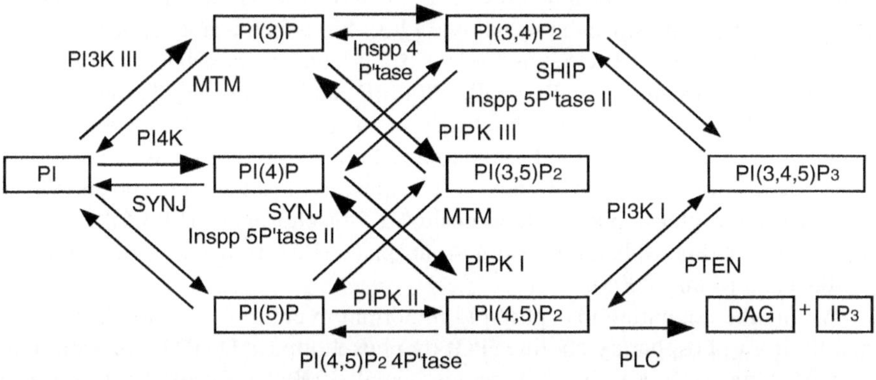

Fig. 1 Phosphoinositide metabolism. Enzymes catalyzing major reactions in animals and their metabolites (boxed) are schematically shown. *Thick arrows* and *boxes* indicate pathways and metabolites present in both plants and animals, respectively. *Inspp 4P'tase* inositol polyphosphate 4-phosphatase, *Inspp 5P'tase II* inositol polyphosphate 5-phosphatase II, *MTM* myotubularin, *PI3K I* class I PI3K, *PI3K III* class III PI3K, *PIPK I* type I PIPK, *PIPK II* type II PIPK, *PI(4,5)P_2 4P'tase* PI(4,5)P_2-inositol 4-phosphatase, *SYNJ* synaptojanin (Martin 1998; Takenawa and Itoh 2001; Mueller-Roeber and Pical 2002; Ercetin and Gillaspy 2004; Hawkins et al. 2006)

$PI(4,5)P_2$ plays a crucial role also in membrane traffic by binding to regulatory proteins of SNARE complexes, CAPS, and clathrin-associated proteins (Itoh et al. 2001; Olsen et al. 2003; Grishanin et al. 2004; Li et al. 2006a). In addition, $PI(4,5)P_2$ interacts with regulatory proteins of small GTPases (e.g., GEFs and GAPs; Paris et al. 1997; Das et al. 2000; Kam et al. 2000; Russo et al. 2001; Nie et al. 2002), leading to the regulation of the actin cytoskeleton and membrane traffic. Another widely studied role of $PI(4,5)P_2$ is the direct interaction with plasma-membrane-integrated proteins that may help establish cell polarity (Suh and Hille 2005). These proteins include voltage-gated potassium and calcium channels and sensory transduction channels (Runnels et al. 2002; Wu et al. 2002; Gamper et al. 2004; Oliver et al. 2004).

Phosphatidylinositol 3,4-bisphosphate ($PI(3,4)P_2$) and phosphatidylinositol 3,4,5-trisphosphate ($PI(3,4,5)P_3$) are known as second messengers for responses to extracellular signals (Di Paolo and De Camilli 2006; Hawkins et al. 2006), and promote cell growth and survival via direct interaction with such protein kinases as Akt/PKB, PDK, and SGK (Alessi et al. 1997; Stokoe et al. 1997; Stephens et al. 1998; Chan et al. 1999; Pao et al. 2007). In addition, $PI(3,4,5)P_3$ regulates the actin cytoskeleton and membrane traffic through binding to GEFs and GAPs for small GTPases belonging to Rho and Arf families (Klarlund et al. 1998; Das et al. 2000; Macia et al. 2000; Russo et al. 2001; Mertens et al. 2003; Cote et al. 2005). Recently, the interconversion between $PI(4,5)P_2$ and $PI(3,4,5)P_3$ by phosphoinositide 3-kinase (PI3K) and 3-phosphatase (PTEN) was shown to regulate cell polarity in several systems (Comer and Parent 2007). In animal epithelial cells, for example, $PI(4,5)P_2$ and $PI(3,4,5)P_3$ are localized to the apical and basolateral plasma membrane, respectively, where they recruit proteins necessary for the identity of each membrane (Gassama-Diagne et al. 2006; Martin-Belmonte et al. 2006). Contrary to PTEN, 5-phosphatases such as SHIP may function to prolong the $PI(3,4,5)P_3$ signaling since the metabolite $PI(3,4)P_2$, which is not dephosphorylated by PTEN, has a same function as $PI(3,4,5)P_3$ in many cases (Majerus et al. 1999). Phosphoinositides other than $PI(3,4)P_2$, $PI(4,5)P_2$, and $PI(3,4,5)P_3$ also act as signaling molecules via their specific binding proteins (Di Paolo and De Camilli 2006; Hawkins et al. 2006). Especially, some phosphoinositides are known to contribute to the generation of organelle identity: phosphatidylinositol 3-phosphate (PI(3)P) for early endosomes, phosphatidylinositol 3,5-bisphosphate ($PI(3,5)P_2$) for late endosomes, and phosphatidylinositol 4-phosphate (PI(4)P) for the Golgi apparatus (Behnia and Munro 2005).

PA, another signaling phospholipid important in cell polarization, is produced mainly from phosphatidylcholine (PC) by phospholipase D (PLD) or from DAG by DAG kinase (DGK) (Fig. 2; Liscovitch et al. 2000; Cockcroft 2001; Testerink and Munnik 2005; Wang et al. 2006; Topham 2006; Oude Weernink et al. 2007). PA is involved in the signal transduction pathways of a vast number of phenomena, including responses to extracellular stimuli, mitogenesis, and cell survival through direct interactions with its effector proteins (e.g., protein kinases mTOR and Raf1, and NADPH oxidase; Ghosh et al. 1996; Rizzo et al. 1999; Fang et al. 2001; Palicz et al. 2001; Chen and Fang 2002). Also in cell morphogenesis, PA

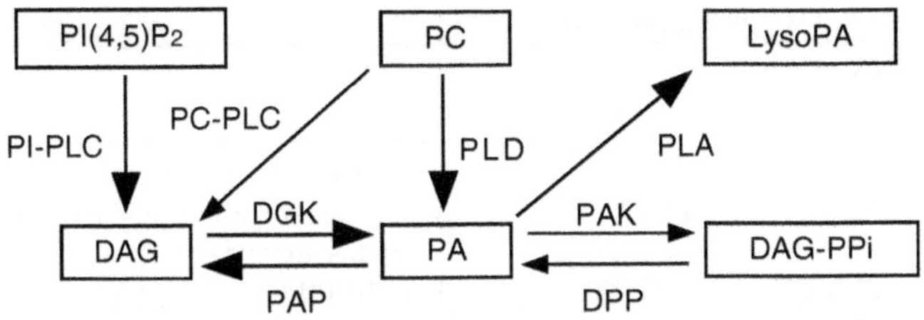

Fig. 2 Phospholipid metabolism around PA signaling. Enzymes catalyzing reactions in plants and their metabolites (boxed) are schematically shown. *Bold arrows* and *boxes* indicate pathways and metabolites present in both plants and animals, respectively. *DAG-PPi* diacylglycerol pyrophosphate, *DPP* diacylglycerol pyrophosphate phosphatase, *LysoPA* lysophosphatidic acid, *PAK* phosphatidic acid kinase, *PAP* phosphatidic acid phosphatase, *PC-PLC* phosphatidylcholine-hydorolyzing phospholipase C, *PI-PLC* phosphoinositide-specific phospholipase C, *PLA* phospholipase A (Testerink and Munnik 2005; Wang et al. 2006)

enhances actin cytoskeletal reorganization and membrane traffic mainly through the activation of type I phosphatidylinositol phosphate kinase (PIPK), a key enzyme in the production of PI(4,5)P_2 (Moritz et al. 1992; Jenkins et al. 1994; Ishihara et al. 1998; Kam and Exton 2001), and modulation of the small GTPase signaling (Manifava et al. 2001; Nie et al. 2002; Lindsay and McCaffrey 2004). In addition to its signaling function, PA is thought to function as a fusogenic lipid during vesicle fission and fusion because of its small anionic head group and its association with divalent cations, which are thought to lower the activation energy required for negative membrane curvature (Scales and Scheller 1999; Kozlovsky et al. 2002).

The signal transduction pathways involving phospholipids are connected to diverse upstream factors via the enzymes producing or consuming them. Both PLD and type I PIPK are directly activated by small GTPases, including Rho and Arf, which are, in turn, regulated by PI(4,5)P_2 via their GEFs and GAPs (Oude Weernink et al. 2004; Jenkins and Frohman 2005; Santarius et al. 2006). In addition, type I PIPK and PLD are mutually activated via their respective products, PI(4,5)P_2 and PA (Oude Weernink et al. 2007). As for the downstream factors of the phospholipid signaling, phospholipid-binding proteins with various molecular functions have been identified (Table 1). Together, these data indicate that phospholipid signaling pathways are tightly interconnected and that they constitute a broad network together with other signaling pathways, especially those of the small GTPases. A much more complicated feature of phospholipid signaling is that the consumption of a signal leads to the production of another signal, and vice versa. For example, PI(4,5)P_2 is not only a signaling molecule by itself but also the precursor of IP_3, DAG, and PI(3,4,5)P_3, and it may lead to PA (see Figs. 1 and 2). Owing to these features, phospholipids are thought to cooperatively endow the intracellular signal transduction network with complicated spatial codes on membranes.

Table 1 Phospholipids-binding proteins

Phospholipid	Binding modules	Binding proteins	Protein functions
PI(3)P	FYVE	EEA1	Early endosomal antigen
	FYVE	SARA	Smad anchor for receptor activation, endosomal protein
	FYVE	PIKfyve	Type III PIPK
	PX	SNX	Sorting nexin, membrane traffic regulation
	PX	P40phox	NADPH oxidase regulatory subunit
	–	Pf1	PHD-containing transcription factor
PI(4)P	A/ENTH	EpsinR	Clathrin-interacting protein
	PH	FAPP	TGN-associated 4-phosphate-adaptor protein
PI(5)P	PHD	ING2	Chromatin regulatory protein
PI(3,4)P_2	PH	TAPP1	Tandem PH-domain-containing protein, protein recruitment
	PH	AKT/PKB	Serine/threonine kinase
	PH	DAPP1	PH-SH2 domain protein, protein recruitment
	PX	P47phox	NADPH oxidase regulatory subunit
PI(3,5)P_2	A/ENTH	Ent3p, Ent5p	Clathrin-interacting protein
	GRAM	Myotubularin	Phosphoinositide 3-phosphatase
	–	SVP1	Swollen vacuole phenotype 1, membrane recycling
PI(4,5)P_2	A/ENTH	AP180	Clathrin-interacting protein
	A/ENTH	Epsin	Clathrin-interacting protein
	A/ENTH	HIP1	Huntingtin-interacting protein, clathrin interaction
	C2	Synaptotagmin	Regulator of SNARE
	FERM	Ezrin	Actin-interacting protein, actin–plasma-membrane linkage
	FERM	Talin	Actin-interacting protein, actin–integrin linkage
	MORN	**PIP5K1**	**Type I PIPK**
	PDZ	Syntenin	Component of signal transduction complex
	PH	VAV, SOS	RhoGEF
	PH	ARNO	ArfGEF
	PH	ASAP1	ArfGAP
	PH	PLCd1	PLC
	PH	Dynamin	GTPase, endocytosis
	PTB	SHC	Component of signal transduction complex
	PX	PI3K II	Class II PI3K
	–	Tubby	Component of signal transduction complex
	–	**AtCP**	**Heterodimeric capping protein, actin binding**
	–	Gelsolin	Actin-capping protein
	–	Profilin	Actin nucleotide exchanging protein
	–	ADF/cofilin	Actin-severing protein
	–	WASP	Actin polymerization activator
	–	CAPS	Ca^{2+}-dependent activator protein for secretion

(continued)

Table 1 (continued)

Phospholipid	Binding modules	Binding proteins	Protein functions
PI(3,4,5)P_3	C2	Rab11-FIP I	Class I Rab11 family interacting protein
	PH	BTK	Protein tyrosine kinase
	PH	AKT/PKB	Serine/threonine kinase
	PH	PDK1	3'-Phosphoinositide-dependent kinase, serine/threonine
	PH	DAPP1	PH-SH2 domain protein, protein recruitment
	PH	VAV, SOS	RhoGEF
	PH	ARNO	ArfGEF
	PH	GRP1	ArfGEF
	PTB	SHC	Component of signal transduction complex
	PX	PLD1	PLD
	PX	CISK	SGK, protein serine/threonine kinase
	–	DOCK180	RhoGEF
PA	C2	PKCε	Protein kinase C, serine/threonine kinase
	C2	Rab11-FIP I	Class I Rab11 family interacting protein
	FRB	mTOR	Serine/threonine kinase
	MORN	**PIP5K1**	**Type I PIPK**
	PH	**AtPDK1**	**PDK1 homolog**
	PX	P47phox	NADPH oxidase regulatory subunit
	PX	PLD1	PLD
	Q2	Opi1	Transcription factor, repressor
	TAPAS	PDE4A1	cAMP phosphodiesterase
	–	PIPK I	Type I PIPK
	–	RafI	MAPKKK, serine/threonine kinase
	–	PP1Cg	Protein phosphatase
	–	AGAP1	ArfGAP
	–	ARF	Small GTPase
	–	SK1	Sphingosin kinase
	–	NSF	*N*-ethylmaleimide-sensitive factor
	–	**ABI1**	**Protein phosphatase**
	–	**PEPC**	**Phosphoenolpyruvate carboxylase**
	–	**AtCP**	**Heterodimeric capping protein, actin binding**

Proteins that bind to phosphoinositides or PA, their binding modules, and their functions are listed. Proteins originated from plants are indicated by bold characters (Suh and Hille 2005; Testerink and Munnik 2005; Di Paolo and De Camilli 2006; Hawkins et al. 2006; Wang et al. 2006)

3 Phosphoinositide Signaling in Root Hair Development

Most of the phosphoinositides and the enzymatic activities shown in Fig. 1 have been detected in plants, suggesting that the knowledge about phosphoinositide signaling pathways in animals are basically applicable to plants, although existence of PI(3,4,5)P_3 signaling pathways is unclear in plants (Mueller-Roeber and Pical

2002; Zonia and Munnik 2006). The first evidence for phosphoinositide involvement in root hair development was provided by the observation that $PI(4,5)P_2$ is specifically localized to root hair bulges and elongating root hairs in maize (Braun et al. 1999). Also, in elongating root hairs of *Arabidopsis*, strong $PI(4,5)P_2$ signals were detected at the plasma membrane and the apical cytoplasmic space using $PH_{PLC\delta1}$-YFP, a chimeric fluorescence protein that specifically binds $PI(4,5)P_2$ in vivo (Vincent et al. 2005).

As for protein factors, genetic studies have revealed that defects in PI transfer protein COW1/AtSFH1 cause short and distorted root hairs (Bohme et al. 2004; Vincent et al. 2005). COW1/AtSFH1 contains an N-terminal lipid-binding domain similar to the yeast PI transfer protein Sec14 and a C-terminal domain similar to the *Lotus japonicus* nodulin Nij16. Its GFP fusion protein showed the same localization pattern as $PI(4,5)P_2$ in elongating root hairs (i.e., the plasma membrane and apical cytoplasmic space; Vincent et al. 2005). The loss of proper $PI(4,5)P_2$ localization in a T-DNA insertion mutant is closely linked to degeneration of the fine F-actin configuration at the tip. This fact, together with the expected function of $PI(4,5)P_2$, strongly suggests that COW1/AtSFH1 generates phosphoinositide landmarks that are coupled to components of the F-actin cytoskeleton via the production of $PI(4,5)P_2$. Interestingly, in *Lotus* nodulogenesis, *LjPLP-IV*, the *AtSFH1* ortholog, is transcriptionally downregulated, and Nij16 nodulin, the C-terminal portion of LiPLP-IV, is highly expressed, through the activation of a bidirectional promoter located in an intron of the *LjPLP-IV* gene (Kapranov et al. 2001). This suggests that the COW1/AtSFH1 class PI transfer proteins can act as a master switch that controls polarization and depolarization in root hair development.

The *Arabidopsis* PI4K PI-4Kβ1 binds the GTP form of RabA4b in yeast and in vitro, and they colocalize to the membrane fraction of growing root hair tips (Preuss et al. 2006). PI-4Kβ1 physically interacts also with the calcium ion sensor protein AtCBL1, suggesting that the ability of PI-4Kβ1 to produce PI4P is regulated by calcium ion concentration. Mutant plants in which both *PI-4Kβ1* and its closest relative *PI-4Kβ2* are disrupted exhibit aberrant root hair morphologies. Since RabA4b-labeled *trans*-Golgi network (TGN) compartments are morphologically altered in the mutant root hair cells, it is thought that PI-4Kβ1/β2 activity and hence PI4P production in these compartments is necessary for proper TGN organization and post-Golgi secretion (Preuss et al. 2006).

In addition to *cow1/atsfh1* mutations, a *pi-4kβ1 pi-4kβ2* double mutation may affect root hair morphogenesis by altering the level of $PI(4,5)P_2$ in the root hair apex. Because the involvement of $PI(4,5)P_2$ in actin cytoskeletal reorganization and membrane traffic has been demonstrated in animals and fungi, and because many effector proteins are conserved in plants, it is reasonable to assume that $PI(4,5)P_2$ is an indispensable regulatory factor for root hair tip growth. Supporting this, mutations in the *Arabidopsis PIP5K3* gene, which encodes PI(4)P 5-kinase (PIP5K) and is expressed preferentially in root hair cells, cause significantly shorter root hairs when compared with wild-type root hairs (Kusano and Aoyama, unpublished data).

$PI(4,5)P_2$ has been observed also in the apical plasma membrane of growing pollen tubes, another tip-growing cellular structure in plants (Kost et al. 1999).

Tobacco and petunia PLCs accumulate laterally on the plasma membrane at the growing pollen tube tip, and suppression of PLC activity leads to PI(4,5)P_2 delocalization at the tip and pollen tube growth depolarization, while its overexpression moderately reduces the growth rate (Dowd et al. 2006; Helling et al. 2006). Based on these results, it has been suggested that PLC restricts PI(4,5)P_2 to the pollen tube apex by digesting PI(4,5)P_2 in the lateral region (Dowd et al. 2006; Helling et al. 2006), although its products, DAG and IP_3, may act as signals during tip growth.

4 Phosphatidic Acid Signaling in Root Hair Development

Plants have two types of PLDs, a C2-domain-containing type that is unique to plants and a PX-PH-domain-containing type that is common to most eukaryotes (Qin and Wang 2002). Studies of PLD-mediated signal transduction in plants have mainly focused on the C2-type PLDs in response to varied environmental stimuli, including wound, cold, and drought stresses (Testerink and Munnik 2005; Wang 2005; Wang et al. 2006). On the other hand, PLDζ1, an *Arabidopsis* PX-PH-type PLD, is involved in root hair cell patterning and morphogenesis (Ohashi et al. 2003). The transcription factor GL2 negatively regulates the *PLDζ1* gene in atrichoblasts during root hair cell pattern formation. Ectopic overexpression of PLDζ1 causes root hair bulges in both atrichoblasts and trichoblasts. Hence, PLDζ1 is thought to be a positive regulator of root hair initiation. PLDζ1 overexpression causes branched root hairs at a high frequency, while reduced PLDζ1 expression disrupts the bulge positioning on a trichoblast and the polarity of root hair expansion. A PLDζ1–GFP fusion protein localized to the apical region of elongating root hairs as well as to vesicle-like compartments in the cortical regions. These findings indicate the involvement of PLDζ1 in regulating the initiation and maintenance of root hair tip growth. T-DNA insertion mutants of the *PLDζ1* gene, however, do not exhibit obvious changes in root hair morphology, suggesting that other *PLD* genes can compensate for the mutant defects (Li et al. 2006b).

The *Arabidopsis* 3'-phosphoinositide-dependent kinase PDK1 is involved in PA signaling in root hair cells (Anthony et al. 2004). Animal PDK1 acts downstream in PI3K signaling via the binding to 3'-phosphoinositide such as PI(3,4,5)P_3 (Alessi et al. 1997; Stephens et al. 1998; Currie et al. 1999). *Arabidopsis* PDK1, however, preferentially binds PA in vitro, and is activated by PA and PI(4,5)P_2 in *Arabidopsis* protoplasts (Anthony et al. 2004). Moreover, PA-dependent, but not PI(4,5)P_2-dependent, activity leads to the activation of the *Arabidopsis* AGC kinase AGC2-1 (Anthony et al. 2004). A loss of AGC2-1 function results in reduced root hair length, and a GFP–AGC2-1 fusion protein showed a dynamic localization pattern during root hair development with localization in the apical region at some stages (Anthony et al. 2004). Interestingly, OXI1, a protein kinase mediating oxidative stress responses, is identical to AGC2-1 (Rentel et al. 2004). These findings indicate that this protein kinase cascade (PKD1-AGC2-1/OXI1) mediates the PA signal and possibly the ROS signal during root hair development. In addition, mutations in the *Arabidopsis IRE* gene, which encodes another member of the AGC kinase family, cause a short root hair phenotype similar to that of the *agc2-1* mutant (Oyama et al. 2002). This suggests that these kinases share a downstream target for regulating root hair elongation while their upstream regulators are likely different from each other.

Pharmacological evidence has also shown that the PA produced by the PLD and PLC-DGK pathways is involved in plant cell morphogenesis. Experiments with 1-butanol and neomycin, which are specific inhibitors of PLD and PLC, respectively, revealed the involvement of both PLD and PLC in root hair deformation during nodule development (den Hartog et al. 2001; Charron et al. 2004). Both pollen germination and tube growth were arrested in the presence of 1-butanol. The inhibition was overcome by addition of exogenous PA-containing liposomes, and partially overcome by addition of taxol, a microtubule-stabilizing agent, indicating that the PA produced by PLD regulates pollen tube development partly via microtubule reorganization (Potocky et al. 2003). Supporting this, PLD , one of C2-type PLDs, binds to microtubules and is involved in cortical microtubule organization (Gardiner et al. 2001, 2003).

5 Other Lipid-Related Signals in Root Hair Development

DAG and IP_3, both of which are crucial second messengers in animals, may also act in intracellular signaling in plants. So far, however, orthologs of the effector proteins found in animals, such as PKC and IP_3 receptors, have not been identified in plants (Meijer and Munnik 2003); hence, their involvement in root hair development remains unclear. Since DAG is rapidly converted to PA by DGK, PA likely acts as a second messenger downstream of PLC rather than DAG (Munnik 2001). As for IP_3, high-affinity IP_3-binding sites have been detected in internal membranes, and IP_3 can trigger the release of calcium ions from internal stores (Alexandre et al. 1990; Meijer and Munnik 2003). In an experiment using caged probes, IP_3 release, as well as that of $PI(4,5)P_2$, resulted in an increase of pollen tube cytosolic-free calcium and reorientation of pollen tube growth (Monteiro et al. 2005).

The mechanisms by which proteins are targeted to and anchored at the plasma membrane in the root hair apical region involve posttranslational modifications, such as *S* acylation and the addition of glycosylophosphatidylinositol (GPI) anchors. The *Arabidopsis TIP1* gene, mutations of which affect tip growth in both root hairs and pollen tubes, encodes an *S*-acyl transferase (Hemsley et al. 2005). Acylation has been implicated in protein sorting into lipid rafts at the plasma membrane, and is known to be involved in determining cell polarity (Bagnat and Simons 2002) and signal transduction (Lai 2003). Proteins modified by acylation include type-II ROP GTPases (Lavy et al. 2002), suggesting that acylation by TIP1 may help recruit these regulatory proteins to the tip. GPI anchor modification takes place in the endoplasmic reticulum, and GPI-anchored proteins are transported via the TGN to the outer leaflet of the plasma membrane (Ikezawa 2002). *Arabidopsis*

SETH1 and *SETH2* encode different putative subunits of the enzyme complex involved in GPI biosynthesis, and mutations in either gene cause defects in pollen germination and pollen tube growth (Lalanne et al. 2004), suggesting that the GPI anchor modification is also critical for polarized cell growth.

6 A Hypothesis of the Signaling Network Sustaining Root Hair Tip Growth

Once root hair tip growth begins, the polarity is strictly sustained, resulting in straight root hairs without branches. The mechanism for sustaining this highly polarized growth, called tip-growth LENS (localization-enhancing network, self-sustaining; Cole and Fowler 2006), has been hypothesized not only to localize events underlying cell expansion, for example, membrane traffic and cytoskeletal reorganization, to the tip continuously, but also to maintain its own localization at the tip against the diffusion concomitant with cell expansion. In this section, I present a hypothesis of the signaling network sustaining root hair tip growth, focusing on the roles of phospholipid signaling.

The tip growth LENS is assumed to use positive feedback loop(s) for self-sustenance, to which phospholipid signaling may contribute significantly. Phospholipid signaling can provide a robust positive feedback loop consisting of two major second messengers, $PI(4,5)P_2$ and PA, and the enzymes that produce them, PIP5K and PLD (Fig. 3). In animal systems, this phospholipid signaling loops have been assumed to regulate actin cytoskeletal reorganization and membrane trafficking (Oude Weernink et al. 2007). Many *Arabidopsis* PLDs, including PLDζ1, are activated by $PI(4,5)P_2$ (Qin and Wang 2002). Since PA activates PIP5K1 through interaction with the MORN domain (Im et al. 2007), PA likely activates also other MORN-containing PIP5Ks, including PIP5K3, which is involved in root hair development (Kusano and Aoyama, unpublished data). Hence, a phospholipid signaling loop likely functions in root hair cells. The loop critically helps to sustain not only the level of signaling via reciprocal activation, but also the peaky spatial pattern of signaling via reciprocal recruitment between PLD and PIP5K at the root hair tip.

The regulatory factors that interact with the phospholipid signaling loop most likely include small GTPases. By analogy with animal and fungal systems, both plant PLDs and PIP5Ks are possibly activated by ROPs and Arfs; in turn, PA and $PI(4,5)P_2$ may regulate GEFs and GAPs (Oude Weernink et al. 2004; Jenkins and Frohman 2005; Santarius et al. 2006). Hence, another signaling loop is expected to occur between phospholipids and small GTPases. Although direct evidence for these interactions has not been obtained, this idea is supported by evidence indicating that the *Arabidopsis* ROP7/Rac2 physically interacts with PIP5K activity in tobacco pollen tubes (Kost et al. 1999) and that plants have small GTPase regulatory proteins with phospholipid-binding domains (e.g., the *Arabidopsis* DOCK180 homolog SPIKE1; Qiu et al. 2002; van Leeuwen et al. 2004).

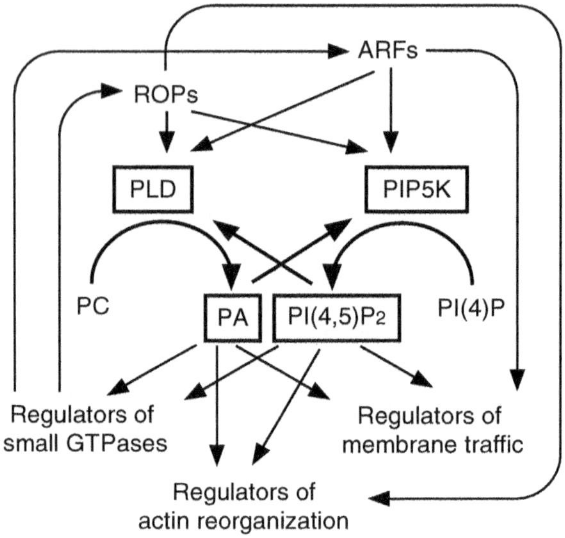

Fig. 3 A schematic representation of a hypothesis of the phospholipid signaling loop sustaining root hair tip growth. A positive feedback loop consisting of PI(4,5)P_2, PA, PIP5K, and PLD, and its upstream and downstream pathways are schematically shown. Other possible regulators and effectors of the signaling loop (e.g., reactive oxygen species and Ca^{2+}) are omitted for simplification

In pathways downstream of the signaling loops, effector proteins of PI(4,5)P_2, PA, and small GTPases are thought to be involved in root hair tip growth. By analogy with animal and fungal systems, effectors of PI(4,5)P_2 likely include several actin-regulatory proteins such as gelsolin, profilin, ADF/cofilin, and WASP for F-actin formation (Yin and Janmey 2003; Logan and Mandato 2006); SNARE and SNARE-regulating proteins such as synaptotagmins for exocytosis (Li et al. 2006a; Nakanishi et al. 2004); and clathrin-associated proteins with the ENTH domain for endocytosis (Grishanin et al. 2004) in membrane recycling between the apical and subapical plasma membrane. In addition, PI(4,5)P_2 might recruit a calcium channel to the root tip plasma membrane by direct interaction. Downstream pathways of ROPs and Arfs are also expected to involve cytoskeletal reorganization and membrane trafficking, respectively, via a wide variety of their effectors (D'Souza-Schorey and Chavrier 2006; Ridley 2006; Santarius et al. 2006; see Zarsky and Fowler 2008).

PA may also recruit and regulate proteins with varied functions required for tip growth. Evidence obtained from *Arabidopsis* and rice indicates that PA produced by PLD increases both NADPH oxidase activity and ROS levels (Sang et al. 2001; Park et al. 2004; Yamaguchi et al. 2004); however, factors mediating these processes are unknown. In animal phagocytic cells, PA binds to the $p47^{phox}$ regulatory subunit of NADPH oxidase and activates the $p91^{phox}$ catalytic subunit (Palicz et al. 2001; Karathanassis et al. 2002). RHD2 is an *Arabidopsis* homolog of the $p91^{phox}$

catalytic subunit (Foreman et al. 2003). Although a plant homolog of $p47^{phox}$ is not known, PA may activate NADPH oxidase through the interaction with its regulatory subunit to activate the signaling pathway via the ROS production to the calcium ion gradient at root hair apices (Foreman et al. 2003).

7 Conclusion

Phospholipid signals are distinct from other second messengers in that they act as site-specific signals on membranes to regulate localized intracellular events. In addition, the production and consumption of one signal can lead to changes in the levels of other signals, indicating that phospholipids are contextual signals. Moreover, their signaling pathways are connected not only to each other (e.g., between $PI(4,5)P_2$ and PA), but also to those of other signaling systems (e.g., small GTPases), constituting a complex signaling network. These features likely allow the robust and sophisticated regulation of root hair tip growth, but hinder a simple understanding of the controlling mechanism. Although information on the molecular basis of the mechanism is still fragmented at present, $PI(4,5)P_2$ and PA most likely play pivotal roles in regulating the highly polarized cell expansion, together with other signaling factors, including small GTPases. For a further clarification of the mechanism regulating root hair tip growth, it will be important to identify protein factors working at the root hair tip, and investigate their interactions with phospholipids, phospholipid-metabolizing enzymes, and other signaling factors in living root hairs.

References

Alessi DR, Deak M, Casamayor A, Caudwell FB, Morrice N, Norman DG, Gaffney P, Reese CB, MacDougall CN, Harison D, Ashworth A, Bownes M (1997) 3-phosphoinositide-dependent protein kinase-1 (PDK1): structural and functional homology with the drosophila DSTPK61 kinase. Curr Biol 7:776–789

Alexandre J, Lassalles JF, Kado RD (1990) Opening of Ca^{2+} channels in isolated red beet vacuole membrane by inositol 1,4,5-trisphosphate. Nature 343:567–570

Anderson RA, Boronenkov IV, Doughman SD, Kunz J, Loijens JC (1999) Phosphatidylinositol phosphate kinase, multifaceted family of signaling enzymes. J Biol Chem 274:17794–17805

Anthony RG, Henriques R, Helfer A, Mezaros T, Rios G, Testerink C, Munnik T, Deak M, Koncz C, Bogre L (2004) A protein kinase target of a PDK1 signalling pathway is involved in root hair growth in *Arabidopsis*. EMBO J 23:572–581

Assaad FF (2008) The membrane dynamics of root hair morphogenesis. In: Emons AMC, Ketelaar T (eds) Root hairs: excellent tools for the study of plant molecular cell biology. Springer, Berlin Heidelberg New York. doi:10.1007/7089_2008_2

Bagnat M, Simons K (2002) Lipid rafts in protein sorting and cell polarity in budding yeast *Saccharomyces cerevisiae*. Biol Chem 383:1475–1480

Behnia R, Munro S (2005) organelle identity and the signposts for membrane traffic. Nature 438:597–604

Berridge MJ, Irvine RF (1984) Inositol triphosphate, a novel second messenger in cellular signal transduction. Nature 312:315–321

Bibikova T, Gilroy S (2008) Calcium in root hair growth. In: Emons AMC, Ketelaar T (eds) Root hairs: excellent tools for the study of plant molecular cell biology. Springer, Berlin Heidelberg New York. doi:10.1007/7089_2008_3

Bohme K, Li Y, Charlot F, Grierson C, Marrocco K, Okada K, Laloue M, Nogue F (2004) The *Arabidopsis COW1* gene encodes a phosphatidylinositol transfer protein essential for root hair tip growth. Plant J 40:686–698

Braun M, Baluska F, von Witsch M, Menzel D (1999) Redistribution of actin, profiling and phosphatidylinositol-4,5-bisphosphate in growing and maturing root hairs. Planta 209:435–443

Carol RJ, Dolan L (2006) The role of reactive oxygen species in cell growth: lessons from root hairs. J Exp Bot 57:1829–1834

Carol RJ, Takeda S, Linstead P, Durrant MC, Kakesova H, Derbyshire P, Crea S, Zarsky V, Dolan L (2005) A RhoGDP dissociation inhibitor spatially regulates growth in root hair cells. Nature 438:1013–1016

Chan TO, Rittenhouse SE, Tsichlis PN (1999) AKT/PBK and other D3 phosphoinositide-regulated kinase: kinase activation by phosphoinositide-dependent phosphorylation. Annu Rev Biochem 68:965–1014

Charron D, Pingret JL, Chabaud M, Journet EP, Barker DG (2004) Pharmacological evidence that multiple phospholipid signaling pathways link Rhizobium nodulation factor perception in *Medicago truncatula* root hairs to intracellular responses, including Ca^{2+} spiking and specific *ENDO* gene expression. Plant Physiol 136:3582–3593

Chen J, Fang YM (2002) A novel pathway regulating the memmalian target of rapamycin (mTOR) signaling. Biochem Pharmacol 64:1071–1077

Cockcroft S (2001) Signalling roles of mammalian phospholipase D1 and D2. Cell Mol Life Sci 58:1674–1687

Cole RA, Fowler JE (2006) Polarized growth: maintaining focus on the tip. Curr Opin Plant Biol 9:579–588

Comer FI, Parent CA (2007) Phosphoinositides specify polarity during epithelial organ development. Cell 128:239–240

Cote JF, Motoyama AB, Bush JA, Vuori K (2005) A novel and evolutionarily conserved PtdIns(3,4,5)P_3-binding domain is necessary for DOCK180 signalling. Nat Cell Biol 7:797–807

Currie RA, Walker KS, Gray A, Deak M, Casamayor A, Downes CP, Cohen P, Alessi DR, Lucocq J (1999) Role of phosphatidylinositol 3,4,5-trisphosphate in regulating the activity and localization of 3-phosphoinositide-dependent protein kinase-1. Biochem J 337:575–583

Das B, Shu XD, Day GJ, Han J, Krishna UM, Falck JR, Broek D (2000) Control of interamolecular interactions between the pleckstrin homology and Db1 homology domains of Vav and Sos1 regulates Rac binding. J Biol Chem 275:15074–15081

de Keijzer MN, Emons AMC, Mulder BM (2008) Modeling tip growth: pushing ahead. In: Emons AMC, Ketelaar T (eds) Root hairs: excellent tools for the study of plant molecular cell biology. Springer, Berlin Heidelberg New York. doi:10.1007/7089_2008_7

den Hartog M, Musgrave A, Munnik T (2001) Nod factor-induced phosphatidic acid and diacylglycerol pyrophosphate formation: a role for phospholipase C and D in root hair deformation. Plant J 25:55–65

Di Paolo G, De Camilli P (2006) Phosphoinositides in cell regulation and membrane dynamics. Nature 443:651–675

Dowd PE, Coursol S, Skirpan AL, Kao TH, Gilroy S (2006) Petunia phospholipase c1 is involved in pollen tube growth. Plant Cell 18:1438–1453

D'Souza-Schorey C, Chavrier P (2006) ARF proteins: roles in membrane traffic and beyond. Nat Rev Mol Cell Biol 7:347–358

Emons AM, Ketelaar T (2008) Intracellular organization: a prerequisite for root hair elongation and cell wall deposition. In: Emons AMC, Ketelaar T (eds) Root hairs: excellent tools for the study of plant molecular cell biology. Springer, Berlin Heidelberg New York. doi:10.1007/7089_2008_4

Ercetin ME, Gillaspy GE (2004) Molecular characterization of an Arabidopsis gene encoding a phospholipid-specific inositol polyphosphate 5-phosphatase. Plant Physiol 135:938–946

Fang Y, Vilella-Bach M, Bachmann R, Flanigan A, Chen A (2001) Phosphatidic acid-mediated mitogenic activation of mTOR signaling. Science 294:1942–1945

Fischer U, Shuzhen M, Grebe M (2004) Lipid function in plant cell polarity. Curr Opin Plant Biol 7:670–676

Foreman J, Demidchik V, Bothwell JH, Mylona P, Miedema H, Torres MA, Linstead P, Costa S, Brownlee C, Jones JD, Davies JM, Dolan L (2003) Reactive oxygen species produced by NADPH oxidase regulate plant cell growth. Nature 422:442–446

Fruman DA, Meyers RE, Cantley LC (1998) Phosphoinositide kinases. Annu Rev Biochem 67:481–507

Gamper N, Shapiro MS (2007) Target-specific PIP_2 signaling: How might it work? J Physiol, 10.1113/jphysiol.2007.132787

Gamper N, Reznikov V, Yamada Y, Yang J, Shapiro MS (2004) Phophatidylinositol 4,5-bisphosphate signals underlie receptor-specific $G_{q/11}$-mediated modulation of N-type Ca^{2+} channels J Neurosci 24:10980–10992

Gardiner JC, Harper JD, Weerakoon ND, Collings DA, Ritchie S, Gilroy S, Cyr RJ, Marc J (2001) A 90-kD phospholipase D from tobacco binds to microtubules and the plasma membrane. Plant Cell 13:2143–2158

Gardiner JC, Collings DA, Harper JD, Marc J (2003) The effects of the phospholipase D-antagonist 1-butanol on seedling development and microtubule organization in *Arabidopsis*. Plant Cell Physiol 44:687–696

Gassama-Diagne A, Yu W, ter Beest M, Martin-Belmonte F, Kierbel A, Engel J, Mostov K (2006) Phosphatidylinositol-3,4,5-trisphosphate regulates the formation of the basolateral plasma membrane in epithelial cells. Nat Cell Biol 8:963–970

Gilroy S, Jones DL (2000) Through form to function: root hair development and nutrient uptake. Trends Plant Sci 5:56–60

Ghosh S, Strum JC, Sciorra VA, Daniel L, Bell RM (1996) Raf-1 kinase possesses distinct binding domains for phosphatidylserine and phosphatidic acid. Phosphatidic acid regulates the translocation of Raf-1 in 12-O-tetradecanoylphorbol-13-acetate-stimulated Madin-Darby canine kidney cells. J Biol Chem 271:8472–8480

Grierson C, Schiefelbein J (2008) Genetics of root hair formation. In: Emons AMC, Ketelaar T (eds) Root hairs: excellent tools for the study of plant molecular cell biology. Springer, Berlin Heidelberg New York. doi:10.1007/7089_2008_15

Grishanin RN, Kowalchyk JA, Klenchin VA, Ann K, Earles CA, Chapman ER, Gerona RR, Martin TF (2004) CAPS acts at a prefusion step in dens-core vesicle exocytosis as a PIP_2 binding protein Neuron 43:551–562

Hawkins PT, Anderson KE, Davidson K, Stephens LR (2006) Signalling through class I PI3Ks in mammalian cells. Biochem Soc Trans 34:647–662

Hemsley PA, Kemp AC, Grierson CS (2005) The tip growth defective1 *S*-acyl transferase regulates plant cell growth in *Arabidopsis*. Plant Cell 17:2554–2563

Hepler PK, Vidali L, Cheung AY (2001) Polarized cell growth in higher plants. Annu Rev Cell Dev Biol 17:159–187

Helling D, Possart A, Cottier S, Klahre U, Kost B (2006) Pollen tube tip growth depends on plasma membrane polarization mediated by tobacco PLC3 activity and endocytic membrane recycling. Plant Cell 18:3519–3534

Ikezawa H (2002) Glycosylphosphatidylinositol (GPI)-anchored proteins. Biol Pharm Bull 25:409–417

Im YJ, Perera IY, Brglez I, Davis AJ, Stevenson-Paulik J, Phillippy BQ, Johannes E, Allen NS, Boss WF (2007) Increasing plasma membrane phosphatudylinositol(4,5)Bisphosphate biosynthesis increases phosphoinositide metabolism in *Nicotiana tabacum*. Plant Cell 19:1603–1616

Ishihara H, Shibasaki Y, Kizuki N, Wada T, Yazaki Y, Asano T, Oka Y (1998) Type I phosphatidylinositol-4-phosphate 5-kinase. Cloning of the third isoform and deletion/substitution analysis of members of this novel lipid kinase family. J Biol Chem 273:8741–8748

Itoh T, Koshiba S, Kigawa T, Kikuchi A, Yokoyama S, Takenawa T (2001) Role of the ENTH domain in phosphatidylinositol-4,5,-bisphosphate binding and endoctosis. Science 291:1047–1051

Jenkins GM, Frohman MA (2005) Phospholipase D: a lipid centric review. Cell Mol Life Sci 62:2305–2316

Jenkins GH, Fisette PL, Anderson RA (1994) Type I phosphatidylinositol 4-phosphate 5-kinase isoforms are specifically stimulated by phosphatidic acid. J Biol Chem 269:11547–11554

Jones MA, Shen JJ, Fu Y, Li H, Yang Z, Grierson CS (2002) The Arabidopsis Rop2 GTPase is a positive regulator of both root hair initiation and tip growth. Plant Cell 14:763–776

Kam Y, Exton JH (2001) Phospholipase D activity is required for actin stress fiber formation in fibroblasts. Mol Cell Biol 21:4055–4066

Kam JL, Miura K, Jackson TR, Gruschus J, Roller P, Stauffer S, Clark J, Aneja R, Randazzo PA (2000) Phosphoinositide-dependent activation of the ADP-ribosylation factor GTPase-activating protein ASAP1. Evidence for the pleckstrin homology domain functioning as an allosteric site. J Biol Chem 275:9653–9663

Kapranov P, Routt SM, Bankaitis VA, de Bruijn FJ, Szczglowski K (2001) Nodule-specific regulation of phosphatidylinositol transfer protein expression in *Lotus japonicus*. Plant Cell 13:1369–1382

Karathanassis D, Stahelin RV, Bravo J, Perisic O, Pacold CM, Cho W, Williams RL (2002) Binding of the PX domain of $p47^{phox}$ to phosphatidylinositol 3,4-bisphosphate and phosphatidic acid is marked by an intramolecular interaction EMBO J 21:5057–5068

Ketelaar T, Emons AM (2008) The actin cytoskeleton in root hairs: a cell elongation device. In: Emons AMC, Ketelaar T (eds) Root hairs: excellent tools for the study of plant molecular cell biology. Springer, Berlin Heidelberg New York. doi:10.1007/7089_2008_8

Klarlund JK, Rameh LE, Cantley LC, Buxton JM, Holik JJ, Sakelis C, Patki V, Corvera S, Czech MP (1998) Regulation of GRP1-catalyzed ADP rebosylation factor guanine nucleotide exchange by phosphatidylinositol 3,4,5-trisphosphate. J Bio Chem 273:1859–1862

Kost B, Lemichez E, Spielhofer P, Hong Y, Tolias K, Carpenter C, Chua N-H (1999) Rac homologues and compartmentalized phosphatidylinositol 4,5-bisphosphate act in a common pathway to regulate polar pollen tube growth. J Cell Biol 145:317–330

Kozlovsky Y, Chernomordik LV, Kozlov MM (2002) Lipid intermediates in membrane fusion: foration, structure, and decay of hemifusion diaphragm. Biophys J 83:2634–2651

Krauss M, Haucke V (2007) Phosphoinositides: Regulators of membrane traffic and protein function. FEBS Lett 581:2105–2111

Lai EC (2003) Lipid rafts make for slippery platforms. J Cell Biol 162:365–370

Lalanne E, Honys D, Johnson A, Borner GH, Lilley KS, Dugree P, Grossniklaus U, Twell D (2004) *SETH1* and *SETH2*, two components of the glycosylphosphatidylinositol anchor biosynthetic pathway, are required for pollen germination and tube growth in Arabidopsis Plant Cell 16:229–240

Lavy M, Bracha-Drori K, Sternberg H, Yalovsky S (2002) A cell-specific, prenylation-independent mechanism regulates targeting of type II RACs. Plant Cell 14:2431–2450

Li L, Shin O-H, Rhee J-S, Arac D, Rah J-C, Rizo J, Sudhof T, Rosenmund C (2006a) Phosphatidylinositol phosphates as co-activators of Ca^{2+} binding to C_2 domains of synaptotagmin 1. J Biol Chem 281:15845–15852

Li M, Qin C, Welti R, Wang X (2006b) Double knockouts of phospholipase Dζ1 and Dζ2 in Arabidopsis affect elongation during phosphate-limited growth but do not affect root hair patterning. Plant Physiol 140:761–770

Lindsay AJ, McCaffrey MW (2004) The C2 domains of the class I Rab11 family of interacting proteins target recycling vesicles to the plasma membrane. J Cell Sci 117:4365–4375

Liscovitch M, Czarny M, Fiucci G, Tang X (2000) Phospholipase D: molecular and cell biology of a novel gene family. Biochem J 345:401–415

Logan MR, Mandato CA (2006) Regulation of the actin cytoskeleton by PIP2 in cytokinesis. Biol Cell 98:377–388

Macia E, Paris S, Chabre M (2000) Binding of the PH and polybasic C-terminal domains of ARNO to phosphoinositides and to acidic lipids. Biochemistry 39:5893–5901

Majerus PW, Kisseleva MV, Norris FA (1999) The role of phosphatases in inositol signaling reactions. J Biol Chem 274:10669–10672

Manifava M, Thuring JW, Lim ZY, Packman L, Holmes AB, Ktistakis NT (2001) Differential binding of traffic-related proteins to phosphatidic acid- or phosphatidylinositol (4,5)-bisphosphate-coupled affinity reagents. J Biol Chem 276:8987–8994

Martin TF (1998) Phosphoinositide lipids as signaling molecules: common themes for signal transduction, cytoskeletal regulation, and membrane trafficking. Annu Rev Cell Dev Biol 14:231–264

Martin-Belmonte F, Gassama-Diagne A, Datta A, Yu W, Rescher U, Gerke V, Mostov K (2006) PTEN-mediated apical segregation of phosphoinositides controls epithelial morphogenesis through Cdc42. Cell 128:383–397

Meijer HJG, Munnik T (2003) Phospholipid-based signaling in plants. Annu Rev Plant Biol 54:265–306

Mertens AE, Roovers RC, Collard JG (2003) Regulation of Tiam-Rac signaling. FEBS Lett 546:11–16

Molendijk AJ, Bischoff F, Rajendrakumar CS, Friml J, Braun M, Gilroy S, Palme K (2001) *Arabidopsis thaliana* Rop GTPases are localized to tips of root hairs and control polar growth EMBO J. 20:2799–2788

Monteiro D, Liu Q, Lisboa S, Sherer GE, Quader H, Malho R (2005) Phosphoinositides and phosphatidic acid regulate pollen tube growth and reorientation through modulation of $[Ca^{2+}]_c$ and membrane secretion J. Exp. Bot. 56:1665–1674

Moritz A, De Graan PN, Gispen WH, Wirtz KW (1992) Phosphatidic acid is a specific activator of phosphatidylinositol-4-phosphate kinase. J. Biol. Chem. 267:7207–7210

Mueller-Roeber B, Pical C (2002) Inositol phospholipid metabolism in Arabidopsis. Characterized and putative isoforms of inositol phospholipid kinase and phosphoinositide-specific phospholipase C. Plant Physiol 130:22–46

Munnik T (2001) Phosphatidic acid: an emerging plant lipid second messenger. Trends Plant Sci 6:227–233

Nakanishi H, de los Santos P, Neiman AM (2004) Positive and negative regulation of a SNARE protein by control of intracellular localization. Mol Biol Cell 15:1802–1815

Nie Z, Stanley KT, Stauffer S, Jacques KM, Hirsch DS, Takei J, Randazzo PA (2002) AGAP1, an endosome-associated, phosphoinositide-dependent ADP-ribosylation factor GTPase-activating protein that affects actin cytoskeleton. J Biol Chem 277:48965–48975

Nielsen E (2008) Plant cell wall biogenesis during tip growth in root hair cells. In: Emons AMC, Ketelaar T (eds) Root hairs: excellent tools for the study of plant molecular cell biology. Springer, Berlin Heidelberg New York. doi:10.1007/7089_2008_11

Niggli V (2005) Regulation of protein activities by phosphoinositide phosphates. Annu Rev Cell Biol 21:57–79

Oliver D, Lien CC, Soom M, Baukrowitz T, Jonas P, Fakler B (2004) Functional conversion between A-type and deleyed rectifier K^+ channels by membrane lipids. Science 304:265–270

Olsen HL, Hoy M, Zhang W, Bertorello AM, Bokvist K, Capito K, Efanov AM, Meister B, Thams P, Yang SN, Rorsman P, Berggren PO, Gromada J (2003) Phosphatidylinositol 4-kinase serves as a metabolic sensor and regulates priming of secretory granules in pancreatic beta cells. Proc Natl Acad Sci USA 100:5187–5192

Ohashi Y, Oka A, Rodrigues-Pousada R, Possenti M, Ruberti I, Morelli G, Aoyama T (2003) Modulation of phospholipid signaling by GLABRA2 in root-hair pattern formation. Science 300:1427–1430

Oude Weernink PA, Schmidt M, Jakobs KH (2004) Regulation and cellular roles of phosphoinositide 5-kinases. Eur J Pharmacol 500:87–99

Oude Weernink PA, Lopez de Jesus M, Schmidt M (2007) Phospholipase D signaling: orchestration by PIP_2 and small GTPases. Naunyn Schmiedeberg's Arch Pharmacol 374:399–411

Oyama T, Shimura Y, Okada K (2002) The *IRE* gene encodes a protein kinase homologue and modulates root hair growth in *Arabidopsis*. Plant J 30:289–299

Palicz A, Foubert TR, Jesaitis AJ, Marodi L, McPhail LC (2001) Phosphatidic acid and diacylglycerol directly activate NADPH oxidase by interacting with enzyme components. J Biol Chem 276:3090–3097

Pao AC, McCormick JA, Li H, Siu J, Govaerts C, Bhalla V, Soundararajan R, Pearce D (2007) NH_2 terminus of serum and glucocorticoid-regulated kinase 1 binds to phosphoinositides and is essential for isoform-specific physiological functions. Am J Physiol Renal Physiol 292: F1741–F1750

Paris S, Beraud-Dufour S, Robineau S, Bigay J, Antonny B, Chabre M, Chardin P (1997) Role of protein-phospholipid interactions in the activation of ARF1 by the guanine nucleotide exchange factor Arno. J Bio Chem 272:22221–22226

Park J, Gu Y, Lee Y, Yang Z, Lee Y (2004) Phosphatidic acid induces leaf cell death in Arabidopsis by activating the Rho-related small G protein GTPase-madiated pathway of reactive oxygen-species generation. Plant Physiol 134:129–136

Peterson RL, Farquhar ML (1996) Root hairs: specialized tubular cells extending root surfaces. Bot Rev 62:2–33

Potocky M, Elias M, Profotova B, Novotna Z, Valentova O, Zarsky V (2003) Phosphatidic acid produced by phospholipase D is required for tobacco pollen tube growth. Planta 217:122–130

Preuss ML, Schmitz AJ, Thole JM, Bonner HKS, Otegui MS, Nielsen E (2006) A role for the RabA4b effector protein PI-4Kβ1 in polarized expansion of root hair cells in *Arabidopsis thaliana*. J Cell Biol 172:991–998

Qin C, Wang X (2002) The Arabidopsis phospholipase D family. Characterization of a calcium-independent and phosphatidylcholine-selective PLDζ1 with distinct regulatory domains. Plant Physiol 128:1057–1068

Qiu JL, Jilk R, Marks MD, Szymanski DB (2002) The Arabidopsis *SPIKE1* gene is required for normal cell shape control and tissue development. Plant Cell 14:101–118

Rameh LE, Cantley LC (1999) The role of phosphoinositide 3-kinase lipid products in cell function. J Biol Chem 274:8347–8350

Rentel MC, Lecourieux D, Ouaked F, Usher SL, Petersen L, Okamoto H, Knight H, Peck SC, Grierson CS, Hirt H, Knight MR (2004) OXI1 kinase is necessary for oxidative burst-mediated signaling in *Arabidopsis*. Nature 427:858–861

Ridley AJ (2006) Rho GTPases and actin dynamics in membrane protrusions and vesicle trafficking. Trends Cell Biol 16:522–529

Rizzo MA, Shome K, Vasudevan C, Stolz DB, Sung TC, Frohman MA, Watkins SC, Romero G (1999) Phospholipase D and its product, phosphatidic acid, mediate agonist-dependent raf-1 translocation to the plasma mmembrane and the activation of the mitogen-activated protein kinase pathway. J Biol Chem 274:1131–1139

Runnels LW, Yue L, Clapham DE (2002) The TRPM7 channel is inactivated by PIP_2 hydrolysis. Nat Cell Biol 4:329–336

Russo C, Gao Y, Mancini P, Vanni C, Porotto M, Falasca M, Torrisi MR, Zheng Y, Eva A (2001) Modulation of oncogenic DBL activity by phosphoinositol phosphate binding to pleckstrin homology domain. J Biol Chem 276:19524–19531

Ryan E, Steer M, Dolan L (2001) Cell biology and genetics of root hair formation in *Arabidopsis thaliana*. Protoplasma 215:140–149

Sang Y, Cui D, Wang X (2001) Phospholipase D and Phosphatidic acid-mediated generation of superoxide in Arabidopsis. Plant Physiol 126:1449–1458

Santarius M, Lee CH, Anderson RA (2006) Supervised membrane swimming: small G-protein lifeguards regulate PIPK signalling and monitor intracellular PtdIns(4,5)P_2 pools. Biochem J 398:1–13

Scales SJ, Scheller RH (1999) Cell biology-lipid membranes shape up. Nature 401:123–124

Smith LG, Oppenheimer DG (2005) Spatial control of cell expansion by the plant cytoskeleton. Annu Rev Cell Dev Biol 21:271–295

Stephens LR, Anderson K, Stokoe D, Erdjument-Bromage H, Painter GF, Holmes AB, Gaffney PR, Reese CB, McCormick F, Tempst P, Coadwell J, Hawkins PT (1998) Protein kinase B kinases that mediate phosphatidylinositol 3,4,5-trisphosphate-dependent activation of protein kinase B. Science 279:710–714

Stokoe D, Stephens LR, Copeland T, Gaffney PR, Reese CB, Painter GF, Holmes AB, McCormick F, Hawkins PT (1997) Dual role of phosphatidylinositol-3,4,5-trisphosphate in the activation of protein kinase B. Science 277:567–570

Suh BC, Hille B (2005) Regulation of ion channels by phosphatidylinositol 4,5-bisphosphate. Curr Opin Neurobiol 15:370–378

Takenawa T, Itoh T (2001) Phosphoinositides, key molecules for regulation of actin cytoskeletal organization and membrane traffic form the plasma membrane. Biochim Biophys Acta 1533:190–206

Testerink C, Munnik T (2005) Phosphatidic acid: a multifunctional stress signaling lipid in plants. Trends Plant Sci 10:368–375

Topham MK (2006) Signaling roles of diacylglycerol kinases. J Cell Biochem 97:474–484

van Leeuwen W, Okresz L, Bogre L, Munnik T (2004) Learning the lipid language of plant signaling. Trends Plant Sci 19:378–384

Vincent P, Chua M, Nogue F, Fairbrother A, Mekeel H, Xu Y, Allen N, Bibikova TN, Gilroy S, Bankaitis VA (2005) A Sec 14p-nodulin domain phosphatidylinositol transfer protein polarizes membrane growth of *Arabidopsis thaliana* root hairs. J Cell Biol 168:801–812

Wang X (2005) Regulatory functions of phospholipase D and phosphatidic acid in plant growth, development, and stress responses. Plant Physiol 139:566–573

Wang X, Devaiah SP, Zhang W, Welti R (2006) Signaling functions of phosphatidic acid. Prog Lipid Res 45:250–278

Wu L, Bauer CS, Zhen XG, Xie C, Yang J (2002) Dual regulation of voltage-gated calcium channels by PtdIns(4,5)P_2. Nature 419:947–952

Yamaguchi T, Tanabe S, Minami E, Shibuya N (2004) Activation of phospholipase D induced by hydrogen peroxide in suspension-cultured rice cells. Plant Cell Physiol 45:1261–1270

Yin HL, Janmey PA (2003) Phosphoinositide regulation of the actin cytoskeleton. Annu Rev Physiol 65:761–789

Zarsky V, Fowler J (2008) ROP (Rho-related protein from Plants) GTPases for spatial control of root hair morphogenesis. In: Emons AMC, Ketelaar T (eds) Root hairs: excellent tools for the study of plant molecular cell biology. Springer, Berlin Heidelberg New York. doi:10.1007/7089_2008_14

Zonia L, Munnik T (2006) Cracking the green paradigm: functional coding of phosphoinositide signals in plant stress responses. Subcell Biochem 39:207–237

ROP (Rho-Related Protein from Plants) GTPases for Spatial Control of Root Hair Morphogenesis

V. Žárský(✉) and J. Fowler

Abstract Cell polarity control is inherently a complex process based on the feedback loops where it is difficult to distinguish cause–effect relationships and identify "master regulators." However, small GTPases from the Rac/Rho family are certainly an important part of polar growth core regulatory circuit/network also in plants. ROPs (Rac/Rho of plant) involvement in root hair polar tip-growth is best documented by the loss of polarity in plants overexpressing specific ROP GTPases, loss of polarized cell expansion in root hairs of RopGDI mutant and root hair tip localization of ROP-GFP fusions. Rho/Rac GTPases switch directs cell growth via a plethora of regulatory and effector proteins known from Opisthokonts, and only recently some of them have been identified also in plant cells. We will review and discuss their root hair function in relation to the cytoskeleton dynamics, Ca^{2+} gradient, NADPH oxidase activity/ROS (reactive oxygen species), cell wall dynamics, phospholipid signaling, auxin signaling, and polarized/targeted secretory vesicle transport and fusion.

1 Introduction

Tip growth of plant cells relies upon an intricate and dynamic network of processes connecting cell wall dynamics/mechanics with signal transduction, membrane lipid modifications, changes in ion transport, and regulation of the secretory pathway and actin cytoskeleton (recently termed the LENS for Localization Enhancing Network, Self-sustaining; Cole and Fowler 2006). Current data suggest that the LENS functions at the growing tip in both root hairs and pollen tubes, and comprises many homologous components in the two cell types. For this chapter it is necessary to

V. Žárský
Department of Plant Physiology, Faculty of Sciences, Charles University, Vinicna 5, 128 44 Prague 2, Czech Republic and Institute of Experimental Botany, Academy of Sciences of the Czech Republic, Rozvojova 135, 165 00 Prague 6, Czech Republic
e-mail: zarsky@ueb.cas.cz

Plant Cell Monogr, doi:10.1007/7089_2008_14

separate out one of the LENS components – ROP GTPase signalling – from the totality of this system; it makes sense partly because of the relatively central position the ROPs (Rho-related protein from plants) occupy in functioning of the LENS.

The evolution of regulatory circuits in eukaryotic cells selected the GTP nucleotide to participate in a mechanistically distinct fashion from the closely related ATP nucleotide. Whereas ATP is involved in signal transduction pathways by covalent modification of target substrates (i.e., phosphorylation), noncovalent binding of GTP to a great array of small GTPase proteins imposes conformational changes to those proteins, making such *GTPases ideal molecular switches* for a plethora of regulatory and signal transduction pathways. Low intrinsic GTPase activity (explaining why these proteins are also called GTP-binding proteins) allows a relatively long life time for their active (GTP-bound) conformation. Generally, only after the intervention of a GTPase-activating protein (GAP) is the γ-phosphate of GTP molecule hydrolyzed, returning the GTPase to the inactive, GDP-bound state.

Along with the founding member of the superfamily, the Ras proto-oncogene, most small GTPases are posttranslationally modified at their C-terminus by the addition of hydrophobic prenyl (farnesylation or geranylgeranylation) or acyl moeities (palmitoylation), thus functioning as peripheral membrane proteins that cycle between membranes and cytoplasm. This cycling is regulated by GDP dissociation inhibitors (GDIs), which are thought to extract GDP-bound GTPases from the target membranes for possible recycling through GDP/GTP exchange and reactivation. This exchange is catalyzed by guanine nucleotide exchange factors (GEFs; for a review on small GTPases, see e.g. Molendijk et al. 2004). The specificity with which these proteins can be localized to particular membranes, and their ability for local activation/deactivation make small GTPases an ideal device to define specific membrane domains, helping to initiate and maintain vectorial/targeted processes of cell morphogenesis and signalling within the cell (e.g., Zerial and McBride 2001).

Over the last years it has become obvious that small GTPases are crucial components of the eukaryotic cell morphogenetic machinery not only in animals and fungi, but also in plants. They are divided into several well-defined (and some less-defined) families or subfamilies, and for our perspective on the functions of plant Rho-related GTPases (ROPs) in root hair initiation and polar growth, it is necessary to mention small GTPases of the Arf and Rab families, in addition to the heterotrimeric GTPases (see other chapters of this volume). Current data and models suggest that they are collectively involved in the regulation of the secretory pathway (e.g., the identity of endomembrane compartments, formation and transport of vesicles and fusion with the target membrane, and membrane recycling), actin cytoskeleton dynamics, cell wall synthesis, activity of membrane-bound enzymes and transporters (e.g., ion channels and NADPH oxidases), signal transduction pathways, and possibly some metabolic pathways. All of these activities also bear on the development of root hairs in Angiosperm plants.

In this chapter we will focus on the ROP GTPase family (also sometimes called RAC), as it is more and more obvious from several model eukaryotes that Rho-type GTPases play particularly important roles in the molecular machines and networks orchestrating cytoskeletal, secretory, and signal transduction pathways from within

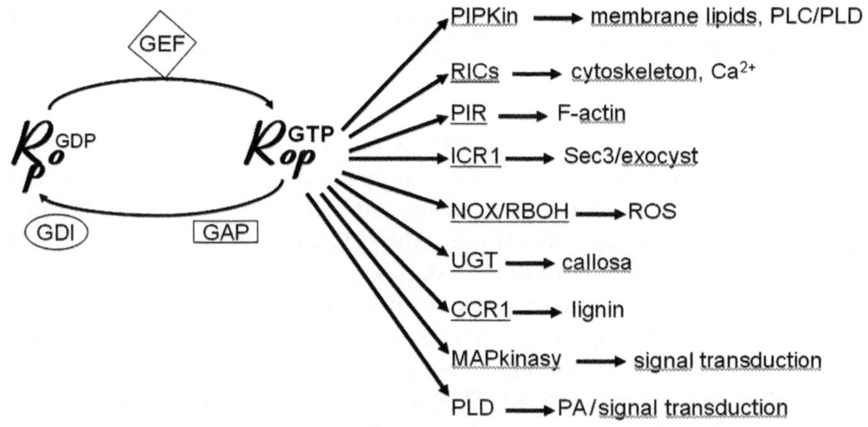

Fig. 1 Regulators and effectors of ROP GTPases (for details – explanation of names – see the text)

the cell to achieve morphogenetic processes. In case of root hairs, we are challenged with the problem to understand polarized cell morphogenesis that occurs in two steps: (1) growth site selection and bulge formation (planar polarization) in trichoblast cells and (2) initiation and maintenance of tip growth. We will first summarize current understanding of the mechanisms of ROP regulation by GEFs, GAPs, and GDIs (Fig. 1), as well as our fragmentary knowledge of ROP downstream effectors. However, our main focus will be on ROP functions in the two major phases of root hair ontogenesis defined in this paragraph.

2 The ROP GTPase Family

The ROP group of plant-specific GTPases diverged from the closely related Rac GTPases, the most basal subgroup of Rho-type GTPases, most probably in algal predecessors before plants conquered the land (Brembu et al. 2006). Phylogenetic analysis suggests that the plant ROP genes underwent rapid evolution prior to the emergence of the embryophyta, creating a group that is distinct from the Rac genes in other eukaryotes. In embryophyta, ROP genes have undergone an expansion through gene duplications (Winge et al. 2000; Christensen et al. 2003; Brembu et al. 2006).

There are two major subfamilies within the ROP family, each with distinct structures, localization mechanisms, and (apparently) functions. The primary structural distinction (and likely the basis for many of the functional differences) between the two lies at the protein C-terminus, affecting their posttranslational modification. Subfamily I (or type I) ROPs (represented in *Arabidopsis thaliana* by ROP1 through ROP8) contain the conventional C-terminal motif (present in the majority of nonplant Rho GTPases) for prenylation. Subfamily II (or type II) ROPs (represented in *A. thaliana* by ROP9 through ROP11) are not prenylated, but rather palmitoylated at a cysteine-containing motif that appears to be a plant-specific innovation (Ivanchenko et al. 2000; Lavy et al. 2002; Lavy and Yalovsky 2006).

Major progress in the ROP stucture–function understanding was achieved recently by resolving the crystal structure of GDP-bound ROP9 protein (Sormo et al. 2006). It was known that plant ROPs not only have slightly different protein composition, but also have evolved different modes of regulation when compared to other eukaryotic Rho homologues (Zheng and Yang 2000). Amino acid residues participating in the interactions between Rho GTPases and GAP, GDI and CRIB domains are well conserved in ROPs. As expected, also GTP-binding domain is very conserved, but important structural differences, compared with other Rho proteins, were described in the insert and switch II regions of plant ROP GTPase (switch I region is well conserved between plant and other Rho proteins). Rho-specific insert region of ROP GTPases is smaller as it lacks two or four amino acids. In switch I region there is one of the discriminatory amino acid positions between group I and group II ROP s – while in group I there is a threonine residue at position 30 (Thr30), in group II ROPs there is a lysine residue at this position (Lys30). In switch II region serine 68 residue in ROPs (compared to aspartic acid at the same position in human Rho/Rac GTPases) increases flexibility of the whole region and possibly facilitates novel interaction/regulation/signal transduction mechanisms for Rops in plants.

Localization of ROPs to the plasma membrane (PM) is dependent on the C-terminal hydrophobic modifications, and over the past few years great progress has been achieved in understanding how these modifications affect localization, particularly by the Yalovsky group. Type-II ROPs are membrane localized by virtue of palmitoylation on two or three cysteines in the C-terminal hypervariable region and a proximal polybasic protein domain (Lavy et al. 2002; Lavy and Yalovsky 2006). However, the *Arabidopsis* type-II ROP9 and ROP10 are not localized to the root hair PM, leading to a focus on type-I ROPs (as exemplified by *Arabidopsis* ROP2) as the major players in organization of the root hair tip (Lavy et al. 2002; Jones et al. 2002). Surprisingly, the type-I ROP6 GTPase (and possibly other type-I ROPs) is also transiently acylated by palmitic and stearic acid when in the active, GTP-bound conformation (Sorek et al. 2007). This modification stabilizes their membrane localization and induces partitioning into detergent-resistant membrane (DRM) domains (also called lipid rafts; Sorek et al. 2007); the wider relevance of this mechanism is supported by the observation that type-I ROPs are uncovered also in the tobacco DRM proteome (Morel et al. 2006). The *tip1* mutant in *Arabidopsis*, which acts early in the bulge and root hair formation pathway (Parker et al. 2000), has disrupted *S*-acyl transferase activity (Hemsley et al. 2005) and thus might affect ROP palmitoylation and DRM partitioning. Recently, Jones et al. (2006) performed an interesting transcriptomic comparison of *Arabidopsis* root hair differentiation primary root zone between WT and root-hair-less mutant rhd2, allowing to pin-point genes expressed specifically in root hairs. This analysis uncovered clear-cut overrepresentation of lipid raft markers – e.g., GPI-anchored proteins – in root hairs; furthermore, two mutants in such proteins are associated with very short, bursting root hairs (Jones et al.

2006). It is possible that activated type-I ROPs are recruited into and help organize DRM domains at the tip of elongating root hairs.

3 ROP Regulators: The Roles of GDIs, GAPs, and GEFs in the Regulation of Specific PM Domains

3.1 GDP Dissociation Inhibitors

RhoGDIs are evolutionarily well conserved, and have very little diversity when compared with other Rho regulators or Rho GTPases themselves. They are fairly promiscuous in their interactions with Rho family members, apparently acting as universal cytosolic chaperones for the GDP-bound form of the GTPase (reviewed in DerMardirossian and Bokoch 2006). Until recently, they were considered as mere housekeepers, functioning to assist in the even distribution of GTPases to membranes. New data however open a different perspective on GDIs, in which, by specific interaction with receptors or displacement factors (GDFs; inducers of GTPase/GDI dissociation) or through regulation by phosphorylation, they actively contribute to the targeting of Rho proteins to specific subcellular membranes (Dransart et al. 2005; Carol et al. 2005; Klahre et al. 2006a).

For mammalian Rho GTPases, the binding sites for RhoGDIs and GEFs overlap, and thus only one partner can be accommodated at a time – in this way, RhoGDI might efficiently inhibit (and so regulate) GEF-stimulated activation of Rho (Dransart et al. 2005). An attractive hypothesis proposed by Dransart et al. (2005) and reviewed in Dovas and Couchman (2005) suggests that RhoGDI–Rho GTPase complexes associate transiently with localized multiprotein-signalling complexes that lead to dissociation and activation of Rho, thus controlling local abundance of active Rho and possibly establishing feedback loops in coordination with Rho effectors, as possible components of the signalling complex. Displacement factors (GDFs) – functioning as quasi-GDI receptors and possible members of such signalling complexes – have only recently been discovered in mammals, and have not yet been described for Angiosperms. Current data from mammals suggest that they are a diverse set of molecules, acting in different compartments (reviewed in DerMardirossian and Bokoch 2006), and thus, plant GDFs (if they exist) may also be diverse and specific for particular membrane domains and functions. In plants, overexpression of an *Arabidopsis* RhoGDI in tobacco pollen tubes inhibits activated ROP-induced "ballooning"/depolarized growth (Fu et al. 2005), consistent with the idea that plant RhoGDIs can limit the abundance of ROPs at the PM. More recent data suggest that a tobacco RhoGDI provides a crucial function that allows ROP to localize to the pollen tube tip (Klahre et al. 2006a). Consistent with this idea that RhoGDIs assist in the polarization of active ROP at the PM, Carol et al. (2005) showed that loss of a particular RhoGDI via the *supercentipede* (*scn1*) mutant in *Arabidopsis* leads to formation of multiple bulges in root hair cells, correlated with ectopic localization of ROP2 in the bulges.

3.2 GTPase-Activating Proteins

Unlike RhoGAPs in other eukaryotes, plant RhoGAPs characterized to date (in *Arabidopsis*, five genes) are equipped with a CRIB (Cdc42/Rac-interactive binding) domain, which not only assists GAP-ROP binding, but is also involved in the GAP activity itself (Wu et al. 2000). In pollen tubes, the domain of active ROP at the tip is limited by a RhoGAP localized subapical to the tip, where it appears to be regulated by interactions with a 14-3-3 protein (Klahre and Kost 2006b). Recently, also NOX interaction with 14-3-3 proteins was demonstrated by Elmayan et al. (2007) opening up a new putative link between ROP signalling module and NOX (see further). Although no data have yet directly addressed whether RhoGAPs play a similar role in root hairs, our analysis of the root hair transcriptome data generated by Jones et al. (2006) suggests candidates for such a function. A partial RopGAP-like mRNA signal (At2g27440; predicted as a pseudogene in TAIR) is twofold increased in root hairs, suggesting that it is in fact an expressed locus and that this partial RhoGAP may have specific functions in the root hair.

3.3 Guanine Nucleotide Exchange Factors

The first candidate for a RopGEF was identified within the context of genetic analysis of leaf trichome development: the *spike1* mutant of *Arabidopsis* is seedling lethal and forms simple nonbranched trichomes (Qiu et al. 2002). *SPIKE1* encodes a member of a new family of RhoGEFs (the CZH family – CDM-Zizimin-homology; Meller et al. 2005), although evidence for GEF activity for SPIKE1 has not yet been published.

A major breakthrough in our understanding of ROP signalling was the discovery of a new class of plant-specific GEFs (encoded by 14 genes in *Arabidopsis*) that activate ROP (Berken et al. 2005; Gu et al. 2006). Now designated the PRONE (plant-specific ROP nucleotide exchanger) family, based on a conserved RopGEF domain (Berken et al. 2005), the family was actually first described in tomato, based on recovery of one member (KPP, kinase partner protein) as an interactor with the cytoplasmic domain of the pollen-specific receptor-like kinases LePRK1 and LePRK2 (Kaothien et al. 2005). As expected for their biochemical activity, overexpression in pollen tubes of either KPP (Kaothien et al. 2005) or one of several *Arabidopsis* PRONE genes (Gu et al. 2006) causes depolarized growth. A most exciting possibility, implied by the KPP–PRK interaction, is that this interaction regulates GEF activity, perhaps by releasing it from the autoinhibitory domain at the PRONE family C-terminus (Gu et al. 2006). Furthermore, given that ROP has been recovered in a complex with another RLK (CLV1; Trotochaud et al. 1999), RopGEF–RLK interactions could be a general mechanism to activate ROP signalling in multiple contexts, including in the root hair. Even if they are unrelated to other known GEFs, it was shown that the basic mechanism of GEF catalysis is similar as in other Rho GEFs (Thomas et al. 2007).

Jones et al. (2006) reported 29 differentially overexpressed RLKs in root hairs; one of these root-hair-specific RLKs causes short, straight root hairs when mutated (At4g18640). Our analysis of the same data set identified two upregulated PRONE RopGEFs: RopGEF11 (At1g522540) shows at least a threefold increase, whereas RopGEF10 (At5g19560) is elevated more that 11-fold in root hairs. Upregulation of these RopGEFs and RLKs (along with RICs and PLDζ, see later) might be a crucial piece of a competence mechanism, making trichoblast cell files sensitive to an auxin/ethylene signal that initiates and positions root hair morphogenesis, in contrast to nontrichoblast cell files.

4 ROP Effectors: RICs (Actin and Calcium), PIP Kinase (PLC and PLD), NADPH Oxidase (ROS), and ICR1–Exocyst

4.1 RICs, Arp2/3, and Actin

It is well established that highly dynamic fine F-actin cytoskeleton configurations are important in the transition from bulge formation into root hair tip growth proper, where actin is a central constituent of the LENS (Miller et al. 1999; Ketelaar et al. 2003). ROP GTPases are crucial for proper actin organization: in transgenic plants expressing a constitutively active (always GTP-bound) mutant of ROP2 (CA-rop2), the extra-wide root hairs contain many actin cytoskeleton bundles throughout, apparently taking the place of the fine actin network at the growing tip in wild type; in plants expressing a dominant negative mutant (DN-rop2), actin cables protrude into the extreme apex of the shorter root hairs, and the tip-localized fine actin meshwork is absent (Jones et al. 2002). Similar phenomena in pollen tubes have now been attributed, at least in part, to RICs (ROP-interactive CRIB-motif-containing proteins) as downstream effectors of ROP GTPases (reviewed in Smith and Oppenheimer 2005). Specifically, RIC4 regulates actin at the pollen tube tip, whereas RIC3 provides counter-acting regulation of the tip-high Ca^{2+} gradient. The first ROP effectors to be described, RICs are relatively small proteins, comprising 11 members of diverse sequence in *Arabidopsis* that contain no recognizable features other than the CRIB domain (Wu et al. 2001). Again, there are no experimental data available on RICs interacting with ROP to fulfil tasks specific for root hair development, although we expect mechanisms similar to the pollen tube to operate during tip growth. The transcriptomic data of Jones et al. (2006) can once more provide candidates: both RIC1 (At2g33460) and RIC10 (At4g04900) are several times more expressed in the root hair zone. (For possible cross-talk with the microtubular cytoskeleton, see the chapter by Sieberer and Timmers in this volume.)

Analysis of "distorted"-type trichome mutants in *Arabidopsis* indicates that the Arp2/3 complex, which provides branched actin nucleation activity, is required for wild-type cell morphogenesis in many plant cell types (reviewed in Smith and Oppenheimer 2005). Furthermore, Arp2/3 activity is regulated by the SCAR/WAVE complex, which appears to contain the SPIKE1 putative GEF, as well as SCAR proteins that are direct targets of activated ROPs (Uhrig et al. 2007). However, in contrast to leaf trichomes, the participation of Arp2/3 and SCAR/WAVE in root hair development seems minimal or nonexistent, as mutants in the constituent proteins have either no effect or only impact root hairs under rapid growth conditions (e.g., mutants in the *ARP2* and *ARP3* genes – Mathur et al. 2003). This implicates other actin nucleators – for example, formins (Yi et al. 2005; Deeks et al. 2005) – as likely major players in the root hair, although the expected regulatory link to activated ROPs is unknown in plants (see also the chapter on actin by Ketelaar and Emons in this volume).

4.2 Reactive Oxygen Species and NADPH Oxidases (NOX); Phosphoinositides, PIP Kinases, and Phospholipases

NADPH oxidases (NOX and their activity-dependent charge transport and reactive oxygen species (ROS) production) belong among the best characterized effectors of Rac GTPases in animal cells, and the same is true in plants (Foreman et al. 2003; Carol et al. 2005; Jones and Smirnoff 2006; reviewed in Carol and Dolan 2006), despite the fact that the nature of the ROP–NOX regulatory link is not known. Activated ROP stimulates NOX activity and ROS production not only in root hairs (Carol et al. 2005; Jones and Smirnoff 2006), but also during plant–pathogen interaction and leaf development (Ono et al. 2001; Park et al. 2004). In root hairs, mutants in *root hair defective 2* (*RHD2*)/*AtrbohC* can suppress growth phenotypes associated with ROP overactivation (either by loss of RhoGDI/SCN1 activity or expression of CA-rop2), indicating that RHD2/AtrbohC is the major NOX target of ROP in this cell type (Carol et al. 2005; Jones and Smirnoff 2006). However, initiation bulges form in the *rhd2* mutant, indicating that its activity is not required for the earliest stages of root hair development; rather, ROP-RHD2/AtrbohC activation is associated with the tip growth phase (reviewed in Carol and Dolan 2006). Because plant NOX proteins are endowed with two calcium-binding EF hands at their N-terminus, RHD2/AtrbohC could also be integrated into LENS signalling via the tip-high calcium gradient (reviewed in Sagi and Fluhr 2006).

Ohashi et al. (2003) uncovered phospholipase D (PLD)ζ1 as one of the targets of the negative transcriptional regulator GLABRA2 in non-root-hair cells. Furthermore, ectopic expression of PLDζ1 (and presumably concomitant generation of its product, phosphatidic acid (PA)) initiates ectopic bulge formation in non-root-hair cell files (Ohashi et al. 2003). As PA was shown to coactivate ROP-mediated ROS production in a cell death response (Park et al. 2004), we speculate that PA derived from PLDζ1 might similarly coactivate ROP-mediated ROS production (e.g., through RHD2/Atrboh2) during root hair initiation or tip growth. As in animals, ROP might stimulate PLD and PLC activity via phosphoinositol 4,5-biphosphate (PIP2) synthesis by direct activation of phosphatidylinositol 4-phosphate 5-kinase (PIP5K; Kost et al. 1999), which may also be stimulated by PA (reviewed in Oude Weernink et al. 2007). This reciprocal stimulation of PLD and PIP5K has been hypothesized to generate rapid feed-forward loops for localized and explosive generation of PA and PIP2, which may then govern the recruitment and activation of proteins to a membrane domain to execute specific tasks, e.g., membrane-trafficking and actin dynamics (reviewed in Oude Weernink et al. 2007). The very tip of growing root hairs and pollen tubes could be such a domain, as PM in this region is PIP2-enriched (Vincent et al. 2005; Kost et al. 1999); type-I ROPs show similar localization patterns (reviewed in Yang 2002; see also the chapter on phospholipids by Aoyama in this volume). Thus, known mechanisms in plant cells are capable of building a specific PM domain based on reciprocal activation of PIP5K, PLD/PA, and ROP activities. As NOX-produced ROS are able to stimulate calcium influx (Foreman et al. 2003), and calcium is able to stimulate NOX activity, it is reasonable to assume a contribution of this positive feedback mechanism to the creation of the tip-focused calcium gradient (Zarsky et al. 2006; Potocky et al. 2007). We hypothesize that both these positive feedback loops help define the tip growth domain, within the framework of the LENS.

4.3 ICR1–Exocyst Complex

The exocyst is an octameric effector complex of both Rho and Rab GTPases, integrating signals at the final stages of polarized secretory pathways in yeast and animal cells (reviewed in Novick et al. 2006; Hsu et al. 2004). We (Cvrckova et al. 2001; Elias et al. 2003; Cole et al. 2005; Synek et al. 2006; Hála et al. 2008) and others (Wen et al. 2005) have accumulated evidence that the exocyst is active in regulation and development of polarity in plants as well. This includes growth of the root hair tip, which is aberrant in *exo70a1* (Synek et al. 2006) and *sec8* (Cole, Kulich, Zarsky, and Fowler, unpublished) mutants. Now a crucial observation made by the Yalovsky group (Lavy et al. 2007) provides the first link from GTP-charged ROP to the SEC3a subunit of the exocyst via the scaffold protein ICR1. In contrast to the situation in fungi and animals, neither type-I nor type-II ROPs interact directly with SEC3a, but do so indirectly via ICR1 at the PM (Lavy et al. 2007). Both *icr1* mutants and *ICR1-RNAi* plants have abnormal roots, root meristems, and adaxial leaf cell shapes; furthermore, double mutant plants *icr1* x CA-rop6 also indicate that formation of rectangular leaf cells by CA-rop6 overexpression is dependent on ICR1, placing the protein downstream of ROP (Lavy et al. 2007). Although *icr1* plants produce apparently wild-type root hairs, overexpression of ICR1 induced aberrant, swollen root hairs, similar to the root hair phenotype caused by activated ROPs (Lavy et al. 2007). Thus, an ROP–ICR1–exocyst linkage may be important in root hairs, obscured in *icr1* plants by redundancy in the ICR family.

The functions of the exocyst, although originally related almost exclusively to exocytosis, are now being put into the broader context as it is now clear in animals that the exocyst functions in membrane recycling via recycling endosome (reviewed in Somers and Chia 2005). In yeast it is the secretory-vesicle-specific Rab GTPase

(Sec4p) known to directly interact with the exocyst, whereas in animals cells it is the recycling endosome-specific GTPase, Rab11 (reviewed in Novick et al. 2006; Somers and Chia 2005). In *Arabidopsis* cell extracts, we find that both ROP and RabA (closely related to Rab11) to some extent coelute with the exocyst in partially purified high molecular weight fractions (Hala, Cole, Synek, Drdova, Fowler, and Zarsky, in preparation).

Intriguingly, localized exocytosis, coupled with endocytosis and recycling, can be a mechanism to maintain dynamic polarization of membrane proteins, if diffusion to equilibrium distribution in the membrane is slow (Valdez-Taubas and Pelham 2003). Such domains of localized exocytosis/rapid recycling may be found in plant cells during cytokinesis, as well as at the tips of root hairs and pollen tubes (Baluska et al. 2005; Ovecka et al. 2005; Dhonukshe et al. 2006). Given that overexpression of an activated type-II ROP (CA-rop11) blocks membrane recycling in *Arabidopsis* root hairs (Bloch et al. 2005), it seems likely that ROPs will also be involved in the regulation of membrane recycling (The graphic presentation of known ROP effectors is shown in Fig. 1)

5 Speculative Model of ROP Function at the Growth Domain Selection and Bulge Formation

The first markers of root hair initiation to be discovered were cell wall proteins – xyloglucan endotransglycosylase (XET; Vissenberg et al. 2001) and expansins (Baluska et al. 2000) – that accumulate locally at the surface-facing basal cell pole in *Arabidopsis* trichoblasts. This domain of accumulation predicts the site of future bulge formation and root hair growth, and is assembled independent of both the microtubule and actin cytoskeleton, based on inhibitor studies (Baluska et al. 2000; Vissenberg et al. 2001). Both XET and expansins are well-represented in the root hair transcriptome (Jones et al. 2006). This implicates localized changes in the PM, cell wall, and possibly cortical secretory pathway (including membrane recycling) as primary targets for the signals that specify root hair position (Baluska et al. 2000; Vissenberg et al. 2001; Fischer et al. 2006; see also the chapter on cell wall by Nielsen in this volume).

It now seems obvious (Molendijk et al. 2001; Jones et al. 2002, 2007; Carol et al. 2005; Fischer et al. 2006) that another very early event of root hair initiation in trichoblast cell files is local activation of ROP GTPases in what appears to be this same basal domain of the trichoblast PM, well before bulge formation. The timing of this event is about the same as the localization of expansins and XET mentioned earlier. Jones et al. (2002, 2006) hint that, in *Arabidopsis*, there are at least three ROP paralogues (with overlapping functions) that participate in the root hair morphogenesis; no single ROP GTPase knock-out has a disturbed root hair development (Jones and Smirnoff 2006). Current data indicate that ROP2 (Jones et al. 2002), ROP4, and ROP6 (Molendijk et al. 2001) are the best candidates, although expression in trichoblasts has only been confirmed for *ROP2*. Signal from both

GFP::Rop2 GTPase and an antibody raised against ROP4 (which likely recognizes other ROPs) is clearly focused to the position of future bulge formation, which persists and becomes focused to a smaller domain during the initiation of root hair tip growth proper (Molendijk et al. 2001; Jones et al. 2002; Carol et al. 2005; Fischer et al. 2006). Overexpression of ROP2 induces ectopic multiple bulges on the rhizodermis, as well as formation of branched root hairs (Jones et al. 2002); as mentioned earlier, this is similar to the phenotype of *scn1/AtRhoGDI1* mutant (Carol et al. 2005). All this implies that localized ROP activity is crucial for defining the site of bulge formation.

What might be the trigger of this precisely targeted ROP activation and localization? To get this specific localization (planar polarization), it seems reasonable to assume that it is a result of intersection of several independent positional signals which certainly include or are enhanced by auxin and ethylene as positive factors (Masucci and Schiefelbein 1996; Fischer et al. 2006, 2007). Indeed, recently Fischer et al. (2006) convincingly demonstrated that auxin (based on the effect of *aux1* lesion, which could be further enhanced by the ethylene signalling mutant *ein1*) is a dominant signal in positioning of ROP and initiation bulges at the base of the trichoblast (see also the chapter on IAA by Lee and Cho in this volume). Furthermore, ROP localization during root hair initiation and bulge positioning is sensitive both to BFA treatment and to reduced function of the BFA-sensitive GNOM ARF GEF, indicating that it is dependent on the ongoing secretory, endocytic, and/or endosome recycling pathways (Molendijk et al. 2001; Fischer et al. 2006, 2007). This also implies that some part of the secretory/endocytic apparatus is polarized well before the visible bulge formation commences. Random, but still either apical or basal, positioning of root hairs in *aux1*, *ein1*, *gnom* triple mutants (Fischer et al. 2006) implies that a cell-junction/"corner" effect can have a default role in providing positional information for this process (Baluska and Hlavacka 2005).

What is then the crucial polarizing signal/landmark, and how is it related to the membrane trafficking pathways? Here we are left with the necessity to speculate – but, owing to recent developments in our knowledge of ROP signalling, these speculations can be at least partly based on pieces of puzzle already known. As stressed earlier, current evidence points to the dominant function of polarized auxin transport, along with ethylene and possible "cell junction" signals, in ROP localization and root hair initiation. Auxin treatment can stimulate activation of ROP in whole seedlings (Tao et al. 2002). Given that PRONE domain RopGEFs interact with (and are possibly regulated by) RLKs (Kaothien et al. 2005), we speculate that some members from the plethora of RLKs present in trichoblasts are sensitive to auxin – i.e., serve as a bona fide auxin receptor on the cell surface. In this scenario, polarized auxin maxima at the bottom of trichoblasts, accompanied by an ethylene signal, might be part of the landmark for root hair initiation by directly activating this hypothetical new type of auxin receptors – IAA-sensitive RLKs; these in turn directly locally activate ROPs via interacting PRONE RopGEFs. As CA-rop11 was able to shut down membrane cycling, including endocytosis (Bloch et al. 2005; this might in fact be part of the mechanism of recently observed IAA-induced endocytosis inhibition (Paciorek et al. 2005)), local endocytosis inhibition might result in a transient stimulation of exocytosis, leading to initiation of bulge formation. Local acidification and relaxation of the cell wall (an effect of localized expansin and XET) might also stimulate cell-wall-interacting RLKs and provide additional local positive feedback to ROP. Additionally, active ROP promotes its own polarized localization, at least in pollen tubes (Gu et al. 2003); a similar mechanism in trichoblast would further enhance positive polarization feedback at the bulge domain. Thus, minimal local activation of ROP by auxin and mechanical stimuli (via RLKs and PRONE GEFs) would be able to switch on an amplifying positive feed back process, resulting in the strongly localized polarization of the actin cytoskeleton and secretory pathway. As suggested by Fischer et al. (2006), an equally plausible scenario would be that nonlocalized auxin/ethylene signalling is superimposed on a weak, preexisting polarity signal, amplifying the weak positional signal, and eventually setting up a similar RLK/RopGEF/ROP positive feedback loop. For instance, auxin/ethylene could stimulate cell-junction-regulated RLKs above some threshold level of activation, leading to polarization of ROP through a RopGEF. This hypothetical model of role for IAA in the activation of ROP GTPase domain before and during bulge formation is represented in Fig. 2.

Local activation of a Rho GTPase is enough to polarize the secretory pathway in yeast (Roumanie et al. 2005). How might locally activated ROP be similarly

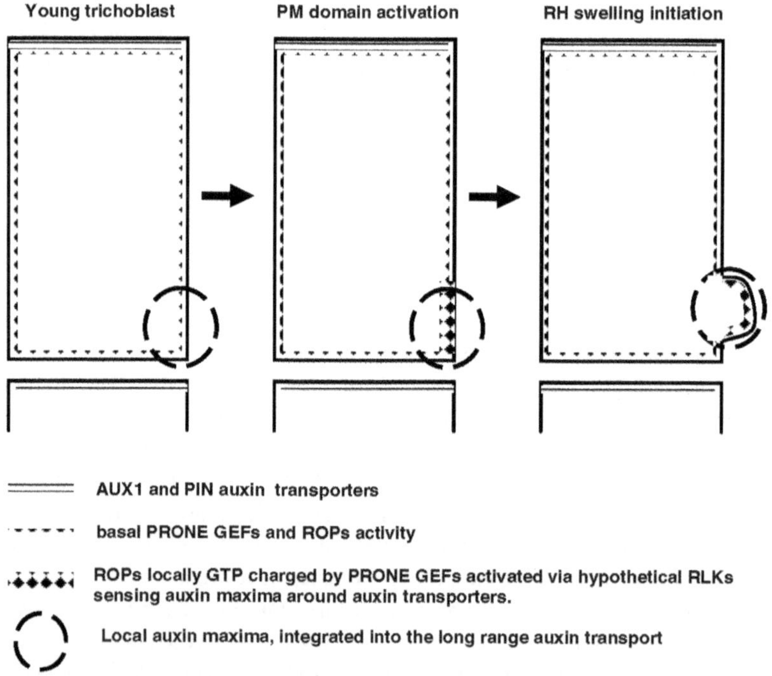

Fig. 2 Speculative hypothesis on the localization and initiation of root hair formation by the activation of ROP via PORNE GEFs stimulated by hypothetical auxin-sensing RLKs

translated into the polarized secretion necessary for bulge formation, as well as for sustained root hair growth in plants? Activated ROP will have as first targets membrane-localized or proximal proteins, for example, PIP5K (Kost et al. 1999). As hypothesized earlier, stimulation of local PIP2 accumulation could directly activate PLD and PA formation, which in turn could feed back to PIP5K/PIP2, as well as further ROP activation, creating self-sustaining membrane domain. Local stimulation of ion (including proton) transporters could also play a role, further stimulating expansin- and XET-assisted cell wall relaxation (Mathur 2006). This locally relaxed cell wall would yield to turgor pressure, resulting in the initial bulge formation and opening of the cell wall (Baluska et al. 2000; Miller et al. 2000; Vissenberg et al. 2001; Carol and Dolan 2002; Mathur 2006).

To achieve a sharp boundary for this activated domain, inhibitory controls exerted by RopGAP and RhoGDI proteins seem inevitable (Carol et al. 2005; Klahre et al. 2006a,b). SCN1/AtRhoGDI1 controls both the position and the shape of RH outgrowths by controlling the RHD2/AtrbohC activity, apparently through ROP (Carol et al. 2005; Carol and Dolan 2006). The fact that RHD2 NOX activity is necessary for formation of multiple bulges in the *scn1* mutant, but is not necessary for the single, correctly positioned bulge in either the *rhd2* or *scn1*, *rhd2* double mutant, suggests that either NOX activity does not play a role in the initial selection process, or a strong, site-specific mechanism can overcome lack of NOX activity in this process.

6 ROPs in Initiation and Maintenance of Root Hair Tip Growth

Creation of the initiation bulge through activation of an RLK/RopGEF/ROP positive feedback loop by auxin and/or mechanical stimuli might in itself be able to amplify to a point at which polarization of cytoskeleton and secretory pathway is focused tightly, allowing initiation of root hair tip growth proper. In the models of root hair tip growth, a central function is attributed to both the actin and microtubular cytoskeletons, with actin important for sustaining polarized growth, and microtubules helping maintain a straight orientation (Bibikova et al. 1999; Baluska et al. 2000; Ketelaar et al. 2003; Sieberer et al. 2005). It seems likely that actin is a primary target of ROP-RIC regulation during this phase of development, in which the root hair tip growth LENS plays the defining role (reviewed in Cole and Fowler 2006). Similarly, the self-sustained domain of ROP-phospholipid kinases/PIP2-phospholipases/PA may act in the LENS to define the PM domain associated with the growing tip. Although ROP-regulated NADPH oxidases are apparently not involved in bulge formation, RHD2/AtrbohC is required to sustain root hair growth itself, possibly forming a regulatory feedback loop with calcium channels to maintain the tip-high gradient of cytoplasmic calcium (Foreman et al. 2003; Carol and Dolan 2006; Potocky et al. 2007). Finally, the activity of these signalling components may also oscillate in phase with growth of the tip, as has been demonstrated for ROP activity in the pollen tube (Hwang et al. 2005).

In light of recent progress in understanding the role of the exocyst in plant cell polarity, particularly given the linkage of the SEC3 subunit to activated ROP at the PM via ICR1 (Lavy et al. 2007), and a ROP function in regulating membrane recycling (Bloch et al. 2005), we will emphasize in our model the activation of a localized secretory/recycling pathway as an equally primary process in root hair out-growth. It seems probable that, as is the case in yeast and animals, the exocyst acts as a Rho effector complex in plant cells, and is intimately involved in secretory pathway polarization and tip growth (Cole et al. 2005; Wen et al. 2005; Synek et al. 2006; Lavy et al. 2007; Hála et al. 2008). The domain of activated ROP at the PM in trichoblasts could stimulate exocyst complex assembly and/or recruitment via the SEC3/ICR1/ROP interaction (Lavy et al. 2007). Supportive of this notion is the observation that all three known exocyst mutants are associated with defects in root hair growth: in *Arabidopsis*, this includes both *exo70a1* (Synek et al. 2006) and *sec8* (Cole et al. 2005; and unpublished observations) mutants; in maize, *roothairless 1* encodes a SEC3 subunit (Wen et al. 2005). As the exocyst was recently shown to be important for membrane recycling in animal cells (reviewed in Somers and Chia 2005), it could also help to stabilize the asymmetric, activated ROP domain, as discussed earlier. We propose that exocyst complex activation via ROP is not only important at the outset of root hair tip growth proper, but is also important for the stabilized polarity of the growing root hair tip. As the tip-high calcium gradient is connected to the tip growth, and calcium is a well-established stimulator of membrane fusion, it is to be expected that ROP may also coordinate membrane transport via both calcium signalling and regulation of the exocyst. Thus, the exocyst will provide another mechanism to link various important components of the LENS (ROP, calcium, membrane trafficking) in a system that allows perpetuation of growth while remaining responsive to developmental and environmental cues.

7 Conclusions

It is now slightly over 5 years since it became clear that ROP GTPases are central regulators of both root hair initiation and root hair tip growth (Molendijk et al. 2001; Jones et al. 2002). Since then, additional work has revealed some parts of the mechanisms that allow ROP signalling to help coordinate the multiple cellular components (e.g., NOX/ROS) that carry out bulge formation and tip growth, as well as those that control ROP itself (e.g., RhoGDI, auxin/ethylene). However, many important links and details of these mechanisms still need experimental data to clearly define how root hair development occurs. What molecule(s) directly activate ROP prior to bulge formation? Do some of the effectors known to be important in pollen tube tip growth – e.g., RICs – also act in the root hair? How are feedback loops maintained at the constantly growing tip, with new membrane and cell wall being repeatedly laid down? We hope that some of the speculative hypotheses and models that we have offered help to motivate additional experimentation to tackle these topics, in hopes that 5 years hence the models describing the role of ROP in root hair development will better answer these questions.

Acknowledgments We are grateful to Hana Soukupova for the help with the finalization of this chapter. The related work in the author's laboratories was supported by the Academy of Sciences of the Czech Republic Grant Agency grant IAA6038410 to V.Z., NSF grant IBN-0420226 to J.F., and KONTAKT MSMT CR – ME841 and MSM0021620858 and MSMTLC06034 to V.Z.

References

Baluska F, Hlavacka A (2005) Plant formins come of age: something special about cross-walls. *New Phytol* 168:499–503

Baluska F, Salaj J, Mathur J, Braun M, Jasper F, Samaj J, Chua NH, Barlow PW, Volkmann D (2000) Root hair formation: F-actin-dependent tip growth is initiated by local assembly of profilin-supported F-actin meshworks accumulated within expansin-enriched bulges. *Dev Biol* 227:618–632

Baluska F, Liners F, Hlavacka A, Schlicht M, Van Cutsem P, McCurdy DW, Menzel D (2005) Cell wall pectins and xyloglucans are internalized into dividing root cells and accumulate within cell plates during cytokinesis. *Protoplasma* 225:141–155

Berken (2006) A ROPs in the spotlight of plant signal transduction. *Cell Mol Life Sci* 63:2446–2459

Berken A, Thomas C, Wittinghofer A (2005) A new family of RhoGEFs activates the Rop molecular switch in plants. *Nature* 436:1176–1180

Bibikova TN, Blancaflor EB, Gilroy S (1999) Microtubules regulate tip growth and orientation in root hairs of Arabidopsis thaliana. *Plant J* 17:657–665

Bloch D, Lavy M, Efrat Y, Efroni I, Bracha-Drori K, Abu-Abied M, Sadot E, Yalovsky S (2005) Ectopic expression of an activated RAC in Arabidopsis disrupts membrane cycling. *Mol Biol Cell* 16:1913–1927

Brembu T, Winge P, Bones AM, Yang Z (2006) A RHOse by any other name: a comparative analysis of animal and plant Rho GTPases. *Cell Res* 16:435–445

Carol RJ, Dolan L (2002) Building a hair: tip growth in Arabidopsis thaliana root hairs. *Philos Trans R Soc Lond B Biol Sci* 357:815–821

Carol RJ, Dolan L (2006) The role of reactive oxygen species in cell growth: lessons from root hairs. *J Exp Bot* 57:1829–1834

Carol RJ, Takeda S, Linstead P, Durrant MC, Kakesova H, Derbyshire P, Drea S, Zarsky V, Dolan L (2005) A RhoGDP dissociation inhibitor spatially regulates growth in root hair cells. *Nature* 438:1013–1016

Cole RA, Fowler JE (2006) Polarized growth: maintaining focus on the tip. *Curr Opin Plant Biol* 9:597–588

Cole RA, Synek L, Zarsky V, Fowler JE (2005) SEC8, a subunit of the putative Arabidopsis exocyst complex, facilitates pollen germination and competitive pollen tube growth. *Plant Physiol* 138:2005–2018

Christensen TM, Vejlupkova Z, Sharma YK, Arthur KM, Spatafora JW, Albright CA, Meeley RB, Duvick JP, Quatrano RS, Fowler JE (2003) Conserved subgroups and developmental regulation in the monocot rop gene family. *Plant Physiol* 133:1791–1808

Cvrckova F, Elias M, Hala M, Obermeyer G, Zarsky V (2001) Small GTPases and conserved signalling pathways in plant cell morphogenesis: from exocytosis to the Exocyst. In: Geitmann A, Cresti M, Heath B (eds) Cell biology of plant and fungal tip growth. IOS Press, Amsterdam, pp 105–122

Deeks MJ, Cvrckova F, Machesky LM, Mikitova V, Ketelaar T, Zarsky V, Davies B, Hussey PJ (2005) Arabidopsis group Ie formins localize to specific cell membrane domains, interact with actin-binding proteins and cause defects in cell expansion upon aberrant expression. *New Phytol* 168:529–540

DerMardirossian CM, Bokoch GM (2006) Phosphorylation of RhoGDI by p21-activated kinase 1. *Methods Enzymol* 406:80–90

Dhonukshe P, Baluska F, Schlicht M, Hlavacka A, Samaj J, Friml J, Gadella TW Jr (2006) Endocytosis of cell surface material mediates cell plate formation during plant cytokinesis. *Dev Cell* 10:137–150

Dovas A, Couchman JR (2005) RhoGDI: multiple functions in the regulation of Rho family GTPase activities. *Biochem J* 390:1–9

Dransart E, Olofsson B, Cherfils J (2005) RhoGDIs revisited: novel roles in Rho regulation. *Traffic* 6:957–966

Elias M, Drdova E, Ziak D, Bavlnka B, Hala M, Cvrckova F, Soukupova H, Zarsky V (2003) The exocyst complex in plants. *Cell Biol Int* 27:199–201

Elmayan T, Fromentin J, Riondet C, Alcaraz G, Blein JP, Simon-Plas F (2007) Regulation of reactive oxygen species production by a 14–3–3 protein in elicited tobacco cells. *Plant Cell Environ* 30:722–732

Fischer U, Ikeda Y, Ljung K, Serralbo O, Singh M, Heidstra R, Palme K, Scheres B, Grebe M (2006) Vectorial information for Arabidopsis planar polarity is mediated by combined AUX1, EIN2, and GNOM activity. *Curr Biol* 16:2143–2149

Fischer U, Ikeda Y, Grebe M (2007) Planar polarity of root hair positioning in Arabidopsis. *Biochem Soc Trans* 35:149–151

Foreman J, Demidchik V, Bothwell JH, Mylona P, Miedema H, Torres MA, Linstead P, Costa S, Brownlee C, Jones JD, Davies JM, Dolan L (2003) Reactive oxygen species produced by NADPH oxidase regulate plant cell growth. *Nature* 422:442–446

Fu Y, Gu Y, Zheng Z, Wasteneys G, Yang Z (2005) Arabidopsis interdigitating cell growth requires two antagonistic pathways with opposing action on cell morphogenesis. *Cell* 120:687–700

Gu Y, Li S, Lord EM, Yang Z (2006) Members of a novel class of Arabidopsis Rho guanine nucleotide exchange factors control Rho GTPase-dependent polar growth. *Plant Cell* 18:366–381

Gu Y, Vernoud V, Fu Y, Yang Z (2003) ROP GTPase regulation of pollen tube growth through the dynamics of tip-localized F-actin. *J Exp Bot* 54:93–101

Hála M, Cole R, Synek L, Drdová E, Pečenková T, Nordheim A, Lamkemeyer T, Madlung J, Hochholdinger F, Fowler JE, Zarsky V (2008) An exocyst complex functions in plant cell growth in *Arabidopsis* and Tobacco. Plant Cell 20

Hemsley PA, Kemp AC, Grierson CS (2005) The TIP GROWTH DEFECTIVE1 S-acyl transferase regulates plant cell growth in Arabidopsis. *Plant Cell* 17:2554–2563

Hsu SC, TerBush D, Abraham M, Guo W (2004) The exocyst complex in polarized exocytosis. *Int Rev Cytol* 233:243–265

Hwang JU, Gu Y, Lee YJ, Yang Z (2005) Oscillatory ROP GTPase activation leads the oscillatory polarized growth of pollen tubes. *Mol Biol Cell* 16:5385–5399

Ivanchenko M, Vejlupkova Z, Quatrano RS, Fowler JE (2000) Maize ROP7 GTPase contains a unique, CaaX box-independent plasma membrane targeting signal. *Plant J* 24:79–90

Jones MA, Raymond MJ, Smirnoff N (2006) Analysis of the root-hair morphogenesis transcriptome reveals the molecular identity of six genes with roles in root-hair development in Arabidopsis. *Plant J* 45:83–100

Jones MA, Raymond MJ, Yang Z, Smirnoff N (2007) NADPH oxidase-dependent reactive oxygen species formation required for root hair growth depends on ROP GTPase. *J Exp Bot* 58:1261–1270

Jones MA, Shen JJ, Fu Y, Li H, Yang Z, Grierson CS (2002) The Arabidopsis Rop2 GTPase is a positive regulator of both root hair initiation and tip growth. *Plant Cell* 14:763–776

Jones MA, Smirnoff N (2006) Nuclear dynamics during the simultaneous and sustained tip growth of multiple root hairs arising from a single root epidermal cell. *J Exp Bot* 57:4269–4275

Kaothien P, Ok SH, Shuai B, Wengier D, Cotter R, Kelley D, Kiriakopolos S, Muschietti J, McCormick S (2005) Kinase partner protein interacts with the LePRK1 and LePRK2 receptor kinases and plays a role in polarized pollen tube growth. *Plant J* 42:492–503

Ketelaar T, de Ruijter NC, Emons AM (2003) Unstable F-actin specifies the area and microtubule direction of cell expansion in Arabidopsis root hairs. *Plant Cell* 15:285–292

Klahre U, Becker C, Schmitt AC, Kost B (2006a) Nt-RhoGDI2 regulates Rac/Rop signaling and polar cell growth in tobacco pollen tubes. *Plant J* 46:1018–1031

Klahre U, Kost B (2006b) Tobacco RhoGTPase ACTIVATING PROTEIN1 spatially restricts signaling of RAC/Rop to the apex of pollen tubes. *Plant Cell* 18:3033–3046

Kost B, Lemichez E, Spielhofer P, Hong Y, Tolias K, Carpenter C, Chua NH (1999) Rac homologues and compartmentalized phosphatidylinositol 4, 5-bisphosphate act in a common pathway to regulate polar pollen tube growth. *J Cell Biol* 145:317–330

Lavy M, Bloch D, Hazak O, Gutman I, Poraty L, Sorek N, Sternberg H, Yalovsky S (2007) A Novel ROP/RAC Effector Links Cell Polarity, Root-Meristem Maintenance, and Vesicle Trafficking. Curr Biol May 8 [Epub ahead of print]

Lavy M, Bracha-Drori K, Sternberg H, Yalovsky S (2002) A cell-specific, prenylation-independent mechanism regulates targeting of type II RACs. *Plant Cell* 14:2431–2450

Lavy M, Yalovsky S (2006) Association of Arabidopsis type-II ROPs with the plasma membrane requires a conserved C-terminal sequence motif and a proximal polybasic domain. *Plant J* 46:934–947

Masucci JD, Schiefelbein JW (1996) Hormones act downstream of TTG and GL2 to promote root hair outgrowth during epidermis development in the Arabidopsis root. *Plant Cell* 8:1505–1517

Mathur J (2006) Local interactions shape plant cells. *Curr Opin Cell Biol* 18:40–46

Mathur J, Mathur N, Kernebeck B, Hulskamp M (2003) Mutations in actin-related proteins 2 and 3 affect cell shape development in Arabidopsis. *Plant Cell* 15:1632–1645

Meller N, Merlot S, Guda C (2005) CZH proteins: a new family of Rho-GEFs. *J Cell Sci* 118:4937–4946

Miller DD, De Ruijter NCA, Bisselign T, Emons AMC (1999) The role of actin in root hair morphogenesis: studies with lipochito-oligosaccharide as a growth stimulator and cytochalasin as an actin perturbing drug. *Plant J.* 17:141–154

Miller DD, Klooster L-t, Emons AMC (2000) Lipochito-oligosaccharide nodulation factors stimulate cytoplasmic polarity with longitudinal endoplasmic reticulum and vesicles at the tip in vetch root hairs. *Mol Plant-Microbe Interact* 13:1385–1390

Molendijk AJ, Bischoff F, Rajendrakumar CS, Friml J, Braun M, Gilroy S, Palme K (2001) Arabidopsis thaliana Rop GTPases are localized to tips of root hairs and control polar growth. *EMBO J* 20:2779–2788

Molendijk AJ, Ruperti B, Palme K (2004) Small GTPases in vesicle trafficking. *Curr Opin Plant Biol* 7:694–700

Morel J, Claverol S, Mongrand S, Furt F, Fromentin J, Bessoule JJ, Blein JP, Simon-Plas F (2006) Proteomics of plant detergent-resistant membranes. *Mol Cell Proteomics* 5:1396–1411

Novick P, Medkova M, Dong G, Hutagalung A, Reinisch K, Grosshans B (2006) Interactions between Rabs, tethers, SNAREs and their regulators in exocytosis. *Biochem Soc Trans* 34:683–686

Ohashi Y, Oka A, Rodrigues-Pousada R, Possenti M, Ruberti I, Morelli G, Aoyama T (2003) Modulation of phospholipid signaling by GLABRA2 in root-hair pattern formation. *Science* 300:1427–1430

Ono E, Wong HL, Kawasaki T, Hasegawa M, Kodama O, Shimamoto K (2001) Essential role of the small GTPase Rac in disease resistance of rice. *Proc Natl Acad Sci USA* 98:759–764

Oude Weernink PA, Han L, Jakobs KH, Schmidt M (2007) Dynamic phospholipid signaling by G protein-coupled receptors. *Biochim Biophys Acta* 1768:888–900

Ovecka M, Lang I, Baluska F, Ismail A, Illes P, Lichtscheidl IK (2005) Endocytosis and vesicle trafficking during tip growth of root hairs. *Protoplasma* 226:39–54

Paciorek T, Zazimalova E, Ruthardt N, Petrasek J, Stierhof YD, Kleine-Vehn J, Morris DA, Emans N, Jurgens G, Geldner N, Friml J (2005) Auxin inhibits endocytosis and promotes its own efflux from cells. *Nature* 435:1251–1256

Park J, Gu Y, Lee Y, Yang Z, Lee Y (2004) Phosphatidic acid induces leaf cell death in Arabidopsis by activating the Rho-related small G protein GTPase-mediated pathway of reactive oxygen species generation. *Plant Physiol* 134:129–136

Parker JS, Cavell AC, Dolan L, Roberts K, Grierson CS (2000) Genetic interactions during root hair morphogenesis in Arabidopsis. *Plant Cell* 12:1961–1974

Potocky M, Jones MA, Bezvoda R, Smirnoff N, Zarsky V (2007) Reactive oxygen species produced by NADPH oxidase are involved in pollen tube growth. *New Phytol* 174:742–751

Qiu JL, Jilk R, Marks MD, Szymanski DB (2002) The Arabidopsis SPIKE1 gene is required for normal cell shape control and tissue development. *Plant Cell* 14:101–118

Roumanie O, Wu H, Molk JN, Rossi G, Bloom K, Brennwald P (2005) Rho GTPase regulation of exocytosis in yeast is independent of GTP hydrolysis and polarization of the exocyst complex. *J Cell Biol* 170:583–594

Sagi M, Fluhr R (2006) Production of reactive oxygen species by plant NADPH oxidases. *Plant Physiol* 141:336–340

Sieberer BJ, Ketelaar T, Essling JJ, Emons AMC (2005) Microtubules guide root hair tip growth. *New Phytol* 167:711–719

Smith LG, Oppenheimer DG (2005) Spatial control of cell expansion by the plant cytoskeleton. *Annu Rev Cell Dev Biol* 21:271–295

Somers WG, Chia W (2005) Recycling polarity. *Dev Cell* 9:312–313

Sorek N, Poraty L, Sternberg H, Bar E, Lewinsohn E, Yalovsky S (2007) Activation status-coupled transient S acylation determines membrane partitioning of a plant Rho-related GTPase. *Mol Cell Biol* **27:**2144–2154

Sormo CG, Leiros I, Brembu T, Winge P, Os V, Bones AM (2006) The crystal structure of Arabidopsis thaliana RAC7/ROP9: the first RAS superfamily GTPase from the plant kingdom. *Phytochemistry* 67:2332–2340

Synek L, Schlager N, Elias M, Quentin M, Hauser MT, Zarsky V (2006) AtEXO70A1, a member of a family of putative exocyst subunits specifically expanded in land plants, is important for polar growth and plant development. *Plant J* 48:54–72

Tao LZ, Cheung AY, Wu HM (2002) Plant Rac-like GTPases are activated by auxin and mediate auxin-responsive gene expression. *Plant Cell* 14:2745–2760

Thomas C, Fricke I, Scrima A, Berken A, Wittinghofer A (2007) Structural evidence for a common intermediate in small G protein-GEF reactions. *Mol Cell* 25:141–149

Trotochaud AE, Hao T, Wu G, Yang Z, Clark SE (1999) The CLAVATA1 receptor-like kinase requires CLAVATA3 for its assembly into a signaling complex that includes KAPP and a Rho-related protein. *Plant Cell* 11:393–406

Uhrig JF, Mutondo M, Zimmermann I, Deeks MJ, Machesky LM, Thomas P, Uhrig S, Rambke C, Hussey PJ, Hulskamp M (2007) The role of Arabidopsis SCAR genes in ARP2-ARP3-dependent cell morphogenesis. *Development* 134:967–977

Valdez-Taubas J, Pelham HR (2003) Slow diffusion of proteins in the yeast plasma membrane allows polarity to be maintained by endocytic cycling. *Curr Biol* 13:1636–1640

Vincent P, Chua M, Nogue F, Fairbrother A, Mekeel H, Xu Y, Allen N, Bibikova TN, Gilroy S, Bankaitis VA (2005) A Sec14p-nodulin domain phosphatidylinositol transfer protein polarizes membrane growth of Arabidopsis thaliana root hairs. *J Cell Biol* 168:801–812

Vissenberg K, Fry SC, Verbelen JP (2001) Root hair initiation is coupled to a highly localized increase of xyloglucan endotransglycosylase action in Arabidopsis roots. *Plant Physiol* 127:1125–1135

Winge P, Brembu T, Kristensen R, Bones AM (2000) Genetic structure and evolution of RAC-GTPases in Arabidopsis thaliana. *Genetics* 156:1959–1971

Wen TJ, Hochholdinger F, Sauer M, Bruce W, Schnable PS (2005) The roothairless1 gene of maize encodes a homolog of sec3, which is involved in polar exocytosis. *Plant Physiol* 138:1637–1643

Wu G, Gu Y, Li S, Yang Z (2001) A genome-wide analysis of Arabidopsis Rop-interactive CRIB motif-containing proteins that act as Rop GTPase targets. *Plant Cell* 13:2841–2856

Wu G, Li H, Yang Z (2000) Arabidopsis RopGAPs are a novel family of rho GTPase-activating proteins that require the Cdc42/Rac-interactive binding motif for rop-specific GTPase stimulation. *Plant Physiol* 124:1625–1636

Yang Z (2002) Small GTPases: versatile signaling switches in plants. *Plant Cell* 14:S375–S388

Yi K, Guo C, Chen D, Zhao B, Yang B, Ren H (2005) Cloning and functional characterization of a formin-like protein (AtFH8) from Arabidopsis. *Plant Physiol* 138:1071–1082

Zarsky V, Potocky M, Baluska F, Cvrckova F (2006) Lipid metabolism, compartmentalization and signalling in the regulation of pollen tube growht. In: Malhó R (ed) The pollen tube - A cellular perspective, Plant cell monographs, vol 3. Springer, Berlin, Heidelberg New York, pp 117–138

Zerial M, McBride H (2001) Rab proteins as membrane organizers. *Nat Rev Moll Cell Biol* 2:107–117

Zheng ZL, Yang Z (2000) The Rop GTPase: an emerging signaling switch in plants. *Plant Mol Biol* 44:1–9

The Actin Cytoskeleton in Root Hairs: A Cell Elongation Device

T. Ketelaar(✉) and A.M. Emons

Abstract The actin cytoskeleton plays an important role in root hair development. It is involved in both the delivery of growth materials to the expanding tip of root hairs and the regulation of the area of tip growth. This review starts with a discussion of the techniques that are available to visualize the actin cytoskeleton in fixed and in live root hair cells, including their advantages and drawbacks. We discuss the function that the actin cytoskeleton performs during tip growth of root hairs, focusing first on filamentous actin organization during root hair development and the response of root hairs to the rhizobial signal molecule Nod factor, which reveals the function of actin in root hair elongation. In addition, we discuss the role of actin binding proteins in organizing the actin cytoskeleton.

1 Introduction

A large body of evidence shows that the actin cytoskeleton plays an important role in the regulation and execution of cell expansion. Root hairs are ideal cells to study the role of the actin cytoskeleton in cell expansion, since they grow locally and rapidly. Besides, since they develop on the root surface, they are easily accessible and easy to manipulate. For these reasons, the knowledge about the actin cytoskeleton in root hairs is much more detailed than in the growth phase of other plant cell types. In this chapter, we will describe the techniques that are available to visualize the actin cytoskeleton in root hairs, both in fixed cells and in live cells, followed by a description of the actin organization and dynamics in root hairs. Together with accumulation of Ca^{2+} in the root hair tip, Ca^{2+} spiking around the nucleus, and microtubule reorganization, fine filamentous actin remodeling is one of the known early (<3 min) responses to *Rhizobium* bacteria.

T. Ketelaar
Laboratory of Plant Cell Biology, Wageningen University, Arboretumlaan 4, 6703 BD Wageningen The Netherlands
e-mail: tijs.ketelaar@wur.nl

Plant Cell Monogr, doi:10.1007/7089_2008_8

This response has given important insight into the role of the actin cytoskeleton in cell elongation. We review this work, as well as the mechanisms of actin remodeling by actin binding proteins.

2 Actin Visualization in Root Hairs

Owing to the dynamic nature of actin filaments, actin visualization is challenging. In growing root hairs, where actin filaments in the apical region have a very high turnover, it is particularly important to follow tested, reliable protocols and use rigorous controls to achieve artifact-free imaging. First, we will discuss visualization of actin filaments in fixed root hair cells, and then we will focus imaging of actin filaments in live cells with fluorescent proteins, fused to (domains of) actin binding proteins.

2.1 Actin Visualization in Fixed Root Hairs

The first step towards actin visualization in fixed root hairs is the fixation. Owing to the dynamicity of the subapical actin filaments, fixation should to be as rapid as possible. The most reliable method to achieve rapid fixation of root hairs is rapid freeze fixation followed by freeze substitution (Emons 1987). During rapid freeze fixation, root hairs are plunged into liquid propane, so that they are instantly frozen and no ice crystals are formed. Ice-crystal-free preservation can be achieved only in the area just below the sample surface that is in direct contact with the liquid propane (Kiss and McDonald 1993; Nicolas and Bassot 1993). Since root hairs are sticking out from the root surface and are generally thin (*Arabidopsis* root hairs are ~8 μm in width), they are well suited for plunge freezing. For freeze fixation of structures that are further away from the surface, high pressure freezing is recommended (Müller and Moor 1984; Kiss et al. 1990; Thijssen et al. 1998), although for light microscopic immunocytochemistry, ice crystal sizes throughout the sample after plunge freezing are far below the resolution limit (≤100 nm) (Ryan 1992) and the method has been used successfully for light microscopical analysis of tissues as large as whole *Arabidopsis* roots (Baskin et al. 1996). After rapid freeze fixation, the watery content of the root hairs is substituted for acetone or methanol at –90°C for 24 h or longer, after which the material is warmed up to room temperature over a period of hours to days. In the freeze substitution medium, water poor, aldehyde-based fixatives can be added to chemically stabilize the freeze-fixed root hairs. We have used 0.5–1% glutaraldehyde (Emons and Derksen 1986; Miller et al. 1999; Sieberer and Emons 2000; Ketelaar et al. 2002, 2003). After freeze substitution, the material can be infiltrated with a resin such as butyl methyl methacrylate (BMM) (Baskin et al. 1992) or Steedmans wax (Steedman 1957). After polymerization, the resin containing the sample is sectioned, whereafter the resin can be washed away with acetone (BMM) or ethanol (Steedmans wax). This makes the sectioned material well accessible for immunolabeling procedures. A disadvantage of sectioning is that it is difficult to reconstruct a three-dimensional image of the actin cytoskeleton of cells having the size of root hairs, since sections are thinner than the root hair tube and it is difficult to position root hairs in such a way that they are parallel to the direction of sectioning. Alternatively, the material can be rehydrated, postfixed with aldehydes, permeabilized with cell-wall-digesting enzymes, and labeled with antibodies without sectioning (Miller et al. 1999; Ketelaar et al. 2002, 2003; Sieberer et al. 2002) and immunolabeled entirely.

Fixation methods that are not based on rapid freeze fixation, such as aldehyde-based chemical fixations, take longer since the chemicals have to penetrate before they can fix cell structures. During a standard chemical fixation (4% paraformaldehyde and 0.1% glutaraldehyde), cytoplasmic rearrangements can still be observed seconds after application of the fixative (Esseling et al. 2000), and changes in cytoplasmic organization before and after fixation are eminent (Ketelaar and Emons 2001). To avoid cytoplasmic reorganizations during chemical fixation of root hairs, prefixations have been performed with, e.g., 400 μM *m*-maleimido benzoyl *N*-hydroxy succinimide (MBS) ester (Sonobe and Shibaoka 1989). MBS-ester stabilizes actin filaments (F-actin) and consequently inhibits cytoplasmic streaming during the subsequent aldehyde fixation (Esseling-Ozdoba et al. 2008), which prevents rearrangements of the cytoplasm and actin filaments during the fixation. The actin organization in root hairs that have been pretreated with MBS-ester is similar to that in root hairs that have been fixed by rapid freeze fixation (Miller et al. 1999; de Ruijter et al. 2000; Ketelaar et al. 2002, 2003). To judge the quality of the fixation method, the cytoarchitecture of root hairs should be checked before and after fixation (Esseling et al. 2000). Many inaccurate actin staining and immunolabeling results, caused by bad fixations, are easily detectable by comparing the cytoarchitecture before and after the fixation procedure. If the cytoarchitecture has been altered during the fixation procedure, the actin preservation is inferior and its localization incorrect, in root hairs a plug of actin is seen at the very tip (Esseling et al. 2000).

After the fixation procedure, there are different methods to fluorescently label the actin cytoskeleton. Broadly, actin can be visualized by immunolabeling (Miller et al. 1999) or by application of fluorescent phalloidin derivates (Emons 1987). Fluorescent phalloidins and phallacidins (which are smaller than phalloidins) have as the advantage over antibody staining that they specifically label F-actin, whereas antibodies against actin label both actin filaments and monomeric actin, which masks the F-actin organization (see, e.g., Vitha et al. 1997, 2000; Ketelaar et al. 2003). For penetration of fluorescently labeled phalloidins into the root hair, treatment with the plasma membrane permeabilizer L-α-lyso-phosphatidylcholine (Lee and Chan 1977; Morris et al. 1980) is sufficient (LPC, 100 mg mL^{-1}; Miller et al. 1999; de Ruijter et al. 2000; Ketelaar et al. 2002, 2003), whereas immunodetection of actin requires enzymatic cell wall digestion in whole-mount immunolabeling procedures, or else sectioning of the embedded material. Cell wall digestion is an elaborate process; we found that the types and concentrations of cell-wall-digesting enzymes are species-dependent (Ketelaar et al. 2002, 2003; Sieberer et al. 2002) even in *Arabidopsis* lines with a different genetic background, large variations in cell wall digestibility can be found (Barriere et al. 2006). Procedures such as freeze shattering (Wasteneys et al. 1997) are not recommended for actin visualization in root hairs, as the shattering tends to fragment the F-actin (our unpublished results). For a good immunocytochemistry protocol for structures other than actin filaments, see Nielsen (2008).

Disadvantages of phalloidin derivates over immunolabeling are that phalloidin staining allows imaging only for several hours before the staining pattern disintegrates, and more importantly, phalloidin does not stain F-actin that has been through a rapid freeze fixation/freeze substitution procedure. After this procedure, F-actin appears to have lost the ability to bind phalloidins. In addition, fluorescently labeled phalloidin derivates have been reported to share the binding location on actin with actin-depolymerizing factor (Nishida et al. 1987; Ressad et al. 1998), and thus not all actin filaments may be decorated with fluorescent phalloidins.

2.2 Actin Visualization in Live Cells

Several methods have been described to grow root hairs so that they are suitable for microscopic imaging of the actin cytoskeleton (Baluska et al. 2000b; Ketelaar et al. 2002, 2004b; Voigt et al. 2005), and it is likely that also other methods that have been used to observe fluorescent probes in live, growing root hairs are suitable to image GFP-decorated F-actin in growing root hairs.

The actin cytoskeleton in live plant cells has been visualized by three different probes: microinjection of fluorescently labeled phalloidin (Esseling et al. 2000) and the genetic probes GFP-mTn (Kost et al. 1998) and GFP:FABD2 (Ketelaar et al. 2004b). All these probes have been used to visualize the actin cytoskeleton in root hairs. Fluorescent phalloidins have been used at different concentrations to visualize the actin cytoskeleton in live root hairs. Cardenas et al. (1998) microinjected root hairs of *Phaseolus* with 20 μM FITC phalloidin (needle concentration). This method is known as the "snap shot method" (Esseling et al. 2000). Although the actin cytoskeleton is visible after microinjection of 20 μM phalloidin, this concentration freezes actin dynamics, induces actin bundling, and causes the cell to die. This makes the snap shot method unsuitable to study actin reorganizations in live cells. Microinjections of lower concentrations of fluorescent phalloidin do allow observation of actin dynamics (Valster et al. 1997), also in root hairs (Esseling et al. 2000). Esseling et al. (2000) report that injection of 0.66 μM Alexa 488 phalloidin (needle concentration) results in actin staining in *Arabidopsis* root hairs. In these root hairs, actin reorganization over time was visible, but only the first 10 min after injection. Thereafter, fluorescence dissociates from actin filaments and accumulates in aggregates. Besides this problem, microinjection of fluorescent phalloidin causes many root hairs to stop growing and is time-consuming. It is therefore not a suitable method for large-scale studies of F-actin organization in root hairs.

In plants that express genetic fusions of GFP to actin-binding domains GFP-mTn and GFP:FABD2, filamentous, fluorescent structures are clearly visible, also in root hairs (Baluska et al. 2000b; Ketelaar et al. 2004b). Both fusion proteins consist of a well-defined actin-binding domain (GFP-mTn: amino acids 2345–2541 from mouse talin (McCann and Craig 1997); and GFP:FABD2: the second actin-binding domain of *Arabidopsis* fimbrin (Klein et al. 2004)). These actin-binding domains bind to actin filaments and thus localize GFP to the actin filaments. As a consequence, these fusion proteins compete for actin binding with endogenous actin-binding proteins (Ketelaar et al. 2004a; Holweg 2007) and they are in a constant flux on and off the actin filaments (Ketelaar et al. 2004a; Sheahan et al. 2004). This could indicate that actin filaments that are heavily decorated with endogenous actin-binding proteins are not, or are weakly, fluorescently labeled. In addition, the rapid flux of ABD:GFP on and off the F-actin causes an inability to follow actin dynamics by fluorescence recovery after photobleaching experiments. The association and dissociation rates of GFP:FABD2 are much more rapid than the actin dynamics themselves (Ketelaar et al. 2004a). Finally, labeling actin filaments that rapidly turn over may be difficult as they depolymerize before the fusion proteins can bind.

Besides the properties of the probe, the sensitivity and speed of microscopic imaging play a role in the visualization of actin filaments, especially in the visualization of fine actin networks and rapidly turning over actin filaments. This is demonstrated when images of growing root hairs, expressing equal levels of GFP: FABD2, that are collected on a conventional point scanning confocal microscope are compared to images collected on a spinning disk confocal microscope, whose speed and sensitivity are superior. Figure 1 shows that in an image collected on a conventional confocal microscope the bundles of actin filaments in the root hair tube are clearly visible, but the subapical network of fine F-actin is not visible, whereas an image collected on a spinning disk confocal microscope does show a network of GFP:FABD2 decorated fine F-actin in the subapical area of the root hair. In the extreme apex, no F-actin is detectable, even under the spinning disk microscope, similar to data obtained by immuno- and phalloidin-staining experiments (Miller et al. 1999; Ketelaar et al. 2003) (see later).

3 Actin Organization and Function in Root Hairs

3.1 Root Hair Development, Actin Organization, and the Effects of Actin Depolymerization

Plant cells are sheathed inside a cell wall and require localized exocytosis for polarized cell growth. Exocytosis involves the fusion of Golgi-derived vesicles with the plasma membrane. This process delivers growth materials – membrane and transmembrane proteins, such as ion pumps and cellulose synthases, to the location of cell growth, as well as cell wall matrix materials. The force that is required for the

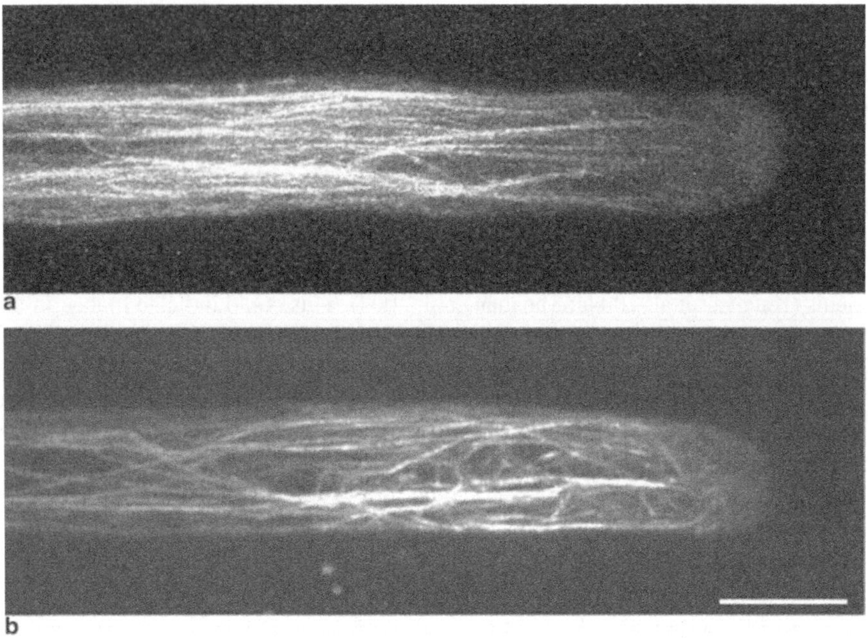

Fig. 1 The actin cytoskeleton in growing *Arabidopsis* root hairs expressing 35S::GFP:FABD2, visualized with a point scanning confocal laser scanning microscope (**a**) and with a spinning disk confocal microscope (**b**). The combination of increased sensitivity and decreased imaging time of spinning disk confocal microscope, equipped with a Yokogawa CSU10 spinning disk unit and a Hamamatsu ORCAII-BT-512G digital camera, allows the detection of GFP:FABD2-decorated actin filaments in the subapex of growing root hairs that cannot be visualized with a point scanning confocal laser scanning microscope (besides this image, see, e.g., Ketelaar et al. (2004b)). The actin cytoskeleton configuration is similar to those seen in images prepared with immunocytochemistry, including an area devoid of filamentous actin at the extreme root hair tip. Bar, 10 μm

stretching of existing cell walls is generated by turgor pressure (Cosgrove 2005), which is outside the scope of this review. Here, we will focus on the organization of the actin cytoskeleton and the role it plays during root hair development. Root hairs expand by tip growth. Tip growth is expansion over a small surface area of the cell, combined with cell wall stiffening in the area just behind the expanding hemisphere, so that a tubular cell is generated, of which the tip expands. Owing to the polar organization of tip-growing cells, the actin cytoskeleton in the tip-growing plant cells, root hairs, and pollen tubes has been the interest of many studies (for reviews, see for example Dai and Sheetz 1995; de Ruijter and Emons 1999; Hepler et al. 2001; Feijo et al. 2004; Grierson and Ketelaar 2004; Smith and Li 2004).

3.1.1 Role of F-Actin in Root Hair Development

Root hairs originate from root epidermal cells that are designated to form root hairs, the so-called trichoblasts. Root hair initiation starts by the appearance of a bulge on the exterior surface of trichoblasts (Emons and Derksen 1986; Dolan et al. 1994; Wymer et al. 1997; de Ruijter et al. 1998; Sieberer and Emons 2000). The actin-depolymerizing drugs latrunculin B and cytochalasin D have no effect on the formation of bulges, nor on their initiation site (Miller et al. 1999; Ovecka et al. 2000; Molendijk et al. 2001), indicating that the actin cytoskeleton is not involved in this process. However, during or just after bulge formation, actin filaments accumulate in the bulge (Emons and Derksen 1986; Baluska et al. 2000b). The appearance of actin filaments correlates with the accumulation of cytoplasm into the bulge (Sieberer and Emons 2000). Once an accumulation of cytoplasm is present, tip growth initiates (Dolan and Roberts 1995). For the initiation of tip growth, a functional actin cytoskeleton is a prerequisite.

Tip growth is characterized by an accumulation of cytoplasm in the tip region of the growing root hair (see Emons and Ketelaar 2008). Further down the root hair tube, only a thin layer of cortical cytoplasm is present, which surrounds a central vacuole, and occasionally, mostly close to the subapical area, transvacuolar cytoplasmic strands are present (Sieberer and Emons 2000). In transmission electron microscopic (TEM) images an accumulation of vesicles is observed in the extreme apex of growing root hairs (Emons 1987; Ridge 1988; Sherrier and Vandenbosch 1994; Galway et al. 1997; Galway 2000; Miller et al. 2000). During root hair growth, the nucleus follows the root hair tip at a fixed distance, which is an actin-based process in *Arabidopsis* (Ketelaar et al. 2002; Van Bruaene et al. 2003). In contrast, nuclear positioning is at least partially mediated by microtubules in the legume *Medicago truncatula* (Sieberer et al. 2002). For more details on root hair cytoarchitecture, see Emons and Ketelaar (2008).

Cytoplasmic streaming is a myosin-based motility of organelles over actin bundles accompanied by hydrodynamic flow (Houtman et al. 2007; Esseling-Ozdoba et al. 2008), which occurs in all plant cells. In growing root hairs, cytoplasmic streaming occurs in a "reversed-fountain-like" pattern. The streaming in the cortex is directed towards the tip. In the center of the root hair tube close to the growing tip, the streaming direction is reversed and directed away from the tip (Emons 1987; Sieberer and Emons 2000).

Like in all plant cells, the F-actin in root hairs is the backbone around which the cytoplasm is localized; accumulations of cytoplasm or cytoplasmic strands can only exist around an actin backbone (Valster et al. 1997; van der Honing et al. 2007). In the cortical cytoplasm of the root hair tube and in the transvacuolar strands, thick bundles of F-actin are present (de Ruijter et al. 1999; Miller et al. 1999; Ketelaar et al. 2003), clearly visible in a transmission electron microscope (Emons 1987). Towards the tip, these bundles flare out into a dense array of net-axial actin filaments, the fine F-actin (Miller et al. 1999; Ketelaar et al. 2003), which is responsible for the accumulation of cytoplasm in the subapex of growing root hairs. In the extreme apex, the small area that correlates with the vesicle-rich area in TEM images, very little F-actin can be detected with fluorescence techniques (Miller et al. 1999; Ketelaar et al. 2003), and also from electron microscopy research F-actin has not been reported to be present in this area. This area is devoid of organelles. Therefore, no light microscopically visible cytoplasmic streaming can be observed, as has been reported for the extreme apex of growing *Equisetum hyemale* root hairs (Emons 1987). Thus, even if there is F-actin in the extreme apex of growing root hairs, which cannot be detected with fluorescence techniques, this pool of actin is not involved in the transport of organelles, but could be involved in vesicle movement. In the apex, only vesicles of ~100 nm in width are present, which have been observed to move, using UV microscopy (Sieberer and Emons 2000) and total internal reflection microscopy (Wang et al. 2006). In Emons and Ketelaar (2008), we argue that delivery of exocytotic vesicles to the vesicle-rich area by fine F-actin from the subapex and their consumption by exocytosis at the plasma membrane are sufficient for directional movement towards the tip by mere diffusion.

By microinjection of DNAse I, which specifically binds monomeric actin, He et al. (2006) conclude that monomeric actin accumulates in the apex of growing root hairs. The authors suggest that there is an active mechanism that targets monomeric actin to the apical region of growing root hairs and prevents the actin from polymerization in the extreme apex. Actin-binding proteins could play a role in this process (see Sect. 3.3). Alternatively, the increased amount of monomeric actin in the apex may not be caused by a locally increased concentration of actin monomers, but may correlate with the larger amount of cytoplasm in the apex, due to the absence of the central vacuole and organelles from this area. In this case the higher amount of monomeric actin would not be caused by an increase in concentration, but an increase in the amount of cytoplasm that contains the same concentration of monomeric actin. An apical preferential localization of a protein can only be proven conclusively by ratio-imaging with another dye that uniformly labels the cytoplasm.

The subapical fine F-actin in growing root hairs is more sensitive to actin-depolymerizing drugs than are the more basal bundles of F-actin in the root hair tube; low concentrations (0.5–1 μM) of cytochalasin D cause the specific disappearance of fine F-actin, whereas thick bundles of F-actin in the root hair tube remain intact and cytoplasmic streaming continues (Miller et al. 1999). The presence of fine F-actin strongly correlates with root hair growth. When the fine F-actin is depolymerized, the delivery of exocytic vesicles that is required for root hair growth is inhibited (Miller et al. 1999; Ketelaar et al. 2003). The indolic compound hypaphorine, which is secreted by the fungus *Pisolithus microcarpus* (establishes an ectomycorrhiza with *Eucalyptus globules*), mimics the treatment with low concentrations of cytochalasin D, namely depolymerization of the subapical actin cytoskeleton and inhibition of cell growth (Ditengou et al. 2003; Dauphin et al. 2006). Miller et al. (1999) hypothesized that the fine F-actin is not only involved in delivering exocytic vesicles to the tip, but also in filtering them so that no larger organelles reach the tip, and in retaining them in the tip.

By application of actin-depolymerizing drugs in even lower concentration (0.01–0.5 μM cytochalasin D) as used by Miller et al. (1999), we were able to show that fine F-actin performs an additional function: it delimits the root hair surface area over which exocytosis takes place (Ketelaar et al. 2003). When *Arabidopsis* root hairs are treated with very low concentrations of actin-depolymerizing drugs, the surface area over which exocytosis takes place increases in a concentration-dependent fashion, resulting in thicker root hairs, whereas the total amount of expansion (in terms of increase in surface area) remains equal (root hair width gradually increases between 0 and 0.1 μM cytochalasin D, where the swelling is maximal). These observations correlate with a decrease in, but not a total disappearance of, fine F-actin and an increase in the size of the apical area with little F-actin, suggesting that this area plays a role in determining the surface area over which exocytosis occurs (Ketelaar et al. 2003).

The final length of root hairs is dependent on the conditions in which roots are cultured, and is species-specific. For screening of hair development mutants, it is important to know that maximal length of root hairs is reached when root hairs are being grown in air on the surface of agar plates. When the cytoplasmic accumulation in the tip disappears, root hairs terminate growth (de Ruijter et al. 1998; Miller et al. 1999; Ketelaar et al. 2003), and the nucleus obtains a random position in the cytoplasm, indicating that the cytoplasm is no longer polarized (Sieberer and Emons 2000). The subapical fine F-actin disappears and thick bundles of F-actin loop through the tip of the hair (Miller et al. 1999; Ketelaar et al. 2003).

3.1.2 The Actin Cytoskeleton in the Response of Legume Root Hairs to *Rhizobium* Bacteria

The actin cytoskeleton of legumes is a target for the signaling cascade that is induced by Nod factor, a lipochitooligosaccharide signaling molecule from *Rhizobium* bacteria (de Ruijter et al. 1999; Miller et al. 1999). Application of Nod factor to the liquid medium in which root hairs grow causes an increase in the density of the subapical fine F-actin within minutes in both growing and fully grown root hairs (de Ruijter et al. 1999). In root hairs that are terminating growth and where the density of fine F-actin is decreasing naturally, this increase in fine F-actin to a density that is seen in growing root hairs causes a reinitiation of tip growth. This reinitiation of tip growth is preceded by the formation of a swelling (Miller et al. 1999). The correlation between the amount of fine F-actin and tip growth is in line with our hypothesis that fine F-actin is essential for cell growth.

During natural symbiosis of legumes with *Rhizobium* bacteria, growing root hairs curl around the bacteria to entrap them. When the bacteria are fully enclosed by the root hair curl, an infection thread is formed which translocates the bacteria to the cortical cells, where nodules are formed that contain the nitrogen-fixing *Rhizobium* bacteroids (see Limpens and Bisseling 2008). Until recently, it was thought that Nod factor by itself was not sufficient to reorient root hair growth, since root hairs do not curl when Nod factor is applied to the liquid medium in which root hairs grow (Heidstra et al. 1994, 1997; de Ruijter et al. 1998, 1999; Miller et al. 1999). By locally applying small drops of Nod factor to one side of the tips of air-grown root hairs, Esseling (Esseling and Emons 2004; Esseling et al. 2004a) was able to show that this is not true, since local Nod factor application is sufficient to cause growing root hairs to reorient their direction and even curl around such a drop of required size and applied at required location. To achieve root hair curling, the location where exocytosis of vesicles takes place should be shifted towards the area with the highest concentration of Nod factor (Esseling and Emons 2004; Esseling et al. 2004a). It is likely that the actin cytoskeleton is involved in this process, since it delivers vesicles (Miller et al. 1999), and is involved in determining the size of the area where exocytosis occurs (Ketelaar et al. 2003). However, microtubules are involved in maintaining a straight growth direction of root hairs, which makes them a more likely primary target of the Nod factor than the actin cytoskeleton (see Sieberer and Timmers 2008). Since it is not possible to image the actin organization at sufficient resolution in air-grown root hairs yet, direct imaging of fine F-actin in air-grown hairs is technically challenging and remains to be published.

3.2 Actin Expression Levels

The data obtained with actin-depolymerizing drugs at low concentrations (0.01–0.5 μM cytochalasin D or 0.001–0.05 μM latrunculin A; see Sect. 3.1) suggest that fine F-actin polymerization is essential for maintenance of fine F-actin and continued cell growth in a fixed diameter. In these experiments, the actin-depolymerizing drugs serve as a tool to decrease the available amount of monomeric actin for polymerization (latrunculin B) or the available barbed ends for polymerization (cytochalasin D). Knocking out actin genes, so that actin expression levels are lower, is a different way to decrease the amount of available actin for polymerization. This should lead to phenotypes similar to those obtained by treatment with actin-depolymerizing drugs. In root hairs, at least three actin genes, *ACT2*, *ACT7*, and *ACT8*, are expressed, as demonstrated by promoter-reporter gene studies (An et al. 1996; McDowell et al. 1996). The different actin isoforms that are involved in root hair development do not seem to have a specific function during root hair development. Overexpression of ACT7 is sufficient to rescue mutations in the *ACT2* gene (Gilliland et al. 2002). Even overexpression of the reproductive actin ACT1, which is the most divergent actin from ACT2, is sufficient to rescue mutations in *ACT2* (Gilliland et al. 2002).

Mutations in *ACT2* and *ACT7* cause a decrease in the total amount of actin protein (Ringli et al. 2002; Gilliland et al. 2003). This reduction in the amount of available actin protein in T-DNA insertion lines in *ACT7* (act7-1 and act7-4) causes a strong reduction of root elongation (Gilliland et al. 2003). Root hair development in these mutants has not been properly analyzed, but reduction in the amount of actin protein, caused by mutations and T-DNA insertions in *ACT2* (mutations: der1-1, der1-2, and der1-3; T-DNA insertion line: act2-1), does cause root hair phenotypes. In these mutants, bulges can swell strongly (Gilliland et al. 2002; Ringli et al. 2002). Also after bulge formation and initiation of tip growth, mutants in the *ACT2* gene produce root hairs with a large variation in width. The changes in root hair diameter are similar in plants with mutations in *ACT2* and plants that have been treated with low concentrations of actin-depolymerizing drugs, albeit more variable in the act2 mutant. Actin localization studies that reveal defects in actin organization or dynamics, or both, in these mutants remain to be carried out. Besides a variation in width, mutations in the *ACT2* gene give rise to multiple root hairs that emerge from a single bulge (Gilliland et al. 2002; Ringli et al. 2002). Treatment with actin-depolymerizing drugs does not produce multiple root hairs from one bulge, suggesting that mutations in the *ACT2* gene do not merely reduce the amount of available actin for polymerization as drug treatments do.

3.3 Actin-Binding Proteins and Their Roles in Root Hair Development and Actin Organization

One can use knocking out, underexpressing, overexpressing, mutating and truncating specific proteins to study the function of these proteins in cellular development and actin organization. Since root hair growth is very sensitive to actin disturbance, it is an excellent system to study the role of actin-binding proteins on cell growth and actin organization. Here, we will discuss the consequences of changing expression levels and/or sequences of several well-studied actin-binding proteins and their consequences on actin organization and root hair growth.

3.3.1 Myosins

Myosins are motor proteins that move over F-actin. In higher plants, two classes of myosins are present: myosin XI and myosin VIII (Jiang and Ramachandran 2004). Four class VIII myosin genes are present in the *Arabidopsis* genome, whereas 13 genes that encode myosin XI are present (Reddy 2001). Class VIII myosins localize to plasmodesmata and young cross walls (Reichelt et al. 1999; Baluska et al. 2000a, 2001). Class XI myosins are thought to be responsible for cytoplasmic streaming, the movement of organelles, and hydrodynamic flow produced by material moved by this action (Houtman et al. 2007; Esseling-Ozdoba et al. 2008). Some of these myosins localize to specific organelles, indicating that different myosin XI proteins are responsible for the transport of specific organelles (Hashimoto 2003; Hashimoto et al. 2005; Reisen and Hanson 2007). Hoffmann and Nebenfuhr (2004) and Sheahan et al. (2007) show that myosins are also involved in determining the organization of the cytoplasm, the cytoarchitecture. It is likely that class XI myosins are responsible for this process, since myosin VIII proteins only localize to areas of the cell cortex.

In T-DNA insertion lines in the Myosin XIk gene, root hair length was reduced to approximately half that of the wild type. Root hair growth velocity was ~30% decreased in this mutant (Ojangu et al. 2007). As the root hair length was reduced more strongly than the growth rate, this suggests that root hairs grow over a shorter time frame. These lines were also defective in trichome development (Ojangu et al. 2007).

3.3.2 Profilin

Profilin is a G-actin-binding protein. Profilin-bound G-actin does not nucleate spontaneously, and incorporation at the pointed end of actin filaments is inhibited, but not incorporation at the barbed end, which continues at the normal velocity (Pollard et al. 2000; McCurdy et al. 2001). *Arabidopsis* possesses five genes that encode profilin, at least one of which is expressed in root hairs (Huang et al. 1996; Ramachandran et al. 2000). In pollen tubes, which are tip-growing cells like root hairs, profilin is evenly distributed through the cytoplasm (Vidali and Hepler 1997; Hepler et al. 2001). In root hairs, immunolabeling experiments have shown an apparent accumulation of profilin in the tips (Braun et al. 1999; Baluska et al. 2000b), but these measurements were not performed with the ratio-imaging technique. Therefore, it cannot be concluded that the profilin accumulation in the apex of root hairs reflects a higher concentration of profilin. Alternatively, the increased signal most probably reflects the accumulation of cytoplasm in the apex of root hairs, which leads to an increased accessible volume of cytoplasm in the root hair apex, and causes a local increase in signal of all cytoplasmic molecules, even if their concentration is not increased (Vos and Hepler 1998). To be conclusive, the accessible volume of the cytoplasm would have to be taken into account in a ratio-imaging experiment.

Although root hairs of plants that overexpress the profilin PFN-1 are twice as long as wild-type hairs (Ramachandran et al. 2000), they do not vary in diameter. The increased length of root hairs indicates that profilin is important for determining the total amount of cell growth. To determine whether the growth velocity is larger or whether root hairs continue to grow for a longer period of time in PFN-1 overexpressing plants, growth velocity measurements will have to be performed. Since actual cell elongation is dependent on the rate of exocytosis combined with force of osmotic pressure of the vacuole and with the cell wall expandability at the hair tip, it is expected that in these mutants root hairs grow not faster but for a longer time.

3.3.3 Actin-Depolymerizing Factor

Actin-depolymerizing factor (ADF), also called cofilin, is an evolutionary conserved protein that binds both to monomeric and F-actin, has the ability to sever F-actin at high pH (8.0), and enhances depolymerization at the pointed end of F-actin (Carlier et al. 1997). At lower pH (6.0) it binds actin filaments, but does not sever them (Gungabissoon et al. 1998). An ADF, ZmADF3, that is expressed in maize root hairs shares these properties: it severs F-actin at high pH (8.0), whereas it binds F-actin at lower pH 6.0 (Jiang et al. 1997). In vitro, it enhances the actin bundling activity of Zm elongation factor 1α in vitro (Gungabissoon et al. 2001). Phosphorylation of Ser-6 of ZmADF3 by a calmodulin-like domain protein kinase inhibits the activity of this ADF (Smertenko et al. 1998; Allwood et al. 2001), as does binding to the phospohlipids phosphatidylinositol 4,5-bisphosphate or phosphatidylinositol 4-monophosphate (Gungabissoon et al. 1998).

By immunolabeling after chemical fixation, Jiang et al. (1997) showed that ZmADF3 is present in the apical region of growing maize root hairs. This suggests that ZmADF3 may be involved in organizing the fine F-actin, which should also be present in this region but is not clearly visible in the presented micrographs, probably because of the suboptimal fixation conditions. Dong et al. (2001) produced *Arabidopsis* plants that overexpress AtADF1 and used expression of AtADF1 in the antisense orientation to decrease ADF levels. Plants overexpressing AtADF1 formed root hairs that have an increased radial diameter and are shorter (128.3 μm, compared with 173.6 μm in wild-type plants). This correlates with a reduction of the amount of longitudinal actin bundles and an irregular actin organization in the root hair tube (not quantified). The actin organization was determined by using GFP-mTn, which is now known to compete for actin binding with ADF (Ketelaar et al. 2004a). Thus, the reported F-actin organization may not be completely reliable. Root hairs of plants in which ADF levels were decreased did not show a change in diameter, but their average length was increased (225.0 μm, compared with 173.6 μm in wild-type plants). This correlated with an increased number of longitudinal bundles of F-actin in the root hair tubes (not quantified), as determined by GFP-mTn-based actin visualization (Dong et al. 2001).

Since GFP-mTn is not suitable to study the configuration of the fine F-actin, the actin organization in the subapex of the ADF over- and underexpressing root hairs cannot be deduced from the images that are included in the work of Dong et al. (2001). Only when the altered actin organization in the nonexpanding root hairs as observed by Dong et al. (2001) becomes a limiting factor in the delivery of exocytotic vesicles to the expanding root hair tip, it would affect root hair growth. Thus, the observed changes in actin organization by Dong et al. (2001) are not likely to be the direct cause of the observed defects in root hair development.

3.3.4 Actin Interacting Protein 1

Actin interacting protein 1 (AIP1) is an actin filament capping protein that enhances the activity of ADF in vitro (Allwood et al. 2002; Ono 2003). In *Arabidopsis*, two genes encode AIP1, one of which is expressed in floral tissues, whereas the other is expressed in vegetative tissues (Allwood et al. 2002). A T-DNA insertion line in AIP1 fails to produce homozygous plants, which suggests that AIP1 is essential for plant viability (Ketelaar et al. 2004b). To get an impression about the function of AIP1 in plant development, Ketelaar et al. (2004b) introduced an ethanol-inducible RNAi construct into *Arabidopsis*. Upon induction with ethanol, AIP1 levels in different lines were differentially reduced. In all lines, the root hair length was reduced, which correlated with a reduced growth velocity of root hairs after induction of the AIP1 RNAi with ethanol. In the most severe lines, hardly any root hair elongation took place. When the actin cytoskeleton was visualized by GFP:FABD2 expression, in the subapex of root hairs of the induced AIP1 RNAi lines, net-axial fine F-actin was visible. In control root hairs, the subapical fine F-actin shown after immunocytochemistry and fluorescent phalloidin staining procedures is not detectable with this construct on the confocal microscope that we used (see Sect. 2.2). This suggests that inhibition of AIP1 expression leads to a reduced dynamicity of F-actin in the subapex of growing root hairs to such an extent that sufficient delivery of exocytotic vesicles does not take place, which explains the reduced growth velocities and the reduced root hair lengths. Strangely, the reduction of root hair length in induced AIP1 RNAi lines is the opposite response than the longer root hairs that are produced when ADF is knocked down (see Sect. 3.3.3). This suggests that although ADF and AIP1 synergistically increase actin dynamics in vitro, in root hair development their activity may be antagonistic (Ketelaar et al. 2004b).

AIP1 levels in plants require tight regulation. Ethanol-inducible overexpression of AIP1 causes *Arabidopsis* plants to form swollen root hairs that occasionally branch (Ketelaar et al. 2007). When fully grown, in these root hairs many actin bundles are present that loop through the tip, far more bundles than in fully grown wild-type root hairs. Since AIP1 increases the turnover of F-actin, one would expect that AIP1 overexpression leads to an increase in actin depolymerization. The data of Ketelaar et al. (2007), more thick bundles of actin filaments, suggest that the opposite occurs.

3.3.5 The Actin-Related Protein 2/3 Complex

The ARP2/3 complex consists of the actin-related proteins 2 and 3 (ARP2/3), and five other subunits (ArpC1/p41, ARPC2/p31, ARPC3/p21, ARPC4/p20, and ARPC5/p16) (Higgs and Pollard 2001). In yeast and animal cells, activated Arp2/3 complex nucleates F-actin from the side of existing actin filaments in a fixed angle of 70° (Higgs and Pollard 2001). Hence, ARP2/3 complex-mediated actin nucleation results in the formation of a branched network of F-actin. Genes that encode the ARP2/3 complex subunits are present in the *Arabidopsis* genome (McKinney et al. 2001). Plant Arp2/3 subunits are able to complement yeast and mammalian Arp2/3 subunits and vice versa (Le et al. 2003; Mathur et al. 2003b; El-Din El-Assal et al. 2004). However, it remains to be biochemically proven whether the plant Arp2/3 complex performs an actin nucleating function similar to that of the Arp2/3 complex in other eukaryotes.

T-DNA insertions into genes that encode several Arp2/3 subunits (*ARP2*, *ARP3*, *ARPC5/p16*) and inhibit their expression have been reported (Li et al. 2003; Mathur et al. 2003a,b). These lines show developmental defects in a variety of cell types in *Arabidopsis*, such as trichomes, leaf pavement cells, and hypocotyl cells. Defects have also been reported in root hair development in Arp2/3 complex knockouts (Mathur et al. 2003a,b). Root hairs are wavy, have an increased width, and occasionally form multiple tips. These observations indicate that the Arp2/3 complex is involved in an actin-cytoskeleton-related process, probably the nucleation of the fine F-actin in growing root hairs. It is unlikely though that the Arp2/3 complex is the only nucleator that is involved in root hair growth, as one would expect more drastic developmental effects, such as the complete absence of root hairs, if the Arp2/3 complex would be solely responsible for actin nucleation in root hairs. More than being a nucleator, Arp2/3 complex appears to organize nucleation, giving rise to a certain F-actin configuration. This suggests that other actin-nucleating proteins are active in growing root hairs.

3.3.6 Formin

Formins are actin-nucleating proteins that consist of several conserved domains, including the formin homology 1 and 2 (FH1 and FH2) domains. Biochemical experiments show that the FH2 domain is involved in actin binding, whereas the FH1 domain is capable of binding to profilin and profilin-bound monomeric actin (Pruyne et al. 2002; Sagot et al. 2002; Li and Higgs 2003). Animal and yeast formins have the ability to bind to the barbed end of actin filaments and allow actin elongation while remaining associated with the barbed end due to the processive activity of the formin (Pruyne et al. 2002; Sagot et al. 2002; Kovar et al. 2003; Li and Higgs 2003; Higashida et al. 2004; Kovar and Pollard 2004; Romero et al. 2004). In the *Arabidopsis* genome, 21 genes that encode formin isoforms are present. These isoforms are separated into two distinct phylogenetic classes (Cvrckova 2000; Deeks et al. 2002; Cvrckova et al. 2004). Class I formins contain a transmembrane domain at their amino-terminus, which targets them to the plasma membrane (Cheung and Wu 2004; Favery et al. 2004; Ingouff et al. 2005). Class II formins remain to be characterized.

Michelot et al. (2005) showed that a truncated version of the *Arabidopsis* group I formin AFH1 can use profilin-bound actin to nucleate F-actin. Unlike yeast and animal formins, the active domain of AFH1 does not remain attached to the barbed end of an actin filament after nucleation, but stays attached to the side of actin filaments (Michelot et al. 2006).

Deeks et al. (2005) used ethanol-inducible overexpression of AtFH8 without the FH2 domain, fused to GFP to assess the role of formins in plant development. They found that upon induction of AtFH8 expression without the FH2 domain, root hair growth was inhibited. The inhibition occurred during or just after bulge formation. This suggests that AtFH8 and/or other formins may be key players in the building of the fine F-actin array, the prerequisite for root hair elongation.

3.3.7 Rho of Plants

Rho of plants (ROP) proteins are members of the Rho family of small GTPases. Strictly speaking, ROP proteins currently cannot be characterized as actin-binding proteins, as no direct interaction with actin has been reported. However, this major class of signaling proteins clearly is involved in regulating actin organization. For a detailed discussion of the ROP proteins and their (putative) effectors, see Zarsky and Fowler (2008).

4 Prospects

Besides the actin-binding proteins that have been found to affect the actin cytoskeleton in root hairs and affect root hair morphology, many other actin-binding proteins have been characterized. Their influence on root hair development has not been characterized, or root hair development is not affected when these proteins are inhibited or overexpressed. Besides continuing to study the role on root hair development of individual proteins, the next challenge will be to integrate the contributions of all the individual proteins and signals into a network of interactions and activities, which results in the formation of a proper actin cytoskeleton required for the formation of a root hair.

The advantages of root hairs as a model system for cell growth have made them an often used system to study the role of actin-binding proteins in cell growth. However, the currently available live cell actin visualization methods are not very well suitable to study the organization and dynamics of fine F-actin. More detailed analysis of how specific proteins contribute to the organization of fine F-actin in tips of growing root hairs will require an actin visualization method that stains fine F-actin in live cells and does not affect the actin dynamics or root hair morphology. The development of improved live cell actin visualization techniques that are able to study the dynamics of fine F-actin is likely to be a prerequisite for continued progress towards understanding the regulation and function of the actin cytoskeleton in root hairs in the near future.

Acknowledgments T.K. was supported by VENI fellowship 863.04.003 from the Dutch Science Foundation (NWO). A.M.C.E. thanks the FOM Institute for Atomic and Molecular Physics (AMOLF), Amsterdam, The Netherlands, for financial support for this project. We thank the FOM Institute for Atomic and Molecular Physics (AMOLF), Amsterdam, for allowing us to use their spinning disk microscope.

References

Allwood EG, Smertenko AP, Hussey PJ (2001) Phosphorylation of plant actin-depolymerising factor by calmodulin-like domain protein kinase. *FEBS Lett* 499:97–100

Allwood EG, Anthony RG, Smertenko AP, Reichelt S, Drobak BK, Doonan JH, Weeds AG, Hussey PJ (2002) Regulation of the pollen-specific actin-depolymerizing factor LIADF1. *Plant Cell* 14:2915–2927

An YQ, Huang SR, McDowell JM, McKinney EC, Meagher RB (1996) Conserved expression of the Arabidopsis ACT1 and ACT3 actin subclass in organ primordia and mature pollen. *Plant Cell* 8:15–30

Baluska F, Polsakiewicz M, Peters M, Volkmann D (2000a) Tissue-specific subcellular immunolocalization of a myosin-like protein in maize root apices. *Protoplasma* 212:37–145

Baluska F, Salaj J, Mathur J, Braun M, Jasper F, Samaj J, Chua NH, Barlow PW, Volkmann D (2000b) Root hair formation: F-actin-dependent tip growth is initiated by local assembly of profilin-supported F-actin meshworks accumulated within expansin-enriched bulges. *Dev Biol* 227:618–632

Baluska F, Cvrckova F, Kendrick-Jones J, Volkmann D (2001) Sink plasmodesmata as gateways for phloem unloading. Myosin VIII and calreticulin as molecular determinants of sink strength? *Plant Physiol* 126:39–46

Barriere Y, Denoue D, Briand M, Simon M, Jouanin L, Durand-Tardif M (2006) Genetic variations of cell wall digestibility related traits in floral stems of Arabidopsis thaliana accessions as a basis for the improvement of the feeding value in maize and forage plants. *Theor Appl Genet* 113:163–175

Baskin T, Busby C, Fowke L, Sammut M, Gubler F (1992) Improvements in immunostaining samples embedded in methacrylate: localization of microtubules and other antigens throughout developing organs in plants of diverse taxa. *Planta* 187:405–413

Baskin TI, Miller DD, Vos JW, Wilson JE, Hepler PK (1996) Cryofixing single cells and multicellular specimens enhances structure and immunocytochemistry for light microscopy. *J Microsc-Oxford* 182:149–161

Braun M, Baluska F, von Witsch M, Menzel D (1999) Redistribution of actin, profilin and phosphatidylinositol-4,5-bisphosphate in growing and maturing root hairs. *Planta* 209:435–443

Cardenas L, Vidali L, Dominguez J, Perez H, Sanchez F, Hepler PK, Quinto C (1998) Rearrangement of actin microfilaments in plant root hairs responding to Rhizobium etli nodulation signals. *Plant Physiol* 116:871–877

Carlier MF, Laurent V, Santolini J, Melki R, Didry D, Xia GX, Hong Y, Chua NH, Pantaloni D (1997) Actin depolymerizing factor (ADF/cofilin) enhances the rate of filament turnover: implication in actin-based motility. *J Cell Biol* 136:1307–1322

Cheung AY, Wu HM (2004) Overexpression of an Arabidopsis formin stimulates supernumerary actin cable formation from pollen tube cell membrane. *Plant Cell* 16:257–269

Cosgrove D (2005) Growth of the plant cell wall. *Nat Rev Mol Cell Biol* 6:850–861

Cvrckova F (2000) Are plant formins integral membrane proteins? Genome Biol 1:RESEARCH001

Cvrckova F, Rivero F, Bavlnka B (2004) Evolutionarily conserved modules in actin nucleation: lessons from Dictyostelium discoideum and plants. *Rev Artic Protoplasma* 224:15–31

Dai J, Sheetz MP (1995) Mechanical properties of neuronal growth cone membranes studied by tether formation with laser optical tweezers. *Biophys J* 68:988–996

Dauphin A, De Ruijter NC, Emons AM, Legue V (2006) Actin organization during eucalyptus root hair development and its response to fungal hypaphorine. *Plant Biol (Stuttg)* 8:204–211

de Ruijter NCA, Emons AMC (1999) Actin-binding proteins in plant cells. *Plant Biol* 1:26–35

de Ruijter NCA, Rook MB, Bisseling T, Emons AMC (1998) Lipochito-oligosaccharides re-initiate root hair tip growth in Vicia sativa with high calcium and spectrin-like antigen at the tip. *Plant J* 13:341–350

de Ruijter NCA, Bisseling T, Emons AMC (1999) Rhizobium Nod factors induce an increase in sub-apical fine bundles of actin filaments in Vicia sativa root hairs within minutes. *Mol Plant-Microbe Interact* 12:829–832

de Ruijter NCA, Ketelaar T, Blumenthal SSD, Emons AMC, Schel JHN (2000) Spectrin-like proteins in plant nuclei. *Cell Biol Int* 24:427–438

Deeks MJ, Hussey PJ, Davies B (2002) Formins: intermediates in signal-transduction cascades that affect cytoskeletal reorganization. *Trends Plant Sci* 7:492–498

Deeks MJ, Cvrckova F, Machesky LM, Mikitova V, Ketelaar T, Zarsky V, Davies B, Hussey PJ (2005) Arabidopsis group Ie formins localize to specific cell membrane domains, interact with actin-binding proteins and cause defects in cell expansion upon aberrant expression. *New Phytol* 168:529–540

Ditengou FA, Raudaskoski M, Lapeyrie F (2003) Hypaphorine, an indole-3-acetic acid antagonist delivered by the ectomycorrhizal fungus Pisolithus tinctorius, induces reorganisation of actin and the microtubule cytoskeleton in Eucalyptus globulus ssp bicostata root hairs. *Planta* 218:217–225

Dolan L, Roberts K (1995) The development of cell pattern in the root epidermis. *Philos Trans R Soc Lond B Biol Sci* 350:95–99

Dolan L, Duckett CM, Grierson C, Linstead P, Schneider K, Lawson E, Dean C, Poethig S, Roberts K (1994) Clonal relationships and cell patterning in the root epidermis of arabidopsis. *Dev Biol* 120:2465–2474

Dong CH, Xia GX, Hong Y, Ramachandran S, Kost B, Chua NH (2001) ADF proteins are involved in the control of flowering and regulate F-actin organization, cell expansion, and organ growth in Arabidopsis. *Plant Cell* 13:1333–1346

El-Din El-Assal S, Le J, Basu D, Mallery EL, Szymanski DB (2004) DISTORTED2 encodes an ARPC2 subunit of the putative Arabidopsis ARP2/3 complex. *Plant J* 38:526–538

Emons AMC (1987) The cytoskeleton and secretory vesicles in root hairs of equisetum and limnobium and cytoplasmic streaming in root hairs of equisetum. *Ann Bot* 60:625–632

Emons AMC, Derksen J (1986) Microfibrils, microtubules and microfilaments of the trichoblast of *equisetum hyemale. Acta Bot Neerl* 35 311–320

Emons AMC, Ketelaar T (2008) Intracellular organization: a prerequisite for root hair elongation and cell wall deposition. In: Emons AMC, Ketelaar T (eds) Root hairs: excellent tools for the study of plant molecular cell biology. Springer, Berlin Heidelberg New York. doi:10.1007/7089_2008_4

Esseling JJ, Emons AMC (2004) Dissection of Nod factor signalling in legumes: cell biology, mutants and pharmacological approaches. *J Microsc-Oxford* 214:104–113

Esseling JJ, de Ruijter NCA, Emons AMC (2000) The root hair actin cytoskeleton as backbone, highway, morphogenetic instrument and target for signalling. In: Ridge RW and Emons AMC (eds). Root Hairs: Cell and Molecular Biology. Springer, Tokyo, Japan. pp 29–52

Esseling JJ, Lhuissier FGP, Emons AMC (2004a). A nonsymbiotic root hair tip growth phenotype in dulation NORK-mutated legumes: Implications for nodulation factor-induced signaling and formation of a multifaceted root hair pocket for bacteria. *Plant Cell* 16:933–944

Esseling-Ozdoba et al. (2008) J Microsc (in press)

Favery B, Chelysheva LA, Lebris M, Jammes F, Marmagne A, Almeida-Engler J, De Lecomte P, Vaury C, Arkowitz RA, Abad P (2004) Arabidopsis formin AtFH6 is a plasma membrane-associated protein upregulated in giant cells induced by parasitic nematodes. *Plant Cell* 16:2529–2540

Feijo JA, Costa SS, Prado AM, Becker JD, Certal AC (2004) Signalling by tips. *Curr Opin Plant Biol* 7:589–598

Galway ME (2000) Root hair ultrastructure and tip growth. In: Ridge RW, Emons AMC (eds) Root hairs, cell and molecular biology, pp 1–15

Galway ME, Heckman JW Jr, Schiefelbein JW (1997) Growth and ultrastructure of Arabidopsis root hairs: the rhd3 mutation alters vacuole enlargement and tip growth. *Planta* 201:209–218

Gilliland LU, Kandasamy MK, Pawloski LC, Meagher RB (2002) Both vegetative and reproductive actin isovariants complement the stunted root hair phenotype of the Arabidopsis act2-1 mutation. *Plant Physiol* 130:2199–2209

Gilliland LU, Pawloski LC, Kandasamy MK, Meagher RB (2003) Arabidopsis actin gene ACT7 plays an essential role in germination and root growth. *Plant J* 33:319–328

Grierson C, Ketelaar T (2004) Development of Root hairs. In: Hussey PJ (ed) The plant cytoskeleton in cell differentiation and development. Annual plant reviews series, Blackwell, London

Gungabissoon RA, Jiang CJ, Drobak BK, Maciver SK, Hussey PJ (1998) Interaction of maize actin-depolymerising factor with actin and phosphoinositides and its inhibition of plant phospholipase C. *Plant Journal* 16:689–696

Gungabissoon RA, Khan S, Hussey PJ, Maciver SK (2001) Interaction of elongation factor 1 alpha from Zea mays (ZmEF-1 alpha) with F-actin and interplay with the maize actin severing protein, ZmADF3. *Cell Motil Cytoskeleton* 49:104–111

Hashimoto T (2003) Dynamics and regulation of plant interphase microtubules: a comparative view. *Curr Opin Plant Biol* 6:568–576

Hashimoto K, Igarashi H, Mano S, Nishimura M, Shimmen T, Yokota E (2005) Peroxisomal localization of a myosin XI isoform in Arabidopsis thaliana. *Plant Cell Physiol* 46:782–789

He X, Liu YM, Wang W, Li Y (2006) Distribution of G-actin is related to root hair growth of wheat. *Ann Bot (Lond)* 98:49–55

Heidstra R, Geurts R, Franssen H, Spaink HP, Van Kammen A, Bisseling T (1994) Root hair deformation activity of nodulation factors and their fate on vicia sativa. *Plant Physiol* 105:787–797

Heidstra R, Yang WC, Yalcin Y, Peck S, Emons A, Van KA, Bisseling T (1997) Ethylene provides positional information on cortical cell division but is not involved in Nod factor-induced root hair tip growth in Rhizobium-legume interaction. *Development Cambridge* 124:1781–1787

Hepler PK, Vidali L, Cheung AY (2001). Polarized cell growth in higher plants. *Annu Rev Cell Dev Biol* 17:159–187

Higashida C, Miyoshi T, Fujita A, Oceguera-Yanez F, Monypenny J, Andou Y, Narumiya S, Watanabe N (2004) Actin polymerization-driven molecular movement of mDia1 in living cells. *Science* 303:2007–2010

Higgs HN, Pollard TD (2001) Regulation of actin filament network formation through ARP2/3 complex: activation by a diverse array of proteins. *Annu Rev Biochem* 70:649–676

Hoffmann A, Nebenfuhr A (2004) Dynamic rearrangements of transvacuolar strands in BY-2 cells imply a role of myosin in remodeling the plant actin cytoskeleton. *Protoplasma* 224:201–210

Holweg CL (2007) Living markers for actin block myosin-dependent motility of plant organelles and auxin. *Cell Motil Cytoskeleton* 64:69–81

Houtman D, Pagonabarraga I, Lowe CP, Esseling-Ozdoba A, Emons AMC, Eiser E (2007) Hydrodynamic flow caused by active transport along cytoskeletal elements. *Eur Phys Lett* 78:18001

Huang SR, McDowell JM, Weise MJ, Meagher RB (1996) The Arabidopsis profilin gene family - Evidence for an ancient split between constitutive and pollen-specific profilin genes. *Plant Physiol* 111:115–126

Ingouff M, Fitz Gerald JN, Guerin C, Robert H, Sorensen MB, Van Damme D, Geelen D, Blanchoin L, Berger F (2005) Plant formin AtFH5 is an evolutionarily conserved actin nucleator involved in cytokinesis. *Nat Cell Biol* 7:374–380

Jiang S, Ramachandran S (2004) Identification and molecular characterization of myosin gene family in Oryza sativa genome. *Plant Cell Physiol* 45:590–599

Jiang CJ, Weeds AG, Hussey PJ (1997) The maize actin depolymerizing factor, ZmADF3, redistributes to the growing tip of elongating root hairs and can be induced to translocate into the nucleus with actin. *Plant J* 12:1035–1043

Ketelaar T, Emons AMC (2001) The cytoskeleton in plant cell growth: lessons from root hairs. *New Phytol* 152:409–418

Ketelaar T, Faivre-Moskalenko C, Esseling JJ, de Ruijter NC, Grierson CS, Dogterom M, Emons AM (2002) Positioning of nuclei in Arabidopsis root hairs: an actin-regulated process of tip growth. *Plant Cell* 14:2941–2955

Ketelaar T, de Ruijter NC, Emons AM (2003) Unstable F-actin specifies the area and microtubule direction of cell expansion in Arabidopsis root hairs. *Plant Cell* 15:285–292

Ketelaar T, Anthony RG, Hussey PJ (2004a) Green fluorescent protein-mTalin causes defects in actin organization and cell expansion in Arabidopsis and inhibits actin depolymerizing factor's actin depolymerizing activity in vitro. *Plant Physiol* 136:3990–3998

Ketelaar T, Allwood EG, Anthony R, Voigt B, Menzel D, Hussey PJ (2004b) The actin-interacting protein AIP1 is essential for actin organization and plant development. *Curr Biol* 14:145–149

Ketelaar T, Allwood EG, Hussey PJ (2007) Actin organization and root hair development are disrupted by ethanol-induced overexpression of Arabidopsis actin interacting protein 1 (AIP1). *New Phytol* 174:57–62

Kiss JZ, McDonald K (1993) Electron microscopy immunocytochemistry following cryofixation and freeze substitution. *Methods Cell Biol* 37:311–341

Kiss JZ, Giddings TH Jr, Staehelin LA, Sack FD (1990) Comparison of the ultrastructure of conventionally fixed and high pressure frozen/freeze substituted root tips of Nicotiana and Arabidopsis. *Protoplasma* 157:64–74

Klein MG, Shi W, Ramagopal U, Tseng Y, Wirtz D, Kovar DR, Staiger CJ, Almo SC (2004) Structure of the actin crosslinking core of fimbrin. *Structure* 12:999–1013

Kost B, Spielhofer P, Chua NH (1998) A GFP-mouse talin fusion protein labels plant actin filaments in vivo and visualizes the actin cytoskeleton in growing pollen tubes. *Plant J* 16:393–401

Kovar DR, Pollard TD (2004) Progressing actin: formin as a processive elongation machine. *Nat Cell Biol* 6:1158–1159
Kovar DR, Kuhn JR, Tichy AL, Pollard TD (2003) The fission yeast cytokinesis formin Cdc12p is a barbed end actin filament capping protein gated by profilin. *J Cell Biol* 161:875–887
Le J, El-Assal Sel D, Basu D, Saad ME, Szymanski DB (2003) Requirements for Arabidopsis ATARP2 and ATARP3 during epidermal development. *Curr Biol* 13, 1341–1347
Lee Y, Chan SI (1977) Effect of lysolecithin on the structure and permeability of lecithin bilayer vesicles. *Biochemistry* 16:1303–1309
Li F, Higgs HN (2003) The mouse Formin mDia1 is a potent actin nucleation factor regulated by autoinhibition. *Curr Biol* 13:1335–1340
Li S, Blanchoin L, Yang Z, Lord EM (2003) The putative Arabidopsis arp2/3 complex controls leaf cell morphogenesis. *Plant Physiol* 132:2034–2044
Limpens E, Bisseling T (2008) Nod factor signal transduction in the Rhizobium-Legume symbiosis. In: Emons AMC, Ketelaar T (eds) Root hairs: excellent tools for the study of plant molecular cell biology. Springer, Berlin Heidelberg New York. doi:10.1007/7089_2008_10
Mathur J, Mathur N, Kernebeck B, Hulskamp M (2003a) Mutations in actin-related proteins 2 and 3 affect cell shape development in Arabidopsis. *Plant Cell* 15:1632–1645
Mathur J, Mathur N, Kirik V, Kernebeck B, Srinivas BP, Hulskamp M (2003b) Arabidopsis CROOKED encodes for the smallest subunit of the ARP2/3 complex and controls cell shape by region specific fine F-actin formation. *Dev Biol* 130:3137–3146
McCann RO, Craig SW (1997) The I/LWEQ module: a conserved sequence that signifies F-actin binding in functionally diverse proteins from yeast to mammals. *Proc Natl Acad Sci USA* 94:5679–5684
McCurdy DW, Kovar DR, Staiger CJ (2001) Actin and actin-binding proteins in higher plants. *Protoplasma* 215:89–104
McDowell JM, An YQ, Huang SR, McKinney EC, Meagher RB (1996) The arabidopsis ACT7 actin gene is expressed in rapidly developing tissues and responds to several external stimuli. *Plant Physiol* 111:699–711
McKinney EC, Kandasamy MK, Meagher RB (2001) The Arabidopsis genome contains ancient classes of extremely complex and differentially expressed actin-related protein (ARP) genes. *Mol Biol Cell* 12:32a–33a
Michelot A, Guerin C, Huang S, Ingouff M, Richard S, Rodiuc N, Staiger CJ, Blanchoin L (2005) The formin homology 1 domain modulates the actin nucleation and bundling activity of arabidopsis FORMIN1. *Plant Cell* 17:2296–2313
Michelot A, Derivery E, Paterski-Boujemaa R, Guerin C, Huang S, Parcy F, Staiger CJ, Blanchoin L (2006) A novel mechanism for the formation of actin-filament bundles by a nonprocessive formin. *Curr Biol* 16:1924–1930
Miller DD, de Ruijter NCA, Bisseling T, Emons AMC (1999) The role of actin in root hair morphogenesis: studies with lipochito-oligosaccharide as a growth stimulator and cytochalasin as an actin perturbing drug. *Plant J* 17:141–154
Miller DD, Leferink-ten Klooster HB, Emons AM (2000) Lipochito-oligosaccharide nodulation factors stimulate cytoplasmic polarity with longitudinal endoplasmic reticulum and vesicles at the tip in vetch root hairs. *Mol Plant Microbe Interact* 13 :1385–1390
Molendijk AJ, Bischoff F, Rajendrakumar CS, Friml J, Braun M, Gilroy S, Palme K (2001) Arabidopsis thaliana Rop GTPases are localized to tips of root hairs and control polar growth. *Embo J* 20:2779–2788
Morris D, McNeil R, Castellino F, Thomas J (1980) Interaction of lysophosphatidylcholine with phosphatidylcholine bilayers. A photo-physical and NMR study. *Biochim Biophys Acta* 599:380–390
Müller M, Moor H (1984) Cryofixation of thick specimens by high pressure freezing. In: Revel JP, Barnard T, Haggins GC (eds) The science of biological specimen preparation. SEM, AMF O'Hare, Chicago, pp 131–138
Nicolas MT, Bassot JM (1993) Freeze substitution after fast-freeze fixation in preparation for immunocytochemistry. *Microsc Res Tech* 24:474–487

Nielsen E (2008) Plant cell wall biogenesis during tip growth in root hair cells. In: Emons AMC, Ketelaar T (eds) Root hairs: excellent tools for the study of plant molecular cell biology. Springer, Berlin Heidelberg New York. doi:10.1007/7089_2008_11

Nishida E, Iida K, Yonezawa N, Koyasu S, Yahara I, Sakai H (1987) Cofilin is a component of intranuclear and cytoplasmic actin rods induced in cultured cells. *Proc Natl Acad Sci USA* 84:5262–5266

Ojangu EL, Jarve K, Paves H, Truve E (2007) Arabidopsis thaliana myosin XIK is involved in root hair as well as trichome morphogenesis on stems and leaves. *Protoplasma* 230:193–202

Ono S (2003) Regulation of actin filament dynamics by actin depolymerizing factor/cofilin and actin-interacting protein 1: new blades for twisted filaments. *Biochemistry* 42:13363–13370

Ovecka M, Nadubinska M, Volkmann D, Baluska F (2000) Actomyosin and exocytosis inhibitors alter root hair morphology in Poa annua. *Biologia* 55:105–114

Pollard TD, Blanchoin L, Mullins RD (2000) Molecular mechanisms controlling actin filament dynamics in nonmuscle cells. *Annu Rev Biophys Biomol Struct* 29:545–576

Pruyne D, Evangelista M, Yang C, Bi E, Zigmond S, Bretscher A, Boone C (2002) Role of formins in actin assembly: nucleation and barbed-end association. *Science* 297:612–615

Ramachandran S, Christensen HE, Ishimaru Y, Dong CH, Chao-Ming W, Cleary AL, Chua NH (2000) Profilin plays a role in cell elongation, cell shape maintenance, and flowering in Arabidopsis. *Plant Physiol* 124:1637–1647

Reddy AS (2001) Molecular motors and their functions in plants. *Int Rev Cytol* 204:97–178

Reichelt S, Knight AE, Hodge TP, Baluska F, Samaj J, Volkmann D, Kendrick-Jones J (1999) Characterization of the unconventional myosin VIII in plant cells and its localization at the post-cytokinetic cell wall. *Plant J* 19:555–567

Reisen D, Hanson MR (2007) Association of six YFP-myosin XI-tail fusions with mobile plant cell organelles. *BMC Plant Biol* 7:6

Ressad F, Didry D, Xia GX, Hong Y, Chua NH, Pantaloni D, Carlier MF (1998) Kinetic analysis of the interaction of actin-depolymerizing factor (ADF)/cofilin with G- and F-actins. Comparison of plant and human ADFs and effect of phosphorylation. *J Biol Chem* 273:20894–20902

Ridge RW (1988) Freeze-substitution improves the ultrastructural preservation of legume root hairs. *Bot J Tokyo* 101:427–441

Ringli C, Baumberger N, Diet A, Frey B, Keller B (2002) ACTIN2 is essential for bulge site selection and tip growth during root hair development of Arabidopsis. *Plant Physiol* 129:1464–1472

Romero S, Le Clainche C, Didry D, Egile C, Pantaloni D, Carlier MF (2004) Formin is a processive motor that requires profilin to accelerate actin assembly and associated ATP hydrolysis. *Cell* 119:419–429

Ryan KP (1992) A simple apparatus and technique for the rapid-freeze and freeze-substitution of single-cell algae. *J Electron Microsc* 39:120–124

Sagot I, Rodal AA, Moseley J, Goode BL, Pellman D (2002) An actin nucleation mechanism mediated by Bni1 and profilin. *Nat Cell Biol* 4:626–631

Sheahan MB, Staiger CJ, Rose RJ, McCurdy DW (2004) A green fluorescent protein fusion to actin-binding domain 2 of Arabidopsis fimbrin highlights new features of a dynamic actin cytoskeleton in live plant cells. *Plant Physiol* 136:3968–3978

Sheahan MB, Rose RJ, McCurdy DW (2007) Actin-filament-dependent remodeling of the vacuole in cultured mesophyll protoplasts. *Protoplasma* 230:141–152

Sherrier DJ, Vandenbosch KA (1994) Secretion of cell wall polysaccharides in Vicia root hairs. *Plant J* 5:185–195

Sieberer B, Emons AMC (2000) Cytoarchitecture and pattern of cytoplasmic streaming in root hairs of Medicago truncatula during development and deformation by nodulation factors. *Protoplasma* 214:118–127

Sieberer BJ, Timmers ACJ (2008) Microtubules in Plant Root Hairs and Their Role in Cell Polarity and Tip Growth. In: Emons AMC, Ketelaar T (eds) Root hairs: excellent tools for the study of plant molecular cell biology. Springer, Berlin Heidelberg New York. doi:10.1007/7089_2008_13

Sieberer BJ, Timmers AC, Lhuissier FG, Emons AM (2002) Endoplasmic microtubules configure the subapical cytoplasm and are required for fast growth of Medicago truncatula root hairs. *Plant Physiol* 130:977–988

Smertenko AP, Jiang CJ, Simmons NJ, Weeds AG, Davies DR, Hussey PJ (1998) Ser6 in the maize actin-depolymerizing factor, ZmADF3, is phosphorylated by a calcium-stimulated protein kinase and is essential for the control of functional activity. *Plant J* 14:187–193

Smith LG, Li R (2004) Actin polymerization: riding the wave. *Curr Biol* 14:R109–R111

Sonobe S, Shibaoka H (1989) Cortical fine actin filaments in higher plant cells visualised by rhodamine-phalloidin after pretreatment with m-maleimidobenzoyl N-hydroxysuccinimide ester. *Protoplasma* 148:80–86

Steedman HF (1957) Polyester wax; a new ribboning embedding medium for histology. *Nature* 179:1345

Thijssen M, Van Went J, Van Aelst A (1998) Heptane and isooctane as embedding fluids for high-pressure freezing of Petunia ovules followed by freeze-substitution. *J Microsc* 192:228–235

Valster AH, Pierson ES, Valenta R, Hepler PK, Emons AMC (1997) Probing the plant actin cytoskeleton during cytokinesis and interphase by profilin microinjection. *Plant Cell* 9:1815–1824

Van Bruaene N, Joss G, Thas O, Van Oostveldt P (2003) Four-dimensional imaging and computer-assisted track analysis of nuclear migration in root hairs of Arabidopsis thaliana. *J Microsc* 211:167–178

van der Honing HS, Emons AM, Ketelaar T (2007) Actin based processes that could determine the cytoplasmic architecture of plant cells. *Biochim Biophys Acta* 1773:604–614

Vidali L, Hepler PK (1997) Characterization and localization of profilin in pollen grains and tubes of Lilium longiflorum. *Cell Motil Cytoskeleton* 36:323–338

Vitha S, Baluska F, Mews M, Volkman D (1997) Immunofluorescence detection of F-actin on low melting point wax sections from plant tissues. *J Histochem Cytochem* 45:89–95

Vitha S, Baluska F, Braun M, Samaj J, Volkmann D, Barlow PW (2000) Comparison of cryofixation and aldehyde fixation for plant actin immunocytochemistry: aldehydes do not destroy F-actin. *Histochem J* 32:457–466

Voigt B, Timmers AC, Samaj J, Hlavacka A, Ueda T, Preuss M, Nielsen E, Mathur J, Emans N, Stenmark H, Nakano A, Baluska F, Menzel D (2005) Actin-based motility of endosomes is linked to the polar tip growth of root hairs. *Eur J Cell Biol* 84:609–621

Vos JW, Hepler PK (1998) Calmodulin is uniformly distributed during cell division in living stamen hair cells of Tradescantia virginiana. *Protoplasma* 201:158–171

Wang X, Teng Y, Wang Q, Li X, Sheng X, Zheng M, Samaj J, Baluska F, Lin J (2006) . Imaging of dynamic secretory vesicles in living pollen tubes of Picea meyeri using evanescent wave microscopy. *Plant Physiol* 141:1591–1603

Wasteneys GO, Willingale Theune J, Menzel D (1997) Freeze shattering: a simple and effective method for permeabilizing higher plant cell walls. *J Microsc-Oxford* 188:51–61

Wymer CL, Bibikova TN, Gilroy S (1997) Cytoplasmic free calcium distributions during the development of root hairs of Arabidopsis thaliana. *Plant J* 12:427–439

Zarsky V, Fowler J (2008) ROP (Rho-related protein from Plants) GTPases for spatial control of root hair morphogenesis. In: Emons AMC, Ketelaar T (eds) Root hairs: excellent tools for the study of plant molecular cell biology. Springer, Berlin Heidelberg New York. doi:10.1007/7089_2008_14

Microtubules in Plant Root Hairs and Their Role in Cell Polarity and Tip Growth

B.J. Sieberer and A.C.J. Timmers (✉)

Abstract The cytoskeleton is part of the tip-growth machinery that assembles in the subapex of elongating root hairs of plant roots. The role of actin in the tip-growth process is well studied and understood (see Ketelaar and Emons 2008), whereas the function of microtubules (MTs) is less clear. However, recent studies demonstrate that MTs also play a crucial role in the tip-growth process. Here we discuss recent techniques to visualize MTs in root hairs, give an overview on current knowledge of structure, dynamics, and function of the MT cytoskeleton in elongating root hairs of different plant species, and give an outlook on future perspectives.

1 Methods to Visualize MTs in Root Hairs

In the past, chemical fixation, in combination with transmission electron microscopy (TEM) (Newcomb and Bonnett 1965; Emons 1982; Emons and Wolters-Arts 1983; Bakhuizen 1988) or immuno fluorescence microscopy (*Equisetum hyemale*: Traas et al. 1985; *Medicago sativa*: Weerasinghe et al. 2003; *Vicia hirsuta*: Lloyd et al. 1987), was used to study the microtubule (MT) cytoskeleton in root hairs. Despite the potential of these methods to visualize MTs, it was not possible to obtain detailed information either on whole-cell MT organization during root hair development or on MT dynamics. Over the past few years, new tools in cell biology have allowed new insights into the nature of the MT cytoskeleton in root hairs. In particular, improved fixation methods (rapid freeze fixation followed by freeze substitution), which had earlier been used for electron microscopy of root hairs (Emons 1987), were used for light microscopy imaging of MTs (Sieberer et al. 2002). In addition, advances in microscopy, combined with in vivo labelling of MTs using green fluorescent protein (GFP) technology, made the study of MT dynamics possible. GFP, either fused to the MT-binding domain (MBD) of mouse

A.C.J. Timmers
Laboratory of Plant–Microorganism Interactions, CNRS INRA, UMR2594, 24 Chemin de Borde Rouge, BP52627, F-31326 Castanet-Tolosan, France
ton.timmers@toulouse.inra.fr

Plant Cell Monogr, doi:10.1007/7089_2008_13

microtubule-associated protein MAP4 (GFP:MBD; Marc et al. 1998; Sieberer et al. 2002; van Bruaene et al. 2004), or to tubulin-a6 of *Arabidopsis* (GFP:TUA6; Vassileva et al. 2005), or yellow fluorescent protein fused to the MT end binding protein 1 (EB1:YFP; Timmers et al. 2007) has made it possible to obtain a clear picture of the structure of the MT array in root hairs of plants and to follow MT dynamics during root hair development.

Rapid freeze fixation is the method of choice when preserving ultrastructural cellular components and the cytoskeleton and minimizing fixation artifacts. For root hairs we have developed a rapid-freeze whole mount procedure, in which whole roots or plant seedlings were rapidly plunged into liquid propane cooled to –180°C by liquid nitrogen and kept there for at least 20 s. The excised roots were freeze substituted in water-free methanol containing 0.05% glutaraldehyde for 48 h at –90°C and allowed to reach room temperature over a 24-h period. Samples were rehydrated in a graded series of methanol in phosphate-buffered saline containing fixative (0.1% glutaraldehyde and 4% paraformaldehyde). After rehydration, a partial cell wall digestion was carried out and the samples were probed with antibodies. This method allowed detection of MTs in root hairs with a high spatial resolution and revealed details that went unnoticed with other techniques, such as chemical fixation and in vivo GFP technology.

For in vivo decoration of MTs with GFP:MBD in *Medicago* root hairs we used an *Agrobacterium-rhizogenes*-mediated transformation strategy (Boisson-Dernier et al. 2001; for details and procedures, see Chabaud et al. 2006). This transformation method yields composite plants with nontransformed shoots and transformed roots that are expressing the reporter protein of interest. The composite plants can be kept for several weeks; hairy root cultures derived from these composite plants can be maintained for several months. Compared with *A. tumefaciens*-mediated transformation, where it can take up to 18 months to obtain stably expressing lines (for *Medicago*), it only takes 3–4 weeks with *A. rhizogenes*. This allows in a relatively time-efficient manner insertion of various reporter proteins (e.g., double transformation with GFP:MBD and EB1:YFP; Timmers et al. 2007) in *Medicago* wild-type or mutant roots.

In our hands, an advanced fixation protocol and in vivo decoration of MTs with GFP reporter proteins are techniques which are complementing each other and minimize the chance of observing artifacts induced by either technique. Furthermore, fixed samples allow one to resolve spatial details of a three-dimensional (3D) MT array spanning throughout a cell, whereas in vivo GFP technology enables us to follow rearrangements of this array and the dynamics of individual MTs. The study of the MT dynamics has been improved considerably by the development of powerful tracking software, such as DiaTrack, which enables the tracking of the complete population of MTs in a single cell (Timmers et al. 2007).

2 MT Organization in Developing Root Hairs

In root hairs, MTs occur in two different configurations: cortical microtubules (CMTs) and endoplasmic microtubules (EMTs) as shown for *Medicago truncatula* in Fig. 1. CMTs are present in all stages of root hair development, whereas EMTs

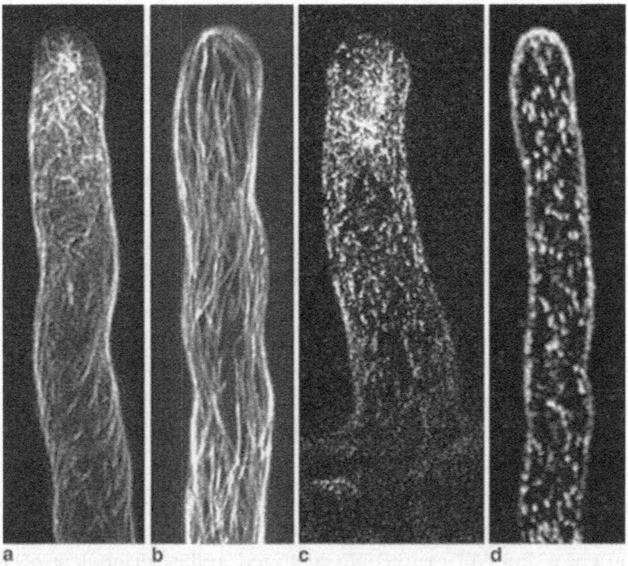

Fig. 1 The organization of the microtubular cytoskeleton in root hairs of *Medicago truncatula*. Growing root hairs (**a** and **c**), fully grown root hairs (**b** and **d**). In (**a**) and (**b**) the cytoskeleton is visualized with GFP:MBD; in (**c**) and (**d**) with EB1:YFP. See the text for the description. (Reproduced with permission from the European Journal of Cell Biology 86 (2007) 69–83)

appear exclusively during initiation of tip growth and remain present until the growth ceases.

CMTs are present in root hairs of all plant species examined (for review, see Ketelaar and Emons 2000, 2001). They form dense arrays in growing root hairs and their overall orientation is either net-axial (*Equisetum hyemale*: Emons 1982; Emons and Wolters-Arts 1983; Traas et al. 1985; *Medicago*: Sieberer et al. 2002; *M. sativa*: Weerasinghe et al. 2003; *Arabidopsis*: van Bruaene et al. 2004) or helical (*Allium cepa*: Lloyd 1983; Traas et al. 1985). The very tip of an elongating hair – the site of actual cell growth – appears to be less dense or devoid of CMTs (evidence from TEM studies – *Raphanus sativus:* Newcomb and Bonnett 1965; *E. hyemale:* Emons 1989; evidence from light microscopy studies – *E. hyemale:* Traas et al. 1985; *Lotus japonicus*: Vassileva et al. 2005; *M. sativa*: Weerasinghe et al. 2003; *M. truncatula*: Sieberer et al. 2002; *Arabidopsis*: van Bruaene et al. 2004). This is different from fully grown root hairs, where CMTs converge at the very tip. After tip growth has ceased, CMT density decreases over time, while their overall orientation remains longitudinal (*Medicago*: Sieberer et al. 2002) or helical (*Arabidopsis*: van Bruaene et al. 2004). In fully grown root hairs of *Arabidopsis*, CMTs are more aligned in parallel than they are in growing hairs (Van Bruaene et al. 2004). Double-labelling experiments of MTs and MT plus-ends in *M. truncatula* root hairs indicate that most of the CMTs could be organized into bundles: MT plus-ends labelled with EB1:YFP show movement along existing MTs decorated with DsRed:MBD (Timmers et al.

2007, and supplementary online material therein). This is in contrast with observations in *Arabidopsis* where CMTs in fully grown root hairs were considered as single MTs (Van Bruaene et al. 2004). The question whether CMTs in root hairs exclusively appear either as bundles or as single MTs, or as a mixture of both should be addressed in a high-resolution electron microscopy study.

Root hairs acquire EMTs when tip growth begins and retain them in their subapical, cytoplasmically dense region until growth stops. Subapical EMTs have been found in several plant species examined – though with differences in patterning and abundance. In *Medicago* and *Arabidopsis*, EMTs were observed around and close to the nucleus at high density (Sieberer et al. 2002; van Bruaene et al. 2004). Van Bruaene et al. (2004) described distinct sites in the vicinity of the surface of the nucleus where EMTs appear in young tip-growing *Arabidopsis* hairs. The authors hypothesize that these locations are MT nucleation sites and that in elongating *Arabidopsis* root hairs EMTs may be involved in the transport of new MT nucleation complexes from the nucleus to other sites in the root hair. As such, these structures resemble so-called satellite microtubule organizing centers, moving particles containing γ-tubulin identified in fission yeast. As in plants, interphase MT organizing centers have not been identified as discrete structures in fission yeast, but MT origins are scattered in the perinuclear region and along existing MTs (Höög et al. 2007). In elongating root hairs of the legume species *L. japonicus* (Sieberer et al. 2005a), *M. truncatula* (Sieberer et al. 2002; Timmers et al. 2007), *M. sativa* (Weerasinghe et al. 2003), and *V. hirsuta* (Lloyd et al. 1987) EMTs have been visualized with different methods. From recent studies with confocal laser scanning microscopy on fixed and living material EMTs appear to be organized into a vast 3D array that spreads throughout the entire subapex between the nucleus and the tip region of elongating hairs, with a few EMTs extending into the very tip. In contrast to legumes, EMTs in growing *Arabidopsis* root hairs do not form a dense array in the subapex, but consist, at most, of a few bundles of EMTs between hair tip and the nucleus, as revealed by studies with GFP:MBD-labelled MTs (Van Bruaene et al. 2004). Using immunocytochemistry after rapid freeze fixation, EMTs could not be detected in *Arabidopsis* hairs (Sieberer et al. 2005a; Ketelaar et al. 2002), or possibly this method gave a much weaker signal for EMTs than in legume root hairs.

3 MT Dynamics in Root Hairs

So far, MT dynamics in root hairs have been studied in detail with quantitative live-cell imaging techniques only in three plant species – *A. thaliana* (Van Bruaene et al. 2004), *L. japonicus* (Vassileva et al. 2005), and *M. truncatula* (Timmers et al. 2007). In the first two studies, only CMTs were analyzed because of technical limitations, whereas the latter study also addressed the dynamic behavior of EMTs.

In brief, all three studies come to similar conclusions concerning general features of MT dynamics in root hairs. Growth and shrinkage rates of root hair MTs are within the range published for other plant systems (Dhonukshe and Gadella

2003; Shaw et al. 2003; Vos et al. 2004; for a review, see Erhardt and Shaw 2006), but seem to depend on the developmental stage of the hair. The average growth rates of MTs in elongating root hairs (*L. japonicus*, 4.62 mm min^{-1}; *M. truncatula*, 3.24 mm min^{-1}) appear to be higher than in fully grown root hairs (*M. truncatula*: 2.88 mm min^{-1}), with peak values in the root hair tip (Timmers et al. 2007, and supplementary online material therein).

In a quantitative analysis of CMT dynamics in fully grown root hairs of *Arabidopsis*, van Bruaene et al. (2004) suggest a process called hybrid tread milling with dynamic instability at the MT plus-ends that mainly point towards the hair tip and slow depolymerization at the MT minus-ends. This is similar to what has been described for CMTs in *Arabidopsis* epidermal cells (Shaw et al. 2003) and in tobacco bright yellow 2 cells (Vos et al. 2004). Van Bruaene et al. (2004) further suggest that this process would allow root hair MTs first to nucleate and then to get ordered. Upon contact with other MTs, MTs partly depolymerise, followed by polymerization/reorientation, finally resulting in alignment of CMTs. Van Bruaene et al. (2004) identified, in elongating *Arabidopsis* root hairs, dispersed nucleation sites of MTs in the subapical cytoplasmic dense region and the cortex, and based on their observations in fully grown hairs the authors suggest a similar MT organizing mechanism in elongating hairs. However, Vassileva et al. (2005) reported that in *Lotus* root hairs only MT plus-ends contribute to the overall CMT dynamics, and not MT minus-ends.

In contrast to the rather stable overall appearance of the CMT array, the EMT array of an elongating root hair constantly changes in density and spatial configuration. From observations with the fluorescent markers GFP:MBD and EB1:YFP it is clear that EMTs are highly dynamic and that these dynamics are caused by rapidly alternating phases of MT growth and shrinkage (*Arabidopsis*: van Bruaene et al. 2004; *Medicago*: Vos et al. 2003; Timmers et al. 2007). Using real-time 4D confocal microscopy, Timmers et al. (2007) followed EMT behavior in detail, and it emerged that EMTs polymerize in several directions with a final net alignment towards the root hair tip. EMTs do not follow predefined tracks but progress in a wavy, searchlike pattern. In elongating root hairs, EMT plus-ends grow towards, and extend into, the very tip, upon which they either shrink back or get relocated into the apical hair cortex without extending into the very tip any further (Timmers et al. 2007).

It is likely that MT dynamics are at least partially under control of MAPs; however, the regulation of MT dynamics by specific proteins and cellular signalling events has so far not been studied in root hairs.

4 The Function of MTs in Elongating Root Hairs

To date, most of the evidence that MTs have a function in the tip-growth process of root hairs comes from pharmacological studies with MT antagonists. *Arabidopsis* root hairs treated with either the MT-depolymerizing drug oryzalin or the MTstabilizing drug paclitaxel (taxol) continue to grow at a normal growth rate, but either in a wavy pattern, or having problems with restricting growth to one single point (Bibikova et al. 1999; Ketelaar et al. 2003). For *Arabidopsis*, it is not clear whether the drug-induced disturbance of EMTs, CMTs, or both is responsible for these deviations from normal hair elongation.

In elongating root hairs, a tip-focussed cytoplasmic calcium-ion $[Ca^{2+}]_c$ gradient is associated with growth (Schiefelbein et al. 1992; Bibikova et al. 1997; Felle and Hepler 1997; Herrmann and Felle 1995; Wymer et al. 1997; de Ruijter et al. 1998; for the role of calcium during root hair tip growth, see Bibikova and Gilroy 2008). Bibikova et al. (1999) showed that in *Arabidopsis* root hairs MTs are involved in regulating directionality and stability of tip growth, possibly through interactions with the cellular machinery that maintains the $[Ca^{2+}]_c$ gradient at the tip. By using UV activation of a caged calcium-ionophore and touch stimulus, they created artificial intracellular $[Ca^{2+}]_c$ gradients, and found that in root hairs pretreated with taxol new growth points and subsequent tip growth occurred at the site of artificially elevated calcium, whereas in control hairs this only led to a transient reorientation of growth. The mechanism through which MTs interact with the tip-focussed calcium gradient is unknown, but MTs might play a role in locally depositing calcium-dependent elements needed for exocytosis, such as annexins, protein kinases, and calcium channels at the very tip of an elongating hair. A high calcium concentration at the hair tip could in turn have a regulatory feedback effect on MT stability/dynamics, since MT depolymerization in plant cells can be triggered by elevated cytoplasmic Ca^{2+} levels (Cyr 1991, 1994).

Legume root hairs respond somewhat differently to treatment with MT antagonists, but experiments in such plants also support the idea that MTs are involved in regulating tip growth. Within minutes after application of 1 μM oryzalin, EMTs, but not CMTs, disappear, which causes a subsequent loss of the typical cytoarchitecture of growing root hairs: the subapical cytoplasmic dense region disappears and the distance between root hair tip and nucleus, which is maintained during root hair elongation at 30–40 mm, increases up to 250 mm (Sieberer et al. 2002). Also Lloyd et al. (1987) have shown in a pharmacological study that in growing *V. hirsuta* root hairs nuclear positioning, but not nuclear movement, depends on EMTs. However, elongating root hairs of *M. truncatula* treated with oryzalin do not completely lose their cytoplasmic polarity since the zone at the very tip containing (exocytotic) vesicles remains present and consequently the tip growth continues, though the rate is reduced by 60% (Sieberer et al. 2002). Taxol, a drug that stabilizes MTs, does not interfere with cytoarchitecture and nuclear position, but causes a similar drop in growth rate as does oryzalin (Sieberer et al. 2002). These experiments indicate that EMTs contribute to cell polarity and that their dynamic properties are responsible for the maintenance of a high growth rate of *Medicago* root hairs.

In *Arabidopsis* root hairs, which have only few EMTs, neither changes in cytoarchitecture nor a drop in growth rate was observed after treatment with MT antagonists (Bibikova et al. 1999; Ketelaar et al. 2002; van Bruaene et al. 2004).

Young, growing root hairs of *Medicago* do not show a wavy growth pattern, like *Arabidopsis* root hairs, upon treatment with oryzalin or taxol (*M. truncatula*:

Sieberer et al. 2002; *M. sativa*: Weerashinge et al. 2003), whereas hairs that are terminating growth show a distinct phenotype after application of oryzalin (see below). Growth-terminating legume root hairs – hairs that still elongate but have almost lost their growth potential – respond to purified rhizobacterial signal molecules, the nodulation factors or Nod factors, with root hair deformation, that is swelling of the hair tips (i.e., isodiametric growth), after which tip growth resumes, albeit in a different direction (Heidstra et al. 1994; reviewed in Miller et al. 1997). Oryzalin causes a deformation phenotype similar to Nod factors in these root hairs; when oryzalin is combined with Nod factors, it increases Nod-factor-induced growth deviation, while taxol suppresses this effect (Sieberer et al. 2005b). Taken together, these results strongly support the idea that a functional MT cytoskeleton is needed to direct tip growth in legume root hairs.

5 MTs in Legume Root Hairs During Early Stages of Infection by Rhizobia

Legume roots infected with rhizobia form new plant organs, the root nodules, in which rhizobia fix atmospheric nitrogen and make it accessible to the plant (see Limpens and Bisseling 2008). One of the first steps in this plant–bacteria interaction is the curling of root hairs around the bacteria (Kijne 1992; Hadri and Bisseling 1998; Esseling et al. 2004; for rhizobacterial attachment to legume roots, see Hirsch et al. 2008). Root hair curling has been described as iterative growth reorientation (Esseling et al. 2003) with two major prerequisites: (1) continuous delivery and insertion of exocytotic vesicles into the plasma membrane of the hair tip (i.e., tip growth) and (2) a constant reorientation of the site where exocytosis takes place without interfering with tip growth itself. With the actin cytoskeleton being the key player in vesicle transport and targeting, MTs could be the cytoskeletal compound that mediates growth reorientation in a curling root hair. Indeed it has been shown that in addition to the actin cytoskeleton (*M. sativa*: Allen et al. 1994; *Phaseolus vulgaris*: Cárdenas et al. 1998, 2003; *Vicia sativa*: De Ruijter et al. 1998; Miller et al. 1999; Ridge 1992), MTs are also a direct or indirect target of Nod factor signalling (*L. japonicus*: Weerasinghe et al. 2005; Vassileva et al. 2005; *Medicago*: Weerasinghe et al. 2003; Sieberer et al. 2005b).

Using immunocytochemistry after chemical fixation, Weerasinghe et al. (2003) observed a complete disintegration of the MT cytoskeleton within 30 min of Nod factor treatment in elongating *M. sativa* root hairs. The MT network reforms 60 min after Nod factor exposure, though in longer hairs recovery is incomplete. However, these findings were not confirmed in *M. truncatula* (Sieberer et al. 2005b) or *L. japonicus* (Vassileva et al. 2005; Weerasinghe et al. 2005) root hairs, where changes in the MT cytoskeleton are more subtle after Nod factor application. Results from *L. japonicus* (Weerasinghe et al. 2005) and *M. truncatula* (Sieberer et al. 2005b) indicate that Nod factors cause a subtle and short-term shortening of the EMT array in elongating hairs. Vassileva et al. (2005) describe a decrease in

CMT growth rate at specific windows of up to 18 h after Nod factor treatment and up to 60 h after inoculation with symbiotic rhizobia, but found no evidence of CMT depolymerization after Nod factor treatment, which is similar to observations made in *M. truncatula* (Sieberer et al. 2005b). In growth-arresting *Medicago* hairs, which respond with hair deformation after Nod factor application, EMTs are also a target of Nod factor signalling. The short EMT array in these hairs completely disappears within minutes after application, but reappears after 20–30 min, whereas CMTs are not obviously affected. The Nod-factor-induced depolymerization of EMTs correlates with a loss of polar cytoarchitecture and straight growth, while the site of EMT reappearance predicts the new direction of hair elongation (Sieberer et al. 2005b).

Further evidence that MTs could be crucial to the infection process of roots by rhizobia comes from *Medicago* root hairs that have already trapped *Sinorhizobium meliloti* bacteria in their curl. In such a curl, EMTs are concentrated at the infection site where bacteria enter the root hair, and are present between the nucleus and the tip of the infection thread during infection thread growth in the root hair shank (Timmers et al. 1999; for infection thread development during rhizobial infection of legume roots, see Gage 2008). This MT distribution does not form in root hairs of the *Medicago hcl* mutant after inoculation with rhizobia. The *hcl* mutant is defective in root hair curling and shows multiple hair tip formation or continuous hair curling in response to *S. meliloti* (Catoira et al. 2001). The authors suggest that the *hcl* mutation alters the formation of signalling centers that normally provide positional information for the reorganization of the MT cytoskeleton in root hairs.

The question remains as to how changes in the MT cytoskeleton are triggered after perception of the Nod factor signal by the root hair. In *Medicago* root hairs, Nod factor treatment leads to elevated cytosolic Ca^{2+} levels (Allen et al. 1994; de Ruijter et al. 1998; Cárdenas et al. 1999; Felle et al. 1999a,b), which in turn lead to several downstream events affecting the cytoskeleton (for a review, see Lhuissier et al. 2001). Hence, Nod factors may indirectly affect MTs in *Medicago* root hairs via transiently increased Ca^{2+} levels. However, it is more likely that a signalling cascade with several molecular protagonists transmits the incoming Nod factor signal towards the MT cytoskeleton. A possible candidate involved in transmitting the Nod factor signal to MTs may be phospholipase D (PLD), since activation of PLD causes MT disruption (Dhonukshe and Gadella 2003; Gardiner et al. 2003) and a PLD signalling pathway is involved in Nod-factor-induced root hair deformation (Den Hartog et al. 2001). In this volume, Aoyama (2008) gives an overview on how phospholipids signalling could regulate tip growth and cytoskeletal elements in root hairs.

The observations reviewed in Paras. 4 and 5 of this chapter indicate that in both legumes and *Arabidopsis*, MTs are involved in determining growth direction, possibly by defining, in an as-yet-unknown manner, the place where exocytosis takes place. In doing so, MTs could steer polar cell expansion without interfering with actin-based tip growth itself. It is tempting to speculate that the specific function of EMTs in the symbiotic process may explain the more extended EMT array in legume root hairs, compared to those of *Arabidopsis*, and that the differential response of legume and *Arabidopsis* root hairs to MT antagonists could be caused by the evolution of the EMT array to facilitate symbiosis. However, there may be specific properties of the EMT array for different plant species, independent on whether they are legumes or not.

6 MTs and Their Putative Role in Targeting Polarity Markers to the Very Tip of Elongating Root Hairs

Interesting observations concerning the role of MTs in determining the site of cell growth were made in fission yeast cells (*Schizosaccharomyces pombe*). Such cells elongate by tip growth and their shape is determined by interphase MTs. These MTs are organized in 3–4 bundles that radiate from the nucleus into the cell tip and carry polarity markers to the cell ends to determine the position of the growth zone. In the absence of functional MTs or polarity markers, tip growth proceeds, but the sites where growth occurs are wrongly positioned and this leads to bent or wavy cells or cells with multiple tips (Beinhauer et al. 1997; Browning et al. 2000; Brunner and Nurse 2000; Behrens and Nurse 2002). Mutant studies, combined with GFP techniques, have shown that it is the linear transport and the delivery of MAPs via polymerizing MTs towards the cell tips that is the crucial process in setting up and maintaining cell polarity. In fission yeast, such an MT plus-end tracking system consists of Mal3 (an EB1 homologue), Tea2 (a kinesin-like motor protein), and Tip1 (a cytoplasmic linker protein-170 homolog and cargo of Tea2). Tea2 moves towards the growing MT plus-ends (Browning and Hackney 2005) and interacts along the MTs with at least Tip1 (Busch and Brunner 2004) and Tea1, a large multi-domain protein that localizes to the cell ends and plays a key role in directing the growth machinery to the cell tip (Mata and Nurse 1997). Mal3 labels MTs and also moves, independently of Tea2, to the MT plus-ends. Once there, Mal3 directly interacts with Tip1 and associates with Tea2, and therefore, Tea2 and Tea1 stay associated with growing MT plus-ends. A recent in vitro study shows that interaction of Mal3, Tea2, and Tip1 is required for the selective tracking of growing MT plus-ends (Bieling et al. 2007). Mal3 also stimulates ATPase activity of Tea2 by recruiting it to MTs (Browning and Hackney 2005). When the plus-ends with the proteins reach the tip of the cell via MT polymerization, Mal3 plus-end labelling of the MTs disappears (Busch and Brunner 2004), the MTs undergo shrinkage, and Tip1, Tea1, and Tea2 stay in the growing tip. Tea1 has been observed to attach to the membrane-associated protein Mod5 (Morphology defective 5; Snaith and Sawin 2003), and forms a complex with Bud6 and For3, a formin that assembles actin cables (Feierbach et al. 2004). This could be a direct link between the polarity complex and actin polymerization, and thus, between the determination of the location of exocytosis, and hence the direction of cell growth (MTs) and the delivery of the growth components (actin cytoskeleton).

A homology search in *Arabidopsis* for the proteins involved in setting up polarity in *S. pombe* reveals that the *Arabidopsis* genome contains three EB1 homologues At3g47690 (AtEB1a), At5g62500 (AtEB1b), and At5g67270 (AtEB1c).

In *M. truncatula*, expressed sequence tags have also been found for EB1 homologues. EB1 reporter fusion proteins (GFP:AtEB1, EB1:YFP) have been shown to decorate the plus-ends of growing MTs in *Arabidopsis* (Chan et al. 2003; Mathur et al. 2003) and *Medicago* (Timmers et al. 2007). Even though not all proteins in the EB1 pathway of fission yeast may be present in plants, EB1 homologues have been found in plant cells and their localization to MT plus-ends is similar to that in *S. pombe*. Therefore, EB1 proteins may perform the same function in plants as in *S. pombe*. Dhonukshe et al. (2005) described MT plus-ends to be essential for intracellular polarization during division-plane establishment in plant cells and suggested that in mitotic plant cells, EMT plus-ends may act as cell shape/polarity sensing and orientating machines by their sustained cortical targeting. Hemsley et al. (2005) found that *TIP1* (tip growth defective 1) encodes for an *S*-acyl transferase in *Arabidopsis* which could possibly regulate root hair tip-growth by directing proteins, vesicle traffic, and the assembly of the cytoskeleton in discrete areas of the plasma membrane. Since growing *M. truncatula* root hairs have an array of highly dynamic EMTs, which can grow from the subapical region into the very root hair tip (Vos et al. 2003; Timers et al. 2007), and determine root hair elongation direction (Sieberer et al. 2005b), linear transport and targeting of polarity markers via polymerizing MTs could exist in plant root hairs as it does in *S. pombe*.

An example of plant-specific polarity markers that might be localized by MTs are small GTPases of the Rho family, named ROPs (Rho-like proteins of plants; see Žárský and Fowler 2008). Molendijk et al. (2001) observed that *Arabidopsis* ROP proteins are polarly localized at the plasma membrane of the trichoblast before the root hair bulge becomes visible. They remain concentrated at the apical membrane of root hair bulges and at the tip of growing root hairs, but they are absent from apical membranes of fully grown root hairs. Seedlings expressing the constitutively active GFP:AtRop4 G15V construct show strong swelling of epidermal cells of the hypocotyl and cotyledons. In root hairs, the polarity of growth is lost. Instead, the root hair tips expand isodiametrically, leading to swelling. Swellings of root hairs from GFP:Rop6 Q64L roots have delocalized Ca^{2+} gradients focused on where growth is occurring (Molendijk et al. 2001). These root hair phenotypes resemble those obtained upon experimental ablation of the MT cytoskeleton. However, it should be noted that root hair swelling is not exclusively triggered by MT antagonists. The regulation of tip growth is a highly regulated process that requests the finely tuned interaction of many cellular components among which MTs are only one.

7 Remaining Questions and Future Work

Most information about the structure of the MT array in root hairs comes from studies with fluorescent markers at the light microscopical level. Fine ultrastructural details are not visible with these techniques and in order to obtain high-resolution structural information concerning MT ends, crossings between bundles or individual MTs, and the association of MTs with the plasma membrane or other membrane structures, an electron tomography study, combined with 3D reconstruction of the entire MT network in root hairs, is necessary.

Tip-growing root hairs develop on epidermal cells which expand by diffuse growth. The transition from one growth form to the other is accompanied by a reorientation of the CMT array (Baluška et al. 2000; Sieberer et al. 2002), and the subsequent formation of an EMT array. How this change is set up and maintained remains unknown. A role of the cell nucleus as a surface for MT initiation has been proposed, but direct evidence is lacking. Further studies to unravel the ontogeny of the MT cytoskeleton in root hairs should include the visualization of γ-tubulin in order to identify MT organizing centers, and analysis of the interaction of MTs with the plasma membrane and other dynamic systems such as the ER and the actin cytoskeleton.

The most direct evidence for a role of MTs in root hair development comes from genetic studies. *Arabidopsis* plants with a reduced level of expression of α-tubulin (TUA6) show dramatically swollen root cells and branched root hairs (Bao et al. 2001). Moreover, the *Arabidopsis* mutant ectopic root hair 3 (botero1/flat root/fragile fiber2), which is affected in a katanin-like protein, develops ectopic root hairs and has swollen roots and defective cell walls. In this mutant the cortical/perinuclear MT organization is also aberrant (Webb et al. 2002). In *Arabidopsis*, additional root hair mutants have been identified with mutations in genes involved in every stage of root hair development (Parker et al. 2000; Schiefelbein 2000; Grierson et al. 2001; Grierson and Schiefelbein 2008). Examples are genes with roles in root hair site initiation, bulge formation, transition from bulge to tip growth, root hair elongation by tip growth, and final root hair length. Whether and how the action of these genes is related to the function of the MT cytoskeleton is unclear. A possibility is that the action of these genes on the cytoskeleton is related to the action of the plant hormones auxin and ethylene, both stimulating root hairs to elongate during tip growth. Examples are axr2 (auxin resistant 2), axr3, and ctr1 (constitutive triple response 1). Interestingly, the action of hypaphorine, an indole alkaloid from the ectomycorrhizal fungus *Pisolithus tinctorius*, suppresses root hair elongation in *Eucalyptus globus* (Ditengou et al. 2003), *A. thaliana* (Reboutier et al. 2002), and Poplar (Valérie Legué, Unité Mixte de Recherche UMR INRA-UHP 1136, Université Nancy, B.P. 239, 54506 Vandoeuvre Cedex, France, 2007, personal communication). At high concentrations (500 μM and above) root hair elongation stops within 15 min after application, but root hair initiation from epidermal cells is not affected. At inhibitory concentrations (100 μM), hypaphorine also induces a transitory root hair swelling and deformation. Since hypaphorine treatment induces a configuration transition of the actin filament configuration from that seen in growing root hairs to the bundled one seen in fully grown hairs (Dauphin et al. 2006), the target of this molecule seems to be the actin cytoskeleton.

The identification of MAPs in plants has turned out to be a difficult task. Only in 2000 the first nonmotor protein plant MAP, MAP-65, was cloned (Smertenko et al. 2000). This protein cross-links MTs and stabilizes the MT array. Since then, others have been identified (Lloyd et al. 2004) but have not been characterized. No data are yet available on the nature or the function of MAPs in root hairs. All plant species so far examined appear to have a large number of genes coding for cytoskeletal proteins. *Arabidopsis*, for example, possesses six α-tubulin, nine β-tubulin, and two γ-tubulin genes (Liu et al. 1994). The number of motor proteins, such as kinesins, is even much higher. The importance of all these putative components of the MT cytoskeleton in root hairs is completely unknown.

Thus, so far, genetic methods have not been fully exploited to decipher MT function, regulation, and interaction with other cytoskeletal components. Future genetic and molecular approaches, in combination with state-of-the-art microscopy, should help to decipher the role of MAPs in regulating MTs and polarity marker deposition in elongating root hairs.

Acknowledgments The authors kindly thank Clare Gough from the Laboratory of Plant–Microorganism Interactions (Castanet-Tolosan, France) for the constructive comments and for English editing.

References

Allen NS, Bennett MN, Cox DN, Shipley A, Ehrhardt DW, Long SR (1994) Effects of Nod factors on alfalfa root hair Ca^{++} and H^+ currents on cytoskeleton behavior. In: Daniels MJ, Downie JA, Osbourn AE (eds) Advances in molecular genetics of plant microbe interactions, vol 3. Kluwer, Dordrecht, pp 107–114

Aoyama T (2008) Phospholipid signaling in root hair development. In: Emons AMC, Ketelaar T (eds) Root hairs: excellent tools for the study of plant molecular cell biology. Springer, Berlin Heidelberg New York. doi:10.1007/7089_2008_1

Bakhuizen R (1988) The plant cytoskeleton in the rhizobium-legume symbiosis. PhD Thesis, Leiden University, The Netherlands

Baluška F, Salaj J, Mathur J, Braun M, Jasper F, Samaj J, Chua N-H, Barlow PW, Volkman D (2000) Root hair formation: F-actin-dependent tip growth is initiated by local assembly of profilin-supported F-actin meshworks accumulated within expansin-enriched bulges. *Devel Biol* 227:618–632

Bao YQ, Kost B, Chua NH (2001) Reduced expression of α-tubulin genes in *Arabidopsis thaliana.* specifically affects root growth and morphology, root hair development and root gravitropism. *Plant J* 28:145–157

Behrens R, Nurse P (2002) Roles of fission yeast tea1p in the localization of polarity factors and in organizing the microtubular cytoskeleton. *J Cell Biol* 157:783–793

Beinhauer JD, Hagan IM, Hegemann JH, Fleig U (1997) Mal3, the fission yeast homolog of the human APC-interacting protein EB-1 is required for micorotubule integrity and the maintenance of cell form. *J Cell Biol* 139:717–728

Bibikova T, Gilroy S (2008) Calcium in root hair growth. In: Emons AMC, Ketelaar T (eds) Root hairs: excellent tools for the study of plant molecular cell biology. Springer, Berlin Heidelberg New York. doi:10.1007/7089_2008_3

Bibikova TN, Zhigilei A, Gilroy S (1997) Root hair growth in *Arabidopsis thaliana* is directed by calcium and an endogenous polarity. *Planta* 203:495–505

Bibikova TN, Blancaflor EB, Gilroy S (1999) Microtubules regulate tip growth and orientation in root hairs of *Arabidopsis thaliana. Plant* 17:657–665

Bieling P, Laan L, Schek H, Munteanu EL, Sandblad L, Dogterom M, Brunner D, Surrey T (2007) Reconstitution of a microtubule plus-end tracking system in vitro. *Nature* 450:1100–1105

Boisson-Dernier A, Chabaud M, Garcia F, Bécard G, Rosenberg C, Barker DG (2001) *Agrobacterium rhizogenes*-transformed roots of *Medicago truncatula* for the study of nitrogen-fixing and endomycorrhizal symbiotic associations. *Mol Plant Microbe Interact* 14:695–700

Browning H, Hayles J, Mata J, Aveline L, Nurse P, McIntosh JR (2000) Tea2p is a kinesin-like protein required to generate polarized growth in fission yeast. *J Cell Biol* 151:15–28

Browning H, Hackney DD (2005) The EB1 homolog Mal3 stimulates the ATPase of the kinesin Tea2 by recruiting it to the microtubule. *J Biol Chem* 280:12299–12304

Brunner D, Nurse P (2000) CLIP170-like tip1p spatially organizes microtubular dynamics in fission yeast. *Cell* 102:695–704

Busch KE, Brunner D (2004) The microtubule plus-end-tracking proteins mal3p and tip1p cooperate for cell-end targeting of interphase microtubules. *Current Biol* 14:548–559

Cárdenas L, Vidali L, Domínguez J, Pérez H, Sánchez F, Hepler PK, Quinto C (1998) Rearrangement of actin microfilaments in plant root hairs responding to *Rhizobium etli* nodulation signals. *Plant Physiol* 116:871–877

Cárdenas L, Feijo JA, Kunkel JG, Sanchez F, Holdaway-Clarke T, Hepler PK, Quinto C (1999) Rhizobium nod factors induce increases in intracellular free calcium and extracellular calcium influxes in bean root hairs. *Plant J* 9:347–352

Cárdenas L, Thomas-Oates JE, Nava N, Lopez-Lara IM, Hepler PK, Quinto C (2003) The role of nod factor substituents in actin cytoskeleton rearrangements in *Phaseolus vulgaris. Mol Plant Microbe Interact* 16:326–334

Catoira R, Timmers ACJ, Maillet F, Galera C, Penmetsa RV, Cook D, Dénarié J, Gough C (2001) The HCL gene of *Medicago truncatula* controls *Rhizobium*-induced root hair curling. *Development* 128:1507–1518

Chabaud M, Boisson-Dernier A, Zhang J, Taylor CG, Yu O, Barker DG (2006) *Agrobacterium rhizogenes*-mediated root transformation. In: Mathesius U, Journet EP, Sumner LW (eds) The *Medicago truncatula* handbook. ISBN 0–9754303–1–9. http://www.noble.org/MedicagoHandbook

Chan J, Calder GM, Doonan JH, Lloyd CW (2003) EB1 reveals mobile microtubule nucleation sites in *Arabidopsis. Nat Cell Biol* 5:967–971

Cyr RJ (1991) Calcium/calmodulin affects microtubule stability in lysed protoplasts. *J Cell Sci* 100:311–317

Cyr RJ (1994) Microtubules in plant morphogenesis: role of the cortical array. *Annu Rev Cell Biol* 10:153–180

Dauphin A, De Ruijter NC, Emons AM, Legué V (2006) Action organization during eucalyptus root hair development and its response to fungal hypaphorine. *Plant Biol (Stuttg)* 8:204–211

Den Hartog M, Musgrave A, Munnik T (2001) Nod factor-induced phosphatidic acid and diacylglycerol pyrophosphate formation: a role for phospholipase C and D in root hair deformation. *Plant J* 25:55–65

De Ruijter NCA, Rook MB, Bisseling T, Emons AMC (1998) Lipochito-oligosaccharides reinitiate root hair tip growth in *Vicia sativa*, with high calcium and spectrin-like antigen at the tip. *Plant J* 13:341–350

Dhonukshe P, Gadella TW Jr (2003) Alteration of microtubule dynamic instability during preprophase band formation revealed by yellow fluorescent protein-CLIP170 microtubule plus-end labeling. *Plant Cell* 15:597–611

Dhonukshe P, Mathur J, Hülskamp M, Gadella Jr TWJ (2005) Microtubule plus-ends reveal essential links between intracellular polarization and localized modulation of endocytosis during division-plane establishment in plant cells. *BMC Biol* 3:11. doi:10.1186/1741–7007–3–11

Ditengou FA, Raudaskoski M, Lapeyrie F (2003) Hyaphorine, an indole-3-acetic acid antagonist delivered by the ectomycorrhizal fungus *Pisolithus tinctorius*, induces reorganisation of actin and the microtubule cytoskeleton in *Eucalyptus globus* ssp *biostata* root hairs. *Planta* 218:217–225

Emons AMC (1982) Microtubules do not control microfibril orientation in a helicoidal cell wall. *Protoplasma* 113:85–87

Emons AMC, Wolters-Arts AMC (1983) Cortical microtubules and microfibril deposition in the wall of root hair of *Equisetum hyemale. Protoplasma* 117:68–81

Emons AMC (1989) Helicoidal microfibril deposition in a tip-growing cell and microtubule alignment during tip morphogenesis: a dry-cleaving and freeze-substitution study. *Can J Bot* 67:2401–2408

Emons AMC (1987) The cytoskeleton and secretory vesicles in root hairs of equisetum and Limnobium and cytoplasmic streaming in root hairs of *Equisetum. Ann Bot* 60:625–632

Erhardt DW, Shaw SL (2006) Microtubule dynamics and organization in the plant cortical array. *Annu Rev Plant Biol* 57:859–875

Esseling JJ, Lhuissier FGP, Emons AMC (2003) Nod factor-induced root hair curling: continuous polar growth towards the point of Nod factor application. *Plant Physiol* 132:1982–1988

Esseling JJ, Lhuissier FGP, Emons AMC (2004) A nonsymbiotic root hair tip growth phenotype in NORK-mutated legumes: implications for nodulation factor–induced signaling and formation of a multifaceted root hair pocket for bacteria. *Plant Cell* 16:933–944

Feierbach B, Verde F, Chang F (2004) Regulation of a formin complex by the microtubule plus-end protein tea1p. *J Cell Biol* 165:697–707

Felle HH, Hepler PK (1997) The cytosolic Ca^{2+} concentration gradient of *Sinapis alba* root hairs as revealed by Ca^{2+}-selective microelectrode tests and fura-dextran ratio imaging. *Plant Physiol* 114:39–45

Felle HH, Kondorosi E, Kondorosi A, Schultze M (1999a) Elevation of the cytosolic free [Ca^{2+}] is indispensable for the transduction of the Nod factor signal in alfalfa. *Plant Physiol* 121:273–279

Felle HH, Kondorosi E, Kondorosi A, Schultze M (1999b) Nod factors modulate the concentration of cytosolic free calcium differently in growing and non-growing root hairs of *Medicago sativa* L. *Planta* 209:207–212

Gage TJ (2008) Architecture of infection thread networks in nitrogen-fixing root nodules. In: Emons AMC, Ketelaar T (eds) Root hairs: excellent tools for the study of plant molecular cell biology. Springer, Berlin Heidelberg New York. doi:10.1007/7089_2008_5

Gardiner J, Collings DA, Harper JD, Marc J (2003) The effects of the phospholipase D-antagonist 1-butanol on seedling development and microtubule organisation in *Arabidopsis. Plant Cell Physiol* 44:687–696

Grierson CS, Parker JS, Kemp AC (2001) *Arabidopsis* genes with roles in root hair development. *J Plant Nutr Soil Sci* 164:131–140

Grierson C, Schiefelbein J (2008) Genetics of root hair formation. In: Emons AMC, Ketelaar T (eds) Root hairs: excellent tools for the study of plant molecular cell biology. Springer, Berlin Heidelberg New York. doi:10.1007/7089_2008_15

Hadri AE, Bisseling T (1998) Responses of the plant to Nod factors. In: Spaink HP, Kondorosi A, Hooykaas PJJ (eds) The Rhizobiaceae: molecular biology of model plant-associated bacteria. Kluwer, Dordrecht, pp 403–416

Heidstra R, Geurts R, Franssen H, Spaink HP, Van Kammen A, Bisseling T (1994) Root hair deformation activity of nodulation factors and their fate on *Vicia sativa. Plant Physiol* 105:787–797

Hemsley PA, Kemp AC, Grierson CS (2005) The tip growth Defective1 S-acyl transferase regulates plant cell growth in *Arabidopsis. Plant Cell* 17:2554–2563

Herrmann A, Felle HH (1995) Tip growth in root hair cells of *Sinapis alba* L.: significance of internal and external Ca^{2+} and pH. *New Phytol* 129:523–533

Hirsch AM, Lum MR, Fujishige NA (2008) Microbial encounters of a symbiotic kind: attaching to roots and other surfaces. In: Emons AMC, Ketelaar T (eds) Root hairs: excellent tools for the study of plant molecular cell biology. Springer, Berlin Heidelberg New York. doi:10.1007/7089_2008_6

Höög JL, Schwartz C, Noon AT, o'Toole ET, Mastronarde DN, McIntosh JR, Antony C (2007) Organisation of interphase microtubules in fission yeqst analysed by electron tomography. *Dev Cell* 12:349–361

Ketelaar T, Emons AMC (2000) The role of microtubules in root hair growth and cellulose microfibril deposition. In: Ridge RW, Emons AMC (eds) Root hairs. Cell and molecular biology. Springer, Berlin Heidelberg New York, pp 17–28

Ketelaar T, Emons AMC (2001) The cytoskeleton in plant cell growth: lessons from root hairs. *New Phytol* 152:409–418

Ketelaar T, Emons AMC (2008) The actin cytoskeleton in root hairs: a cell elongation device. In: Emons AMC, Ketelaar T (eds) Root hairs: excellent tools for the study of plant molecular cell biology. Springer, Berlin Heidelberg New York. doi:10.1007/7089_2008_8

Ketelaar T, Faivre-Moskalenko C, Esseling JJ, de Ruijter NCA, Grierson CS, Dogterom M, Emons AMC (2002) Positioning of nuclei in *Arabidopsis*. root hairs: an actin-regulated process of tip growth. *Plant Cell* 14:2941–2955.

Ketelaar T, de Ruijter NCA, Emons AMC (2003) Unstable F-actin specifies the area and microtubule direction of cell expansion in *Arabidopsis* root hairs. *Plant Cell* 15:285–292

Kijne JW (1992) The *Rhizobium* infection process. In: Stacey G, Burris R, Evans H (eds) Biological nitrogen fixation. Chapman and Hall, New York, pp 349–398

Lhuissier FGP, de Ruijter NCA, Sieberer BJ, Esseling JJ, Emons AMC (2001) Time course of cell biological events evoked in legume root hair by *Rhizobium*. Nod factors: State of the art. *Ann Bot* 87:289–302

Limpens E, Bisseling T (2008) Nod factor signal transduction in the Rhizobium-Legume symbiosis. In: Emons AMC, Ketelaar T (eds) Root hairs: excellent tools for the study of plant molecular cell biology. Springer, Berlin Heidelberg New York. doi:10.1007/7089_2008_10

Liu B, Joshi HC, Wilson TJ, Silflow CD, Palevitz BA, Snustad DP (1994) Gamma-tubulin in *Arabidopsis*: Gene sequence, immunoblot, and immunofluorescence studies. *Plant Cell* 6:303–314

Lloyd C (1983) Helical microtubular arrays in onion root hairs. *Nature* 305:311–313

Lloyd CW, Pearce KJ, Rawlins DJ, Ridge RW, Shaw PJ (1987) Endoplasmic microtubules connect the advancing nucleus to the tip of legume root hairs, but F-actin is involved in basipetal migration. *Cell Motil Cytoskeleton* 8:27–36

Lloyd CW, Chan J, Hussey PJ (2004) Microtubules and microtubule-associated proteins in plants. In: Hussey PJ (ed) The plant cytoskeleton in cell differentiation and development. Blackwell, Oxford, pp 3–27

Marc J, Granger CL, Brincat J, Fisher DD, Kao T, McCubbin AG, Cyr RJ (1998) A GFP:MAP4 reporter gene for visualizing cortical microtubule rearrangements in living epidermal cells. *Plant Cell* 10:1927–1939

Mata J, Nurse P (1997) Tea1 and the microtubular cytoskeleton are important for generating global spatial order within the fission yeast cell. *Cell* 89:939–949

Mathur J, Mathur N, Kernebeck B, Srinivas BP, Hülskamp M (2003) A novel localization pattern for an EB1-like protein links microtubule dynamics to endomembrane organization. *Current Biol* 13:1991–1997

Miller DD, de Ruijter NCA, Emons AMC (1997) From signal to form: aspects of the cytoskeleton-plasma membrane-cell wall continuum in root hair tips. *J Exp Bot* 48:1881–1896

Miller DD, de Ruijter NCA, Bisseling T, Emons AMC (1999) The role of actin in root hair morphogenesis: studies with lipochito-oligosaccharide as a growth stimulator and cytochalasin as an actin perturbing drug. *Plant J* 17:141–154

Molendijk AJ, Bischoff F, Rajendrakumar CS, Friml J, Braun M, Gilroy S, Palme K (2001) *Arabidopsis thaliana* Rop GTPases are localized to tips of root hairs and control polar growth. *EMBO J* 20:2779–2788

Newcomb EH, Bonnett HT Jr (1965) Cytoplasmic microtubules and wall microfibril orientation in root hairs of radish. *J Cell Biol* 27:575–589

Parker JS, Cavell AC, Dolan L, Roberts K, Grierson CS (2000) Genetic interactions during root hair morphogenesis in *Arabidopsis*. *Plant Cell* 12:1961–1974

Reboutier D, Bianchi M, Brault M, Roux C, Dauphin A, Rona J-P, Legué V, Lapeyrie F, Bouteau F (2002) The indolic compound hypaphorine produced by ectomycorrhizal fungus interferes with auxin action and evokes early responses in nonhost *Arabidopsis thaliana Mol Plant Microbe Interact* 15:932–938

Ridge RW (1992) A model of legume root hair growth and *Rhizobium* infection. *Symbiosis* 14:359–373

Schiefelbein JW, Shipley A, Rowse P (1992) Calcium influx at the tip of growing root-hair cells of *Arabidopsis thaliana. Planta* 187:455–459

Schiefelbein JW (2000) Constructing a plant cell. The genetic control of root hair development. *Plant Physiol* 124:1525–1531

Shaw SL, Kamyar R, Erhardt DW (2003) Sustained microtubule treadmilling in *Arabidopsis.* cortical arrays. *Science* 300:1715–1718

Sieberer BJ, Timmers ACJ, Lhuissier FGP, Emons AMC (2002) Endoplasmic microtubules configure the subapical cytoplasm and are required for fast growth of *Medicago truncatula* root hairs. *Plant Physiol* 130:977–988

Sieberer BJ, Ketelaar T, Esseling JJ, Emons AMC (2005a) Microtubules guide root hair tip growth. *New Phytol* 167:711–719

Sieberer BJ, Timmers AC, Emons AMC (2005b) Nod factors alter the microtubule cytoskeleton in *Medicago truncatula* root hair to allow root hair reorientation. *Mol Plant-Microbe Interact* 11:1195–1204

Smertenko A, Saleh N, Igarashi H, Mori H, Hauser-Hahn I, Jiang CJ, Sonobe S, Lloyd CW, Hussey PJ (2000) A new class of microtubule-associated proteins in plants. *Nat Cell Biol* 2:750–753

Snaith HA, Sawin KE (2003) Fission yeast mod5p regulates polarized growth through anchoring of tea1p at cell tips. *Nature* 423:647–651

Timmers ACJ, Auriac M-C, Truchet G (1999) Refined analysis of early symbiotic steps of the Rhizobium-*Medicago* interaction in relationship with microtubular cytoskeleton rearrangements. *Development* 126:3617–3628

Timmers AC, Vallotton P, Heym C, Menzel D (2007) Microtubule dynamics in root hairs of *Medicago truncatula*. Eur J Cell Biol 8669–83

Traas JA, Braat P, Emons AMC, Meeks H, Derksen J (1985) Microtubules in root hairs. *J Cell Sci* 76:303–320

Van Bruaene N, Joss G, Oostveldt P van (2004) Reorganization and in vivo dynamics of microtubules during *Arabidopsis* root hair development. *Plant Physiol* 136:3905–3919

Vassileva VN, Kouchi H, Ridge RW (2005) Microtubule dynamics in living root hairs: transient slowing by lipochitin oligosaccharide nodulation signals. *Plant Cell* 17:1777–1787

Vos JW, Sieberer B, Timmers ACJ, Emons AMC (2003) Microtubule dynamics during preprophase band formation and the role of endoplasmic microtubules during root hair elongation. *Cell Biol Int* 27:295

Vos JW, Dogterom M, Emons AMC (2004) Microtubules become more dynamic but not shorter during preprophase band formation: a possible "search-and-capture" mechanism for microtubule translocation. *Cell Motil Cytoskel* 57:246–258

Webb M, Jouannic S, Foreman J, Linstead P, Dolan L (2002) Cell specification in the *Arabidopsis.* root epidermis requires the activity of *ECTOPIC ROOT HAIR 3* – a katanin-p60 protein. *Development* 129:123–131

Weerasinghe RR, Collings DA, Johannes E, Allen NS (2003) The distributional changes and role of microtubules in Nod factor-challenged *Medicago sativa* root hairs. *Planta* 218:276–287

Weerasinghe RR, Bird DMCK, Allen NS (2005) Root-knot nematodes and bacterial Nod factors elicit common signal transduction events in *Lotus japonicus. Proc Natl Acad Sci USA* 102:3147–3152

Wymer CL Bibikova TN, Gilroy S (1997) Cytoplasmic free calcium distributions during the development of root hairs of *Arabidopsis thaliana. Plant J* 12:427–439

Zarsky V, Fowler J (2008) ROP (Rho-related protein from Plants) GTPases for spatial control of root hair morphogenesis. In: Emons AMC, Ketelaar T (eds) Root hairs: excellent tools for the study of plant molecular cell biology. Springer, Berlin Heidelberg New York. doi:10.1007/7089_2008_14

Nod Factor Signal Transduction in the Rhizobium–Legume Symbiosis

E. Limpens and T. Bisseling (✉)

Abstract The symbiotic interaction between *Rhizobium* bacteria and most legume plants is initiated by the perception of bacterial signal molecules, the nodulation (Nod) factors, at the root hairs of the plant. This induces responses both in the root hairs, leading to infection by the bacteria, as well as at a distance in the root cortex, leading to nodule organ formation. Molecular genetic approaches have been very successful in elucidating the key components essential for this Nod factor signal transduction. Cloning of these key regulators has been possible because of the establishment of two model legumes, *Lotus japonicus* and *Medicago truncatula*, for which extensive molecular genetic tools are available. We discuss the characteristics of the identified epidermal Nod-factor-signaling components from these two legumes and position them in a genetically based signal transduction cascade. To allow a successful rhizobial symbiosis, the responses in the root hairs need to be tightly coordinated with responses in the inner root cells. This is likely achieved through secondary signals that are generated upon Nod factor perception in the epidermis and are transported to the pericycle/cortex. The recent identification of a cytokinin receptor that is essential for the cortical responses supports the involvement of secondary signals, and the possible role of cytokinin as intercellular signal is discussed.

1 Introduction

The interaction of *Rhizobium* bacteria and certain legumes results in the formation of nitrogen-fixing root nodules. The formation of these nodules requires that the bacteria enter the root. In most legumes infection threads are made in root hairs and

T. Bisseling
Laboratory of Molecular Biology, Graduate School of Experimental Plant Sciences (EPS), Wageningen University and Research Center, 6703 HA Wageningen, The Netherlands
ton.bisseling@wur.nl

Plant Cell Monogr, doi:10.1007/7089_2008_10

these allow the bacteria to enter in a manner strictly controlled by the host plant. Further, already differentiated cortical cells are reprogrammed and enter the cell cycle, by which a nodule primordium is formed. The infection threads grow towards this primordium and upon release of the bacteria in cells of the primordium it differentiates into a nodule (Gage 2004; Oldroyd and Downie 2004). The infection process is induced by so-called Nod factors that are secreted by rhizobia when they colonize the roots of their host. Also, the induction of cortical cell divisions is induced by Nod factors, but most likely by a secondary signal generated in the epidermis upon Nod factor perception. Such intercellular communication coordinates infection with nodule formation in both space and time to ensure a successful symbiosis. In this chapter, we will first introduce the symbiotic partners, legumes and rhizobia. Subsequently, we will focus on Nod factor perception and transduction in the root epidermis and will discuss its role in infection as well as in nodule primordium formation.

2 Legumes Can Establish a Rhizobial Symbiosis

The rhizobial root nodule symbiosis is restricted to a single plant family, the *Fabaceae* (legumes). The *Fabaceae* is the third largest taxonomic plant family, encompassing ~18,000 species divided over three subfamilies: *Caesalpinioideae*, *Mimosoideae*, and *Papilionoideae*. Rhizobial symbiosis occurs within most genera belonging to the latter two subfamilies, whereas only ~5% of the genera of the subfamily *Caesalpinioideae* contain species that can form nodules (Sprent 2007). Since these nodulating genera are rather scattered within the phylogenetic tree, it is assumed that the interaction has evolved several times in evolution (Soltis et al. 1995). The multiple origin of rhizobial symbiosis is supported by the occurrence of a rhizobial interaction outside the *Fabaceae*. The genus *Parasponia* (encompassing ~4 species) belonging to the *Ulmaceae* family can also form a root nodule symbiosis with rhizobia (Trinick 1973; Scott 1986; Bender et al. 1987; Lafay et al. 2006).

Most important crop legumes, such as pea, soybean, bean, clover, and alfalfa, are in the *Papilionoideae* subfamily, and experimental research on nodulation is especially done with species belonging to this family. To facilitate molecular genetic studies in legumes, two model systems were developed, namely *Medicago truncatula* (Medicago) and *Lotus japonicus* (Lotus) (Oldroyd and Geurts 2001; Udvardi et al. 2005). Both species have all characteristics essential for a model plant species, such as a relatively short generation time, a small diploid genome, self-pollination, and efficient transformation. Medicago and Lotus are slightly different in their symbiotic properties as Medicago nodules have a persistent meristem supporting an indeterminate growth, whereas Lotus makes determinate nodules in which the meristematic activity disappears at an early stage of development. The description of Nod factor perception and transduction will be based on studies in these two model legumes.

3 Rhizobia and Nod Factors

The bacteria that can form nodules on legume roots are a diverse group of 40–50 species divided in 12 genera within the alpha, beta, and gamma proteobacteria (Moulin et al. 2001; Sawada et al. 2003; Benhizia et al. 2004). The bacterial species that are most studied are collectively known as Rhizobium (or rhizobia) and belong to two taxonomic families within the alpha protobacteria, namely the *Rhizobiaceae* (the genera *Rhizobium, Allorhizobium, Sinorhizoibum*, and *Mesorhizobium*) and *Bradyrhizobiaceae* (the genera *Bradyrhizobium and Azorhizobium*) (Sawada et al. 2003; Gupta 2005). Although the bacterial species involved are rather diverse, the bacterial genes involved in the symbiosis are in general highly conserved, which suggests that they have been obtained by horizontal gene transfer. Since the bacterial symbiosis genes are generally located on an extrachromosomal plasmid (sym plasmid, pSym), conjugational transfer in the rhizosphere seems indeed the most important evolutionary force spreading key genetic information essential for symbiotic nitrogen fixation among different bacterial species (Velazquez et al. 2005).

The rhizobial genes that are essential for the induction of infection as well as nodule primordium formation are the so-called *nod* (*nodulation*) genes and they are part of a molecular dialogue between legume host and *Rhizobium.* This dialogue starts by activation of the rhizobial NodD protein(s). NodD proteins belong to the class of LysR-type transcriptional regulators that become activated upon binding of external signals. In case of NodD this signal is generally a plant-secreted flavonoid (Mulligan and Long 1989; Honma et al. 1990). Flavonoids bind directly to NodD, thereby causing a conformational change resulting in an increased binding affinity for *nod* gene promoters by which the other *nod* genes are activated (Ogawa and Long 1995; Feng et al. 2003; Chen et al. 2005).

The products encoded by the *nod* genes are involved in the production and secretion of Nod factors. The basic structure of Nod factors, lipochitooligosaccharides, produced by different rhizobial species, is very similar (Fig. 1). They consist of an *N*-acetylglucosamine backbone with a fatty acyl chain attached at the nonreducing terminal sugar residue. Depending on the rhizobial species, the structure of the acyl chain can vary, and specific substitutions can be present at the reducing (position R4; Fig. 1) and nonreducing terminal glucosamine residues (positions R2 and R3; Fig. 1). These differences are due to the presence of species-specific *nod* genes or are the result of (allelic) variation causing a slightly different specificity of the encoded enzymes. In general rhizobia produce a population (ranging from 2 to 60 species) of different Nod factors (D'Haeze and Holsters 2002). The resulting variation in Nod factor structure plays an important role in the ability of the bacterium to interact with its specific host plants, where they are active in concentrations as low as 10^{-9} – 10^{-12}M and are major determinants of host specificity.

The microsymbiont of Medicago *Sinorhizobium meliloti* (*S. meliloti*) produces one major Nod factor, which is essential to nodulate *Medicago* species. This Nod factor is tetrameric and contains an acyl chain of 16 carbon atoms in length with two unsaturated bonds (C16:2). Furthermore, the terminal reducing glucosamine

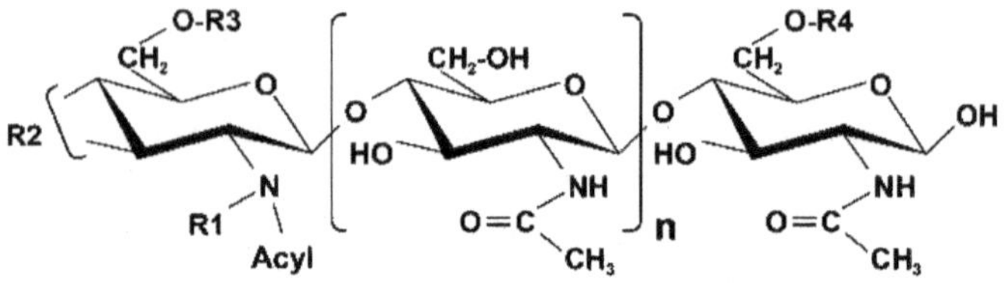

	Acyl	R1	R2	R3	R4	n
S. meliloti	$C_{16:2}$	H	OH	H / Ac	Sulfate	2
M. loti	$C_{18:1}$	Met	Carb	H	AcFuc	3

Fig. 1 Basal *N*-acetylglucosamine backbone of Nod factors. An acyl chain is attached to the terminal nonreducing glucosamine. Substitutions present on position from R1 to R4 are given for main Nod factors produced by *Sinorhizobium meliloti* (*S. meliloti*) and *Mesorhizobium loti* (*M. loti*)

residue of this Nod factor is *O*-sulfated, whereas the other terminal glucosamine contains an *O*-acetyl group (Lerouge et al. 1990; Schultze et al. 1992) (Fig. 1).

The microsymbiont of Lotus *Mesorhizobium Loti* (*M. loti*) mainly synthesizes pentameric Nod factors with an acetylated fucosyl at C-6 (R4) of the reducing sugar, a hydrogen at C-6 (R3), a carbamoyl at C-4 (R2), and an *N*-methylated (R1) 18:1 acyl chain at the nonreducing sugar (López-Lara et al. 1995) (Fig. 1).

4 Infection

Most legumes are infected by rhizobia via their root hairs. Nod-factor-secreting rhizobia get attached to the root hairs and some hairs form a tight curl around these rhizobia. Curling is the result of a continuous redirection of growth towards the side where a bacterium is attached to the root hair. This curling is induced by the recognition of Nod factors since a microdroplet containing Nod factor is sufficient to redirect growth to the side of application (Esseling et al. 2003). By root hair curling, the bacteria become entrapped in the pocket of this curl. Within the pocket of the curl the rhizobia induce a local modification of the cell wall at which site the plasma membrane invaginates and a tubelike structure, the so-called infection thread, is formed (Brewin 2004; Gage 2004). By these infection threads the rhizobia can enter their host. Concomitantly with the infection process, root cortical cells divide and form a nodule primordium. The infection thread grows towards this primordium and there bacteria are released – surrounded by a plant-derived membrane – into nodule primordium cells (Brewin 2004). The primordium then differentiates into a nodule and the released rhizobia differentiate into their symbiotic form and start to fix atmospheric nitrogen.

5 Nod-Factor-Signaling Pathway

5.1 Nod-Factor-Induced Epidermal Responses

Several legume mutants have been isolated that have disturbed Nod factor signaling, as Nod-factor-induced epidermal responses are lost. Studies on these mutants and the cloning of the mutant genes have markedly deepened our insight into mechanisms involved in Nod factor signaling. The first Nod-factor-signaling events take place in the epidermis when bacteria colonize the root. The epidermal Nod factor responses that were especially useful to position genes in a pathway are (1) morphological changes induced in root hairs, namely, root hair deformation and reorientation of root hair growth; (2) electrophysiological changes such as calcium spiking and calcium influx induced in the epidermis; and (3) the induction of expression of nodulin genes. First, these responses are described in more detail.

5.1.1 Root Hair Deformation and Reorientation of Root Hair Growth

Root hair deformation is the earliest morphological response to Nod factors. Nod factors cause deformation of root hair tips within 1 h. The swelling of root hair tips is the result of isotropic growth and occurs primarily in root hair cells that are terminating growth (Heidstra et al. 1994). Subsequently, tip growth is reestablished by which a new root hair tip emerges from the swelling at an angle from the original growth direction (root hair deformation). Alternatively, root hair tip growth can be reinitiated along the shaft of the root hair, causing a branched appearance.

When Nod factors are applied in a small droplet at the surface of a growing root hair, this can also cause a morphological response. This local application can result in redirection of growth towards the site of application and therefore it is reminiscent to root hair curling induced by rhizobial bacteria (Esseling et al. 2003).

5.1.2 Electrophysiological Responses to Nod Factor

Electrophysiological responses (especially changes in [Ca^{2+}]) and induction of gene expression, respectively, have also been used to genetically dissect the Nod factor-signaling pathway. Ca^{2+} influx is one of the earliest responses to Nod factor application (Felle et al. 1998). Ten seconds after addition of Nod factor an increase in cytoplasmic Ca^{2+} can be observed in the root hair tip, which occurs mainly at the periphery of the cell (Shaw and Long 2003). This primary rise in peripheral cytoplasmic $[Ca^{2+}]$ at the root hair tip is followed by a second one that originates from sources around the nucleus. This process is referred to as the secondary calcium flux (Shaw and Long 2003).

Calcium spiking is another electrophysiological response and starts from 5 to 30 min after application of Nod factors (Ehrhardt et al. 1996). Calcium spikes are transient elevations of $[Ca^{2+}]$ that originate from sources around the nucleus. Spiking can sustain for a couple of hours after the initial Nod factor application and is up to four magnitudes more sensitive to Nod factors (10^{-12}M) than the calcium flux response (1×10^{-9} to 10×10^{-9}M) (Shaw and Long 2003).

5.1.3 Transcriptional Responses to Nod Factor

Purified Nod factors are sufficient to activate the expression of several so-called early nodulin genes (*ENOD*) of which *ENOD11* and *ENOD12* are most frequently used to characterize mutants (Horvath et al. 1993; Pichon et al. 1992; Pingret et al. 1998; Journet et al. 2001). Induction of these genes, or reporter genes (such as *GUS* (*glucoronidase*)) making use of their promoters, can be detected in the epidermis within 1 h after Nod factor application.

Table 1 Nod-factor-signaling genes active in the epidermis

Medicago	Lotus	(Putative) function of gene products	Reference
NFP	NFR5	Nod-factor-signaling receptor	Madsen et al. 2003; Radutoiu et al. 2003; Arrighi et al. 2006
–	*NFR1*	Nod-factor-signaling receptor	Radutoiu et al. 2003
DMI2	SYMRK	Leucine-rich repeat receptor kinase	Stracke et al. 2002; Endre et al. 2002
DMI1	Castor	Cation channel	Ané et al. 2004; Imaizumi-Anraku et al. 2005
–	*Pollux*	Cation channel (duplication of Castor)	Imaizumi-Anraku et al. 2005
DMI3	SYM15	Ca^{2+}-/calmodulin-dependent protein kinase	Levy et al. 2004; Gleason et al. 2006
NSP1	NSP1	GRAS-type transcription factor	Smit et al. 2005; Heckmann et al. 2006
NSP2	NSP2	GRAS-type transcription factor	Kalo et al. 2005; Heckmann et al. 2006
NIN	NIN	RWP-RK transcription factor	Schauser et al. 1999; Marsh et al. 2007
LYK3	–	Nod factor entry receptor	Limpens et al. 2003; Smit et al. 2007
–	*Nup133*	Nucleoporin	Kanamori et al. 2006
–	*Nup85*	Nucleoporin	Saito et al. 2007
ERN1	–	AP2-ERF transcription factor	Middleton et al. 2007

–, orthologous gene in this species is not (yet) described

5.2 Genetic Dissection of the Nod-Factor-Signaling Pathway

Mutants that are blocked at a very early stage of Nod factor signaling lose some or all epidermal responses. Such mutants have been made, for example, in the model legumes Medicago and Lotus (Table 1). The mutant phenotypes have been studied in most detail in Medicago; therefore, we will mainly give a description of the genetic dissection of Nod factor signaling in this species. If necessary, additional characterized mutants from Lotus will be addressed. In Table1 we have presented the Lotus and Medicago Nod-factor-signaling genes that have been cloned and are active in the epidermis.

The above-described epidermal Nod factor responses have been used to study the Medicago mutants indicated in Table1. These are *Nod Factor Perception* (*nfp*), *LysM containing receptor kinase* (*Lyk3*)/*Hair Curling* (*hcl*), *Doesn't Make Infections 1* (*dmi1*), *dmi2*, *dmi3*, *Nod factor Signaling Pathway 1* (*nsp1*) and *nsp2*, *Branching Infection Threads 1* (*bit1*)/*ERF required for Nodulation 1* (*ern1*), *Nodule Inception* (*nin*) (Catoira et al. 2000, 2001; Amor et al. 2003; Oldroyd and Long 2003; Marsh et al. 2007; Middleton et al. 2007). In Lotus, two additional early mutants *Nucleoporin 133* (*nup133*) and *nup85* have been characterized (Kanamori et al. 2006; Saito et al. 2007). First, we will discuss the phenotypic characterization of these mutants by which most of the mutated genes could be positioned in a genetically based signal transduction cascade. In a following paragraph the proteins encoded by these genes are described in more detail.

A loss-of-function mutation in the gene(s) active most upstream within the Nod-factor-signaling pathway likely has the most severe effect on Nod factor signaling. Only in *nfp* mutants all Nod factor responses are lost (Amor et al. 2003). This positions NFP at the upstream end of the signaling cascade and this fits well with a proposed function as Nod factor receptor (Amor et al. 2003) (Fig. 2). In Medicago only one mutant was identified that had lost all Nod factor responses. However, in Lotus two loci were identified that are essential for all Nod factor responses (Schauser et al. 1998; Szczyglowski et al. 1998; Madsen et al. 2003; Radutoiu et al. 2003). These are *NFR1* and *NFR5*, both of which encode putative Nod factor receptors and of which the latter is orthologous to Medicago *NFP* (see later). The Medicago gene most homologous to Lotus *NFR1* is *LYK3*. However, in contrast to NFR1, a loss-of-function mutation in LYK3, in the *hcl* mutant, is first blocked at the root hair curling stage (Catoira et al. 2001; Smit et al. 2007).

In the *dmi1* and *dmi2* mutants the primary calcium influx can be induced and also their root hairs can reorient their growth direction in response to Nod factor (Shaw and Long 2003; Esseling et al. 2004). However, the secondary calcium increase around the nucleus as well as the calcium spiking response are lost (Wais et al. 2000; Shaw and Long 2003). In contrast, calcium spiking can still be induced in *dmi3*, *nsp1*, and *nsp2* (Wais et al. 2000; Shaw and Long 2003). This positions DMI1 as well as DMI2 downstream of NFP but upstream of calcium spiking, whereas DMI3, NSP1, and NSP2 are positioned downstream of calcium spiking (Fig.2). Further, *dmi3* mutants have been shown to be more sensitive to Nod-factor-induced calcium spiking (Oldroyd et al. 2001). Therefore, DMI3 activation appears to negatively

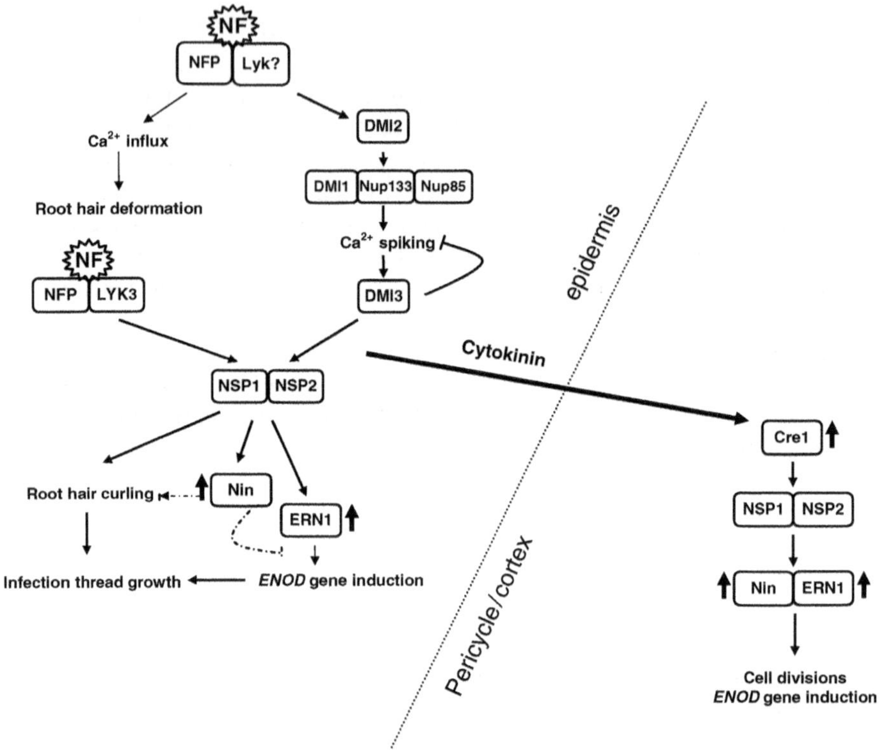

Fig. 2 Schematic representation of the Nod-factor-signaling cascade from the epidermis to the pericycle/cortex. To simplify the model, we only use the Medicago gene names; we refer to the main text for the orthologous genes from Lotus. Nod factors are most likely perceived by LysM domain containing receptors, such as NFP and LYK3, located in the plasma membrane. NFP is absolutely required for all Nod factor responses, whereas LYK3 appears to be only essential for root hair curling, infection, and a subset of Nod-factor-induced genes. NFP likely forms heterodimeric complexes with other LYK genes to activate the signaling cascade. Upon Nod factor perception the DMI pathway is activated. This pathway is not required for calcium influx at the tip of the root hairs or for root hair deformations. However, the DMI proteins do affect root hair growth by controlling the sensitivity of the root hairs to mechanical stress. The LRR receptor kinase DMI2 is located in the plasma membrane whereas DMI1 and the Nup proteins (so far only identified in Lotus) are located in the nuclear envelope. Therefore DMI2 is placed upstream of DMI1 and NUP133/85. All these proteins are required to induce calcium spiking. The resulting calcium signature is decoded by the calcium/calmodulin dependent protein kinase DMI3, which is located in the nucleus. There, DMI3 activates the Nod factor response genes NSP1 and NSP2, putative GRAS-type transcription factors. In *dmi3* mutants the calcium spiking response is more sensitive to Nod factors, which indicates that DMI3 controls calcium spiking via a negative feedback loop. NSP1 and NSP2 are required to trigger (*ENOD*) gene induction and to allow root hair curling around Nod-factor-producing bacteria attached to the root hairs. As root hair curling and a subset of NSP1/2-dependent Nod-factor-induced genes require LYK3, it is likely that signaling from the LYK3 Nod factor receptor complex, parallel to the DMI pathway, converges at or just after NSP1/2 to specify the set of activated genes. Among the genes that are induced by NSP1/2 are the putative transcription factors *NIN* and *ERN1*. NIN is required to allow the formation of tight root hair curls entrapping the bacteria. The AP2-ERF transcription factor ERN1 is not

control the calcium spiking response. In the Lotus temperature-sensitive *nup85* and *nup133* mutants calcium spiking is also blocked, which positions them upstream of DMI3/CCAMK (Kanamori et al. 2006; Saito et al. 2007) (Fig. 2). To date, no orthologs of the Lotus *nup* mutants have been identified in Medicago.

Nod factors are unable to induce the second step of root hair deformation in *dmi1, dmi2, and dmi3* mutants under conditions used in most laboratories (Heidstra et al. 1994; Catoira et al. 2000). So isotropic growth after Nod factor application occurs, but new polar outgrowth from these swellings cannot be initiated. However, when mechanical impulses are minimized (e.g., during replacement of medium), root hairs of *dmi2* mutants deform like those of wild-type plants (Esseling et al. 2004). This indicates that the touch response of *dmi2* mutants is markedly altered (Esseling et al. 2004; Geurts et al. 2005). They appear hypersensitive to touch as a block of root hair growth is induced by very mild mechanical stimuli, and their ability to restart growth is markedly reduced (Esseling et al. 2004). However, *dmi1*, *dmi2*, and *dmi3* mutant root hairs can respond to careful local Nod factor application by reorienting their growth direction just like wild-type root hairs do (Esseling et al. 2004). This suggests that NFP, by a signal transduction pathway independent of *DMI* genes, induces deformation and root hair reorientation. Parallel to this pathway, the DMI pathway influences sensitivity of root hair growth to mechanical stress, and, via the *NSP* genes, is essential for regulation of early nodulin gene expression (Fig. 2). Signaling via NSP1 and NSP2 is required to allow root hair curling in response to rhizobia.

Expression analyses of the *ENOD11/12* reporter genes showed that mutations in all genes positioned upstream of *NSP1* and *NSP2* completely eliminated induction of *ENOD* expression by Nod factors (Catoira et al. 2000; Mitra and Long 2004). Also *nsp2* mutants show no induction of reporter genes, while in *nsp1* mutants *ENOD* genes can be induced albeit at a markedly reduced level (Catoira et al. 2000; Oldroyd and Long 2003). An always-active mutant of DMI3 leads to Nod-factor-independent expression of *ENOD11* and this requires the two *NSP* genes (Gleason

Fig. 2 (continued) essential for root hair curling but is required for a subset of Nod-factor-induced genes, such as *ENOD11* induction. NIN, though not required for *ENOD11* induction, does limit the size of the root zone responsive to Nod factors. Gene induction controlled by ERN1 is required to allow proper infection thread growth. Nod factor signaling in the epidermis also triggers cell activation and (*ENOD*) gene induction in the pericycle and cortex, leading to the formation of a nodule primordium. This is likely achieved via the intercellular transport of a secondary signal molecule(s) from the epidermis to the pericycle. There Nod factor responses require the presence of the cytokinin receptor CRE1. Therefore, it is possible that cytokinin is produced upon Nod factor signaling in the epidermis, and then transported to the inner cell layers where it activates CRE1. However, other or additional secondary signals might also be involved. A gain-of-function mutant of the cytokinin receptor LHK1 in Lotus (orthologous to CRE1) is able to trigger spontaneous nodule formation in the absence of Nod-factor-producing bacteria. This requires the presence of the NSP1/2 genes as well as (upregulation of (*thick arrows*)) both NIN and ERN1. Spontaneous nodule formation can also be triggered by an always-active DMI3 mutant. Such an always-active DMI3 mutant, in contrast to the gain-of-function cytokinin receptor, also induces *ENOD11* expression in the epidermis in the absence of Nod factors

et al. 2006). Therefore DMI3 is positioned upstream of both NSP1 and NSP2 (Fig. 2). Despite the slight difference, we place NSP1 and NSP2 in a similar position downstream of DMI3.

Expression of *ENOD11* is also blocked in the Medicago *bit1–1/ern1* mutant (Middleton et al. 2007). However, a subset of early Nod-factor-induced genes can still be induced in *bit1*. Further, *bit1* mutants do show root hair curling but infection thread growth in the root hairs is impaired. Therefore *bit1* is positioned downstream of NSP1 and NSP2 (Fig. 2).

In the Lotus and Medicago *nin* mutants, excessive root hair curling is triggered; however, tight curls entrapping the bacteria cannot be formed (Schauser et al. 1999; Marsh et al. 2007). As a consequence infection threads cannot be formed. Furthermore, the susceptible zone of root hairs showing root hair deformations and *ENOD11* induction is expanded in the *nin* mutants, which indicates that NIN is involved in a negative feedback to control root hair competence (Schauser et al. 1999; Marsh et al. 2007). It is not known to which extent *nin* mutants affect Nod factor induced expression; however, NIN does not appear to be required for *ENOD11* induction (Marsh et al. 2007). Therefore we position NIN parallel to ERN1 downstream of NSP1 and NSP2.

All the above-mentioned mutants are also impaired in their ability to induce Nod factor responses in the inner cell layers of the root, e.g., pericycle and cortex.

6 The Nod-Factor-Signaling Genes in the Epidermis

6.1 The Nod Factor Receptors

Nod factors are thought to be perceived by LysM-domain-containing receptor kinases that are most likely located in the plasma membrane. In Medicago these are encoded by *NFP* and *LYK3* and in Lotus by *NFR1* and *NFR5* (Amor et al. 2003; Madsen et al. 2003; Limpens et al. 2003; Radutoiu et al. 2003; Arrighi et al. 2006; Smit et al. 2007). Medicago *NFP* is the ortholog of Lotus *NFR5*; both are essential for all Nod factor responses. Their sequences are highly conserved, and also the chromosome regions where these genes are located are syntenic (Amor et al. 2003; Madsen et al. 2003; Radutoiu et al. 2003; Arrighi et al. 2006). The predicted extracellular domain of NFP/NFR5 contains three LysM domains. LysM domains are proposed to bind *N*-acetylglucosamine oligomers as found in bacterial peptidoglycan, chitin, and the Nod factor backbone (Bateman and Bycroft 2000; Limpens et al. 2003; Radutoiu et al. 2003; Steen et al. 2005; Arrighi et al. 2006). In *Arabidopsis* a LysM-domain-containing receptor kinase was shown to be required for chitin-elicitor defense signaling (Miya et al. 2007). Further, a LysM-domain-containing glycoprotein from rice has been shown to directly bind fungal chitin elicitors triggering defense responses (Kaku et al. 2006). This supports the idea that the putative Nod factor receptors directly bind Nod factors via their LysM domains, although physical interaction remains to be demonstrated. It further suggests an evolutionary link between (Nod factor) symbiotic signaling and fungal chitin-elicitor defense signaling. Genetics has shown that components of the Nod-factor-signaling pathway (the DMI and Nup proteins, see later) are also required for the more ancient mycorrhizal fungal symbiosis, which is triggered by a not-yet-identified mycorrhizal signal molecule called Myc factor (Kosuta et al. 2003). It can therefore be expected that the Myc-factor is a chitin-like molecule that is perceived by LysM-domain-containing receptors. For interaction of mycorrhizae and root hair interaction see also Novero et al. (2008).

The kinase domain of NFP, like NFR5, is missing the activation loop, normally present in kinases (Madsen et al. 2003; Arrighi et al. 2006). It also lacks the so-called P-loop and a conserved DFG motif next to the missing activation loop, regions normally highly conserved in protein kinases. These regions play a role in positioning ATP at the active site and its availability to substrates (Johnson and Ingram 2005). It was shown that NFP does not show autophosphorylation activity in vitro, which indicates that the kinase domain is not active, although it cannot be excluded that this is induced upon ligand binding (Arrighi et al. 2006). It might be that NFP/NFR5 needs to heterodimerize with an active kinase.

A good candidate for such an interacting active kinase is NFR1. Like NFR5, NFR1 is required for (almost) all Nod-factor-induced responses in Lotus (Radutoiu et al. 2003). NFR1 contains three LysM domains in its extracellular domain and, in contrast to NFR5, contains all motifs of an active intracellular kinase domain. The fact that both NFR5 and NFR1 are required for all Nod factor responses suggests that they form a heterodimer (Madsen et al. 2003; Radutoiu et al. 2003). Introduction of NFR5 and NFR1 in Medicago allowed an incompatible *M. loti* and DZL strain to nodulate Medicago in a Nod-factor-dependent manner (Radutoiu et al. 2007). It was further shown that the ability to extend the host range is controlled by the extracellular LysM domains, strongly supporting the idea that they directly bind Nod factors (Radutoiu et al. 2007). Lotus NFR1 is highly homologous to Medicago LYK3 and also the Medicago and Lotus chromosomal regions where these genes are located are syntenic (Limpens et al. 2003). However, LYK3, mutated in the *hair curling* (*hcl*) mutant, is not required for the early Nod factor induced responses such as root hair deformation, calcium spiking, and gene induction (Catoira et al. 2001; Mitra and Long 2004; Smit et al. 2007). In contrast, strong and weak alleles of LYK3 show that it is required for root hair curling, infection thread formation and a subset of Nod factor induced genes (Catoira et al. 2001; Limpens et al. 2003; Mitra and Long 2004; Smit et al. 2007). The identification of a Nod factor receptor specifically involved in infection-related steps is very well in line with the two-receptor model that was postulated, based on the finding that the structural demands for different Nod factor responses vary (Debellé et al. 1986; Surin and Downie 1988; Firmin et al. 1993, Ardourel et al. 1994; Journet et al. 1994; Demont-Caulet et al. 1999; Wais et al. 2002; Shaw and Long 2003). The start of infection was shown to have a more stringent demand with respect to Nod factor structure, whereas, for example, root hair deformation, calcium flux, calcium spiking, and induction of early nodulin gene expression are more promiscuous. These studies were the basis for the hypothesis that responses such as root hair deformation and cortical cell division are induced by a so-called Nod-factor-signaling receptor, whereas the start of infection thread formation required a second Nod factor receptor that was named entry receptor (Ardourel et al. 1994). *LYK3* knock-down mutants also show Nod factor structure dependent infection phenotypes, which support a physical interaction with Nod factors (Limpens et al. 2003; Smit et al. 2007). So LYK3 fits the criteria of being an entry receptor (Fig. 2). NFP in this respect represents a signaling receptor. The RNA-interference-mediated knock down of a homologous *LYK* gene (*LYK4*) also causes a Nod factor structure dependent infection phenotype, where infection threads appear to lose polarity as they grow, resulting in sac-like structures (Limpens et al. 2003). Therefore, it might be that in Medicago multiple heterodimers containing NFP and different LYK genes are formed that have different Nod-factor-binding properties or substrates, or both (Cullimore and Dénarié 2003) (Fig. 2). The Medicago genome contains at least ten *LYK* gene members, of which seven form a cluster at the LYK3 locus (Limpens et al. 2003; Arrighi et al. 2006).

6.2 DMI Genes

Genetic analyses showed that the *DMI* genes have a pivotal role in both rhizobial as well as in mycorrhizal symbioses. As the mycorrhizal symbiosis is much more ancient than the rhizobial symbiosis, it is thought that the latter has recruited the ancient DMI pathway in evolution to control nodulation. Here we will briefly describe the characteristics of these three DMI proteins.

6.2.1 DMI2

Medicago DMI2 is orthologous to Lotus SYMRK (Stracke et al. 2002; Endre et al. 2002). *DMI2* encodes a membrane-localized receptor-like kinase, consisting of an N-terminal extracellular domain of 595 amino acids that contains three leucine-rich repeats near the transmembrane domain and an intracellular kinase domain (Endre et al. 2002). A functional DMI2-GFP fusion protein was localized to the plasma membrane (Limpens et al. 2005). Leucine-rich repeats are known to be involved in protein–protein and protein–ligand interactions. Therefore it is possible that DMI2 is activated by a ligand generated upon Nod factor perception or that it is part of a large complex together with the Nod factor receptors. The fact that *dmi2* mutants appear to have a nonsymbiotic enhanced touch response suggests that plant-derived ligands are involved (Esseling et al. 2003). Pharmacological studies have shown that DMI2 signaling leading to calcium spiking and *ENOD* gene induction can be bypassed by application of Mastoparan, an activator of hetero-trimeric G-proteins and MAP kinases (Pingret et al. 1998; Den Hartog et al. 2003; Charron et al. 2004; Esseling et al. 2004; Sun et al. 2007). This step is blocked by inhibitors of phospholipase D and phospholipase C activity, implicating phospholipid signaling in the pathway between DMI2 and calcium spiking. The molecular link between DMI2 and phospholipid signaling is not yet known. The kinase domain of DMI2 was recently shown to interact with a 3-hydroxy-3-methylglutaryl coenzyme A reductase, HMGR1 (Kevei et al. 2007). HMGR1 belongs to a multigene family in Medicago and these proteins are key enzymes in the mevalonate pathway. This mevolanate pathway leads to a range of isoprenoid compounds in plants (Aharoni et al. 2005), such as cytokinins, sterols, phytosteroids, and isoprenyl lipid moieties, that are attached to a range of signaling proteins such as, for example, ROP GTPases and are important for the localized activation of these signal transduction proteins (Yang and Fu 2007). Interestingly, cytokinin plays a key role in the Nod-factor-signaling pathway, triggering cortical cell divisions (see Sect. 7). Pharmacological inhibition of HMGR1 activity as well as RNAi severely impaired nodulation, indicating that HMGR1 is essential for nodulation (Kevei et al. 2007). The exact step at which HMGR1 affects nodulation is not clear and the mechanism by which HMGR1 controls nodulation remains to be unraveled.

6.2.2 DMI1

DMI1, like DMI2, is required to induce calcium spiking in response to Nod factors. *DMI1* encodes a membrane protein, containing five predicted transmembrane domains, and shows strong similarity to a MthK calcium-gated potassium channel (Ané et al. 2004; Peiter et al. 2007). However, DMI1 lacks a typical GYGD motif found in K^+-selective channels and hence it is not clear which ions are actually transported by DMI1. It is unlikely that DMI1 transports calcium, as pharmacological studies have shown that Mastoparan-induced calcium spiking and *ENOD11* gene induction similar to Nod factors do not require DMI1 (Charron et al. 2004; Sun et al. 2007). The cytosolic C terminus of DMI1 contains a RCK (regulator of the conductance of K^+) domain that may be involved in controlling the activity of the channel in response to calcium or other ligands. A functional DMI1-GFP fusion protein was localized to the nuclear envelope, corresponding to the region where the major calcium changes occur (Riely et al. 2007). So by an unknown signaling cascade, most likely involving phospholipid signaling (Den Hartog et al. 2003; Charron et al. 2004; Sun et al. 2007), the plasma-membrane-located DMI2 activates DMI1 in the nuclear envelope. A *dmi1* mutant allele lacking the C-terminal part acts as a dominant-negative form that interferes with calcium ion release (Peiter et al. 2007). Further, Mastoparan is unable to induce calcium ion spikes in this mutant. Therefore it is hypothesized that DMI1 modulates calcium ion channels that are activated by both Mastoparan and Nod factor signaling. This is possibly achieved through modulation of the membrane potential of the nuclear membrane by altering the transport of an unknown cation (Peiter et al. 2007). In contrast to Medicago, Lotus has two *DMI*-like genes, namely *Castor* and *Pollux*, and both are essential for Nod-factor-induced calcium spiking (Imaizumi-Anraku et al. 2005; Miwa et al. 2006). GFP fusion proteins of both proteins localized to plastids in onion cells and pea roots (Imaizumi-Anraku et al. 2005). However, the functionality of these fusion constructs was not tested and the plastid localization is difficult to reconcile with their role in calcium spiking.

6.2.3 DMI3

The calcium spiking response must be transduced to downstream Nod factor responses such as *ENOD* gene expression. This is most likely done by DMI3. *DMI3* encodes a calcium- and calmodulin-dependent protein kinase (CCaMK), which is homologous to a biochemically well characterized lily CCaMK (Levy et al. 2004; Mitra et al. 2004; Gleason et al. 2006). The N-terminal part contains a conserved serine or threonine protein kinase domain in which a putative nuclear localization signal is present. This domain is flanked by an autoinhibitory domain that overlaps with a calmodulin-binding domain and three calcium-binding EF hands reside in the C-terminal part of the protein. Calcium-enhanced calmodulin-binding relieves the autoinhibition. DMI3 has been shown to be nuclear located, where it most likely functions as an integrator of [Ca^{2+}] signals (spiking) (Oldroyd and Downie 2004; Kalo et al. 2005; Smit et al. 2005). Analysis of calcium spiking using a fluorescent reporter has shown that the duration and number of calcium spikes, together with the developmental status of the cell, regulate *ENOD* gene expression (Miwa et al. 2006). The nuclear localization of DMI3 suggests that further downstream signaling leading to gene expression takes place in the nucleus and most likely DMI3 activates transcription factors. The key role for DMI3 in further signal transduction is supported by the identification of always-active alleles in both Lotus (*snf1*) and Medicago (Gleason et al. 2006; Tirichine et al. 2007). Such mutant alleles are mutated in (or lack) the autoinhibition domain, resulting in an active kinase. This induces the formation of spontaneous nodules in the absence of bacteria as well as *ENOD* gene induction (Gleason et al. 2006; Tirichine et al. 2006). These responses require the NSP proteins (see later), indicating that DMI3 activates these. DMI3 likely forms a multimeric complex as has been reported for the homologous animal CaMKII (Oldroyd and Downie 2004). This is supported by the fact that the Lotus *snf1* allele is recessive and negatively affected by the wild-type DMI3 protein (Tirichine et al. 2006). It is likely that the always-active allele can still be negatively regulated by incorporation in a complex with wild-type DMI3 protein.

6.3 Nucleoporins

Calcium spiking is also blocked in the temperature-sensitive Lotus *nup133* and *nup85* mutants. *NUP133* and *NUP85* encode nucleoporin genes homologous to the yeast and mammalian Nup133 and Nup85 genes (Kanamori et al. 2006; Saito et al. 2007). Nucleoporins are components of a nuclear pore complex, containing around 30 proteins, which mediates transport of mRNA and proteins across the nuclear envelope. In mammals Nup133 and Nup85 are both components of a single Nup107–160 subcomplex (Walther et al. 2003). Localization of a functional YFP-Nup133 fusion protein to the nuclear rim in root hair cells supports the idea that it is part of the nuclear pore complex in plants (Kanamori et al. 2006). Both Nup proteins are expressed constitutively in different organs of the plant and no induction is seen upon inoculation with rhizobia (Kanamori et al. 2006; Saito et al. 2007). It is currently not clear how these nucleoporins control calcium spiking. It might be that they affect the membrane potential of the nuclear envelope and thereby interfere with the activity of calcium channels, as was suggested for DMI1, which also localizes to the nuclear envelope. Interestingly, in *Arabidopsis* a nucleoporin (Nup96) has been shown to be required for resistance against pathogens (Zhang and Li 2005). Lotus *Nup133* and *Nup85* are also required to establish a mycorrhizal symbiosis (Kanamori et al. 2006; Saito et al. 2007). This indicates a general role for the nucleoporins in plant–microbe interactions, both pathogenic and symbiotic.

6.4 NSP1 and NSP2

NSP1 and NSP2 have been identified in both Medicago and Lotus and are essential for all reported Nod-factor-induced transcriptional changes (Catoira et al. 2000; Oldroyd and Long 2003; Mitra et al. 2004; Kalo et al. 2005; Smit et al. 2005; Heckmann et al. 2006). *NSP1* and *NSP2* encode putative transcription factors belonging to the plant-specific GRAS (GAI, RGA, SCR)-type protein family (consisting of 33 members in *Arabidopsis*). They contain all five amino acid motifs present in GRAS-type proteins. Medicago NSP1 is constitutively present in the nucleus, where it is likely activated by DMI3 upon Nod factor signaling, either directly or indirectly (Smit et al. 2005). The Medicago NSP2 was located in the ER surrounding the nucleus and upon Nod factor signaling it accumulated in the nucleus (Kalo et al. 2005). However, this localization might be flawed through the use of a strong 35S promoter driving the fusion protein. Since both *NSPs* are constitutively expressed they have been named Nod factor response factors (Smit et al. 2005). Phosphorylation of both proteins by DMI3 in the nucleus might activate them and trigger transcriptional responses. NSP1 and NSP2, in contrast to DMI3, are not required for mycorrhization, which suggests that either Nod factor signaling parallel to the DMI pathway converges at NSP1/2 or that perception of different calcium signatures (nodulation vs. mycorrhization) by DMI3 leads to different outputs.

Both *NSPs* are single copy genes that have putative orthologs in, e.g., rice, poplar, and *Arabidopsis*, suggesting that NSPs have a nonsymbiotic function. However, uninoculated *nsp1* and *nsp2* mutants do not show a phenotype when grown under laboratory conditions. In addition, a tobacco NSP1 homolog was shown to complement the epidermal Nod factor responses in the *nsp1* mutants when expressed under the control of the Medicago *NSP1* promoter, indicating that this function is conserved through evolution (Heckmann et al. 2006).

6.5 Nodule Inception

The *nin* mutants in both Lotus and Medicago show excessive root hair curling, an enlarged root zone that responds to Nod factor and rhizobia, no infections, and no cortical cell divisions (Schauser et al. 1999; Marsh et al. 2007). *nin* encodes an atypical putative transcription factor with a predicted nuclear localization signal as well as two putative membrane-spanning helices (Schauser et al. 1999). Its C-terminal half combines characteristics of both bZIP and bHLH/Z transcription factors and also contains basic DNA-binding domains. Two additional features of NIN are the RWP-RK motif (within the basic region) and the PB1 domain at the C terminus. The RWP-RK motif is conserved in the *Chlamydomona reinhardtii* Minus dominance proteins that regulate sexual reproduction in response to nitrogen starvation. As nodulation is also regulated by nitrogen, this may hint at the evolutionairy conservation of this type of regulators. The PB1 domain mediates protein–protein interactions by interacting with a PC motif present in cell-polarity-determining proteins in yeast (Ito et al. 2001). NIN appears to have both positive (primordium formation and infection) as well as negative (restricting the susceptible zone) activities during nodulation. To explain this, it has been postulated that NIN activity is related to that of transmembrane proteins that are proteolytically cleaved to release a transcriptional regulator intracellularly as well as a signal molecule extracellularly with different function (Schauser et al. 1999; Marsh et al. 2007). However, such a model including the subcellular localization of NIN remains to be demonstrated. Upon Nod factor signaling, *NIN* expression itself is induced in the epidermis and in the forming primordium (Schauser et al. 1999; Radutoiu et al. 2003). This suggests that *NIN* is one of the genes regulated by the Nod-factor-signaling pathway as depicted in Fig. 2. NIN seems to be a "primary" target and it is required for certain Nod factor responses. Homologs of *NIN*, as holds for all Nod-factor-signaling genes, have been identified also in nonleguminous plants (Schauser et al. 2005).

6.6 Ern

The Medicago *bit1–1* mutant was characterized by a block in the initiation and development of infection threads (Middleton et al. 2007). *bit1–1* turned out to be mutated in an AP2 domain containing ERF transcription factor termed ERN1 (ERF Required for Nodulation 1). Introduction of the always-active DMI3 allele into the *bit1–1* mutant further showed that ERN1 is essential for spontaneous nodule formation. Furthermore, ERN1 is required for Nod-factor-induced *ENOD11* induction as well as a subset of early Nod-factor-induced genes (Middleton et al. 2007). This shows that ERN1 functions in the Nod-factor-signaling pathway. *ERN1* expression, like *NIN*, is itself rapidly induced upon Nod factor signaling. It is not known whether *NIN* induction depends on ERN1. The activation of a subset of early Nod-factor-induced genes in the *bit1–1* mutant seems to be sufficient to allow limited infection thread development; however, proper infection seems to require the full set of genes similar to what is observed in the *hcl* (*lyk3*) mutants (Mitra et al. 2004; Smit et al. 2007). However, *ERN1* expression was induced in *hcl* to the same level as in wild type; so, *ERN1* induction is independent of LYK3. A search for transcription factors regulating Nod-factor-induced expression of *ENOD11* in root hairs identified two additional ERN1 homologs, ERN2 and ERN3 (Andriankaja et al. 2007). ERN1 and ERN2 act as transcriptional activators and ERN3 as repressor. This suggests that differential activation of subsets of such activating and repressing factors controls the spatial and temporal expression of target genes.

7 Epidermis to Inner Root Signaling

Infection thread formation and growth needs to be tightly coordinated in both time and space with nodule primordium formation to establish a successful symbiosis. Infection threads initiate in root hairs, while the nodule primordium concomitantly develops from inner cortical cells in, for example, Medicago. To develop a functional nodule, the infection threads need to reach the primordium cells in a timely manner. Both infection thread formation and nodule primordium formation are induced by Nod factors. In all legume mutants characterized so far, a loss of all the Nod factor responses in the epidermis is always correlated with the loss of Nod-factor-induced cortical cell divisions (Stougaard 2001; Tsyganov et al. 2002; Limpens and Bisseling 2003). This observation led to the hypothesis that secondary signals are generated in the epidermis upon Nod factor perception and these subsequently trigger cortical cell divisions (Dudley and Long 1989). The involvement of Nod-factor-induced short distance signaling is supported by studies on Nod-factor-induced early nodulin gene expression in *Vicia sativa* (Vijn et al. 1995). Here, external Nod factor application induced cortical cell divisions and in these cells *ENOD12* expression is induced; however, *ENOD5* is not induced in such nodule primordia. *ENOD5* was only induced in cells that are in direct contact with the Nod factor, e.g., epidermal cells exposed to Nod factor or cells with growing infection threads. Therefore it was hypothesized that *ENOD12* can be induced cell-autonomously by Nod factor and at a distance by a second messenger. In contrast, *ENOD5* cannot be activated by this second messenger, but only cell autonomously by Nod factor. The cell autonomous activity of Nod factors is in line with their strong affinity for plant cell walls, where they become immobilized. This makes it very unlikely that Nod factors themselves can diffuse from epidermis to inner cell layers (Goedhart et al. 2000). These observations fit best with the hypothesis that Nod factor signaling in the epidermis generates a secondary signal(s) that subsequently triggers cortical cell divisions.

Mutants that uncouple the cortical responses from the epidermal responses can give insight into this intercellular communication. Such mutants were recently identified in Lotus and some of them turned out to be mutated in a plasma-membrane-localized cytokinin receptor, Lotus histidine kinase – LHK1 (Murray et al. 2007; Tirichine et al. 2007). Three independent loss-of-function mutations in LHK1 (*from hit1–1* to *hit1–3*) resulted in a loss of nodule formation, while epidermal responses such as infection thread formation were not blocked. This suggests that cytokinin induces cortical responses but not the epidermal responses. These mutants even appeared to be hyperinfected (Murray et al. 2007), which suggests that nodule primordia provide a negative feedback on infection of the root hairs. Infection threads in the *hit* mutant mainly fail to penetrate into the cortex and are arrested at the epidermis/cortex boundary. This is likely due to the absence of cortical cell activation and lack of "cytoplasmic bridges" through which the infection threads are guided (Yang et al. 1994; Murray et al. 2007). In wild-type roots these cytoplasmic bridges are formed in outer cortical cells that are arrested in the G2 phase of the cell cycle upon Nod factor treatment (Yang et al. 1994). In Medicago, silencing of the orthologous cytokinin receptor *CRE1* also blocked nodule formation, but in contrast to the situation in Lotus, hyperinfection did not occur (Gonzalez-Rizzo et al. 2006) (Fig. 2).

It has been shown in the past that application of cytokinin to the outside of legume roots (or *E. coli* secreting cytokinin) can induce several Nod factor responses such as cortical cell divisions and associated early nodulin expression (such as *ENOD40* and *NIN*) (Cooper and Long 1994; Bauer et al. 1996; Fang and Hirsch 1998; Mathesius et al. 2000; Compaan et al. 2001; Lohar et al. 2004; Gonzalez-Rizzo et al. 2006). Therefore it can be hypothesized that cytokinin acts as an intercellular/short distance signal to trigger cortical cell divisions in response to Nod factor signaling at the epidermis. In other words, Nod factor signaling might induce the formation of cytokinin species that as secondary signals trigger responses in the cortex. Alternatively, Nod factor signaling might induce cortical cell divisions via other secondary signals that impinge on cell cycle control by cytokinin (Cooper and Long 1994).

These observations suggest that Nod factor signaling independent activation of the CRE1/LHK1 cytokinin receptor is sufficient to trigger nodule formation. Indeed, a gain-of-function mutation in the extracellular domain of LHK1, in the Lotus *snf2* mutant, results in the spontaneous formation of nodules in the absence of bacteria (Tirichine et al. 2007). Interestingly, the ability to induce spontaneous nodules by this *snf2* allele depends on the putative transcription factors NSP2 and NIN, but does not require functional NFR1, NFR5, SYMRK (DMI2), or CCaMK (DMI3) (Fig. 2).

Cytokinin signaling seems to trigger responses only in the pericycle and cortex. Epidermal Nod-factor-induced responses do not seem to be affected by cytokinin application or in the different CRE/LHK mutants. This may be explained by the absence of *CRE1/LHK1* expression in the epidermis, as is suggested by *CRE1* promoter-GUS studies in Medicago (Lohar et al. 2004). The first nonepidermal responses to Nod factor (or cytokinin application) in Medicago occur in the pericycle, where ENOD40 expression is induced within 3 h and one or two rounds of pericycle divisions precede cortical cell divisions (Timmers et al. 1999; Compaan et al. 2001). This suggests that CRE1 is first active in the pericycle, where we hypothesize that it activates *NIN* via NSP2.

Always-active mutants of the CCaMK (DMI3) in both Lotus (*snf1*) and Medicago are also able to trigger spontaneous nodule formation (Tirichine et al.

2006; Gleason et al. 2006). Double mutants containing both the LHK1 *snf2* allele and the always-active CCaMK (*snf1*) show an additive effect with more spontaneous nodules than the single mutants. This implies two (mutually nonexclusive) possibilities; the always-active CCaMK generates a secondary signal that induces the expression of the cytokinin receptor in the pericycle and cortex, or, the always-active CCaMK elevates cytokinin levels. Transcriptome analysis of early nodulation events, combined with promoter-GUS analysis, showed that *MtCRE1* was significantly up-regulated in roots between 6 and 48 h after inoculation with rhizobia. Strong expression was observed in nodule primordia, which later became restricted to the meristem in mature nodules (Lohar et al. 2006). As cytokinin itself can induce the expression of the cytokinin receptor, it cannot be excluded that this is due to enhanced cytokinin production (Hutchison and Kieber 2002; Gonzalez-Rizzo et al. 2006). The LHK1 *snf2* allele still responds to cytokinin by increasing the expression of response genes upon cytokinin treatment (Tirichine et al. 2007). Therefore, increased levels of the mutated receptor, possibly in combination with increased cytokinin production, would explain the additive effect (Fig. 2).

In case cytokinin is produced upon Nod factor signaling in the epidermis, it is possible that cytokinin signaling already occurs in the epidermis. Studies using an *Arabidopsis* cytokinin response regulator (ARR5 promoter-GUS reporter) in Medicago showed that this response regulator is induced in curled/deformed root hairs by rhizobia (Lohar et al. 2004), supporting the hypothesis that cytokinin species are produced upon Nod factor signaling in the epidermis. It might be that other CRE1 homologs are present in the epidermis to relay the cytokinin signal.

Interestingly, it was recently shown that a Nod-factor-induced AP2-ERF transcription factor, termed ERN1, is required for both infection thread growth and primordium development during nodulation (Middleton et al. 2007). Introduction of the always-active CCaMK in the *ern1* mutant *bit1–1* was unable to induce spontaneous nodules, indicating that ERN1 is required for Nod factor responses in the cortex (Fig. 2). Nod-factor-dependent root hair expression of *ENOD11* was shown to be controlled by both activating (ERN1 and ERN2) and repressive activities (ERN3) of homologous AP2-ERF transcription factors (Andriankaja et al. 2007). In *Arabidopsis* it was shown that a subset of AP2-ERF transcription factors mediates cytokinin responses together with the type-B ARR cytokinin response regulators (Rashotte et al. 2006). This raises the possibility that signaling output in response to cytokinin and Nod factors could be modulated by specific sets of these transcription factors in different cell types.

Other studies suggest that cytokinin might not be the only signal triggering cortical responses, at least in indeterminate nodules. Although application of the cytokinin BAP (benzylamino purine) to white clover roots induces cortical cell divisions, it fails to induce several of the early cortical responses observed in response to rhizobia (Mathesius et al. 2000). Among these differences was the lack of induction of chalcone synthase genes in the inner cortex 24 h after cytokinin treatment. Chalcone synthases are involved in the production of flavonoids, which have been shown to function as auxin transport inhibitors and are required for nodule formation in indeterminate nodules (Mathesius et al. 1998; Wasson et al. 2006). Exogenous application of auxin transport inhibitors (NPA) was also shown to induce nodule-like structures that express early nodulin genes such as *ENOD2* and *ENOD12* (Hirsch et al. 1989; Scheres et al. 1992). However, (iso)flavonoid-induced auxin transport inhibition is not required for determinate nodule formation (Subramanian et al. 2006, 2007). It will be interesting to see whether a *Mt*CRE1 gain-of-function mutant, as the Lotus *snf2* allele, will be sufficient to induce genuine spontaneous nodules in Medicago.

Nodulation is strictly controlled by nitrogen availability. Furthermore, nodule number is controlled in a process called autoregulation, by which the plant controls the balance between energetically costly nodule formation and benefit (nitrogen source) (Limpens and Bisseling 2003). This involves a long distance signaling between root and shoot. A Clavata1-like receptor kinase HAR1 (hypernodulation aberrant root 1) has been identified in Lotus, which is orthologous to Medicago SUNN, and is required only in the shoot to control nodule number (Krusell et al. 2002; Nishimura et al. 2002; Schnabel et al. 2005). HAR1 is likely involved in perceiving a signal from the root and in the generation of an autoregulation signal by which nodulation is stopped (Oka-Kira and Kawaguchi 2006). Autoregulation of nodule numbers and repression of nodulation by fixed nitrogen are likely controlled in part by a similar mechanism, because autoregulation mutants are insensitive to nitrate. It has recently been shown that plant development in response to nitrogen is signaled in part by cytokinin as a local and long-range messenger (Takei et al. 2002; Sakakibara et al. 2006). Nitrate was shown to upregulate specific cytokinin synthesis genes in *Arabidopsis*, leading to cytokinin accumulation (Takei et al. 2002). The nonsymbiotic root phenotype of the *har1* autoregulation mutant can be rescued in part by the addition of the cytokinin zeatin in combination with the ethylene blocker AVG (Wopereis et al. 2000). On the other hand, growing roots in the continued presence of (high) cytokinin blocks nodule formation, which also functions systemically in split-root systems (van Brussel et al. 2002). This shows that cytokinin signaling and autoregulation are interconnected. However, it was shown that the cyokinin (zeatin)-induced systemic block of nodulation in *Vicia* is different from nodule-induced autoregulation, as the former could be prevented by controlling the pH of the growth medium (van Brussel et al. 2002). However, our understanding of the complex cross-talk and fine-tuning of nitrogen and cytokinin signaling as well as specificity towards different nitrogen sources and cytokinin species is still very limited, especially in legumes. It might be that specific production of cytokinin(s) in response to Nod factor signaling acts as the secondary signal that couples cortical cell division to autoregulation of nodule number. Future research on the role of cytokinin in nodule formation will likely shed light on this interesting link and might also give insight into the evolutionary aspects of nodulation.

8 Concluding Remarks

Last decade, major novel insights into Nod factor perception and transduction have been obtained by the cloning and characterization of legume genes that had been identified by a genetic approach. The cloning of these key regulators became possible in legume species for which efficient molecular genetic tools had been developed, namely, the legumes Medicago and Lotus. The cloning of Lotus and Medicago genes facilitates now the cloning of orthologs in important crops such as soybean and pea. In this way these genes will be available from many legume species. It further opens the possibility to identify and study the corresponding genes from nonlegumes and to study how such genes were recruited in evolution to establish the rhizobial symbiosis.

The cloning of symbiotic genes has especially been focused on genes involved in Nod factor signaling occurring during the early steps of the interaction. However, at later stages, for example, when the infection thread traverses the root hair, major cell biological changes occur and it will be important to identify and characterize the key regulators of these steps. Fortunately, several mutants with disturbed infection thread growth have already been identified. It has further become clear that continuous Nod factor signaling is required to facilitate these later steps.

The cloning of Nod-factor-signaling genes now provides a toolbox by which a more detailed understanding of the involved mechanisms can be obtained. For example, the availability of cloned receptors provides the tools to address important research questions, such as the following: Are Nod factor receptors located in special membrane domains? Are receptors internalized by endocytosis and if so is this because signaling takes place at endomembrane compartments or because receptors have to be removed upon local activation? Further, the cloned Nod-factor-signaling genes provide the anchor point of a signaling cascade in which many gaps remain to be filled. For example, how is signaling transferred from plasma-membrane-located DMI2 to the nuclear envelope where DMI1 is located? How is positional information achieved during root hair curling and infection thread formation and how is this communicated to the cytoskeleton and integrated with membrane trafficking? Since most responses occur very locally, e.g., infection threads are only formed in a few root hair cells, it is clear that these questions can only be addressed in the coming years by applying advanced cell biological approaches.

Acknowledgments Erik Limpens is supported by The Netherlands Organization of Scientific Research (NWO) VENI grant 86305023.

References

Aharoni A, Jongsma MA, Bouwmeester HJ (2005) Volatile science? Metabolic engineering of terpenoids in plants. Trends Plant Sci 10:594–602

Amor BB, Shaw SL, Oldroyd GE, Maillet F, Penmetsa RV, Cook D, Long SR, Dénarié J, Gough C (2003) The NFP locus of *Medicago truncatula* controls an early step of Nod factor signal transduction upstream of a rapid calcium flux and root hair deformation. Plant J 34:495–506

Andriankaja A, Boisson-Dernier A, Frances L, Sauviac L, Jauneau A, Barker DG, de Carvalho-Niebel F (2007) AP2-ERF transcription factors mediate Nod factor dependent Mt ENOD11 activation in root hairs via a novel cis-regulatory motif. Plant Cell 19:2866–2885

Ané JM, Kiss GB, Riely BK, Penmetsa RV, Oldroyd GE, Ayax C, Levy J, Debellé F, Baek JM, Kalo P, Rosenberg C, Roe BA, Long SR, Dénarié J, Cook DR (2004) *Medicago truncatula* DMI1 required for bacterial and fungal symbioses in legumes. Science 303:1364–1347

Ardourel M, Demont N, Debellé F, Maillet F, de Billy F, Promé JC, Dénarié J, Truchet G (1994) *Rhizobium meliloti* lipooligosaccharide nodulation factors: different structural requirements for bacterial entry into target root hair cells and induction of plant symbiotic developmental responses. Plant Cell 6:1357–1374

Arrighi JF, Barre A, Amor BB, Bersoult A, Campos Soriano L, Mirabella R, De Carvalho-Niebel F, Journet EP, Ghérardi M, Huguet T, Geurts R, Dénarié J, Rougé P, Gough C (2006) The *Medicago truncatula* lysine motif-receptor-like kinase gene family includes *NFP* and new nodule-expressed genes. Plant Physiol 142:265–279

Bateman A, Bycroft M (2000) The structure of a LysM domain from *E coli* membrane-bound lytic murein transglycosylase D (MltD). J Mol Biol 299:1113–1119

Bauer P, Ratet P, Crespi MD, Schultze M, Kondorosi A (1996) Nod factors and cytokinins induce similar cortical cell division, amyloplast deposition and *Msenod12A* expression in alfalfa roots. Plant J 10:91–105

Bender GL, Goydych W, Rolfe BG, Nayudu M (1987) The role of *Rhizobium* conserved and host specific nodulation genes in the infection of the non-legume *Parasponia andersonii*. Mol Gen Genet 210:299–306

Benhizia Y, Benhizia H, Benguedouar A, Muresu R, Giacomini A, Squartini A (2004) Gamma proteobacteria can nodulate legumes of the genus *Hedysarum* Syst Appl Microbiol 27:462–468

Brewin NJ (2004) Cell wall remodeling in the *Rhizobium*-legume symbiosis. Crit Rev Plant Sci 23:1–24

Catoira R, Galera C, de Billy F, Penmetsa RV, Journet EP, Maillet F, Rosenberg C, Cook D, Gough C, Dénarié J (2000) Four genes of *Medicago truncatula* controlling components of a nod factor transduction pathway. Plant Cell 12:1647–1666

Catoira R, Timmers AC, Maillet F, Galera C, Penmetsa RV, Cook D, Dénarié J, Gough C (2001) The *HCL* gene of *Medicago truncatula* controls *Rhizobium*-induced root hair curling. Development 128:1507–1518

Charron D, Pingret J, Chabaud M, Journet E, Barker DG (2004) Pharmacological evidence that multiple phospholipid signaling pathways link rhizobium nodulation factor perception in *Medicago truncatula* root hairs to intracellular responses, indicating Ca^{2+} spiking and specific *ENOD* gene expression. Plant Physiol 136:3582–3593

Chen XC, Feng J, Hou BH, Li FQ, Li Q, Hong GF (2005) Modulating DNA bending affects NodD-mediated transcriptional control in *Rhizobium leguminosarum*. Nucl Acids Res 33:2540–2548

Compaan B, Yang WC, Bisseling T, Franssen H (2001) *ENOD40* expression in the pericycle precedes cortical cell division in Rhizobium-legume interaction and the highly conserved internal region of the gene does not encode a peptide. Plant Soil 230:1–8

Cooper JB, Long SR (1994) Morphogenetic rescue of *Rhizobium meliloti* nodulation mutants by trans-zeatin secretion. Plant Cell 6:215–225

Cullimore J, Dénarié J (2003) Plant sciences. How legumes select their sweet talking symbionts. Science 302:575–578

D'Haeze W, Holsters M (2002) Nod factor structures, responses, and perception during initiation of nodule development. Glycobiology 12:79R–-105R

Debellé F, Rosenberg C, Vasse J, Maillet F, Martinez E, Dénarié J, Truchet G (1986) Assignment of symbiotic developmental phenotypes to common and specific nodulation (*nod*) genetic loci of *Rhizobium meliloti*. J Bacteriol 168:1075–1086

Demont-Caulet N, Maillet F, Tailler D, Jacquinet JC, Promé JC, Nicolaou KC, Truchet G, Beau JM, Dénarié J (1999) Nodule-inducing activity of synthetic *Sinorhizobium meliloti* nodulation factors and related lipo-chitooligosaccharides on alfalfa Importance of the Acyl Chain Structure. Plant Physiol 120:83–92

Den Hartog M, Verhoef N, Munnik T (2003) Nod factor and elicitors activate different phospholipid signaling pathways in suspension-cultured alfalfa cells. Plant Physiol 132:311–317

Dudley ME, Long SR (1989) A non-nodulating alfalfa mutant displays neither root hair curling nor early cell division in response to *Rhizobium meliloti*. Plant Cell 1:65–72

Ehrhardt DW, Wais R, Long SR (1996) Calcium spiking in plant root hairs responding to *Rhizobium* nodulation signals. Cell 85:673–681

Endre G, Kereszt A, Kevei Z, Mihacea S, Kalo P, Kiss GB (2002) A receptor kinase gene regulating symbiotic nodule development. Nature 417:962–966

Esseling JJ, Lhuissier FGP, Emons AMC (2003) Nod factor-induced root hair curling: continuous polar growth towards the point of Nod factor application. Plant Physiol 132:1982–1988

Esseling JJ, Lhuissier FGP, Emons AMC (2004) A nonsymbiotic root hair tip growth phenotype in NORK-mutated legumes: implications for Nodulation factor–induced signaling and formation of a multifaceted root hair pocket for bacteria. Plant Cell 16:933–944

Fang Y, Hirsch AM (1998) Studying early nodulin gene *ENOD40* expression and induction by nodulation factor and cytokinin in transgenic alfalfa. Plant Physiol 116:53–68

Felle HH, Kondorosi E, Kondorosi A, Schultze M (1998) The role of ion fluxes in Nod factor signaling in *Medicago sativa.* Plant J 13:455–463

Feng J, Li Q, Hu HL, Chen XC, Hong GF (2003) Inactivation of the nod box distal half-site allows tetrameric NodD to activate *nodA* transcription in an inducer-independent manner. Nucl Acids Res 31:3143–3156

Firmin JL, Wilson KE, Carlson RW, Davies AE, Downie JA (1993) Resistance to nodulation of cv Afghanistan peas is overcome by *nodX*, which mediates an O-acetylation of the *Rhizobium leguminosarum* lipo-oligosaccharide nodulation factor. Mol Microbiol 10:351–360

Gage DJ (2004) Infection and invasion of roots by symbiotic, nitrogen-fixing, rhizobia during nodulation of temperate legumes, Microbiol. Mol Biol Rev 68:280–300

Geurts R, Fedorova E, Bisseling T (2005) Nod factor signalling genes and their function in the early stages of *Rhizobium* infection. Curr Opin Plant Biol 8:346–352

Gleason C, Chaudhuri S, Yang T, Muñoz A, Poovaiah BW, Oldroyd GE (2006) Nodulation independent of rhizobia induced by a calcium-activated kinase lacking autoinhibition. Nature 441:1149–1152

Goedhart J, Hink MA, Visser AJ, Bisseling T, Gadella TW Jr (2000) In vivo fluorescence correlation microscopy (FCM) reveals accumulation and immobilization of Nod factors in root hair cell walls. Plant J 21:109–119

Gonzalez-Rizzo S, Crespi M, Frugier F (2006) The *Medicago truncatula.* CRE1 cytokinin receptor regulates lateral root development and early symbiotic interaction with *Sinorhizobium meliloti*. Plant Cell 18:2680–2693

Gupta RS (2005) Protein signatures distinctive of alpha proteobacteria and its subgroups and a model for alpha-proteobacterial evolution. Crit Rev Microbiol 31:101–135

Heckmann AB, Lombardo F, Miwa H, Perry JA, Bunnewell S, Parniske M, Wang TL, Downie JA (2006) Lotus japonicus nodulation requires two GRAS domain regulators, one of which is functionally conserved in a non-legume. Plant Physiol 142:1739–1750

Heidstra R, Geurts R, Franssen H, Spaink HP, van Kammen A, Bisseling T (1994) Root hair deformation activity of nodulation factors and their fate on *Vicia sativa.* Plant Physiol 105:787–797

Hirsch AM, Bhuvaneswari TV, Torrey JG, Bisseling T (1989) Early nodulin genes are induced in alfalfa root outgrowths elicited by auxin transport inhibitors. Proc Natl Acad Sci USA 86:1244–1248

Honma MA, Asomaning M, Ausubel FM (1990) *Rhizobium meliloti nodD* genes mediate host-specific activation of *nodABC*. J Bacteriol 172:901–911

Horvath B, Heidstra R, Miklos L, Moerman M, Spaink HP, Promé J, van Kammen A, Bisseling T (1993) Lipo-oligosaccharides of *Rhizobium* induce infection related early nodulin gene expression in pea root hairs. Plant J 4:727–733

Hutchison CE, Kieber JJ (2002) Cytokinin signaling in Arabidopsis. Plant Cell 14 (Suppl):S47–S59

Imaizumi-Anraku H, Takeda N, Charpentier M, Perry J, Miwa H, Umehara Y, Kouchi H, Murakami Y, Mulder L, Vickers K, Pike J, Downie JA, Wang T, Sato S, Asamizu E, Tabata S, Yoshikawa M, Murooka Y, Wu GJ, Kawaguchi M, Kawasaki S, Parniske M, Hayashi M (2005) Plastid proteins crucial for symbiotic fungal and bacterial entry into plant roots. Nature 433:527–531

Ito T, Matsui Y, Ago T, Ota K, Sumimoto H (2001) Novel modular domain PB1 recognizes PC motif to mediate functional protein-protein interactions. EMBO J 20:3938–3946

Johnson KL, Ingram GC (2005) Sending the right signals: regulating receptor kinase activity. Curr Opin Plant Biol 8:648–656

Journet EP, Pichon M, Dedieu A, de Billy F, Truchet G, Barker DG (1994) *Rhizobium meliloti* Nod factors elicit cell-specific transcription of the *ENOD12* gene in transgenic alfalfa. Plant J 6:241–249

Journet EP, El-Gachtouli N, Vernoud V, de Billy F, Pichon M, Dedieu A, Arnould C, Morandi D, Barker DG, Gianinazzi-Pearson V (2001) *Medicago truncatula ENOD11*: a novel RPRP-encoding early nodulin gene expressed during mycorrhization in arbuscule-containing cells. Mol Plant Microbe Interact 14:737–748

Kaku H, Nishizawa Y, Ishii-Minami N, Akimoto-Tomiyama C, Dohmae N, Takio K, Minami E, Shibuya N (2006) Plant cells recognize chitin fragments for defense signaling through a plasma membrane receptor. Proc Natl Acad Sci USA 103:11086–11091

Kaló P, Gleason C, Edwards A, Marsh J, Mitra RM, Hirsch S, Jakab J, Sims S, Long SR, Rogers J, Kiss GB, Downie JA, Oldroyd GE (2005) Nodulation signaling in legumes requires NSP2, a member of the GRAS family of transcriptional regulators. Science 308:1786–1789

Kanamori N, Madsen LH, Radutoiu S, Frantescu M, Quistgaard EM, Miwa H, Downie JA, James EK, Felle HH, Haaning LL, Jensen TH, Sato S, Nakamura Y, Tabata S, Sandal N, Stougaard J (2006) A nucleoporin is required for induction of Ca2 + spiking in legume nodule development and essential for rhizobial and fungal symbiosis. Proc Natl Acad Sci USA 103:359–364

Kevei Z, Lougnon G, Mergaert P, Horváth GV, Kereszt A, Jayaraman D, Zaman N, Marcel F, Regulski K, Kiss GB, Kondorosi A, Endre G, Kondorosi E, Ané JM (2007) 3-Hydroxy-3-Methylglutaryl Coenzyme A Reductase1 interacts with NORK and is crucial for nodulation in medicago truncatula. Plant Cell 19:3974–3989

Kosuta S, Chabaud M, Lougnon G, Gough C, Dénarié J, Barker DG, Bécard G (2003) A diffusible factor from arbuscular mycorrhizal fungi induces symbiosis-specific *MtENOD11* expression in roots of *Medicago truncatula*. Plant Physiol 131:952–962

Krusell L, Madsen LH, Sato S, Aubert G, Genua A, Szczyglowski K, Duc G, Kaneko T, Tabata S, De Bruijn F, Pajuelo E, Sandal N, Stougaard J (2002) Shoot control of root development and nodulation is mediated by a receptor-like kinase. Nature 420:422–426

Lafay B, Bullier E, Burdon JJ (2006) Bradyrhizobia isolated from root nodules of *Parasponia*. (*Ulmaceae*) do not constitute a separate coherent lineage. Int J Syst Evol Microbiol 56:1013–1018

Lerouge P, Roche P, Faucher C, Maillet F, Truchet G, Promé JC, Dénarié J (1990) Symbiotic host-specificity of *Rhizobium meliloti* is determined by a sulphated and acylated glucosamine oligosaccharide signal. Nature 344:781–784

Levy J, Bres C, Geurts R, Chalhoub B, Kulikova O, Duc G, Journet EP, Ané JM, Lauber E, Bisseling T, Dénarié J, Rosenberg C, Debellé F (2004) A putative Ca^{2+} and calmodulin-dependent protein kinase required for bacterial and fungal symbioses. Science 303:1361–1364

Limpens E, Bisseling T (2003) Signaling in symbiosis. Curr Opin Plant Biol 6:343–350

Limpens E, Franken C, Smit P, Willemse J, Bisseling T, Geurts R (2003) LysM domain recetor kinases regulating rhizobial Nod factor-induced infection. Science 302:630–633

Limpens E, Mirabella R, Fedorova E, Franken C, Franssen H, Bisseling T, Geurts R (2005) Formation of organelle-like N_2-fixing symbiosomes in legume root nodules is controlled by DMI2. Proc Natl Acad Sci USA 102:10375–10380

Lohar DP, Schaff JE, Laskey JG, Kieber JJ, Bilyeu KD, Bird DM (2004) Cytokinins play opposite roles in lateral root formation, and nematode and Rhizobial symbioses. Plant J 38:203–214

Lohar DP, Sharopova N, Endre G, Peñuela S, Samac D, Town C, Silverstein KA, VandenBosch KA (2006) Transcript analysis of early nodulation events in *Medicago truncatula*. Plant Physiol 140:221–234

López-Lara IM, van den Berg JD, Thomas-Oates JE, Glushka J, Lugtenberg BJ, Spaink HP (1995) Structural identification of the lipo-chitin oligosaccharide nodulation signals of *Rhizobium loti*. Mol Microbiol 15:627–638

Madsen EB, Madsen LH, Radutoiu S, Olbryt M, Rakwalska M, Szczyglowski K, Sato S, Kaneko T, Tabata S, Sandal N, Stougaard J (2003) A receptor kinase gene of the LysM type is involved in legume perception of rhizobial signals. Nature 425:637–640

Marsh JF, Rakocevic A, Mitra RM, Brocard L, Sun J, Eschstruth A, Long SR, Schultze M, Ratet P, Oldroyd GE (2007) *Medicago truncatula* NIN is essential for rhizobial-independent nodule organogenesis induced by autoactive calcium/calmodulin-dependent protein kinase. Plant Physiol 144:324–335

Mathesius U, Schlaman HR, Spaink HP, Of Sautter C, Rolfe BG, Djordjevic MA (1998) Auxin transport inhibition precedes root nodule formation in white clover roots and is regulated by flavonoids and derivatives of chitin oligosaccharides. Plant J 14:23–34

Mathesius U, Charon C, Rolfe BG, Kondorosi A, Crespi M (2000) Temporal and spatial order of events during the induction of cortical cell divisions in white clover by *Rhizobium leguminosarum* bv *trifolii* inoculation or localized cytokinin addition. Mol Plant Microbe Interact 13:617–628

Middleton PH, Jakab J, Penmetsa RV, Starker CG, Doll J, Kaló P, Prabhu R, Marsh JF, Mitra RM, Kereszt A, Dudas B, VandenBosch K, Long SR, Cook DR, Kiss GB, Oldroyd GE (2007) An ERF transcription factor in *Medicago truncatula* that is essential for Nod factor signal transduction. Plant Cell 19:1221–1234

Mitra RM, Long SR (2004) Plant and bacterial symbiotic mutants define three transcriptionally distinct stages in the development of the *Medicago truncatula /Sinorhizobium meliloti* symbiosis. Plant Physiol 134:595–604

Mitra RM, Gleason CA, Edwards A, Hadfield J, Downie JA, Oldroyd GE, Long SR (2004) A Ca^{2+} /calmodulin-dependent protein kinase required for symbiotic nodule development Gene identification by transcript-based cloning. Proc Natl Acad Sci USA 101:4701–4705

Miwa H, Sun J, Oldroyd GE, Downie JA (2006) Analysis of calcium spiking using a cameleon calcium sensor reveals that nodulation gene expression is regulated by calcium spike number and the developmental status of the cell. Plant J 48:883–894

Miya A, Albert P, Shinya T, Desaki Y, Ichimura K, Shirasu K, Narusaka Y, Kawakami N, Kaku H, Shibuya N (2007) CERK1, a LysM receptor kinase, is essential for chitin elicitor signaling in Arabidopsis. Proc Natl Acad Sci USA 104:19613–19618

Moulin L, Munive A, Dreyfus B, Boivin-Masson C (2001) Nodulation of legumes by members of the -subclass of proteobacteria. Nature 411:948–950

Mulligan JT, Long SR (1989) A family of activator genes regulates expression of *Rhizobium meliloti* nodulation genes. Genetics 122:7–18

Murray JD, Karas BJ, Sato S, Tabata S, Amyot L, Szczyglowski K (2007) A cytokinin perception mutant colonized by *Rhizobium* in the absence of nodule organogenesis. Science 315:101–104

Nishimura R, Hayashi M, Wu GJ, Kouchi H, Imaizumi-Anraku H, Murakami Y, Kawasaki S, Akao S, Ohmori M, Nagasawa M, Harada K, Masayoshi K (2002) HAR1 mediates systemic regulation of symbiotic organ development. Nature 420:426–429

Novero M, Genre A, Szczyglowski K, Bonfante P (2008) Root hair colonization by mycorrhizal fungi. In: Emons AMC, Ketelaar T (eds) Root hairs: excellent tools for the study of plant molecular cell biology. Springer, Berlin Heidelberg New York. doi:10.1007/7089_2008_12

Ogawa J, Long SR (1995) The *Rhizobium meliloti groELc* locus is required for regulation of early nod genes by the transcription activator NodD. Genes Dev 9:714–729

Oka-Kira E, Kawaguchi M (2006) Long-distance signlaing to control root nodule number. Curr Opin Plant Biol 9:496–502

Oldroyd GE, Geurts R (2001) Medicago truncatula, going where no plant has gone before. Trends Plant Sci 6:552–554

Oldroyd GE, Long SR (2003) Identification and characterization of *nodulation-signaling pathway 2*, a gene of *Medicago truncatula* involved in Nod factor signaling. Plant Physiol 131:1027–1032

Oldroyd ED, Downie JA (2004) Calcium, kinases and nodulation signaling in legumes. Nat Rev 5:566–576

Oldroyd GE, Mitra RM, Wais RJ, Long SR (2001) Evidence for structurally specific negative feedback in the Nod factor signal transduction pathway. Plant J 28:191–199

Peiter E, Sun J, Heckmann AB, Venkateshwaran M, Riely BK, Otegui MS, Edwards A, Freshour G, Hahn MG, Cook DR, Sanders D, Oldroyd GE, Downie JA, Ané JM (2007) The *Medicago truncatula* DMI1 protein modulates cytosolic calcium signaling. Plant Physiol 145:192–203

Pichon M, Journet EP, Dedieu A, de Billy F, Truchet G, Barker DG (1992) Rhizobium meliloti elicits transient expression of the early nodulin gene *ENOD12* in the differentiating root epidermis of transgenic alfalfa. Plant Cell 4:1199–1211

Pingret JL, Journet EP, Barker DG (1998) Rhizobium Nod factor signaling: evidence for a G protein-mediated transduction mechanism. Plant Cell 10:659–671

Radutoiu S, Madsen LH, Madsen EB, Felle HH, Umehara Y, Grønland M, Sato S, Nakamura Y, Tabata S, Sandal N, Stougaard J (2003) Plant recognition of symbiotic bacteria requires two LysM receptor-like kinases. Nature 425:585–592

Radutoiu S, Madsen LH, Madsen EB, Jurkiewicz A, Fukai E, Quistgaard EM, Albrektsen AS, James EK, Thirup S, Stougaard J (2007) LysM domains mediate lipochitin-oligosaccharide recognition and *Nfr* genes extend the symbiotic host range. EMBO J 26:3923–3935

Rashotte AM, Mason MG, Hutchison CE, Ferreira FJ, Schaller GE, Kieber JJ (2006) A subset of Arabidopsis AP2 transcription factors mediates cytokinin responses in concert with a two-component pathway. Proc Natl Acad Sci USA 103:11081–11085

Riely BK, Lougnon G, Ané JM, Cook DR (2007) The symbiotic ion channel homolog DMI1 is localized in the nuclear membrane of *Medicago truncatula* roots. Plant J 49:208–216

Saito K, Yoshikawa M, Yano K, Miwa H, Uchida H, Asamizu E, Sato S, Tabata S, Imaizumi-Anraku H, Umehara Y, Kouchi H, Murooka Y, Szczyglowski K, Downie JA, Parniske M, Hayashi M, Kawaguchi M (2007) NUCLEOPORIN85 is required for calcium spiking, fungal and bacterial symbioses, and seed production in *Lotus japonicus*. Plant Cell 19:610–624

Sakakibara H, Takei K, Hirose N (2006) Interactions between nitrogen and cytokinin in the regulation of metabolism and development. Trends Plant Sci 11:440–448

Sawada H, Kuykendall LD, Young JM (2003) Changing concepts in the systematics of bacterial nitrogen-fixing legume symbionts. J Gen Appl Microbiol 49:155–179

Schauser L, Handberg K, Sandal N, Stiller J, Thykjær T, Nielsen A, Stougaard J (1998) Symbiotic mutants deficient in nodule establishment identified after T-DNA transformation of *Lotus japonicus*. Mol Gen Genet 259:414–423

Schauser L, Roussis A, Stiller J, Stougaard J (1999) A plant regulator controlling development of symbiotic root nodules. Nature 402:191–195

Schauser L, Wieloch W, Stougaard J (2005) Evolution of NIN-like proteins in Arabidopsis, rice, and *Lotus japonicus*. J Mol Evol 60:229–237

Scheres B, McKhann HI, Zalensky A, Löbler M, Bisseling T, Hirsch AM (1992) The *PsENOD12*. gene is expressed at two different sites in Afghanistan pea pseudonodules induced by auxin transport inhibitors. Plant Physiol 100:1649–1655

Schnabel E, Journet PE, Carvalho-Niebel F, De Duc G, Frugoli J (2005) The *Medicago truncatula SUNN* gene encodes a CLV1-like leucine-rich repeat receptor kinase that regulates nodule number and root length. Plant Mol Biol 58:809–822

Schultze M, Quiclet-Sire B, Kondorosi E, Virelizier H, Glushka JN, Endre G, Géro SD, Kondorosi A (1992) *Rhizobium meliloti* produces a family of sulfated lipo-oligosaccharides exhibiting different degrees of plant host specificity. Proc Natl Acad Sci USA 89:192–196

Scott KF (1986) Conserved nodulation genes from the non-legume symbiont *Bradyrhizobium* sp. (*Parasponia*). Nucl Acid Res 14:2905–2919

Shaw SL, Long SR (2003) Nod factor elicits two separable calcium responses in *Medicago truncatula* root hair cells. Plant Physiol 131:976–984

Smit P, Raedts J, Portyanko V, Debellé F, Gough C, Bisseling T, Geurts R (2005) NSP1 of the GRAS protein family is essential for rhizobial Nod factor-induced transcription. Science 308:1789–1791

Smit P, Limpens E, Geurts R, Fedorova E, Dolgikh E, Gough C, Bisseling T (2007) Medicago LYK3, an entry receptor in rhizobial nodulation factor signaling. Plant Physiol 145:183–191

Soltis DE, Soltis PS, Morgan DR, Swensen SM, Mullin BC, Dowd JM, Martin PG (1995) Chloroplast gene sequence data suggest a single origin of the predisposition for symbiotic nitrogen fixation in angiosperms. Proc Natl Acad Sci USA 92:2647–2651

Sprent I (2007) Evolving ideas of legume evolution and diversity: a taxonomic perspective on the occurrence of nodulation. New Phytologist 174:11–25

Steen A, Buist G, Horsburgh GJ, Venema G, Kuipers OP, Foster SJ, Kok J (2005) AcmA of *Lactococcus lactis* is an N-acetylglucosaminidase with an optimal number of LysM domains for proper functioning. FEBS J 272:2854–2868

Stougaard J (2001) Genetics and genomics of root symbiosis. Curr Opin Plant Biol 4:328–335

Stracke S, Kistner C, Yoshida S, Mulder L, Sato S, Kaneko T, Tabata S, Sandal N, Stougaard J, Szczyglowski K, Parniske M (2002) A plant receptor-like kinase required for both bacterial and fungal symbiosis. Nature 417:959–962

Subramanian S, Stacey G, Yu O (2006) Endogenous isoflavones are essential for the establishment of symbiosis between soybean and *Bradyrhizobium japonicum.* Plant J 48:261–273

Subramanian S, Stacey G, Yu O (2007) Distinct, crucial roles of flavonoids during legume nodulation. Trends Plant Sci 12:282–285

Sun J, Miwa H, Downie JA, Oldroyd GE (2007) Mastoparan activates calcium spiking analogous to Nod factor-induced responses in *Medicago truncatula.* root hair cells. Plant Physiol 144:695–702

Surin BP, Downie JA (1988) Characterization of the *Rhizobium leguminosarum* genes *nodLMN* involved in efficient host-specific nodulation. Mol Microbiol 2:173–183

Szczyglowski K, Shaw SS, Wopereis J, Copeland S, Hamburger D, Kasiborski B, Dazzo FB, de Bruijn FJ (1998) Nodule organogenesis and symbiotic mutants of the model legume *Lotus japonicus*. Mol Plant Microbe Interact 11:684–697

Takei K, Takahashi T, Sugiyama T, Yamaya T, Sakakibara H (2002) Multiple routes communicating nitrogen availability from roots to shoots: a signal transduction pathway mediated by cytokinin. J Exp Bot 53:971–977

Timmers ACJ, Auriac MC, Truchet G (1999) Refined analysis of early symbiotic steps of the Rhizobium-Medicago interaction in relationship with microtubular cytoskeleton rearrangements. Development 126:3617–3628

Tirichine L, Sandal N, Madsen LH, Radutoiu S, Albrektsen AS, Sato S, Asamizu E, Tabata S, Stougaard J (2007) A gain-of-function mutation in a cytokinin receptor triggers spontaneous root nodule organogenesis. Science 315:104–107

Trinick MJ (1973) Symbiosis between *Rhizobium* and the non-legume, *Trema aspera.* Nature 244:459–468

Tsyganov VE, Voroshilova VA, Priefer UB, Borisov AY, Tikhonovich IA (2002) Genetic dissection of the initiation of the infection process and nodule tissue development in the *Rhizobium* - pea (*Pisum sativum* L) symbiosis. Ann Bot (Lond) 89:357–366

Udvardi MK, Tabata S, Parniske M, Stougaard J (2005) *Lotus Japonicus*: legume research in the fast lane. Trends Plant Sci 10: 222–228

van Brussel AA, Tak T, Boot KJ, Kijne JW (2002) Autoregulation of root nodule formation: signals of both symbiotic partners studied in a split-root system of *Vicia sativa* subsp *Nigra.* Mol Plant Microbe Interact 15:341–349

Velazquez E, Peix A, Zurdo-Pineiro JL, Palomo JL, Mateos PF, Rivas R, Munoz-Adelantado E, Toro N, Garcia-Benavides P, Martinez-Molina E (2005) The coexistence of symbiosis and pathogenicity-determining genes in *Rhizobium rhizogenes* strains enables them to induce nodules and tumors or hairy roots in plants. Mol Plant-Microbe Interact 12:1325–1332

Vijn I, Martinez-Abarca F, Yang WC, das Neves L, van Brussel A, van Kammen A, Bisseling T (1995) Early nodulin gene expression during Nod factor-induced processes in *Vicia sativa.* Plant J 8:111–119

Wais RJ, Galera C, Oldroyd G, Catoira R, Penmetsa RV, Cook D, Gough C, Dénarié J, Long SR (2000) Genetic analysis of calcium spiking responses in nodulation mutants of *Medicago truncatula*. Proc Natl Acad Sci USA 97:13407–13412

Wais RJ, Keating DH, Long SR (2002) Structure-function analysis of nod factor-induced root hair calcium spiking in *Rhizobium*-legume symbiosis. Plant Physiol 129:211–224

Walther TC, Alves A, Pickersgill H, Loïodice I, Hetzer M, Galy V, Hülsmann BB, Köcher T, Wilm M, Allen T, Mattaj IW, Doye V (2003) The conserved Nup107–160 complex is critical for nuclear pore complex assembly. Cell 113:195–206

Wasson AP, Pellerone FI, Mathesius U (2006) Silencing the flavonoid pathway in Medicago truncatula inhibits root nodule formation and prevents auxin transport regulation by rhizobia. Plant Cell 18:1617–1629

Wopereis J, Pajuelo E, Dazzo FB, Jiang Q, Gresshoff PM, De Bruijn FJ, Stougaard J, Szczyglowski K (2000) Short root mutant of *Lotus japonicus* with a dramatically altered symbiotic phenotype. Plant J 23:97–114

Yang WC, de Blank C, Meskiene I, Hirt H, Bakker J, van Kammen A, Franssen H, Bisseling T (1994) *Rhizobium* nod factors reactivate the cell cycle during infection and nodule primordium formation, but the cycle is only completed in primordium formation. Plant Cell 6:1415–1426

Yang Z, Fu Y (2007) ROP/RAC GTPase signaling. Curr Opin Plant Biol 10:490–494

Zhang Y, Li X (2005) A putative nucleoporin 96 Is required for both basal defense and constitutive resistance responses mediated by suppressor of npr1–1,constitutive 1. Plant Cell 17:1306–1316

Architecture of Infection Thread Networks in Nitrogen-Fixing Root Nodules

D.J. Gage

Abstract During the development of nitrogen-fixing root nodules, symbiotic bacteria are often delivered to the nodule interior by a network of tubes formed by the invagination of plant cell wall and plasma membrane. These tubes, called infection threads, are cooperatively constructed by both the plant host and its symbiotic bacteria. This chapter outlines how infection threads develop in root hairs and in root cortical cells, and how the three-dimensional architecture of infection thread networks in nodules change during the course of nodule development. Three-dimensional reconstructions of infection thread networks inside *M. truncatula* nodules infected with *Sinorhizobium meliloti* show that the infection threads form relatively simple, treelike networks that exhibit changes in growth orientation as nodules mature. Questions concerning possible mechanisms that determine the direction of infection thread development in nodule tissue, and whether or not the mechanisms of infection thread construction in nodule tissue differ from those in root hairs, are discussed.

1 Introduction

Sinorhizobium meliloti, similar to other rhizobial α-proteobacteria of the genera *Rhizobium*, *Mesorhizobium*, *Azorhizobium*, and *Bradyrhizobium*, is able to form nitrogen-fixing root nodules when it infects compatible host plants from the Leguminosae family (Batut et al. 2004; Brewin 2002; Gage 2004). Such symbioses are usually specific in that a particular bacterial species is able to infect a narrow range of host plants, though there are exceptions to this rule (Perret et al. 2000). All known interactions between rhizobial bacteria and their legume hosts require that the bacteria synthesize a lipooligosaccharide molecule called Nod factor. Nod factor is required for many of the changes seen in roots upon inoculation with compatible bacteria (see Limpens and Bisseling 2008). Another feature common to

---

D.J. Gage
Department of Molecular and Cell Biology, University of Connecticut, 91 N. Eagleville Road, U-3125, Storrs, CT 06269-3125, USA
e-mail: daniel.gage@uconn.edu

Plant Cell Monogr, doi:10.1007/7089_2008_5

these symbioses is that bacteria must enter the root before they begin to fix nitrogen in conjunction with their host. There is substantial variation in how such entry can occur. In some cases, bacteria enter root systems through cracks between epidermal cells, or at sites where lateral roots are emerging from developed roots. Bacteria then induce underlying cortical cells to form infection structures, through which the bacteria gain access to nodule tissue and they eventually enter host cell cytoplasm (Chandler 1978, 1982; De Faria et al. 1988; Gonzalez-Sama et al. 2004; Lotocka et al. 2000; Ndoye et al. 1994; Rana and Krishnan 1995). This form of infection is often considered to be more primitive than other forms because the the host root hairs do not seem to undergo any particularly sophisticated cellular differentiation during the process. There are instances in which such a mode of infection may be a secondary adaptation allowing infection of submerged roots lacking root hairs (Goormachtig et al. 2004). Variations on crack entry have been documented. For example, a species of *Bradyrhizobium* can enter lupine roots via junctions between epidermal cells, or where curled root hairs meet the outer wall of epidermal cells and trap bacteria (Lotocka et al. 2000). There have been reports that *Azorhizobium*, which invades the aquatic legume *Sesbania* via crack entry, can enter wheat through this process, grow in the intercellular spaces, and possibly fix nitrogen for the host plant (Sabry et al. 1997).

Two general types of nodules have been identified based on their development. Infection in these cases is more complex than infection through the crack entry described earlier because bacteria enter the plant through modified root hairs. In the first type, nodule cells arise from a series of initial cell divisions that takes place in the outer cortex (Gonzalez-Sama et al. 2004; Lotocka et al. 2000). Those cells expand in size and are concomitantly infected by nitrogen-fixing bacteria. No new cells are added after the initial cell divisions. This type of nodule (determinate) is usually round, does not display an obvious developmental gradient, and lacks a persistent meristem (Fig. 1). Infection structures (infection threads, ITs) develop early to deliver symbionts to cells in the nodule, but after the initial nodule cells are infected, no new infection threads develop because there are no new nodule cells to infect. There is often passage of bacteria from parental cells to daughter cells during division of infected cells. Legumes that form determinate nodules are typically tropical in origin and include *Glycine max* (soybean), *Vicia faba* (bean), and *Lotus japonicus*. *L. japonicus* has characteristics that make it particularly suitable as a model to study the formation of determinate type nodules, and much of our understanding of the process by which determinant nodules form has come from genetic and physiological studies of this plant and its symbiont *Mesorhizobium loti* (Handberg and Stougaard 1992; Udvardi et al. 2005). In contrast to determinate nodules, indeterminate nodules have an elongated shape because of a persistent meristem. The meristem (zone I) is near the growing tip of the nodule and slightly older cells, which undergo infection, are closer to the root (root-proximal, zone II). Proximal to these are the mature cells, which are generally full of nitrogen-fixing bacteria (zone III), and proximal to these is a zone of cells that undergo senescent degeneration (zone IV) (Vasse et al. 1990). This developmental process results in a gradient of developmental stages in mature nodules (Fig. 1). The new, meristem-derived, zone II cells are infected by bacteria

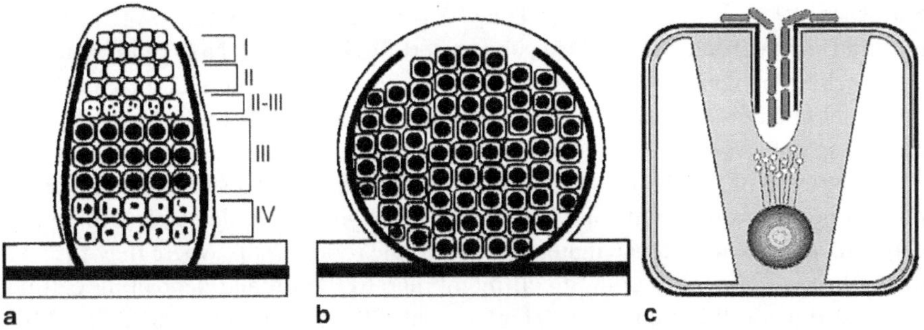

Fig. 1 Nodule types and infection thread development in cortical cells. (**a**) An indeterminate nodule. Zones I–IV are indicated. There is a developmental gradient from meristematic zone I cells to the senescent zone IV cells. *Black* indicates vacuolar space which coalesces and is centrally located in zone III cells. Cells in Zone I are uninfected. (**b**) A determinate nodule. Nodule volume is mostly occupied by tissue that is actively fixing nitrogen and a developmental gradient does not exist. *Black* indicates vacuolar space which is centrally located in nitrogen-fixing nodule cells. (**c**) An infection thread developing a cortical cell in an indeterminate-type nodule. *Light grey* represents cytoplasm, most of which is contained in the centrally located PIT. The *outer, thick, black line* represents plant cell wall. The *inner, thin, black line* around the circumference of the cell represents plant plasma membrane. The nucleus, and the cytoskeletal array between it and the infection thread are also represented. Vesicles are shown on the array. Bacteria are shown entering the developing thread from the intercellular space above the cortical cell

delivered by infection threads that continually develop in this region. *Medicago sativa* (alfalfa), *Pisum sativum* (pea), *Vicia* species (vetches), and *Trifolium* species (clovers) have historically been used as models for studying the formation of indeterminate nodules. *M. truncatula* has recently become a favored model for studies focusing on the genetics and cell biology of indeterminate nodule formation because it is diploid, has a small genome, and can be readily inbred to form genetically homogenous lines (Cook 2000). This chapter will focus on the development of infection thread networks during infection of *M. truncatula* or *M. sativa* L. with *S. meliloti*. Related topics such as root hair growth, plant responses to Nod factor, the role of the cytoskeleton, and signal transduction can be found in Assaad (2008); Limpens and Bisseling (2008); Emons and Ketelaar (2008); Bibikova and Gilroy (2008); Ketelaar and Emons (2008); Nielsen (2008); Sieberer and Timmers (2008); and Zarsky and Fowler (2008).

2 Infection Thread Development in Root Hairs

During growth in the rhizosphere of host plants such as alfalfa and *M. truncatula*, *S. meliloti* senses compounds released by roots and responds by inducing *nod* genes (Denarie et al. 1992; Hirsch 1992; Long et al. 1991; Spaink 1995; Triplett and

Sadowsky 1992; Vijn et al. 1993). The *nod* genes encode about 25 proteins that are required for the bacterial synthesis and export of Nod factor. Nod factor is a lipooligosaccharide signal with mitogenic activity that initiates many of the early developmental changes seen in the host plant early in the nodulation process (Limpens and Bisseling 2008).

The growth of root hairs is one of the few cellular processes known to be accomplished by tip growth in plants. During root hair growth, vesicles containing plant cell wall and cell membrane material move to the root hair tip, where they fuse with the cell membrane, thus adding cell membrane to the tip and depositing cell wall material outside the membrane (Hepler et al. 2001; Lhuisser et al. 2001; Miller et al. 2000; Peterson and Farquhar 1996). Root hair infection by *S. meliloti* causes reorientation of cell wall growth, which can lead to root hair curling and entrapment of the infecting bacteria. It has been suggested that Nod-factor-dependent reorganization of the actin cytoskeleton redirects vesicle traffic from the root hair tip to a new site away from the center of the apical dome of the root hair, thereby causing root hair deformation and curling (Cardenas et al. 1998; Catoira et al. 2001; de Ruijter et al. 1998; Esseling et al. 2003; Miller et al. 1999; Timmers et al. 1999). Following the entrapment of bacteria in a root hair curl, inwardly directed growth of the infection thread results, probably from tip growth (Esseling et al. 2003). The root hair microtubule cytoskeleton also undergoes dynamic changes during the formation of infection threads. Before infection, the microtubule network in growing root hairs is distributed primarily parallel to the long axis of the cells. As infection proceeds, the network becomes asymmetrically distributed and focused on the bacterial microcolony at the center of the curled root hair (Catoira et al. 2001). As the infection thread grows down the root hair, the microtubule network in the root hair becomes distributed between the infection thread tip and the nucleus, which proceeds down the root hair in front of the infection thread tip. In alfalfa and *M. truncatula*, the microtubule arrays likely colocalize with the column of actively streaming cytoplasm located between the nucleus and the thread tip (Gage 2004; Timmers et al. 1999). These endoplasmic microtubules may be involved in growth of the thread tip in much the same way that microtubules are involved in tip growth of root hairs, i.e., in determining the direction of growth (Sieberer and Timmers 2008; Bibikova et al. 1999; Lloyd et al. 1987; Sieberer et al. 2002, 2005). In summary, it is thought that growth of an infection thread as it develops in a root hair may be a modified form of the tip growth that normally elongates root hairs (Gage 2004). Therefore, a better understanding of the mechanisms of root hair growth should contribute to a better understanding of infection thread development.

Because infection threads are ingrowths the wall and membrane are contiguous with plant cell wall and plasma membrane and bacteria inside them are topologically outside the root hair. Because of this continuity with external cell walls, infection thread walls are likely to be similar to root hair, nodule, or cell walls and contain esterified and unesterified pectins, xyloglucans, and cellulose (Cosgrove 2005; Rae et al. 1992). The lumen of the thread contains a mixture of material normally found as part of the extracellular matrix of plant cell walls, combined with extracellular polysaccharides, proteins, and other macromolecules that originate from the bacterial symbiont (Rae et al. 1991; Van den Bosch et al. 1989). As infection threads elongate, bacteria near the tip of the developing infection thread divide and grow, keeping the majority of the tube filled with bacteria (Gage 2004). However, bacteria are absent from the very tip of the developing thread, which appears membranous and is a site where very active cytoplasmic streaming occurs in the infected root hair (Gage 2004). During root hair infection, the infection thread eventually fuses with the far wall of the epidermal cell and bacteria enter the intercellular space between the epidermal cell and the underlying outer cortical cell. Degradation of cell wall, invagination of the plasma membrane, and growth of infection thread occur in the underlying cortical cell, and the bacteria-filled thread is thus propagated further toward the nodule primordium developing root interior (Turgeon and Bauer 1985; van Brussel et al. 1992; van Spronsen et al. 1994).

2.1 Infection Thread Development in Cortical Tissue and Young Indeterminate Nodules

During nodule development in alfalfa and *M. truncatula*, inner cortical cells beneath infection sites reenter the cell cycle, begin dividing, and form a nodule primordium. While the primordium develops, infection threads develop in polarized outer cortical cells and traverse those cells through centrally located cytoplasmic bridges called preinfection threads (PITs, Fig. 1c) (Timmers et al. 1999; van Brussel et al. 1992). PITs in adjacent cells are aligned and mark the future paths through which infection threads will develop (Fig. 2). The polarization of these cells is manifested in the trapezoidal shape of the PITs which are wider at the outer walls, and in the amyloplasts that accumulate in the PITs, against the inner walls (van Brussel et al. 1992). PITs also contain bundles of microtubules that run the length of the cytoplasmic bridge (Timmers et al. 1999). Similar to PITs, these bundles align between adjacent cells through which infection threads will develop. The polarized cytoplasm of the PITs is probably responsible for the polarized growth of the infection thread network at this stage of nodulation (Fig. 2a). After the thread network develops in the root cortex and invades the nodule primordium, some uninfected cells in the middle cortex reenter the cell cycle and form islets of cell division activity. The number of cells in these islets grows and eventually the islets fuse forming a persistent arc of meristematic cells (Timmers et al. 1999). When the meristem first develops, it is located on the distal side of the nodule primordium and it grows away from the interior of the root, toward the root surface (Fig. 2b). After meristem formation, the inwardly developing infection threads in the nascent nodule are located between the meristem and the center of the root (Fig. 2b). As the nodule develops further, the region behind the meristem (the infection zone or zone II) differentiates and contains a branched network of developing infection threads that deliver infecting bacteria to some, but not all, of the nodule cells in the region (Fig. 2c). Release of bacteria from unwalled portions of the developing network into the cytoplasm of nodule cells occurs in zone II. Once inside the nodule cells,

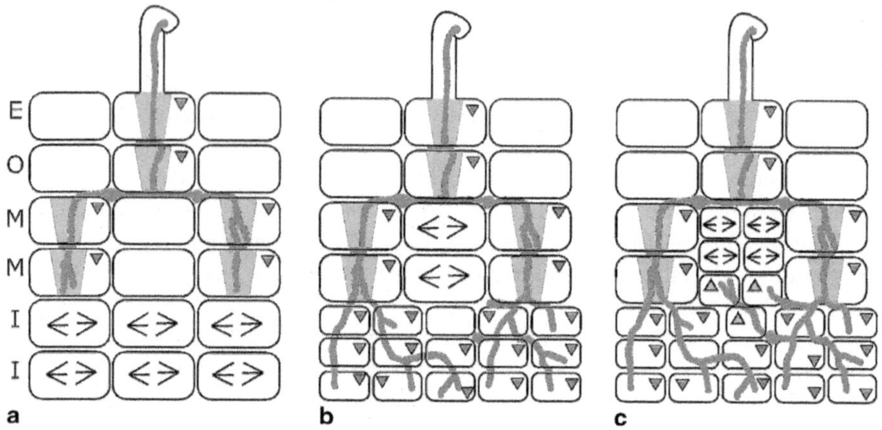

Fig. 2 Directional development of infection threads in polarized plant tissue. (**a**) Stage I of infection thread network development. Threads are shown following development in a root hair and in cells in the outer- and middle cortex. Cells in the inner cortex divide and give rise to the nodule primordium. (**b**) Stage II of infection thread network development. Threads invade the nodule primordium. Uninfected cells in the middle cortex begin to divide and develop into a nascent meristem. (**c**) Stage III of infection thread network development. New infection thread branches develop in nodule cells that are newly derived from meristematic cells. These nodule cells represent the beginning of the infection zone and are polarized such that infection threads which develop in them are oriented toward the nodule meristem. The direction of polarization in nodule cells is indicated by the *triangles*. *E* epidermis, *O* outer cortex, *M* middle cortex, *I* inner cortex

the bacteria differentiate and synthesize proteins required for nitrogen fixation (Vasse et al. 1990).

In indeterminate nodules such as those of alfalfa and *M. truncatula*, the meristem continually grows away from the interior of the root. As the meristem moves outward, the infection thread network must grow into the region behind the meristem that contains new cells that need to be infected. How the network of infection threads which was inwardly directed in order to deliver bacteria to the nodule primordium can later infect cells behind the outwardly growing meristem is a problem with no (or several) obvious solutions. Recent work, summarized later, shows how this topological reordering takes place, and in doing so raises new questions.

2.2 Lessons from Three-Dimensional Reconstructions of Infection Thread Networks

Three-dimensional (3D) representations of infection thread networks were reconstructed from serial sections of 5-, 10-, and 30-day-old *M. truncatula* nodules after infection with wild-type *S. meliloti* strain Rm1021 (Monahan-Giovanelli et al. 2006). In all nodules, the infection thread networks formed open, treelike structures with branches that rarely fused (Fig. 3c, d). Average infection thread width varied along the length of the nodule, with the thinnest parts of the threads near the advancing edge of the networks. The thickest parts of the threads were generally in zone II where bacteria were being released into nodule cells (Fig. 3d). Bacterial release, which is known to require the product of the *NIP* gene in *M. truncatula* (Veereshlingam et al. 2004), was readily observed in nodule cells that were 50–100 mm (2–4 cell layers) from the advancing edge of the IT network in the 10- and 30-day-old nodules. It was also in this zone that thread density was the highest (Fig. 3d). The observed difference in thread thickness between the newly developed threads and those that were releasing bacteria indicated that bacterial populations in the threads continued to increase after the thread had developed. If there is deposition of plant material (nutrients, cell wall precursors) into the matrix of the thread after it has formed, this may allow continued growth of bacterial populations inside nodule infection threads. This differs from what has been seen in infection threads in root hairs where bacteria grow only near the extending tip of the threads, and where the threads are generally uniform in width (Fahraeus 1957; Gage 2002; Gage et al. 1996). IT thickness and spatial density decreased in regions root-proximal to zone II (Fig. 3d) (Monahan-Giovanelli et al. 2006). This is likely explained by the fact that zone II cells are relatively small, but increase in size as they develop further (Truchet 1978; Vasse et al. 1990; Vinardell et al. 2003). As this expansion takes place, the infection threads, which are tunnels through the nodule cells, appear to stretch and straighten out as their host cell expands (Monahan-Giovanelli et al. 2006). This likely occurs because the rate of cell wall expansion of the infection threads is less than that of the host cells in which they are contained. In zone III of mature nodules, threads appear not to develop new branches. Therefore, as those cells increase in size, the number of threads per unit nodule volume necessarily decreases.

The reconstructed infection thread networks occupied only 1.5–3.0% of the total nodule volume (Table 1). The fact that the networks occupy a small fraction of the nodule volume makes it difficult to ascertain infection thread characteristics or qualitative aspects of network organization from single sections of nodule tissue. Therefore, conclusions about the effect of mutations on infection thread phenotype in nodules would best be based on multiple sections from multiple nodules, at a minimum. Ideally, 3D reconstructions would be done because they should contribute greatly to understanding how mutations affect infection thread development, and alter IT network organization, during nodule development.

The 3D reconstructions of nodule tissue described above were initially undertaken to determine how IT networks develop in order to deliver bacteria to the new tissue laid down by the nodule meristem. One possibility was that the inward-growing IT network which developed in nascent nodule cells, between the root center and the meristem, repolarized and developed outwards, following the meristem. The other possibility was that the network developed in nodule tissue in a directionally random fashion, with some of its branches delivering bacteria to the new tissue laid down by the nodule meristem. The reconstructions clearly showed that 10 days after inoculation the IT networks in *M. truncatula* nodules had a strong growth bias toward the meristem and tip of the nodule. This bias strengthened, and

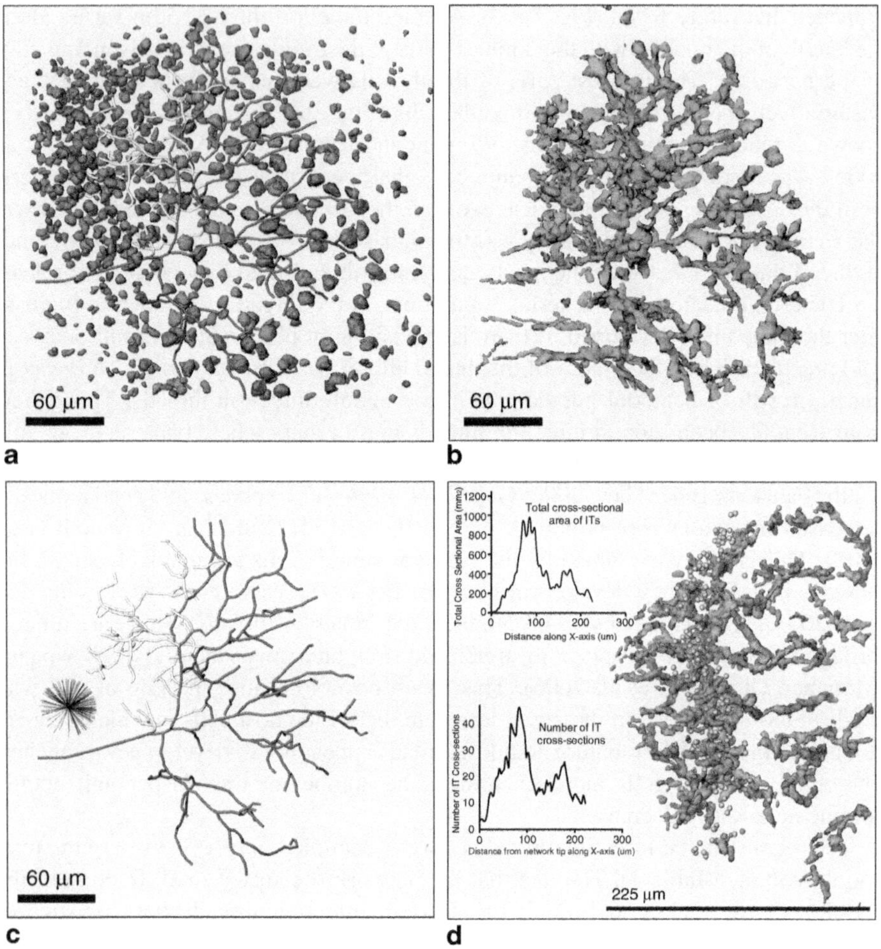

Fig. 3 Three-dimensional reconstructions of infection thread networks in nodules. Nodule tips are to the left in all panels. (**a**) A reconstruction of a 5-day-old nodule showing nuclei and the centerlines of the infection threads that make up the IT network. Inwardly directed branches of the IT network are dark grey, outwardly directed branches are white. Note the high density of nuclei in the *upper left quadrant* of the image indicative of a meristematic islet. (**b**) Reconstructed infection thread volumes from the same 5-day-old nodule shown in (**a**). (**c**) Infection thread centerlines from (**a**). Inwardly directed branches of the IT network are dark grey, outwardly directed branches are white. A sunburst plot showing the growth direction of the network branches with respect to the nodule tip is shown in the left part of the image. Branches growing directly toward the nodule tip are represented by lines pointing directly to the left, and branches pointing away from the nodule tip are represented by lines pointing directly to the right. (**d**) A reconstruction of 30-day-old nodule showing infection thread volumes and the location of the distal-most bacteria released from cells in zone II (*spheres*). The *insets* show the total cross-sectional area of infection threads and the number of infection thread cross-sections in the reconstructed volume

Table 1 Volume of nodule tissue and infection threads in three-dimensional reconstructions (data from Monahan-Giovanelli et al. 2006)

Nodule	Dimensions of tissue occupied by infection threads (μm)			Tissue volume (um³)	Infection thread volume (um³)	Infection thread volume/tissue volume
	X	*Y*	*Z*			
5 Day A	241	282	68	4,633,587	104,272	0.023
5 Day B	127	189	60	1,432,508	43,786	0.031
10 Day A	290	308	62	5,551,209	79,829	0.014
10 Day B	314	278	62	5,421,533	82,142	0.015
30 Day A	228	304	62	4,304,982	87,590	0.020
30 Day B	293	333	48	4,691,173	89,062	0.019

by 30 days the IT network was highly polarized and growing toward the nodule tip (Fig. 3) (Monahan-Giovanelli et al. 2006).

The picture in young, 5-day-old nodules was more complex. In these nodules the overall growth bias of the IT network was toward the root interior. This was likely because the threads had developed to deliver bacteria to the nodule primordium, and the cortical cells from which they were derived were polarized to support inwardly directed infection thread development (Fig. 2b). Importantly, there were well-defined regions in these nodules that showed local growth biases directed toward the nodule tip. In locations root-proximal to regions of cell division activity (i.e. meristematic islets (Timmers et al. 1999)), the inwardly oriented IT networks formed new branches that developed in an outward direction, toward the islets of meristematic activity (Fig. 3a, c). The outwardly growing branches generally pointed toward the region of meristematic activity from their inception, and did not typically result from branches that turned or changed their growth from inward to outward. Interestingly, in 1973 Libbenga mentioned observing such backward branching in young nodules (Libbenga and Harkes 1973).

2.3 *Two Questions Concerning the Development of Infection Threads in Indeterminate Nodules*

Investigations into the development of infection threads in root hairs and in cortical tissue have raised important questions, and progress has been made on many of them. The analysis of infection thread development in nodule tissue is not nearly as developed, but what has been done raises a number of questions. Two of the most interesting ones, from my point of view, are discussed here.

(1) *How is infection thread development targeted toward the nodule meristem?* The requirements for reorientation of IT networks in response to meristem development, or for their progression through the infection zone toward the meristem, are not known.

Polarized, meristem-directed, network growth could be supported if infection zone cells near the meristem are polarized in a fashion similar to the cortical cells that support polarized thread growth early during infection (Timmers et al. 1999; van Brussel et al. 1992). However, unlike IT development in cortical tissue, there were no obvious visual markers, similar to PITs, that indicated the future paths of ITs through the infection zone in the 3D work described earlier. If PIT-like structures were present, it may be that they were not readily visible because infection zone cells in front of extending ITs are cytoplasm-rich and have no large vacuoles to be displaced if polarized cytoplasm moves to a position marking paths of future threads.

Alternatively, the direction of IT network development in nodules may respond to diffusible targeting signals arising from the meristematic cells, or from signals synthesized by bacteria in infection threads. Directional cues from diffusible signals and polarized cells are not mutually exclusive because cells root-proximal to meristematic cells could be polarized in response to diffusible signals arising from the meristematic cells or from signals synthesized by bacteria in infection threads. In this regard, Nod factor is a strong candidate for signal that might be needed for host cell polarization and polarized IT propagation. Nod factor is clearly involved in these processes early in infection, and the genes required for its synthesis and transport are known to be expressed by bacteria in the infection zone of nodules, but not by bacteria in zone III where new ITs do not develop (Sharma and Signer 1990). Thus, Nod factor gradients may play a role in polarizing nodule tissue and establishing directionality for growth of infection threads. However, a Nod factor gradient alone is not likely to be the only signal needed for establishing polarized nodule tissue because if it was sufficient, polarized tissue should form all along the distal edge of the infection zone in young nodules, not only in those areas near meristematic islets.

It was recently shown that developing *M. truncatula* nodules formed symplastic fields, connected to phloem, that were capable of trafficking large molecules. The location of these fields changed as nodules developed, and they corresponded to regions of nodules that supported IT growth. In young nodules, all internal regions formed a phloem-connected symplastic field, whereas only the meristem and infection zones did so in mature nodules (Complainville et al. 2003). Clearly, signals that influence IT targeting could be brought to their correct locations, and distributed to neighboring regions by such transport mechanisms (Crawford and Zambryski 1999).

(2) *Do infection threads in nodules develop in a manner that involves tip growth, similar to that seen in root hairs?* In cells with large amounts of vacuolar space, phragmosomes can be observed prior to cell division. Phragmosomes are accumulations of transvacuolar strands that center the nucleus and mark the location of the future division plane and typically occur in the plane that marks the shortest path to divide the cell volume in two (Sinnott and Bloch 1940; Smith 2001). During normal cell division, after the separation of chromosomes, a microtubule array is assembled perpendicular to the division plane, and this structure, the phragmoplast, is involved in transporting vesicles containing cell wall material to the division

plane in order to assemble the cell plate, which separates the mother cell into two daughter cells (Jurgens 2005). Phragmoplast-dependent cell plate formation can begin in the center of the mother cell and move centripetally toward the walls, or it can demonstrate a polarized growth pattern, start at a wall, and move toward the other side of the cell (Jurgens 2005; Smith 2001). The latter form of cell plate construction occurs when a well-defined phragmoplast assembles between two daughter nuclei, moves across the cell, and synthesizes the plate between the daughter nuclei (Cutler and Ehrhardt 2002).

van Brussel et al. (1992) described the formation of preinfection threads (PITs) that develop in cortical cells in response to rhizobial infection and Nod factor addition to alfalfa roots. PITs are similar to phragmosomes in that they are accumulations of transvacuolar cytoplasm that center the nucleus near the middle of the cell. As mentioned earlier, PITs form in adjacent cells and mark the future path of the infection thread. van Brussel et al., and others, have noted that Nod factor can alter cell wall and cell wall synthesis patterns of cortical cells and cause the formation of root hairs in these cells (Ardourel et al. 1994; van Brussel et al. 1992). These results show that Nod factor can reprogram some nongrowing cells, causing them to reorganize and polarize their cytoplasm and induce new regions of tip growth. These facts, combined with the fact that IT growth in root hairs appears to result from a reorganization of root hair tip growth, suggest that infection thread development in cortical cells is a process of modified tip growth. The fact that PITs are very similar to phragmosomes leaves open the possibility that IT development in cortical cells is more closely related to the process of polarized cell plate formation that proceeds from one cell wall across to the other side of the cell, than it is to tip growth. However, if so, the growth of what is normally a platelike structure would have to be constrained at its edges in order to give rise to the tube shape that is characteristic of an infection thread. In addition, the modified cell plate would have to form in a way that did not depend on there being a standard phragmoplast, given that infection threads develop in cells containing single nuclei. These considerations make it likely that IT development depends on a form of modified tip growth, and not on modified cell plate formation, in root cortical cells. This modified tip growth may be related to the mechanisms used to extend root hairs, or it may be derived from the tip-growth-like mechanisms that plant cells use to wall off invading microorganisms (Schmelzer 2002; Schulze-Lefert 2004). In the latter case, cell cytoplasm becomes polarized and there are cytoskeletal and cytoplasmic rearrangements similar to those seen in cortical cells during IT formation that coincide with cell wall deposition at the site of pathogen invasion (Takemoto and Hardham 2004; Takemoto et al. 2003).

It is difficult to assess the cytoplasmic organization of nodule cells that are undergoing, or will undergo, the process of infection thread development. This is because the cells in the infection zone are not accessible to observation without fixation and sectioning. In addition, they are cytoplasmically dense with little or no vacuolar space, which makes discerning structures such as PITs difficult. However, sectioned material has provided some information on the process of IT formation in these cells (Ardourel et al. 1994). Infection threads in a single cell often meander

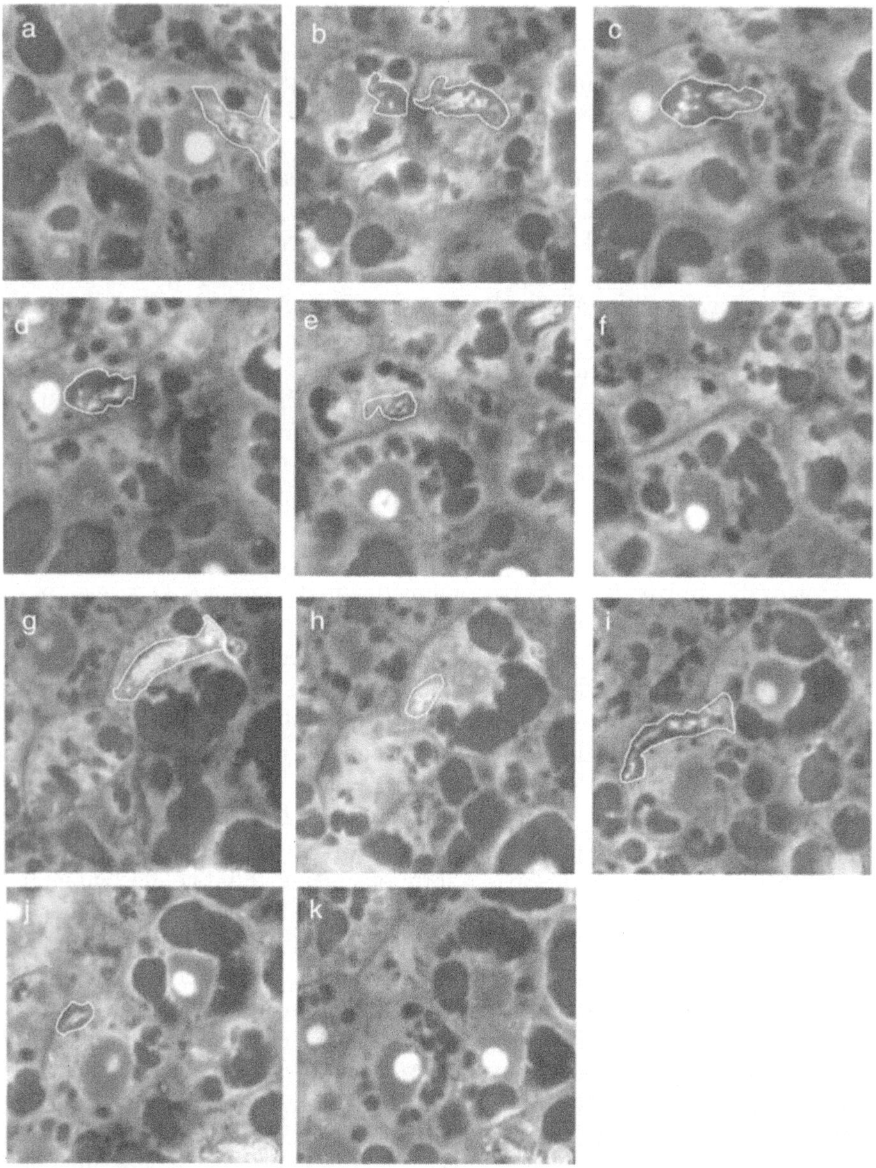

Fig. 4 Structure of infection thread tips in nodule tissue from 10-day-old nodules. The nodule meristems are to the left in all panels. Panels (**a**–**f**) are images of 2-μm-thick serial sections showing the structure of an infection thread tip that first developed in the right-most cell in the images and was developing in the left-most cell when the nodule was harvested. The tip of the thread is wholly contained in the sections shown in panels (**b**) through (**e**) because sections of the left-most cell above and below these do not contain infection threads. Panels (**g**–**k**) are similar to (**a**–**f**), but they show an infection thread tip that skirts along the edge of the nodule cell in which it is developing, and clearly does not take the shortest, unobstructed path across that cell. Similar to the previous series, the tip of the thread is wholly contained in sections shown in panels (**i**) and (**j**). Infection thread structures are outlined in white in all panels. Note that infection thread tips are broad, and enclose relatively few bacteria

and do not correspond to a straight path that efficiently halves the cell (Fig. 4). This raises the question of whether a PIT-like structure similar to those in vacuolated cells is associated with IT development in infection zone cells (Batut et al. 2004). Infection threads clearly initiate at one side of the cell, and the cell wall at the site of their initiation is often highly deformed. This is similar to the cell-wall weakening and deformation seen in cortical cells where infection threads approach underlying cells, and may be related to wall weakening and deformation required for tip growth (Bibikova et al. 1999). The advancing tip of the developing thread is composed of an empty space with few bacteria in it (Fig. 4). This is likely a membranous compartment associated with the coalescence of vesicles that construct the thread. This membrane-delimited compartment is often quite large, rather than very narrow like the membrane-delimited compartments that make up cell plates as they form. These results suggest that in nodule cells infection threads develop in a process related to tip growth, as they probably do in root hairs and cortical cells. If this is the case, then proteins specific for tip growth (Zarsky and Fowler 2008) should be associated with developing infection threads in all three cell types.

3 Materials and Methods

3.1 Fixation, Dehydration, Embedding, and Staining of Nodule Tissue

In this method, sections are dyed with an aqueous mixture of DAPI and acridine orange. The stained sections are brightly colored, and it is easy to differentiate plant cell wall, plant cell nuclei, plant cell vacuoles, bacteroids, and infection threads (see Fig. 4 for an example of stained tissue). These methods are based on a method by Dudley and Long (1987).

3.2 Materials Needed

Phosphate-buffered saline; solid paraformaldehyde; ethanol; 24-well microtiter dishes; Straight Peer or Dumont tweezers (no. 5); JB4 embedding kit from PolySciences (Warrington, PA); Embedding molds such as PolySciences catalog no. 23257; BEEM capsules size 00 PolySciences catalog no. 00224 ; double-edged razor blades; cellophane wrap; desiccator; superglue; microtome; glass knives for the microtome; polylysine-coated slides; NH_4OH; cover slips; 4′,6-Diamidino-2-phenylindole dihydrochloride (DAPI); 3,6-Bis[dimethylamino]acridine hydrochloride hydrate (Acridine orange); and an epifluorescence microscope.

4 Methods

1. Excise nodules and submerge for at least 24 h in phosphate-buffered saline (PBS) containing 4% paraformaldehyde. The 4% solution is made by gently heating 0.4 g of paraformaldehyde in 10 mL PBS for 1 h, with stirring. Fixation and dehydration are conveniently done in 24-well microtiter dishes. Nodules can be pooled at this point, but use less than ten nodules per well to assure that tissue is well infiltrated and dehydrated.
2. Remove PBS + paraformaldehyde solution. Begin dehydration series by adding 1 mL of 30% ethanol to each well; incubate for 30 min. Follow by 30 min in 50%, 70%, and 90% ethanol. Nodules can be left indefinitely in 70% ethanol solution if needed.
3. Remove 90% ethanol and add a 1:1 mixture of 90% ethanol and JB4 solution A plus catalyst (see JB4 instructions). Incubate overnight.
4. Remove the 1:1 mixture and incubate nodules overnight in JB4 solution A plus catalyst.
5. Prepare fresh JB4 embedding solution A plus catalyst.
6. Transfer nodules to an embedding tray. Add JB4 solution B to solution A plus catalyst and mix (avoid introducing excess oxygen – see JB4 instructions). Add the JB4 mixture to the tray wells containing nodules. Also, add some JB4 mixture to BEEM capsules. These will be used to mount the nodules for sectioning. JB4 will not polymerize properly in the presence of oxygen; to help prevent oxygen inhibition, place cellophane wrap on top of the embedding tray wells, and place the trays and the BEEM capsules in a desiccator. Evacuate the desiccator and let JB4 polymerize overnight. If the JB4 polymerizes poorly in the embedding tray, deeper trays can be used to decrease the surface-to-volume ratio, which will lessen oxygen inhibition of polymerization.
7. Remove embedded nodules from trays, and trim them using a double-edged razor blade that has been carefully snapped in two lengthwise. Change blades frequently. Trim such that the nodule is contained in a small block of plastic. This is then glued with superglue onto the sharp end of the JB4 cast in the BEEM capsules. To have a platform large enough to mount a nodule, it is usually necessary to cut off the tip of the capsule.
8. Once the superglue has dried, the capsule and the attached nodule can be mounted on a microtome and sectioned using a glass knife to get 1–4 μm-thick sections. The sections are picked off the knife individually using very fine tweezers and dropped flat onto a drop of 1% solution of NH_4OH placed on a polylysine-coated slide. The section should spread on the surface of the drop and dry flat. Typically, 12–18 sections can be placed on a single slide.
9. Stain the sections by placing a drop of water containing 10 μg mL^{-1} DAPI and 40 μg mL^{-1} acridine orange. Leave the drop for 1 min and destain for 2 min in a slide jar containing water. Remove the slide, dry the back, and place a cover slip over the stained sections. Observe under an epifluorescence microscope using a long pass filter set which excites in the UV (330–380 nm) and emits visible light in the blue to red (all visible wavelengths longer than 435 nm or so).

References

Ardourel M, Demont N, Debelle F, Maillet F, Billy F, de Prome JC, Denarie J, Truchet G (1994) *Rhizobium meliloti* lipooligosaccharide nodulation factors: different structural requirements for bacterial entry into target root hair cells and induction of plant symbiotic developmental responses. *Plant Cell* **6:**1357–1374

Assaad FF (2008) The membrane dynamics of root hair morphogenesis. In: Emons AMC, Ketelaar T (eds) Root hairs: excellent tools for the study of plant molecular cell biology. Springer, Berlin Heidelberg New York. doi:10.1007/7089_2008_2

Batut J, Andersson SGE, O'Callaghan D (2004) The evolution of chronic infection strategies in the α-proteobacteria. *Nat Rev Microbiol* **2:**933–945

Bibikova T, Gilroy S (2008) Calcium in root hair growth. In: Emons AMC, Ketelaar T (eds) Root hairs: excellent tools for the study of plant molecular cell biology. Springer, Berlin Heidelberg New York. doi:10.1007/7089_2008_3

Bibikova TN, Blancaflor EB, Gilroy S (1999) Microtubules regulate tip growth and orientation in root hairs of *Arabidopsis thaliana*. *Plant J* **17:**657–665

Brewin NJ, (2002). Pods and nods: a new look at symbiotic nitrogen fixing. *Biologist* **49:**1–5

Cardenas L, Vidali L, Dominguez J, Perez H, Sanchez F, Hepler PK, Quinto C (1998) Rearrangement of actin microfilaments in plant root hairs responding to *Rhizobium etli* nodulation signals. *Plant Physiol* **116:**871–877

Catoira R, Timmers ACJ, Maillet F, Galera C, Penmetsa RV, Cook D, Denarie J, Gough C (2001) The HCL gene of *Medicago truncatula* controls *Rhizobium*-induced root hair curling. *Development* **128:**1507–1518

Chandler MR (1982) Infection and root-nodule development in *Stylosanthes* species by *Rhizobium*. *J Exp Bot* **33:**47–57

Chandler MR (1978) Some observations on infection of *Arachis hypogaea* L. by *Rhizobium*. *J Exp Bot* **29:**749–755

Complainville A, Brocard L, Roberts I, Dax E, Sever N, Sauer N, Kondorosi A, Wolf S, Oparka K, Crespi M (2003) Nodule initiation involves the creation of a new symplasmic field in specific root cells of *Medicago* species. *Plant Cell* **15:**2778–2791

Cook D (2000). *Medicago truncatula* – a model in the making! *Curr Opin Plant Biol* **2:**301–304

Cosgrove DJ (2005) Growth of the plant cell wall. *Nat Rev Mol Cell Biol* **6:**850–861

Crawford KM, Zambryski PC (1999) Plasmodesmata signaling: many roles, sophisticated statutes. *Curr Opin Plant Biol* **2:**382–387

Cutler SR, Ehrhardt DW (2002) Polarized cytokinesis in vacuolate cells of *Arabidopsis*. *Proc Natl Acad Sci USA* **99:**2812–2817

De Faria SM, Hay GT, Sprent JI (1988) Entry of rhizobia into roots of *Mimosa scabrella* Bentham occurs between epidermal cells. *J Gen Microbiol* **134:**2291–2296

de Ruijter NCA, Rook MB, Bisseling T, Emons AMC (1998) Lipochito-oligosaccharides re-initiate root hair tip growth in *Vicia sativa*. with high calcium and spectrin-like antigen at the tip. *Plant J* **13:**341–350

Denarie J, Debelle F, Rosenberg C (1992) Signaling and host range in nodulation. *Annu Rev Microbiol* **46:**497–525

Dudley ME, Jacobs TW, Long SR (1987) Microscopic studies of cell divisions induced in alfalfa roots by *Rhizobium meliloti*. *Planta* **171:**289–301

Emons AMC, Ketelaar T (2008) Intracellular organization: a prerequisite for root hair elongation and cell wall deposition. In: Emons AMC, Ketelaar T (eds) Root hairs: excellent tools for the study of plant molecular cell biology. Springer, Berlin Heidelberg New York. doi:10.1007/7089_2008_4

Esseling JJ, Lhuissier FG, Emons AM (2003) Nod factor-induced root hair curling: continuous polar growth towards the point of nod factor application. *Plant Physiol* **132:**1982–1988

Fahraeus G (1957) The infection of clover root hairs by nodule bacteria studied by a simple glass slide technique. *J Gen Microbiol* **16:**374–381

Gage DJ (2002) Analysis of infection thread development using Gfp- and DsRed-expressing *Sinorhizobium meliloti*. *J. Bacteiol* **184:**7042–7046

Gage DJ (2004) Infection and invasion of roots by symbiotic, nitrogen-fixing rhizobia during nodulation of temperate legumes. *Microbiol Mol Biol Rev* **68:**280–300
Gage DJ, Bobo T, Long SR (1996) Use of green fluorescent protein to visualize the early events of symbiosis between *Rhizobium meliloti* and alfalfa (*Medicago sativa*). *J Bacteriol* **178:**7159–7166
Gonzalez-Sama A, Lucas MM, Felipe MR, De Peuyo JJ (2004) An unusual infection mechanism and nodule morphogenesis in white lupin (*Lupinus albus*). *New Phytol* **163:**371–380
Goormachtig S, Capoen W, James EK, Holsters M (2004) Switch from intracellular to intercellular invasion during water stress-tolerant legume nodulation. *Proc Natl Acad Sci USA* **101:** 6303–6308
Handberg K, Stougaard JS (1992) *Lotus japonicus*, an autogamous, diplod legume species for classical and molecular genetics. *Plant J* **2:**487–496
Hepler PK, Vidali L, Cheung AY (2001) Polarized cell growth in higher plants. *Annu Rev Cell Dev Biol* **17:**159–187
Hirsch AM (1992) Developmental biology of legume nodulation. *New Phytol* **122:**211–237
Jurgens G (2005) Cytokinesis in higher plants. *Annu Rev Plant Biol* **56:**281–299
Ketelaar T, Emons AMC (2008) The actin cytoskeleton in root hairs: a cell elongation device. In: Emons AMC, Ketelaar T (eds) Root hairs: excellent tools for the study of plant molecular cell biology. Springer, Berlin Heidelberg New York. doi:10.1007/7089_2008_8
Lhuisser FGP, De Ruijter NCA, Sieberer BJ, Esseling JJ, Emons AMC (2001) Time course of cell biological events evoked in legume root hairs by *Rhizobium* Nod factors: state of the art. *Ann Bot* **87:**289–302
Libbenga KR, Harkes PAA (1973) Initial proliferation of cortical cells in the formation of root nodules in *Pisum sativum. Planta* **114:**17–28
Limpens E, Bisseling T (2008) Nod factor signal transduction in the Rhizobium-Legume symbiosis. In: Emons AMC, Ketelaar T (eds) Root hairs: excellent tools for the study of plant molecular cell biology. Springer, Berlin Heidelberg New York. doi:10.1007/7089_2008_10
Lloyd C, Pearce K, Rawlins DJ, Ridge RW, Shaw PJ (1987) Endoplasmic microtubules connect the advancing nucleus to the tip of legume root hairs, but F-actin is involved in basipetal migration. *Cell Motil Cytoskeleton* **8:**27–36
Long SR, Fisher RF, Ogawa J, Swanson J, Ehrhardt DW, Atkinson EM, Schwedock J (1991) *Rhizobium meliloti*, nodulation gene regulation and molecular signals, In: Hennecke H, Verma DPS (ed) Advances in molecular genetics of plant-microbe interactions, vol 1. Kluwer, Dordrecht
Lotocka B, Kopcinska J, Gorecka M, Golinowski W (2000) Formation and abortion of root nodule primordia in *Lupinus luteus* L. *Acta Biol Crac ser Bot* **42:**87–102
Miller DD, de Ruijter NCA, Bisseling T, Emons AMC (1999) The role of actin in root hair morphogenesis: studies with lipochito-oligosaccharide as a growth stimulator and cytochalasin as an actin perturbing drug. *Plant J* **17:**141–154
Miller DD, Leferink-ten Klooster HB, Emons AMC (2000) Lipochito-oligosaccharide nodulation factors stimulate cytoplasmic polarity with longitudinal endoplasmic reticulum and vesicles at the tip in vetch root hairs. *Mol Plant-Microbe Interact* **13:**1385–1390
Monahan-Giovanelli H, Arango-Pinedo C, Gage DJ (2006) Architecture of infection thread networks in developing root nodules induced by the symbiotic bacterium *Sinorhizobium meliloti* on *Medicago truncatula. Plant Physiol* **104:**661–670
Ndoye I, Billy F, de Vasse J, Dreyfus B, Truchet G (1994) Root nodulation of *Sesbania rostrata. J Bacteriol* **176:**1060–1068
Nielsen E (2008) Plant cell wall biogenesis during tip growth in root hair cells. In: Emons AMC, Ketelaar T (eds) Root hairs: excellent tools for the study of plant molecular cell biology. Springer, Berlin Heidelberg New York. doi:10.1007/7089_2008_11
Perret X, Staehelin C, Broughton WJ (2000) Molecular basis of symbiotic promiscuity. *Micro Mol Biol Rev* **164:**180–201
Peterson RL, Farquhar ML (1996) Root hairs: specialized tubular cells extending root surfaces. *Bot Rev* **62:**1–40

Rae AL, Bonfante-Fasolo P, Brewin NJ (1992) Structure and growth of infection threads in the legume symbiosis with *Rhizobium leguminosarum*. *Plant J* **2:**385–395

Rae AL, Perrotto S, Knox JP, Kannenberg EL, Brewin NJ (1991) Expression of extracellular glycoproteins in the uninfected cells of developing pea nodule tissue. *Mol Plant Microbe Interact* **4:**563–570

Rana D, Krishnan HB (1995) A new root-nodulating symbiont of the tropical legume *Sesbania*. , *Rhizobium* sp. SIN-1, is closely related to *R. galegae*, a species that nodulates temperate legumes. *FEMS Microbiol Lett* **134:**19–25

Sabry SRS, Saleh SA, Batchelor CA, Jones JD, Jotham J, Webster G, Kothari SL, Davey MR, Cocking EC (1997) Endophytic establishment of *Azorhizobium caulinodans* in wheat. *Proc R Soc London Ser B* **264:**341–346

Schmelzer E (2002) Cell polarization, a crucial process in fungal defence. *Trends Plant Sci* **7:**411–415

Schulze-Lefert P (2004) Knocking on heaven's wall: pathogenesis of and resistance to biotrophic fungi at the cell wall. *Curr Opin Plant Biol* **7:**377–383

Sharma SB, Signer ER (1990) Temporal and spatial regulation of the symbiotic genes of *Rhizobium meliloti* in planta revealed by transposon Tn5-*gusA*. *Genes Dev* **4:**344–356

Sieberer BJ, Timmers ACJ (2008) Microtubules in Plant Root Hairs and Their Role in Cell Polarity and Tip Growth. In: Emons AMC, Ketelaar T (eds) Root hairs: excellent tools for the study of plant molecular cell biology. Springer, Berlin Heidelberg New York. doi:10.1007/7089_2008_13

Sieberer BJ, Ketelaar T, Esseling JJ, Emons AMC (2005) Microtubules guide root hair tip growth. *New Phytol* **167:**711–719

Sieberer BJ, Timmers ACJ, Lhuissier FGP, Emons AMC (2002) Endoplasmic microtubules configure the subapical cytoplasm and are required for fast growth of *Medicago truncatula* root hairs. *Plant Physiol* **130:**977–988

Sinnott EW, Bloch R (1940) Cytoplasmic behavior during division of vacuolate plant cells. *Proc Natl Acad Sci USA* **26:**223–227

Smith LG (2001) Plant cell division: building walls in the right places. *Nat Rev Mol Cell Biol* **2:**33–39

Spaink HP (1995) The molecular basis of infection and nodulation by rhizobia: the ins and outs of sympathogenesis. *Annu Rev Phytopathol* **33:**345–368

Takemoto D, Hardham AR (2004) The cytoskeleton as a regulator and target of biotic interactions in plants. *Plant Physiol* **136:**3864–3876

Takemoto D, Jones DA, Hardham AR (2003) GFP-tagging of cell components reveals the dynamics of subcellular re-organization in response to infection of *Arabidopsis* by oomycete pathogens. *Plant J* **33:**775–792

Timmers AC, Auriac MC, Truchet G (1999) Refined analysis of early symbiotic steps of the *Rhizobium-Medicago* interaction in relationship with microtubular cytoskeleton rearrangements. *Development* **126:**3617–3628

Triplett E, Sadowsky MJ (1992) Genetics of competition for nodulation of legumes. *Annu Rev Microbiol* **46:**399–428

Truchet G (1978) Sur l'état diploide des cellules du mériteme des nodules radiculaires des légumineuses. *Ann Sci Bot Paris* **19:**3–38

Turgeon BG, Bauer WD (1985) Ultrastructure of infection-thread development during the infection of soybean by *Rhizobium japonicum*. *Planta* **163:**328–349

Udvardi MK, Tabata S, Parniske M, Stougaard J (2005) *Lotus japonicus*: legume research in the fast lane. *Trends Plant Sci* **10:**222–228

van Brussel AAN, Bakhuizen R, van Spronsen PC, Spaink HP, Tak T, Lugtenberg BJJ, Kijne JW (1992) Induction of pre-infection thread structures in the leguminous host plant by mitogenic lipo-oligosaccharidess of *Rhizobium*. *Science* **257:**70–72

Van den Bosch KA, Bradley DJ, Knox JP, Perotto S, Butcher GW, Brewin NJ (1989) Common components of the infectgion thread matrix and intercellular space identified by immunocytochemical analysis of pea nodules and uninfected roots. *EMBO J* **8:**335–342 van Spronsen PC, Bakhuizen R, van Brussel AA, Kijne JW (1994) Cell wall degradation during infection thread formation by the root nodule bacterium *Rhizobium leguminosarum* is a two-step process. *Eur J Cell Biol* **64:**88–94

Vasse J, de Billy F, Camut S, Truchet G (1990) Correlation between ultrastructural differentiation of bacteroids and nitrogen fixation in alfalfa nodules. *J Bacteriol* **172:**4295–4306

Veereshlingam H, Haynes JG, Penmetsa RV, Cook DR, Sherrier DJ, Dickstein R (2004) *nip*, a symbiotic *Medicago truncatula* mutant that forms root nodules with aberrant infection threads and plant defense-like response. *Plant Physiol* **136:**3692–3702

Vijn I, das Neves L, van Kammen A, Franssen H, Bisseling T (1993) Nod factors and nodulation in plants. *Science* **260:**1764–1765

Vinardell JM, Fedorova E, Cebolla A, Kevei Z, Horvath G, Kelemen Z, Tarayre S, Roudier F, Mergaert P, Kondorosi A, Kondorosi E (2003) Endoreduplication mediated by the anaphase-promoting complex activator CCS52A is required for symbiotic cell differentiation in *Medicago truncatula* nodules. *Plant Cell* **15:**2093–2105

Zarsky V, Fowler J (2008) ROP (Rho-related protein from Plants) GTPases for spatial control of root hair morphogenesis. In: Emons AMC, Ketelaar T (eds) Root hairs: excellent tools for the study of plant molecular cell biology. Springer, Berlin Heidelberg New York. doi:10.1007/7089_2008_14

Microbial Encounters of a Symbiotic Kind: Attaching to Roots and Other Surfaces

A.M. Hirsch(✉), M.R. Lum, and N.A. Fujishige

Abstract Attachment of bacteria to plant surfaces is a necessary prelude to the interaction, either pathological or mutualistic, that follows. For symbiotic nitrogen fixation to occur and, in particular, for nodules to develop for housing the nitrogen-fixing bacteria in the legume–*Rhizobium* mutualism, attachment of rhizobia to roots is critical. Nodules form on some legume roots as a consequence of Nod factor perception but if the rhizobia do not attach, they remain uninfected because attachment is needed for infection-thread formation. Numerous studies have shown that rhizobial cell surface components are required for optimal root attachment and colonization. These components include polysaccharides such as exopolysaccharides, lipopolysaccharides, cyclic β-1,2-glucans, and cellulose fibrils; and also proteins, including flagellae, pili, rhicadhesin, and a bacterial lectin known as Bj38. Loss of function of genes encoding exo-, capsular-, and lipopolysacchrides as well as cyclic β-1,2-glucan often result in diminished root attachment and poorly infected nodules. However, no mutant phenotypes have been described for the loss of function of either rhicadhesin or bacterial lectin because genes encoding these traits have not yet been identified. *Rhizobium leguminosarum* mutants defective in cellulose fibril production still induce nitrogen-fixing nodule formation, and moreover, not all rhizobia synthesize cellulose fibrils, strongly suggesting that fibrils are not universally required for attachment to plant roots. On the plant side, very little is known about the factors required for rhizobial attachment. Carbohydrate-binding proteins, particularly lectins, have been implicated, but few other plant proteins have been described. This review describes what is known about the genes and proteins that are involved in attachment and colonization of rhizobia on legumes. We focus not only on attachment to root hairs and epidermal cells, but also on *ex planta* adherence. To that end, we consider rhizobial attachment to the root surface as well as

A.M. Hirsch
Department of Molecular, Cell and Developmental Biology, University of California, Los Angeles, CA, 90095-1606, USA and Molecular Biology Institute, University of California, Los Angeles, CA, 90095-1606, USA
e-mail: ahirsch@ucla.edu

Plant Cell Monogr, doi:10.1007/7089_2008_06

to abiotic surfaces as a biofilm, i.e., a structured community of bacteria adherent to a surface and to each other, and surrounded by exopolymer. We also examine the effects of cell surface mutations on biofilm development in other bacteria with the goal of establishing commonalities with nitrogen-fixing rhizobia.

1 Introduction

Although roots are normally covered with numerous microbes, surprisingly little attention has been paid to them by most biologists other than microbial ecologists. In part, this is because rhizospheric interactions are not only very heterogeneous, but also difficult to study in a soil-based system. The classic studies by Foster et al. (1983) on the ultrastructure of the soil–root interphase illustrate the complexities of the rhizosphere environment. This tour-de-force, electron microscopic analysis shows an amalgam of root cells, bacteria, fungal cells, virus particles, polysaccharides, and soil (Fig. 1). Technical improvements, such as the use of reporter genes, fluorescent antibodies, fluorescence in situ hybridization as well as confocal scanning laser microscopy, have greatly aided our investigation of the rhizosphere (see references in Wagner et al. 2003). Moreover, metagenomic analyses of rhizosphere

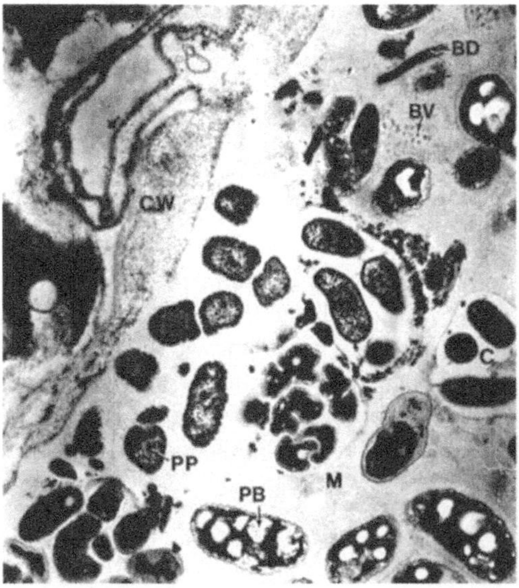

Fig. 1 Ultrastructure of the clover rhizosphere (×25,000). *BD Bdellovibrio* parasitizing various rhizosphere microbes, *BV* bacteriophages, *C* capsule, *CW* cell wall, *M* mucigel, *PB* phosphate bodies in soil microbes, *PP* polyphosphate granules (reprinted with permission, from Foster and Rovira (1978))

organisms are beginning to uncover the incredible diversity that exists underground (Erkel et al. 2006; Gros et al. 2006).

To learn the identity of the factors important for attachment to a plant surface, which is a necessary prelude to the commensal, pathological, or mutualistic plant–microbe interactions that follow, requires a less global view than that of metagenomes. Studies of single species, their biochemistry, and genetics have greatly helped in the identification of genes and gene products needed for bacterial adherence to biotic and abiotic surfaces. However, much of the research so far has focused on clinically relevant or polluting microbes because of their bearing on disease and biofouling. Attachment to plant surfaces has been investigated much less often, but is just as important for a successful interaction as those encounters that result in illness or pollution. For example, nitrogen-fixing root nodules will not develop without rhizobia attaching to and colonizing the root surface. (For more on nodule development, see Limpens and Bisseling 2008). Although a completely attachment-minus (Att$^-$) rhizobial mutant has not been described, numerous mutants affected in attachment to the root surface have been isolated; most infect nodules poorly. Thus, attachment to the root surface is a critical stage for the initiation of infection threads for proper nodule formation and subsequent nitrogen fixation. Two recent reviews address bacterial attachment to plant and other surfaces (Danhorn and Fuqua 2007; Rodríguez-Navarro et al. 2007).

A two-step system of "docking" and "locking" is characteristic of all bacteria that adhere to surfaces (Dunne 2002). Docking is considered to be reversible whereas locking is generally thought of as irreversible. Docking can also be thought of as nonspecific binding whereas locking is species-specific. Attachment follows a similar two-step process for Rhizobiaceae and many other symbiotic bacteria (Matthysse et al. 1981; Dazzo et al. 1984; Smit et al. 1992). Interestingly, most of the components important for rhizobial docking that have been studied are of bacterial origin. Little is known about the corresponding receptors in the plant that recognize the bacterial docking and locking factors. Only two plant factors, i.e., lectins, plant proteins that bind carbohydrates, and a potential receptor for rhicadhesin, have been studied.

This review describes what is known about the genes and proteins that are involved in the attachment and colonization of roots by symbiotic microbes, focusing not only on adherence to root hairs and epidermal cells, but also on *ex planta* adhesion. To that end, we consider the attachment of rhizobia to the root surface to be a biofilm, i.e., "a structured community of bacteria enclosed in a self-produced polymeric matrix and adherent to an inert or living surface" (Costerton et al. 1995) (Fig. 2). Most important, the bacteria in a biofilm also attach to each other. The genes involved in the various stages of biofilm formation have been uncovered mainly based on the study of mutants, particularly of model systems such as *Escherichia coli*, *Pseudomonas aeruginosa*, and *Staphylococcus epidermidis*. For example, flagella are needed for the initial stages of attachment, but pili are required for microcolony formation. Microcolony formation paves the way for a mature biofilm architecture consisting of towers, mushrooms, mounds, or streamers (Fig. 2). However, in addition to the well-studied mushroom-type biofilms of *P. aeruginosa* and other model bacteria, additional types of biofilm architecture develop, some of which are independent of flagella. Microcolonies may be established by clonal propagation, with type IV pili mediating bacterial migration by using twitching motility to extend the microcolonies to form a continuous or ridgelike biofilm (Fig. 3).

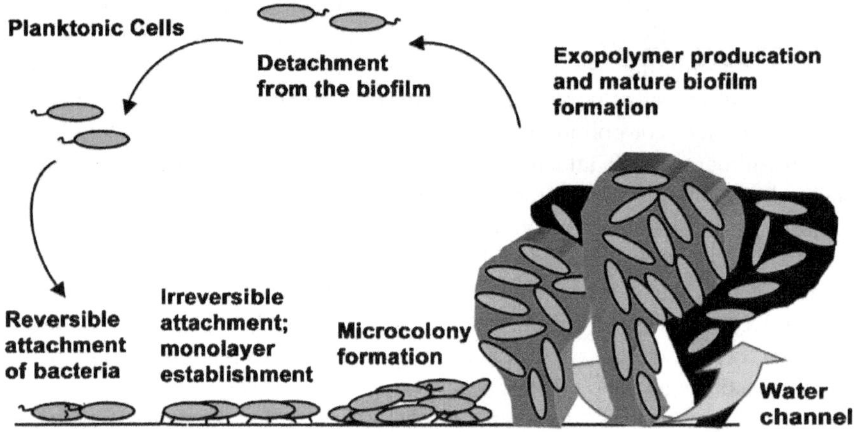

Fig. 2 Generalized diagram of the main stages in biofilm formation. Planktonic, free-swimming bacteria are signaled by environmental cues to attach to a surface. Irreversible attachment follows, and the cells establish a monolayer. Later, microcolonies form on the surface. The mature biofilm develops a distinctive architecture comprising mushrooms, towers, ridges, or streamers. Exopolymer covers the biofilm cells, but cells can detach and become planktonic (reprinted with permission, from Fujishige et al. (2006c))

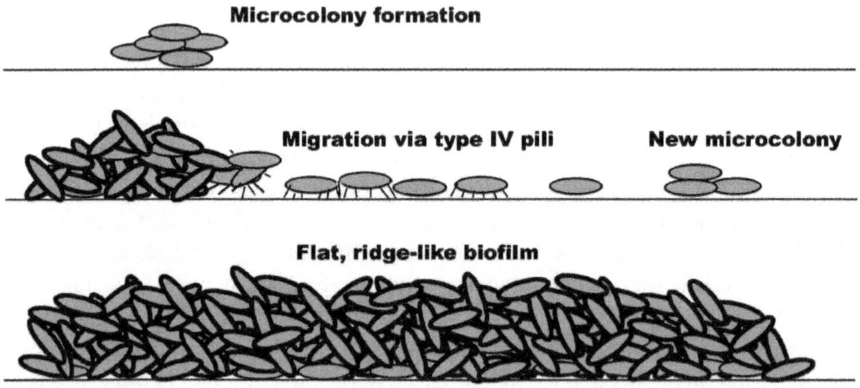

Fig. 3 Generalized diagram of the stages of a flat or ridgelike biofilm whereby microcolony formation is mediated by clonal growth. Bacteria migrate via twitching behavior due to the presence of type IV pili (modified from Klausen et al. (2003)). See also Fig. 4

These differences in biofilm architecture appear to depend both on cultural conditions, such as carbon source, as well as on strain differences (Klausen et al. 2003). Where data are available, we will compare the rhizobial genes critical for attaching to the root surface with those that are important for biofilm formation in other bacteria.

2 Docking Components

Often, bacteria will attach or not attach to a surface based solely on the variabilities of the surface, such as its texture, composition, or whether it is hydrophobic or hydrophilic. These factors will not be discussed in this review, but many parameters, including temperature, pH, electrostatic and hydrophobic interactions, nutrient status, and culture age, condition biofilm formation and attachment to roots (Carpentier and Cerf 1993; Albareda et al. 2006; Rinaudi et al. 2006).

For different strains of *R. leguminosarum*, the docking step appears to be mediated by rhicadhesin and lectins, whereas the locking step is mediated by cellulose fibrils (Smit et al. 1987, 1992). For other rhizobia, additional factors come into play. These will be described.

2.1 Rhicadhesin

Rhicadhesin is a calcium-binding protein of 14–16 kDa (depending on the method of analysis; Smit et al. 1989) that when purified, inhibits attachment to pea roots of not only *R. leguminosarum* strain RBL5523, the strain from which it was originally isolated, but also other Rhizobiaceae, including *Agrobacterium tumefaciens*, *A. rhizogenes*, *Bradyrhizobium japonicum*, and *Phyllobacterium* spp., suggesting that rhicadhesin is common in this bacterial family. A *Bradyrhizobium* calcium-dependent, bacterial surface protein of 14 kDa, assumed to be rhicadhesin, mediates this strain's attachment to peanut, a legume that shows a "crack-entry" rather than infection-thread mode of rhizobial incursion (Dardanelli et al. 2003). Thus, rhicadhesin-mediated attachment is involved in both forms of rhizobial entry into roots. Rhicadhesin is also involved in the attachment of rhizobia to wheat and other non-legume root hairs (Smit et al. 1989), pointing to rhicadhesin's role in non-specific or docking attachment rather than in the second, more specific, locking stage.

In an attempt to find the gene encoding rhicadhesin, Ausmees et al. (2001) uncovered four *Rhizobium*-adhering proteins (Rap) in *Rhizobium leguminosarum* bv. *trifolii* strain R200 by using phage-display cloning techniques. Although these were all secreted calcium-binding proteins, RapA1and the other three were larger (24 kDa) than rhicadhesin and not found in *Rhizobium* species other than *R. leguminosarum* biovars and *R. etli*. Thus, up to now, a gene for a 14-kDa rhicadhesin protein has not been discovered, making it difficult to assess the exact role of this protein in the first phase of attachment. Also, without a gene or protein sequence, it is difficult to determine whether rhicadhesin is present in bacteria other than the Rhizobiaceae.

A legume rhicadhesin receptor had been postulated based on its ability to suppress rhicadhesin activity (Swart et al. 1994). A cell wall component from pea roots was purified and found to be glycosylated. It was estimated to be about 32 kDa when glycosylated and has an isoelectric point of about 6.4. However, only the N-terminal region was sequenced, and no homology was observed to any other proteins other than a germin (Matthysse and Kijne 1998). Germins are common to a wide range of plants and are expressed in all parts of the plant. A signal peptide, a β-barrel structure, and one or two N-glycosylation sites at a constant position characterize them. Germin-like proteins (GLPs), but not germins, have an Arg-Gly-Asp (RGD) tripeptide or related KGD/KDE peptides (Bernier and Berna 2001). An RGD hexapeptide, as well as vitronectin, an extracellular matrix protein in animal cells with an RGD motif, suppressed rhicadhesin-mediated attachment of both rhizobia and agrobacteria to root hairs (Swart et al. 1994), suggesting that an RGD-containing protein functions as a receptor. Recently, a GLP sharing sequence similarity in its N-terminus with the comparable region in the putative receptor for rhicadhesin has been identified from pea. GLP mRNA is expressed in nodules in the expanding cells adjacent to the nodule meristem and to a lesser extent in the nodule epidermis (Gucciardo et al. 2007), locations where a rhicadhesin receptor might not be expected. The newly identified GLP has superoxide dismutase activity, which is resistant to high temperatures among other stresses, suggesting that this GLP may be involved as a target for protein cross-linking. Some germins and GLPs become insoluble in response to stress (Bernier and Berna 2001). Nevertheless, GLP may be a rhicadhesin receptor because these proteins are associated with the plant extracellular matrix and they have the tripeptide RGD, which is found in animal adhesion proteins. More work is needed to validate this hypothesis.

2.2 Flagella

Flagella have been proposed to be a docking step for biofilm formation for a number of bacteria (O'Toole and Kolter 1998a; Pratt and Kolter 1998; Klausen et al. 2003). For example, mutating *flgH* genes in *Caulobacter crescentus* result in biofilms that are thick, homogenous monolayers (Entcheva-Dimitrov and Spormann 2004). Moreover, the transition from microcolonies to the large mushroom-shapes that characterize the mature biofilm in *Caulobacter* is impaired in the *flgH* mutants.

Flagella have also been proposed as the docking step for *Azospirillum* attachment to plant roots (Michiels et al. 1991; Vande Broek and Vanderleyden 1995), although flagella are best known for transporting microbes across distances. For example, when a legume seed is planted, the seed coat and in some cases the roots exude flavonoids and related molecules that attract rhizobia, which are believed to move by flagellar action, towards the legume root. However, it is unclear exactly how far rhizobia actually move in the soil environment because motility is dependent on soil matric potential and composition among other factors. For *Sinorhizobium meliloti* grown under axenic conditions, Fla$^-$ mutants are delayed in forming nodules, but normal nitrogen-fixing nodules develop (Finan et al. 1995; Fujishige et al. 2006a). However, under field conditions, it is likely that motility mutants are at a competitive disadvantage when compared to wild-type, motile *S. meliloti* (Ames and Bergman 1981; Fujishige et al. 2006a). Similarly, *S. meliloti* Fla$^-$ mutants show a more than 50% reduction in biofilm formation when compared with that in the wild-type strain (Fujishige et al. 2006a). Hence, both the biofilm and nodulation phenotypes are likely to be a consequence of impaired motility as well as the reduced docking exhibited by the *S. meliloti fliP* and *flgH* mutants.

2.3 Lectin-Mediated Attachment

The only bacterial lectin studied in detail so far is BJ38, which is organized in a tuftlike mass to one pole of *B. japonicum* cells. This bacterial lectin is important for adherence to other bradyrhizobia, resulting in the formation of "star"-like clusters (Loh et al. 1993). Soybean agglutinin (a plant lectin) binds to the opposite pole that BJ38 binds to, suggesting that soybean agglutinin or soybean lectin is not related to bacterium-to-bacterium attachment, and must bind to a cell surface molecule other than the bacterial lectin BJ38. However, the details of how this bacterial lectin functions in attachment have not been elucidated because no specific gene(s) has been identified, although two non-BJ38-producing bacteria were described. The mutants showed a decreased ability to bind to young emergent root hairs and to nodulate soybean (Ho et al. 1994).

Plant lectins have been implicated in the *Rhizobium*–legume symbiosis ever since a strong correlation between rhizobial cross-inoculation groups and the host legume lectin was noted (see references in Hirsch 1999). Experiments whereby lectins have been transferred from one legume to another, such as the transfer of genes encoding soybean lectin to *Lotus corniculatus* (van Rhijn et al. 1998) and pea lectint tansferred to clover (Díaz et al. 1989) or to soybean and alfalfa (van Rhijn et al. 2001), demonstrated that the "wrong" rhizobia could nodulate the transgenic plants, but only if the rhizobia were producing the Nod factor appropriate to the "new" host. To explain these results, we proposed that the cognate lectin facilitated attachment of rhizobia to the root surface, thereby increasing Nod factor concentration above the threshold required for nodulation (van Rhijn et al. 2001). Binding sites for lectins have been detected in exopolysaccharide (EPS) and lipopolysaccharide (LPS; Mort and Bauer 1980), and we presented data showing that Exo$^-$ mutants were not as efficient as wild-type rhizobia in attaching to transgenic lectin plants (van Rhijn et al. 1998, 2001). However, others have argued that EPSs are not ligands for lectins (Laus et al. 2006).

Laus et al. (2006) reported the isolation and partial characterization of a *R. leguminosarum* glucose–mannose polysaccharide containing minor amounts of galactose and rhamnose. The *Rhizobium* strain studied in this investigation was originally isolated from a clover nodule and thus has the *R. leguminosarum* bv. *trifolii* chromosome. However, its endogenous pSym was replaced by the symbiotic plasmid pRl1JI from *R. leguminosarum* bv. *viciae* so that it produces a Nod factor recognized by pea and vetch. Pea lectin and vetch seed lectin are both glucomannan-binding lectins, and show high affinity for this newly described bacterial polysaccharide. Moreover, pea lectin was shown to localize to one pole of the rhizobial cell, thereby pinpointing the location of this novel polysaccharide. Nevertheless, mutant rhizobia that do not produce the glucomannan polysaccharide are impaired in root attachment and infection only under slightly acidic conditions (pH 5.2) (Laus et al. 2006).

As for rhicadhesin, chromosomal genes encode glucomannan polysaccharide production in *R. leguminosarum* (Laus et al. 2006). Thus, we expected that rhizobial strains deleted of pSym would show the same biofilming activity as does the wild-type parent if these components are primarily responsible for attachment. We tested RlvB151, a symbiotic plasmid-minus derivative of *R. leguminosarum* bv. *viciae* (Brewin et al. 1983) in a biofilm assay and found that the pSym$^-$ strain established biofilms poorly when compared with wild-type (Fujishige et al. 2006b). This result indicates that genes encoding either glucomannan or rhicadhesin are probably not needed for biofilm formation under the conditions studied. A similar result was observed for wild-type *S. meliloti* strain 2011 and its pSymA-cured (SmA818; Oresnik et al. 2000) derivative (Fujishige et al. 2006b). Thus, one can infer from these experiments that genes borne on symbiotic plasmids are important for attachment to surfaces. This result also indicates that chromosomal gene products such as glucomannan or rhicadhesin cannot substitute for the pSym gene products required for normal biofilm formation by *R. leguminosarum* bv. *viciae*.

3 Locking Components

After reaching a surface, bacteria undergo a period of reversible attachment, but this quickly changes to irreversible attachment once the bacteria stop being motile. Again, there is an overlap between those components needed for locking to a root and those for biofilm formation. Both polysaccharides and proteins are involved in the irreversible steps of biofilm formation in a number of plant–microbe interactions.

3.1 Polysaccharides

Several investigators have reported that polysaccharides on the rhizobial cell surface are absolutely required for optimal root attachment and colonization. Some of these cell surface components include EPS and capsular polysaccharide (Becker and Pühler 1998), LPS (Kannenberg et al. 1998), cyclic β-1,2-glucans (Dylan et al. 1990), and cellulose fibrils (Smit et al. 1987). Why so much diversity is required is unclear, but a redundancy of binding components may ensure that a sufficient population of rhizobia becomes firmly attached to the root so that the subsequent stages of nodule initiation are initiated.

3.1.1 Exopolysaccharide

Two EPSs are produced by *S. meliloti*: EPSI, a succinoglycan, and EPSII, a galactoglucan-repeating unit modified with acetyl and pyruvyl residues. The sequenced *S. meliloti* strain, Rm1021, does not synthesize EPSII, but is symbiotically competent. However, *S. meliloti* Exo⁻ cells in the Rm1021 background lack EPSI (and EPSII), and thus do not enter the nodule properly. For example, the *S. meliloti exoY* (*exoY* encodes a galatosyl-1-P-transferase that carries out the first step of synthesis; Reuber and Walker 1993) mutant in the Rm1021 background is impaired in the formation of infection threads, resulting in nodules that are free of rhizobial cells (Finan et al. 1985; Leigh et al. 1985). Alfalfa nodules induced by *exoY* mutant *S. meliloti* fail to form persistent nodule meristems, and remain small, round, and ineffective (Yang et al. 1992). Exopolysaccharide is very important for biofilm formation in many bacteria and loss-of-function mutants generally show impaired biofilm formation (Yildiz and Schoolnik 1999; Danese et al. 2000; Whiteley et al. 2001; Matsukawa and Greenberg 2004). Similarly, the *S. meliloti exoY* loss-of-function mutant shows reduced biofilm formation; the bacteria never make the transition from the microcolony stage to a mature three-dimensional biofilm (Fujishige et al. 2006a).

Two gain-of-function *exo* mutants, *exoS* and *exoR*, have been described in *S. meliloti*; the *exoR* and *exoS-chvI* system regulates EPSI synthesis (Cheng and Walker 1998; Wells et al. 2007). The *exoS* mutant elicits the formation of normal nitrogen-fixing nodules on alfalfa whereas the *exoR* mutant induces a mixture of ineffective and effective nodules (Cheng and Walker 1998). We found that *S. meliloti exoS* mutants develop more extensive biofilms, although the cells are more loosely attached to each other in the biofilm than are wild-type cells (Fujishige et al. 2006a). The *S. meliloti exoR* mutants also produce larger biofilms than do the wild-type cells, but because they do not remain attached after handling, we found the measurements of the amount of biofilm formation to be extremely variable (Fujishige 2005). Nevertheless, Wells et al. (2007) were able to show that *exoR* and *exoS* mutants exhibited increased biofilm formation when compared with wild-type rhizobia, thereby confirming our unpublished results and also those of Fujishige et al. (2006a) with regard to *exoS S. meliloti*.

R. leguminosarum bv. *viciae* mutants (*prsD* and *prsE*) defective in the secretion of the EPS glycanases PlyA and PlyB make biofilms that are not only delayed in timing, compared with wild-type, but also are arrested in their development (Russo et al. 2006). The EPS produced by these mutants is longer than the wild-type EPS because the PlyA and PlyB glycanases normally cleave EPS into shorter lengths. Although *prsE* mutants have no nodulation phenotype, *prsD* mutants elicit nodules that do not fix nitrogen. Completely EPS⁻ mutants (*pssA* and other *pss* genes) show minimal biofilm formation. These mutants are also defective for root infection; empty nodules result. Interestingly, Russo et al. (2006) found that the *R. leguminosarum* bv. *viciae* Raps (*Rhizobium*-adhering proteins) RapA1, RapA2, and RapC are secreted by the PrsD-PrsE type I secretion system and are likely to be important for biofilm formation.

3.1.2 Lipopolysaccharide

Lipopolysaccharides are a major component of the bacterial outer envelope (Campbell et al. 2002). They are involved in the later stages of infection of the root by rhizobia, being responsible for continual infection. In the *S. meliloti*-alfalfa symbiosis, *lps* mutants produce approximately half the number of fully extended infection threads as the wild-type strain, but the rhizobia are still able to enter into the nodule cells. However, once the bacteria are released from the infection thread into the infection droplet, the *lps* mutants cannot sustain the infection, and the nodule usually becomes ineffective. Nodule cells often contain bacteroids with unusual cell morphologies (Campbell et al. 2002).

Few studies have examined the root attachment phenotype of rhizobial *lps* mutants. Using the microtiter plate assay, we examined two *S. meliloti lps* mutants defective in two parts of the LPS structure for their effects on biofilm formation. The *lpsB* mutant is lacking glycosyltransferase I, which is responsible for the biosynthesis of the LPS core (Campbell et al. 2002), and the *bacA* mutant is defective in the distribution of fatty acids on the lipid-A component of LPS (Ferguson et al. 2002). Mutation of *lpsB* resulted in a slight reduction of biofilm formation, whereas mutation of *bacA* resulted in biofilms that were reduced by roughly half when compared with that in the wild-type (Fujishige 2005). Microscopic examination did not show profound differences in the biofilm structures of the *lps* mutant. However, in the *bacA* mutant, microcolonies and towers occurred less frequently and were reduced in size.

3.1.3 Cellulose Fibrils

Cellulose fibrils are part of the locking step for *R. leguminosarum* RBL5523 and *Agrobacterium tumefaciens* attachment (Matthysse and Kijne 1998). However, cellulose fibrils are not universally present in symbiotic rhizobia and interestingly, not all strains of *R. leguminosarum* produce detectable cellulose (Russo et al. 2006). Furthermore, the *R. leguminosarum* RBL5523 cellulose fibril mutant elicits effective nodule formation, demonstrating that the lack of a component for the locking step has minimal effect on nodulation (Smit et al. 1987; Ausmees et al. 1999; Laus et al. 2005). For *A. tumefaciens*, cellulose fibrils bind the bacteria tightly to the plant surface as well as to each other. Mutants deficient in synthesizing these fibrils did not bind as strongly to plant cell walls and were reduced in colonizing roots grown in quartz sand (Matthysse and McMahan 1998). On the other hand, cellulose overproduction increased the ability of *A. tumefaciens* to bind to and establish biofilms on roots (Matthysse et al. 2005). Indeed, cellulose has been described as being involved in several bacterial biofilms such as those of *Acetobacter xylinum* (Cannon and Anderson 1991). However, *A. tumefaciens* cellulose overproduction did not result in better root surface colonization, perhaps because of the slower growth rate of these bacteria (Matthysse et al. 2005).

3.1.4 Cyclic β-1,2-Glucans

Cyclic β-1,2-glucans are periplasmic space components of gram-negative bacteria and are important for microbe–host interactions, whether that interaction be pathogenic (Matthysse et al. 2005; Roset et al. 2004) or symbiotic (Breedveld and Miller 1994). The exact mechanism whereby cyclic β-1,2-glucans function in attachment is unknown in part because mutants show highly pleiotropic phenotypes, and do not always disrupt an interaction with the plant host. For example, for nodulation, it has been suggested that these hydrated polysaccharides increase turgor pressure within the infection thread; this pressure could be required to drive the infection thread growth (Nagpal et al. 1992).

The genes that encode cyclic β-1,2-glucans in *S. meliloti* are *ndvA* and *ndvB*. *ndvB* encodes a cytoplasmic membrane protein that synthesizes cyclic β-glucan from UDP-glucose whereas *ndvA* encodes an ATP-binding transport protein that is responsible for the secretion of cyclic β-glucans to the extracellular space. Comparable genes exist in *A. tumefaciens*. The phenotype resulting from mutating either *chvA* or *chvB* in *A. tumefaciens* (Douglas et al. 1982) or *ndvA* or *ndvB* in *S. meliloti* (Dylan et al. 1990) is poor attachment to plant cells. The *chvA* and *chvB* mutants also formed reduced biofilms when compared with the wild-type *A. tumefaciens* on roots (Matthysse et al. 2005). Perhaps not surprisingly, the *A. tumefaciens chvB* mutant defect was partially complemented by overproduction of cellulose (Matthysse et al. 2005). Although a 41% reduction in biofilm formation, compared with that in wild-type *S. meliloti*, in microtiter plate wells was observed for *ndvB* mutants, we saw no change in the level of biofilm formation, using the same assay for *S. meliloti ndvA* mutants (Fujishige 2005). Previous work showed that mutations in *ndvA* result in increased EPSI production (Breedveld and Miller 1994). An increase in EPSI may compensate for the loss of *ndvA*, or alternatively, cyclic β-glucan may be secreted through a completely different transporter than NdvA.

3.2 *Proteins*

Few proteins have been implicated in the locking stage of rhizobial attachment. It seems logical that pili of various types are involved in the locking steps for the *Rhizobium*–legume symbiosis, not only because of their importance for adherence to animal cells and environmental surfaces, but also because of their involvement in biofilm formation.

3.2.1 Pili/Fimbriae

Pili are important for adherence to animal cells and are also critical for biofilm formation, particularly the early stages. Whereas flagellae are important for transport of microbes across a distance to a surface where they reversibly attach, pili are believed to be involved in the irreversible attachment of the bacteria to a surface and also for the characteristic architecture of the mature biofilm. Pratt and Kolter (1998) found that *E. coli* biofilm formation required flagella for the early stages of biofilm establishment and type I pili for irreversible attachment. Type IV pili, which are important for "twitching" behavior, are also required for biofilm formation in *P. aeruginosa* because *pilB*, *pilC*, and *pilY1* mutants, all of which lack type IV pili, do not establish microcolonies (O'Toole and Kolter 1998b).

For plant-associated interactions, very few investigations have been pursued regarding the role of pili in surface interactions either with biotic or abiotic surfaces. *Xylella fastidiosa* cells, which are nonflagellated, establish biofilms in the xylem cells of their host by using type IV pili for migrating within the plant while type I pili are important for cell-to-cell aggregation in biofilm formation (Li et al. 2007). Type IV pili have been shown to be important for adhesion of *Azoarcus* sp. to plant root and fungal surfaces (Dörr et al. 1998).

There are very few reports in the literature that describe pili/fimbriae in *Rhizobium*. One of the exceptions is the account of Vesper and Bauer (1986), who proposed that pili mediate the locking step between *Bradyrhizobium japonicum* and its soybean host. Piliated cells were found to bind to roots and plastic plates proportionally to their numbers, and an antipilus antibody made to pili from soybean-nodulating rhizobia blocked both attachment and nodulation without affecting bacterial viability. Pilus-minus mutants were also defective in adhesion (Vesper and Bauer 1986; Vesper et al. 1987). However, further research on pili was not pursued, and the antipilus antibody was lost (W.D. Bauer, personal communication).

We revisited this issue by mutagenizing *pilA* in *S. meliloti* (M.R. Lum and A.M. Hirsch, unpublished results). PilA makes up the pilin subunit of the pilus. The mutants were defective in twitching behavior and much less competitive for nodulation. On the other hand, preliminary results suggested that biofilm formation was not reduced when compared with wild-type *S. meliloti*, although differences were observed in biofilm architecture (M.R. Lum and A.M. Hirsch, unpublished results). However, a second copy of *pilA* exists in the *S. meliloti* genome. Studies are in progress to elucidate the role of the *pilA* genes by making a double knock-out mutant.

3.3 Additional Components Involved in Attachment

A number of bacteria use glucosamine oligomers and polymers to adhere to each other in biofilm matrices (Leriche et al. 2000). An excellent example is polysaccharide intercellular adhesin (PIA) in *Staphylococcus aureus* and *S. epidermidis*, two soil-inhabiting bacteria that cause nosocomial infections in humans. PIA, also called "slime," is a polymer of at least 130 β-1,6-*N*-acetylglucosamine (GlcNAc) residues, some of which are deacylated (Götz 2002). In the biofilm, PIA acts as an intercellular adhesin, linking the bacterial cells together in the biofilm. Interestingly, several of the *ica* genes that encode proteins giving rise to PIA show sequence similarity to *Rhizobium nod* genes (Heilmann et al. 1996; Götz 2002). Mutations in *ica* genes yield a biofilm-minus phenotype.

Caulobacter crescentus is an aquatic bacterium characterized by two distinct cell types: a swarmer cell with pili and a polar flagellum; and a stalked cell, which has a holdfast at the distal end. The holdfast facilitates cell-to-cell adherence (rosette formation) as well as adhesion in a biofilm. Mutations in genes in holdfast synthesis adhere poorly and form loosely attached microcolonies (Entcheva-Dimitrov and Spormann 2004). Lectin-binding assays indicate that the adhesive tip of the holdfast contains GlcNAc residues (Merker and Smit 1988; Ong et al. 1990), which are most likely to be β-1,4-linked because chitinase and lysozyme treatment disrupts the rosettes (Merker and Smit 1988). Proteins and uronic acids also make up the holdfast polysaccharide. These components form a dense gel, which has been modeled as an elastic leaf spring, with the GlcNAc oligomers playing a significant role in elasticity (Li et al. 2005).

We found that GlcNAc oligomers (i.e. Core Nod factor) also promote cell adhesion in *S. meliloti* biofilms (Fujishige et al. 2008). Mutations in common *nod* genes (*nodD1ABC*), but not in the host-specific *nod* genes, result in biofilms that remain a monolayer and do not transition into a three-dimensional structure. Moreover, common *nod* gene mutants do not attach as well to the roots of their legume host, indicating that root colonization is also affected. Exactly how the Core Nod factor (product of the common *nod* genes) functions in *S. meliloti* biofilm formation is unknown. There are at least two mechanisms, which are not mutually exclusive. (1) Nod^+ bacteria are more hydrophobic than Nod^- rhizobial cells, allowing them to attach to each other and to roots and other surfaces. (2) Core Nod factor functions as a "glue" similar to the GlcNAc oligomers of *Caulobacter* or *Staphylococcus* (Fujishige et al. 2008). Deciphering the mechanism of Core Nod factor in biofilm formation is a goal for future research.

4 Concluding Remarks

We have approached this review by examining the genes that are involved in attachment, not only to roots, but also to abiotic surfaces via biofilm formation. The parallels between biofilm formation and rhizobial infection are numerous. Ramey et al. (2004) suggested that infection threads are actually biofilms, and we concur with this proposal. The study of biofilms has the potential to unlock many mysteries of infection because they are much simpler to study than the symbiotic root.

Considering the importance of attachment for the initiation of the nitrogen-fixing symbiosis between legumes and rhizobia, it is surprising how little we know of the factors involved from the plant's perspective. Our lack of knowledge contrasts with the studies of attachment of mammalian-associated bacteria, where adherence is a precursor to infection and disease. Numerous receptors, such as pilin-binding proteins, and components of the extracellular matrix such as vitronectin and fibronectin have been shown to bind bacterial components. Although a plant rhicadhesin receptor, which may be analogous to vitronectin, has been uncovered, so far no protein has been definitively identified that specifically binds to rhicadhesin. Plant lectins have been shown to be involved in attachment, and evidence exists for their binding to various polysaccharides on the rhizobial surface. The exact component is still under debate. With evidence now showing that the *nod* genes have a second function, i.e., holding a biofilm together (Fujishige et al. 2008), as well as the fact that receptors for signaling Nod factors have been described (Limpens et al. 2003; Madsen et al. 2003; Radutoiu et al. 2003), the question can be asked whether the Nod factor receptors are also important for rhizobial attachment. This could be a rich area for future study.

5 Method: Growing Rhizobia in Microtiter Plate Wells

Many of the experiments described herein utilize various methods for growing rhizobia in biofilms. One of the most useful is the microtiter plate procedure. We adapted a method published by O'Toole et al. (1999) that facilitates a rapid, high through-put, and quantitative approach to analyzing biofilms. Microtiter plates having either 96 or 20 wells of either poly(vinyl chloride) (PVC) plastic or polystyrene have been used successfully in our laboratory for diverse species of a- and β-rhizobia, including *S. meliloti*, *R. leguminosarum* bv. *viciae* (Fujishige et al. 2006a), *Rhizobium* NGR234, and *Burkholderia tuberum* (A. Maghsoodpour and A.M. Hirsch, unpublished results). Although we routinely use PVC plates with U-bottom wells, plates with flat-bottomed wells are very useful for visualizing the timing of biofilm development under phase contrast or fluorescence optics of an inverted microscope. The timing of incubation in the microtiter plate wells varies depending on the experimental question. We frequently analyze and compare the early stages of biofilm development (<24 h or 24–48 h), but others and we have utilized longer times for certain experiments (Fujishige et al. 2006a; Russo et al. 2006). O'Toole et al. (1999) reported that *P. aeruginosa* grown for extended times in PVC plates start to detach; this possibility must be assessed for each bacterial species studied.

Liquid cultures for biofilm analysis are grown in TY or modified Rhizobium defined medium (RDM) to OD_{600} of 1.5–2.0. The cells are then diluted in modified RDM (Fujishige et al. 2006a) to an OD_{600} of 0.2, and then 100 μL of the diluted cells is added to a minimum of 10 wells in a 96-well PVC plate (Falcon 353911, Becton Dickinson, Franklin Lakes, NY). The same amount of uninoculated medium is used as a control. We found that significantly better biofilm formation is obtained by growing the cells in a minimal medium; Russo et al. (2006) also found this to be the case. The plates are sealed with either sterile rayon adhesive film (AeraSeal, Excel Scientific, Wrightwood, CA) or covered with flexible PVC lids (Falcon 353913, Becton Dickinson), and incubated at 28°C without shaking. Wells et al. (2007) grew *S. meliloti* biofilms in LB at 28°C with shaking whereas Russo et al. (2006) analyzed static biofilms. For our biofilm experiments, we carefully remove the culture medium once a day by aspiration and replace it with 100 μL of fresh modified RDM.

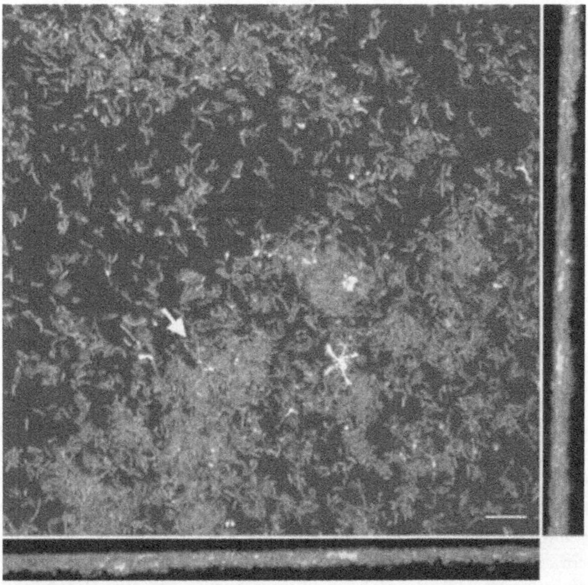

Fig. 4 Confocal scanning microscopic views (*top* and *side*) of a 72-h-old biofilm of wild-type, green-fluorescent protein-expressing *Sinorhizobium meliloti* strain RCR2011. The *arrow* points to a ridgelike structure in the biofilm as seen from the top. Bar, 10 μm

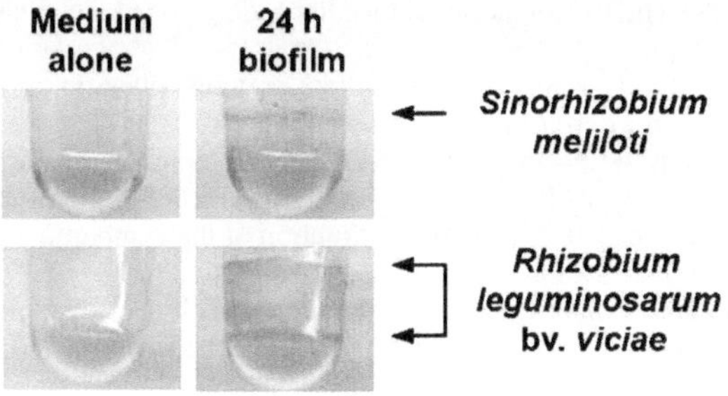

Fig. 5 Side view of microtiter plate wells from a 96-well plate showing the medium controls (*left*) and two biofilms (*right*) 24 h after the start of the experiment of either Rm1021 (*top*) or *R. leguminosarum* bv. *viciae* (*bottom*). The Rm1021 wells have been stained with crystal violet and the *R. leguminosarum* bv. *viciae* wells with safranin (reprinted with permission, from Fujishige et al. (2006a))

At the end of the experiment, the OD_{595} is read in a Bio Rad (Richmond, CA) Microtiter Plate reader (model no. 680) to ascertain the growth rates of the bacteria. Then, the medium is removed by aspiration and the biofilms are stained with 110 μL of 0.3% filtered aqueous crystal violet (this is a change in concentration from our previously published protocol) or 0.1% safranin (Fig. 5) for 10 min. After staining, the biofilms are washed three times with water to remove any residual cell and excess dye; this must be done carefully, otherwise excess dye gets into the wells and skews the OD_{570} (for crystal violet) or OD_{490} (for safranin) reading. Before reading the OD, the stained biofilms are allowed to dry, and then the dye is solubilized in 80% ethanol and 20% acetone. By careful rinsing and pipeting, we have been able to reduce the human-induced variation from well to well significantly. Microsoft Excel is used to calculate the average and standard deviation of the strains tested, the latter value giving us an idea of biological variation. Standard errors of the mean can also be used to show differences in treatments.

References

Albareda M, Dardanelli MS, Sousa C, Megías M, Temprano F, Rodríguez-Navarro DN (2006) Factors affecting the attachment of rhizospheric bacteria to bean and soybean roots. *FEMS Microbiol Lett* 259:67–73

Ames P, Bergman K (1981) Competitive advantage provided by bacterial motility in the formation of nodules by *Rhizobium meliloti*. *J Bacteriol* 148:728–729

Ausmees N, Jacobsson K, Lindberg M (2001) A unipolarly located, cell-surface-associated agglutinin, RapA, belongs to a family of *Rhizobium* adhering proteins (Rap) in *Rhizobium leguminosarum* bv. *trifolii*. *Microbiology* 147:549–559

Ausmees N, Jonsson H, Höglund S, Ljunggren H, Lindberg M (1999) Structural and putative regulatory genes involved in cellulose synthesis in *Rhizobium leguminosarum* bv. *trifolii*. *Microbiology* 145:1253–1262

Becker A, Pühler A (1998) Production of exopolysaccharides. In: Spaink HP, Kondorosi A, Hooykaas PJJ (eds) The Rhizobiaceae: molecular biology of model plant-associated bacteria. Kluwer, Dordrecht, pp 97–118

Bernier F, Berna A (2001) Germins and germin-like proteins: plant do-all proteins. But what do they do exactly? *Plant Physiol Biochem* 39:545–554

Breedveld MW, Miller KJ (1994) Cyclic beta-glucans of members of the family *Rhizobiaceae*. *Microbiol Mol Biol Rev* 58:145–161

Brewin NJ, Wood EA, Young JPW (1983) Contribution of the symbiotic plasmid to the competitiveness of *Rhizobium leguminosarum*. *J Gen Microbiol* 129:2973–2977

Campbell GR, Reuhs BL, Walker GC (2002) Chronic intracellular infection of alfalfa nodules by *Sinorhizobium meliloti* requires correct lipopolysaccharide core. *Proc Natl Acad Sci USA* 99:3938–3943

Cannon RE, Anderson SM (1991) Biogenesis of bacterial cellulose. *Crit Rev Microbiol* 17:435–447

Carpentier B, Cerf O (1993) Biofilms and their consequences, with particular reference to hygiene in the food industry. *J Appl Bacteriol* 75:499–511

Cheng H-P, Walker GC (1998) Succinoglycan is required for initiation and elongation of infection threads during nodulation of alfalfa by *Rhizobium meliloti*. *J Bacteriol* 180:5183–5191

Costerton JW, Lewandowski Z, Caldwell DE, Korber DR, Lappin-Scott HM (1995) Microbial biofilms. *Annu Rev Microbiol* 49:711–745

Danese PN, Pratt LA, Kolter R (2000) Exopolysaccharide production is required for development of *Escherichia coli* K-12 biofilm architecture. *J Bacteriol* 182:3593–3596

Danhorn T, Fuqua C (2007) Biofilm formation by plant-associated bacteria. *Annu Rev Microbiol* 61:401–422

Dardanelli M, Angelini J, Fabra A (2003) A calcium-dependent bacterial surface protein is involved in the attachment of rhizobia to peanut roots. *Can J Microbiol* 49:399–405

Dazzo FB, Truchet GL, Sherwood JE, Hrabak EM, Abe M, Pankratz SH (1984) Specific phases of root hair attachment in the *Rhizobium trifolii*-clover symbiosis. *Appl Environ Microbiol* 48:1140–1150

Díaz CL, Melchers LS, Hooykaas PJJ, Lugtenberg BJJ, Kijne JW (1989) Root lectin as a determinant of host plant specificity in the *Rhizobium*-legume symbiosis. *Nature* 338:579–581

Dörr J, Hurek T, Reinhold-Hurek B (1998) Type IV pili are involved in plant-microbe and fungus-microbe interactions. *Mol Microbiol* 30:7–17

Douglas CJ, Halperin W, Nester EW (1982) *Agrobacterium tumefaciens.* mutants affected in attachment to plant cells. *J Bacteriol* 152:1265–1275

Dunne WM Jr (2002) Bacterial adhesion: seen any good biofilms lately. *Clin Microbiol Rev* 15:155–166

Dylan T, Helinski DR, Ditta GS (1990) Hypoosmotic adaptation in *Rhizobium meliloti* requires β-1 2-glucan. *J Bacteriol* 172:1400–1408

Entcheva-Dimitrov P, Spormann AM (2004) Dynamics and control of biofilms of the oligotrophic bacterium *Caulobacter crescentus. J Bacteriol* 186:8254–8266

Erkel C, Kube M, Reinhardt R, Liesack W (2006) Genome of rice cluster I archaea – the key methane producers in the rice rhizosphere. *Science* 313:370–372

Ferguson GP, Roop RM II, Walker GC (2002) Deficiency of a *Sinorhizobium meliloti bacA* mutant in alfalfa symbiosis correlates with alteration of the cell envelope. *J Bacteriol* 184:5625–5632

Finan TM, Hirsch AM, Leigh JA, Johansen E, Kuldau GA, Deegan S, Walker GC, Signer ER (1985) Symbiotic mutants of *Rhizobium meliloti* that uncouple plant from bacterial differentiation. *Cell* 40:869–877

Finan TM, Gough C, Truchet G (1995) Similarity between the *Rhizobium meliloti fliP* gene and pathogenicity-associated genes from animal and plant pathogens. *Gene* 152:65–67

Foster RC, Rovira AD (1978) The ultrastructure of the rhizosphere of *Trifolium subterraneum* L. [Fig. 1] In: Loutit MW, Miles JAR (eds) Microbial ecology. Springer-Verlag, Berlin

Foster RC, Rovira AD, Cock TW (1983) Ultrastructure of the root-soil interface. St. Paul, The American Phytopathological Society

Fujishige NA (2005) Molecular analysis of biofilm formation by *Rhizobium* species. Ph.D. thesis, University of California-Los Angeles

Fujishige NA, Kapadia NN, De Hoff PL, Hirsch AM (2006a) Investigations of *Rhizobium* biofilm formation. *FEMS Microbiol Ecol* 56:195–205

Fujishige NA, Rinaudi L, Giordano W, Hirsch AM (2006b) Superficial liaisons: colonization of roots and abiotic surfaces by rhizobia. In: Sánchez F, Quinto C, López-Lara IM, Geiger O (ed) Biology of plant-microbe interactions, vol 5. Proceedings of the 12th international congress on molecular plant-microbe interactions. St. Paul, ISMPMI, pp 292–299

Fujishige NA, Kapadia NN, Hirsch AM (2006c) A feeling for the microorganism: Structure on a small scale. Biofilms on plant roots. *Bot J Linn Soc* 150:79–88

Fujishige NA, Lum MR, De Hoff PL, Whitelegge JP, Faull KF, Hirsch AM (2008) *Rhizobium* common *nod* genes are required for biofilm formation. *Mol Microbiol* 67:504–515

Götz F (2002) *Staphylococcus* and biofilms. *Mol Microbiol* 43:1367–1378

Gros R, Jocteur-Monrozier L, Faivre P (2006) Does disturbance and restoration of alpine grassland soils affect the genetic structure and diversity of bacterial and N_2 -fixing populations? *Environ Microbiol* 8:1889–1901

Gucciardo S, Wisniewski J-P, Brewin NJ, Bornemann S (2007) A germin-like protein with superoxide dismutase activity in pea nodules with high protein sequence identity to a putative rhicadhesin receptor. *J Exp Bot* 58:1161–1171

Heilmann C, Schweitzer O, Gerke C, Vanittanakom N, Mack D, Götz F (1996) Molecular basis of intercellular adhesion in the biofilm-forming *Staphylococcus epidermidis. Mol Microbiol* 20:1083–1091

Hirsch AM (1999) Role of lectins (and rhizobial exopolysaccharides) in legume nodulation. *Curr Opin Plant Biol* 2:320–326

Ho S-C, Wang JL, Schindler M, Loh JT (1994) Carbohydrate binding activities of *Bradyrhizobium japonicum.* III. Lectin expression, bacterial binding, and nodulation efficiency. *Plant J* 5:873–884

Kannenberg EL, Reuhs BL, Forsberg LS, Carlson RW (1998) Lipopolysaccharides and K-antigens: their structures, biosynthesis and functions. In: Spaink HP, Kondorosi A, Hooykaas PJJ (ed) The Rhizobiaceae: molecular biology of model plant-associated bacteria. Kluwer, Dordrecht, pp 119–154

Klausen M, Heydorn A, Ragas P, Lambertsen L, Aaes-Jørgensen A, Molin S, Tolker-Nielsen T (2003) Biofilm formation by *Pseudomonas aeruginosa* wild type, flagella, and type IV pili mutants. *Mol Microbiol* 48:1511–1524

Laus MC, van Brussel AAN, Kijne JW (2005) Role of cellulose fibrils and exopolysaccharides of *Rhizobium leguminosarum* in attachment to and infection of *Vicia sativa* root hairs. *Mol Plant-Microbe Interact* 18:533–538

Laus MC, Logman TJ, Lamers GE, van Brussel AAN, Carlson RW, Kijne JW (2006) A novel polar surface polysaccharide from *Rhizobium leguminosarum* binds host plant lectin. *Mol Microbiol* 59:1704–1713

Leigh JA, Signer ER, Walker GC (1985) Exopolysaccharide-deficient mutants of *Rhizobium meliloti* that form ineffective nodules. *Proc Natl Acad Sci USA* 82:6231–6235

Leriche V, Sibille P, Carpentier B (2000) Use of an enzyme-linked lectinsorbent assay to monitor the shift in polysaccharide composition in bacterial biofilms. *Appl Environ Microbiol* 66:1851–1856

Li G, Smith CS, Brun YV, Tang JX (2005) The elastic properties of the *Caulobacter crescentus* adhesive holdfast are dependent on oligomers of *N*-acetylglucosamine. *J Bacteriol* 187:257–265

Li Y, Hao G, Galvani CD, Meng De Y, La Fuente L, Hoch HC, Burr TJ (2007) Type I and type IV pili of *Xylella fastidiosa* affect twitching motility, biofilm formation and cell-cell aggregation. *Microbiology* 153:719–726

Limpens E, Bisseling T (2008) Nod factor signal transduction in the Rhizobium-Legume symbiosis. In: Emons AMC, Ketelaar T (eds) Root hairs: excellent tools for the study of plant molecular cell biology. Springer, Berlin Heidelberg New York. doi:10.1007/7089_2008_10

Limpens E, Franken C, Smit P, Willemse J, Bisseling T, Geurts R (2003) LysM domain receptor kinases regulating rhizobial Nod factor-induced infection. *Science* 302:630–633

Loh JT, Ho S-C, de Feijter AW, Wang JL, Schindler M (1993) Carbohydrate binding activities of *Bradyrhizobium japonicum* : unipolar localization of the lectin BJ38 on the bacterial cell surface. *Proc Natl Acad Sci USA* 90:3033–3037

Madsen EB, Madsen LH, Radutoiu S, Olbryt M, Rakwalska M, Szczyglowski K, Sato S, Kaneko T, Tabata S, Sandal N, Stougaard J (2003) A receptor kinase gene of the LysM type is involved in legume perception of rhizobial signals. *Nature* 425:637–640

Matsukawa M, Greenberg EP (2004) Putative exopolysaccharidesynthesis genes influence *Pseudomonas aeruginosa* biofilm development. *J Bacteriol* 186:4449–4456

Matthysse AG, Kijne JW (1998) Attachment of Rhizobiaceae to plant cells. In: Spaink HP, Kondorosi A, Hooykaas PJJ (ed) The Rhizobiaceae: molecular biology of model plant-associated bacteria. Kluwer, Dordrecht, pp. 235–249

Matthysse AG, McMahan S (1998) Root colonization by *Agrobacterium tumefaciens* is reduced in *cel*, *attB*, *attD*, and *attR* mutants. *Appl Environ Microbiol* 64:2341–2345

Matthysse AG, Holmes KV, Gurlitz RHG (1981) Elaboration of cellulose fibrils by *Agrobacterium tumefaciens* during attachment to carrot cells. *J Bacteriol* 145:583–595

Matthysse AG, Marry M, Krall L, Kaye M, Ramey BE, Fuqua C, White AR (2005) The effect of cellulose overproduction on binding and biofilm formation on roots by *Agrobacterium tumefaciens. Mol Plant-Microbe Interact* 18:1002–1010

Merker RI, Smit J (1988) Characterization of the adhesive holdfast of marine and freshwater caulobacters. *Appl Environ Microbiol* 54:2078–2085

Michiels K, Croes C, Vanderleyden J (1991) Two different modes of attachment of *Azospirillum brasilense* S7 to wheat roots. *J Gen Microbiol* 137:2241–2246

Mort AJ, Bauer WD (1980) Composition of the capsular and extracellular polysaccharides of *Rhizobium japonicum*: changes with culture age and correlations with binding of soybean seed lectin to the bacteria. *Plant Physiol* 66:158–163

Nagpal P, Khanuja SP, Stanfield SW (1992) Suppression of the *ndv* mutant phenotype of *Rhizobium meliloti* by cloned *exo* genes. *Mol Microbiol* 6:479–488

Ong CJ, Wong MLY, Smit J (1990) Attachment of the adhesive holdfast organelle to the cellular stalk of *Caulobacter crescentus*. *J Bacteriol* 172:1448–1456

Oresnik IJ, Liu SL, Yost CK, Hynes MF (2000) Megaplasmid pRme2011a of *Sinorhizobium meliloti* is not required for viability. *J Bacteriol* 182:3582–3586

O'Toole GA, Kolter R (1998a) Flagellar and twitching motility are necessary for *Pseudomonas aeruginosa* biofilm development. *Mol Microbiol* 30:295–304

O'Toole GA, Kolter R (1998b) Initiation of biofilm formation in *Pseudomonas fluorescens* WCS365 proceeds via multiple, convergent signaling pathways: a genetic analysis. *Mol Microbiol* 28:449–461

O'Toole GA, Pratt LA, Watnick PI, Newman DK, Weaver VB, Kolter R (1999) Genetic approaches to study of biofilms. *Methods Enzymol* 310:91–109

Pratt LA, Kolter R (1998) Genetic analysis of *Escherichia coli* biofilm formation: roles of flagella, motility, chemotaxis and type I pili. *Mol Microbiol* 30:285–293

Radutoiu S, Madsen LH, Madsen EB, Felle HH, Umehara Y, Grønlund M, Sato S, Nakamura Y, Tabata S, Sandal N, Stougaard J (2003) Plant recognition of symbiotic bacteria requires two LysM receptor-like kinases. *Nature* 425:585–592.

Ramey BE, Koutsoudis M, von Bodman SB, Fuqua C (2004) Biofilm formation in plant-microbe associations. *Curr Opin Microbiol* 7:602–609

Reuber TL, Walker GC (1993) Biosynthesis of succinoglycan, a symbiotically important exopolysaccharide of *Rhizobium meliloti. Cell* 74:269–280

Rinaudi L, Fujishige NA, Hirsch AM, Banchio E, Zorreguieta A, Giordano W (2006) Effects of nutritional and environmental conditions on *Sinorhizobium meliloti* biofilm formation. *Res Microbiol* 157:867–875

Rodríguez-Navarro DN, Dardanelli MS, Ruíz-Saínz J (2007) Attachment of bacteria to the roots of higher plants. *FEMS Microbiol Lett* 272:127–136

Roset MS, Ciocchini AE, Ugalde RA, Iñón de Iannino N (2004) Molecular cloning and characterization of *cgt*, the *Brucella abortus* cyclic beta-1, 2-glucan transporter gene, and its role in virulence. *Infect Immun* 72:2263–2271

Russo DM, Williams A, Edwards A, Posadas DM, Finnie C, Dankert M, Downie JA, Zorreguieta A (2006) Proteins exported via the PrsD-PrsE type I secretion system and the acidic exopolysaccharide are involved in biofilm formation by *Rhizobium leguminosarum. J Bacteriol* 188:4474–4486

Smit G, Kijne JW, Lugtenberg BJJ (1987) Involvement of both cellulose fibrilsand a Ca^{2+}-dependent adhesin in the attachment of *Rhizobium leguminosarum* to pea root hair tips. *J Bacteriol* 169:4294–4301

Smit G, Logman TJ, Boerrigter ME, Kijne JW, Lugtenberg BJ (1989) Purification and partial characterization of the *Rhizobium leguminosarum* biovar *viciae* Ca^{2+}-dependent adhesin, which mediates the first step in attachment of cells of the family *Rhizobiaceae* to plant root hair tips. *J Bacteriol* 171:4054–4062

Smit G, Swart S, Lugtenberg BJJ, Kijne JW (1992) Molecular mechanisms of attachment of *Rhizobium* bacteria to plant roots. *Mol Microbiol* 6:2897–2903

Swart S, Logman TJJ, Smit G, Lugtenberg BJJ, Kijne JW (1994) Purification and partial characterization of a glycoprotein from pea (*Pisum sativum*) with receptor activity for rhicadhesin, an attachment protein of Rhizobiaceae. *Plant Mol Biol* 24:171–183

Vande Broek A, Vanderleyden J (1995) The role of bacterial motility, chemotaxis, and attachment in bacteria-plant interactions. *Mol Plant-Microbe Interact* 8:800–810

van Rhijn P, Goldberg RB, Hirsch AM (1998) *Lotus corniculatus* nodulation specificity is changed by the presence of a soybean lectin gene. *Plant Cell* 10:1233–1249 van Rhijn P, Fujishige NA, Lim P-O, Hirsch AM (2001) Sugar-binding activity of pea (*Pisum sativum*) lectin enhances heterologous infection of transgenic alfalfa plants by *Rhizobium leguminosarum* biovar *viciae*. *Plant Physiol* 126:133–144

Vesper SJ, Bauer WD (1986) Role of pili (fimbriae) in attachment of *Bradyrhizobium japonicum* to soybean roots. *Appl Environ Microbiol* 52:134–141

Vesper SJ, Malik NSA, Bauer WD (1987) Transposon mutants of *Bradyrhizobium japonicum* altered in attachment to host roots. *Appl Environ Microbiol* 53:1959–1961

Wagner M, Horn M, Daims H (2003) Fluorescence *in situ* hybridisation for the identification and characterisation of prokaryotes. *Curr Opin Microbiol* 6:302–309

Wells DH, Chen EJ, Fisher RF, Long SR (2007) ExoR is genetically coupled to the ExoS-ChvI two-component system and located in the periplasm of *Sinorhizobium meliloti*. *Mol Microbiol* 64:647–664

Whiteley M, Bangera MG, Bumgarner RE, Parsek MR, Teitzel GM, Lory S, Greenberg EP (2001) Gene expression in *Pseudomonas aeruginosa* biofilms. *Nature* 413:860–864

Yang C, Signer ER, Hirsch AM (1992) Nodules initiated by *Rhizobium meliloti* exopolysaccharide mutants lack a discrete, persistent nodule meristem. *Plant Physiol* 98:143–151

Yildiz FH, Schoolnik GK (1999) *Vibrio cholerae* O1 El Tor, identification of a gene cluster required for the rugose colony type, exopolysaccharide production, chlorine resistance, and biofilm formation. *Proc Natl Acad Sci USA* 96:4028–4033

Root Hair Colonization by Mycorrhizal Fungi

M. Novero, A. Genre, K. Szczyglowski, and P. Bonfante (✉)

Abstract Mycorrhizal fungi, i.e., the soil fungi that form mutualistic associations with many terrestrial plants, are provided by the host with carbon sources required to complete their life cycle, whereas they assist the plant in nutrient uptake from soil. Such acquisition is also considered to be one of the primary functions of root hairs. The aim of this chapter is to investigate the importance of root hairs in the establishment of mycorrhizal interactions, to verify whether plant (root hairs) and fungal (extraradical hyphae) structures work synergistically to provide efficient mineral nutrition. Evidence from morphological studies, where the mycorrhizal typologies have been compared, point to the direct involvement of root hairs in ecto- and arbuscular mycorrhizas (AM). Root hairs probably play a role during the first stages of ectomycorrhizal development, being sensitive to diffusible factors released by the symbiotic fungi and acting as a preferential anchorage site. During AM establishment, a variety of interactions are reported. In liverworts, AM fungi often penetrate the rhizoids, but the colonization process in higher plants does not usually involve root hairs. The analysis of mutant plants with impaired root hair development has not demonstrated any discernible impact on their mycorrhizal capacities. Confocal microscopy has recently provided important insight into understanding the plant responses upon encountering mycorrhizal fungi. Root hairs respond to the fungal presence with nuclear movements, although fungal penetration most often occurs through atrichoblasts or, occasionally, at the base of trichoblasts. Taken together, experimental evidence points to a strong difference in root hair involvement during AM and nodulation development. Nitrogen-fixing bacteria may have found in root hairs a specific anatomical niche, often neglected by AM fungi, to achieve tissue colonization.

Mineral nutrient acquisition from soil is considered one of the primary functions of root hairs, together with the anchorage of the plant (Gilroy and Jones 2000). However, these crucial structures are not alone in the acquisition of nutrients, since mycorrhizal fungi,

P. Bonfante
Dipartimento di Biologia Vegetale dell' Università and Istituto per la Protezione delle Piante-CNR, Viale Mattioli 25, Torino, 10125, Italy
p.bonfante@ipp.cnr.it

Plant Cell Monogr, doi:10.1007/7089_2008_12

i.e., the soil fungi that form mutualistic associations with many land plants, also assist their hosts in this way. The mycorrhizal symbiosis is in fact characterized by reciprocal nutrient exchanges between the symbiotic partners: while the fungus obtains photosynthetically derived carbon compounds, the plant receives mineral nutrients. The fungus receives up to 20% of the photoassimilated carbon allocated by the plant to the root (Smith and Read 1997). In exchange, the fungus improves the mineral supply to the plant (mainly phosphate) through the external mycelium, extending through and beyond the nutrient depletion area that surrounds the root (Jakobsen 1995). In fact, mineral nutrients such as phosphorus have very limited mobility in soil, and depletion zones – where the entire available nutrient has been scavenged – quickly appear around roots (Marschner 1995). To obtain phosphorus, plants have to extend their root surface area, and they are helped in this task by mycorrhizal fungal hyphae, which are thinner and more extensive than the root hairs themselves.

The biological meaning of the symbiotic association between the plant roots and some soil fungi, defined for the first time as "mycorrhiza" by Frank (1885), has become more and more apparent in recent years, especially since the corresponding research covers important aspects of ecology, evolutionary biology, genetics, and developmental biology.

Despite their enormous ecological relevance, knowledge of the cellular and molecular mechanisms that control the success of plant–fungal symbiotic associations is still limited, as highlighted by recent reviews and books (Martin et al. 2007; Gianinazzi-Pearson et al. 2007; Pühler and Strack 2007). However, the development of technological platforms in plant genomics is greatly facilitating the comprehensive identification of genes that are activated during mycorrhizal symbiosis. For these reasons, research on mycorrhizas has entered the mainstream of biology, since new tools to uncover symbiont communication and associated developmental mechanisms, diversity, and contributions of symbiotic partners to functioning of mycorrhizal associations are now available. All of these new aspects have been documented in the above-mentioned reviews, as well as in Paszkowski (2006), Bucher (2007), and Genre and Bonfante (2007).

The aim of this chapter is to investigate the importance of root hairs in the establishment of mycorrhizal interactions, and to verify whether plant (root hairs) and fungal (extraradical hyphae) structures work synergistically to provide efficient mineral uptake. For these reasons, a short summary of the main mycorrhizal typologies and morphological features is provided to address the question "what is the involvement of root hairs in mycorrhizal associations?"

1 A Short Overview of Mycorrhizas

Different kinds of mycorrhizas exist; they differ according to their anatomy and the fungi involved. Broadly speaking, mycorrhizas can be assigned to one of the following groups: ectomycorrhizas or endomycorrhizas, depending on whether the fungus colonizes the root intercellular spaces or develops inside plant cells (Fig. 1).

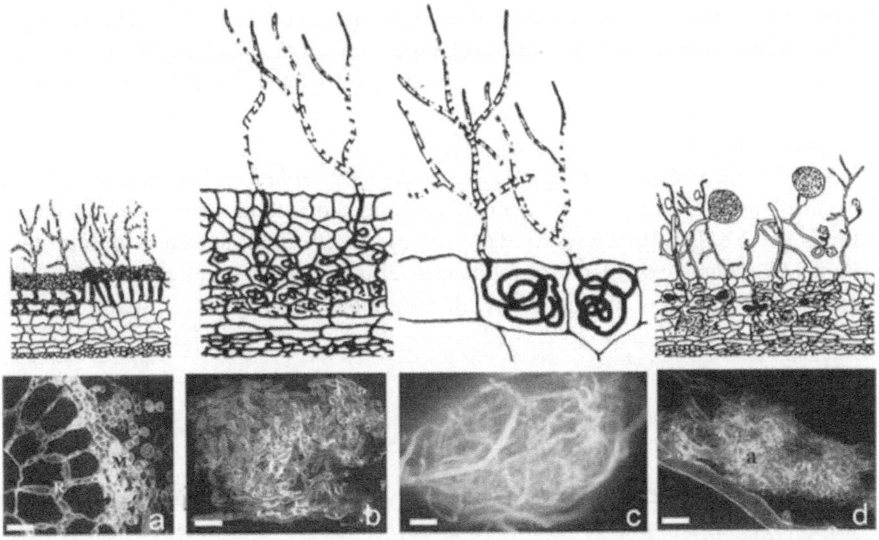

Fig. 1 *Upper part:* the different mycorrhizal types and the main colonization structures. From the left are ectomycorrhizas, arbuscular mycorrhizas, orchid mycorrhizas, and ericoid mycorrhizas. *Lower part:* the fungal mantle and the Hartig net in *Quercus* ectomycorrhiza (**a**); a coil produced by an ericoid fungus in a hair root of *Calluna vulgare* (**b**); a coil produced by an endophytic fungus inside an orchid root (**c**); an arbuscule in the cortical cell of a leek root (**d**). Bars correspond to 30 μm (**a**) and to 15 μm (**b–d**). (After Bonfante 2001)

1.1 Ectomycorrhizas

In this kind of association, the fungus does not enter the host cell. Two features are usually recognized (Fig. 1a): a mantle of fungal hyphae covering thin secondary roots, which gives them the characteristic shape similar to a finger of a glove and a hyphal network that develops between the root cells, called the Hartig net. The plants involved in ectomycorrhiza are usually trees (i.e., members of Fagaceae, Tiliaceae, Salicaceae, Pinaceae, etc.) and very rarely shrubs and herbaceous species (Peterson et al. 2004). Most ectomycorrhizal fungi belong to the Basidiomycete and Ascomycete classes.

As a result of the interaction between root cells and hyphae (see later), the root hairs and the sloughing root cap cells become incorporated in the mantle, and the root appears swollen and totally devoid of root hairs. This is an important feature that allows ectomycorrhizas to be easily identified. The mantle is probably involved in nutrient storage, since the carbohydrates acquired by the fungus from the plant cells may be stored temporarily in this compartment, mainly in the form of trehalose (Lopez et al. 2007). Hartig (1840) first described this network based on his observation on pine mycorrhizas. This hyphal network, known as Hartig net, starts to develop around the wall of the root epidermal and cortical cells. The Hartig net is the preferential site for nutrient exchange, and it is characterized by a complex pattern of hyphal branching, where linear development is replaced by a multi-branched growth. The main biological significance of such a network is to increase the surface area devoted to nutrient exchange.

Some ectomycorrhizal fungi (*Laccaria laccata*, *Tuber melanosporum*, *Paxillus involutus*) are currently the subject of genome sequencing projects (Voegele and Mendgen 2007, http://www.newphytologist.org/fungal-genomics/default.htm). This knowledge is expected to provide crucial information on how fungal genomes mediate the symbiotic interactions with host plants.

1.2 Endomycorrhizas

This kind of association groups together very diverse fungi (Glomero-, Asco-, and Basidiomycota) and plants (Brundrett 2002). Irrespective of this huge biodiversity, the fungus always enters into the epidermal and cortical cells of the host, where it develops hyphae or specialized structures such as arbuscules, coils, or vesicles. Such intracellular structures are always surrounded by an invagination of the host plasma membrane (Bonfante 2001). Unlike ectomycorrhizas, morphological changes in the gross anatomy of the colonized roots are not easy to note.

Different types of endomycorrhizas are currently listed, depending on the type of host–fungus association (Peterson et al. 2004). In the context of this chapter, Ericoid and Orchid mycorrhizas are briefly mentioned, while more attention is devoted to arbuscular mycorrhizas.

1.2.1 Ericoid Mycorrhizas

Ericoid mycorrhizas play a relevant role in several ecosystems characterized by nutrient-poor soils and low temperatures. Fungal symbionts, belonging to a huge number of taxa (Martin et al. 2007), help Ericales, a large and diverse group of dicotyledons represented by many life forms, including trees, bushes, lianas, and herbaceous plants, to acquire both phosphate and nitrogen. Thanks to the secretion of acid phosphatases, ericoid fungi enable their hosts to access phosphate from organic and condensed phosphates, and nitrogen from diverse sources. In certain environments these fungi also have the ability to protect the plant from toxic levels of heavy metals, such as copper and zinc.

A peculiar feature of the Ericales that form this type of mycorrhiza is the formation of specialized lateral roots, called hair roots (Fig. 1b). These are narrow (about 100 μm in diameter), short, with a strikingly simple anatomy, and never undergo secondary growth. In both Ericaceae and Epacridaceae, transverse sections of such roots show a central vascular cylinder surrounded by one or two layers of cortical cells and an epidermal layer, which is usually the preferential niche for mycorrhizal fungi (Bonfante and Perotto 1995). Interestingly, such hair roots do not possess root hairs, since, possibly due to their miniaturized structure, the roots themselves play a direct role in nutrient absorption, along with mycorrhizal fungi.

1.2.2 Orchid Mycorrhizas

Even though orchid mycorrhizas are only found in the Orchidaceae, interest in this symbiosis is becoming more and more widespread because symbiosis is essential for both seed germination and seedling establishment. The fungal partners are Basidio- and Ascomycota, which form hyphal coils (pelotons) usually within cortical cells (Fig. 1c). Interestingly, an increasing number of reports have shown how these fungi may act as a functional bridge between ectomycorrhizal plants and orchids that have low photosynthetic activity. The fungus can transfer the sugars acquired by the photosynthetic tree to the orchid, which is therefore fully dependent on the fungal partner for its carbohydrate nutrition. Such ecological and functional studies have led to a new research field, called mycoheterotrophy, and confirmed the ecological importance of mycorrhizal fungi in nutrient cycles (Selosse et al. 2006). Less attention has been paid to the morphology of such associations. However, based on observations dating back to the 1980s (Scannerini and Bonfante 1983) and more recent samplings (Selosse et al. 2004), large-diameter orchid roots do not seem to possess root hairs. This suggests that, in these particular roots, mycorrhizal fungi might represent a functional replacement in nutrient and water absorption, for the missing root hairs.

1.2.3 Arbuscular Mycorrhizas

One noteworthy aspect of arbuscular mycorrhizas (AM) is their ecological success. They are present in the roots of 80% of vascular plants (both angiosperms and gymnosperms), and also in ferns and bryophytes. It is quite exceptional that one small group of fungi, Glomeromycota (Schüßler et al. 2001), can colonize such diverse plant species irrespective of tissue ploidy, as both gametophytic and sporophytic tissues are involved in the symbiosis. Several features are common to this extremely successful group of mycorrhizas:

1. Land plants and AM fungi share a long coevolutionary history (Rémy et al. 1994); the symbiosis gives benefits to both partners (Smith and Read 1997).
2. Reciprocal nutrient exchange requires close physical contact between the partner cells (Genre and Bonfante 2005).
3. Functional symbiosis requires profound readjustments in plant and fungal cells (Fig. 1d).

Glomeromycota are highly dependent on their hosts. Hyphae germinating from their large asexual spores grow for only a few days in the absence of the plant. Upon the recognition of the host plant, these presymbiotic hyphae develop infection units that colonize the root epidermis and cortex. The sequence of events leading to AM symbiosis is largely conserved among different combinations of fungal and plant species, suggesting the presence of common molecular and genetic determinants across different plant taxa. Broadly speaking, three stages can be identified during the colonization process: (1) the presymbiotic phase; (2) the plant–fungal contact stage, followed by penetration; and (3) the intraradical fungal proliferation stage, leading to the development of arbuscules, which are the preferential sites of nutrient transfer (Paszkowski 2006). However, since root colonization is an asynchronous process, all of these steps can occur concurrently.

During the presymbiotic phase, AM spores germinate spontaneously (asymbiotic stage) in the soil, developing a germ tube that is fed by the carbon stored in the spore (Bianciotto et al. 1995). In the absence of a host plant, hyphal growth stops before the complete depletion of carbon resources (Bago et al. 2000). On the other hand, when a host plant is present, fungal growth is enhanced and hyphal branching is induced, thereby increasing the probability of contact between the two partners. A recent breakthrough discovery demonstrated that plants produce a "branching factor," which was isolated and identified as a strigolactone (Akiyama et al. 2005). The chemical nature of AM fungal signals (myc factor) remains elusive, even though indirect evidence of their existence has been presented (Kosuta et al. 2003; Olah et al. 2005; Navazio et al. 2007). Once the hypha contacts the root epidermis, appressorium-like hyphopodia (i.e., large adhesion structures) develop on its surface (Genre and Bonfante 2007). Following this event, the host plant responds by forming a prepenetration apparatus (Genre et al. 2005), which prepares the epidermal cell for fungal entry by assembling a tunnel-like structure. Fungal hyphae enter and cross the epidermal cell using the preformed tunnel and avoiding a direct contact between the fungal wall and the host cytoplasm, which remain separated by a thin apoplastic compartment called the interface (Bonfante 2001).

Once the fungus has passed the epidermal layer, it grows inter- and intracellularly within the root. During this stage, some fungi (e.g., *Glomus* species) may differentiate particular structures, called vesicles, which occur within exodermal and cortical cells. These circular structures completely fill up the cell lumen and are believed to act as storage sites. Only when the fungus has penetrated a cortical cell does a specialized branching process initiate, leading to the formation of arbuscules. These highly branched structures, formed by recursive dichotomous branching of a hyphal trunk (Bonfante 1984), are the key elements of the symbiosis, since they are considered to be the main site of nutrient exchanges. The development of such a massive intracellular structure changes the architecture of the host cell to a great extent: the nucleus moves from the periphery to the center of the cell, the vacuole is fragmented, plastids change their morphology, and a new apoplastic space, based on membrane proliferation, is built around all the arbuscule branches (Bonfante and Perotto 1995). The construction of this interface compartment, which mediates reciprocal nutrient exchange between the symbiont cells, results from an intense reorganization of the plant cell components and metabolic activity, ranging from specific gene activation (Gianinazzi-Pearson and Brechenmacher 2004) to localized cell wall and membrane deposition (Balestrini and Bonfante 2005), cytoskeleton remodelling (Genre and Bonfante 1998), organelle mobilization (Lohse et al. 2005) and phosphate transport (Balestrini et al. 2007). Arbuscules are ephemeral structures with a life cycle of only a few days (Toth and Miller 1984): after 4–5 days, the arbuscule branches collapse, and the host cell gradually regains its original preinfection form.

To conclude this schematic overview on mycorrhizal types, it should be mentioned that root hairs are apparently not involved in ericoid mycorrhizas, since these tiny hair roots never produce root hairs; nor are they involved in the mycorrhizas of the large, hairless, orchid roots. Hence, only ecto- and arbuscular mycorrhizas are considered in detail in the rest of this chapter.

2 The Role of Root Hairs in Mycorrhizal Establishment

2.1 Root Hairs and Ectomycorrhizal Fungi

Although the transcriptome profile of developing ectomycorrhizas has been thoroughly investigated in many plant/fungus systems (Le Quere et al. 2005), morphological information on the early events is more limited. The signaling processes between ectomycorrhizal fungi and plants are largely unknown, with one noticeable exception: the case of *Eucalyptus globulus* and *Pisolithus tinctorius*. On the one hand, rutin (a flavanol present in *Eucalyptus* root exudates) stimulates fungal growth (Lagrange et al. 2001), while on the other, fungal exudates such as hypaphorine (an indole alkaloid) and auxins trigger morphological changes in the root system, inhibiting or stimulating root hair growth, respectively (Béguiristain and Lapeyrie 1997). Hypaphorine is able to stop root hair elongation changing the actin cytoskeleton of growing hairs with a shift from fine F-actin to F-actin bundles in the subapical region. This configuration differs from that observed during the developmental growth arrest of root hairs, where F-actin bundles extend into the apex and surround completely the central vacuole (Dauphin et al. 2006).

Taking advantage of an in vitro system, *Tuber borchii* hyphae were observed when contacting the root cap or the epidermal cells of *Tilia platyphyllos* lateral roots (Sisti et al. 2003). This experimental system revealed changes in fungal morphology during the interaction: the hyphae branch, increase their diameter, and contact the host cells. At this stage, the root hairs represent a preferential adhesion point. Hyphae surround them, and abundant extracellular material becomes detectable at the contact point (Fig. 2a). When seen under the electron microscope (Fig. 2b, c), this material is found to be of a fibrillar polysaccharidic nature. After contact, the fungus develops between the epidermal cells and in the first layers of the cortex where it produces the Hartig net. Finally, the mantle develops on the root surface.

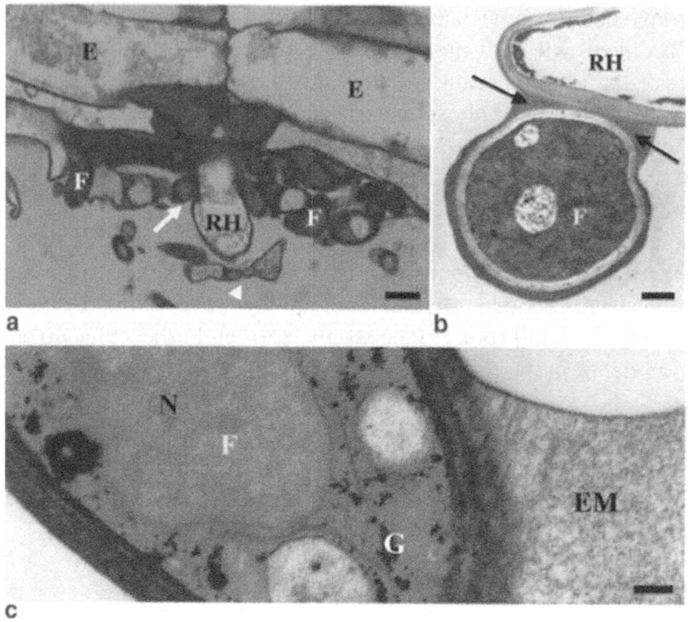

Fig. 2 Root hairs during the development of an ectomycorrhiza: hyphae of *Tuber borchii* grow in the presence of *Tilia platyphyllos* roots. (**a**) Light micrographs of a longitudinal section from a colonized root after 30 days. Hyphae closely surround a root hair, both at the base (*arrow*) and at the very tip (*arrowhead*). *E* epidermal cell, *F* hypha, *RH* root hair. Bar corresponds to 8.8 μm. (**b**) Under electron microscope, abundant extracellular material is seen at the contact point (*arrows*) between a root hair and a hypha. The section is labelled with an antibody against β1–3 glucans. *RH* root hair, *F* hypha. Bar corresponds to 0.5 μm. (**c**) At higher magnification the extracellular material shows a fibrillar structure that reacts to the silver reaction performed according to the method of Thiéry (1967): carbohydrates were oxidized with periodic acid, and visualized by incubation in thiocarbohydrazide and silver proteinate. *EM* extracellular material, *F* hypha, *n* nucleus, *g* glycogen. Bar corresponds to 0.3 μm. See the Appendix for methodological detalis

As a consequence, root hairs and root cap cells are incorporated in the mantle and the roots appear swollen and totally devoid of root hairs.

On the basis of these preliminary observations, it would seem that the root hairs play a role during the first stages of ectomycorrhizal development, being sensitive to diffusible factors released by the symbiotic fungi and acting as a preferential anchorage site. However, a better complementation between molecular and morphological data is required before final conclusions can be drawn.

2.2 *Root Hairs and Arbuscular Mycorrhizas*

The first contact between an AM fungus and its host plant occurs at the root surface: hyphae adhere to epidermal cells and develop a specialized swollen structure called appressorium or *hyphopodium.* Penetrating hyphae originate from hyphopodial branches. Even though direct evidence is not available, some morphological and biochemical data suggest that penetrating hyphae may produce hydrolases or release small molecules that cause the epidermal wall to relax and allow hyphal penetration (Garcia-Romera et al. 1991; Peretto et al. 1995). AM fungi usually enter the root tissue through epidermal cells, crossing their radial or tangential walls (Bonfante et al. 2000; Demchenko et al. 2004; Genre et al. 2005). Penetration through root hairs (Guinel and Hirsch 2000) without the formation of a clear appressorium-like structure is less frequent.

By contrast, nonvascular plants seem to follow the "root hair pathway" to start the colonization process. In natural conditions, liverwort gametophytes are usually colonized by Glomeromycota, which often use the rhizoids as the preferential access to the plant. Recent observations on Marchantiopsida and Metzgeriidae indicate that the fungus penetrated the rhizoids at any point, forming large intracellular hyphae running in both directions (Ligrone et al. 2007).

As shown earlier, the main role of AM fungi in symbiosis is to acquire mineral nutrients by exploring the soil. Their hyphae are more extended and thinner than roots or root hairs: for this reason, they are more efficient in reaching soil interstices. However, not all land plants develop AM symbioses. Members of Chenopodiaceae, Amaranthaceae, Caryophyllaceae, Polygonaceae, Brassicaceae, Scrophulariaceae, Commelinaceae, Juncaceae, and Cyperaceae do not have mycorrhizas (Brundrett 1991). Their root apparatus is probably very efficient in exploring the soil, thus providing the plant with mineral nutrients while saving the resources necessary to maintain a symbiotic fungus. Generally speaking, plants do not usually support high levels of mycorrhizal colonization and the extent of root apparatus is tightly controlled, mainly because of the high metabolic cost that this would require (Brundrett 1991). These considerations lead to the question concerning whether root architecture is directly linked to plant dependence on mycorrhizas.

3 Root Anatomy, Root Hairs, and Mycorrhizas: The Baylis Hypothesis

During one of the first symposia on endomycorrhizas organised by Francis Sanders and Bernard Tinker in Leeds in 1975, Geoffrey T.S. Baylis presented his considerations on the evolutionary trends shown by root architecture and, for the first time, used the definition of "magnolioid root" to describe the little-branched, hairless roots of Magnoliales. He hypothesized the existence of a link between root architecture and mycorrhizal dependency, claiming that root hair length and abundance are good guides to the degree of dependence a plant will have on mycorrhizas (Baylis 1975).

According to many authors (Pirozynski and Malloch 1975; Redecker et al. 2000), Glomeromycota coevolved with land plants, playing a crucial role in land colonization. The earliest land plants did not possess a developed root system and the acquisition of soil nutrients by their unbranched root axes must have been very challenging, as these plants progressively moved from an aquatic environment to the land. Under such conditions, their association with soil-heterotrophic fungi was evolutionarily rewarding. Mathematical models of nutrient uptake from the soil indicate that root growth and extension into unexploited volumes of soil are very important to acquire nutrients that diffuse slowly in the soil, such as phosphorus (Clarkson 1985). As a direct consequence, root architecture is determined not only by genetically inherited developmental programs, but also by external biotic and abiotic stimuli (Zobel 1996).

The architecture of a root apparatus is determined by the root pattern (root length, number of secondary and higher order roots, diameter, etc.) and by the root hair density. Root hairs occur in most vascular plants. Owing to their tubular shape (Gilroy and Jones 2000), the formation of root hairs considerably increases the root surface with relatively little dry matter investment. Root hairs make up 70–90% of the total root surface area (Bates and Lynch 1996) and play a dominant role in a number of root functions. They are among the first cells to come into contact with the soil solution. Several studies have reported the importance of these cells in the acquisition of nutrients (Gilroy and Jones 2000). Root hair length and density increase in response to iron (Fe) and phosphorus (P) deficiencies, enhancing the efficiency of inorganic orthophosphate (Pi) acquisition by the plant (Ma et al. 2001).

Root hairs abound in rape (*Brassica napus*), spinach (*Spinacia oleracea*), and tomato roots at low P concentrations (<10 mM), but they are absent or rudimentary at high P (>100 mM) (Föhse and Jungk 1983). When the Pi or K^+ uptake rates of different plant species are compared, a positive correlation can be seen between the uptake rate per unit root length and the volume of the root hair cylinder (Marschner 1995). A study on hairless-root mutants has revealed the important role played by root hairs in Pi uptake from the soil solution; in a low-P environment, hairs are crucial for P acquisition and plant survival, while they might be dispensable under high-P conditions (Bates and Lynch 2000). A study including barley genotypes with different root hair lengths showed that long-hair genotypes are better adapted to low-P soils and express high yield potentials in both low- and high-P soils (Gahoonia and Nielsen 2004). *Citrus* roots have short and poorly distributed root hairs on large-diameter roots (Menge et al. 1978). They therefore offer another good example of magnolioid roots. Studies performed in pot conditions suggest that there are differences in mycorrhizal dependency among rootstocks. In general, lower mycorrhizal dependency has been attributed to thinner rootstocks (Graham and Syvertsen 1985).

On the basis of such a rationale, plants with large-diameter roots and a few root hairs such as *Citrus* are expected to be strongly mycorrhiza-dependent. On the other hand, plants with thinner roots and a huge number of root hairs, such as Graminaceae, should be less dependent on symbiosis. Over the last 30 years, many field studies have provided support to the Baylis hypotheses. However, in spite of the interest of such pioneering observations, the relationship has not always been confirmed. For example, in laboratory conditions, tomato and carrot hairy roots have numerous and very long root hairs, but they are quite easily colonized by AM fungi (Fig. 3).

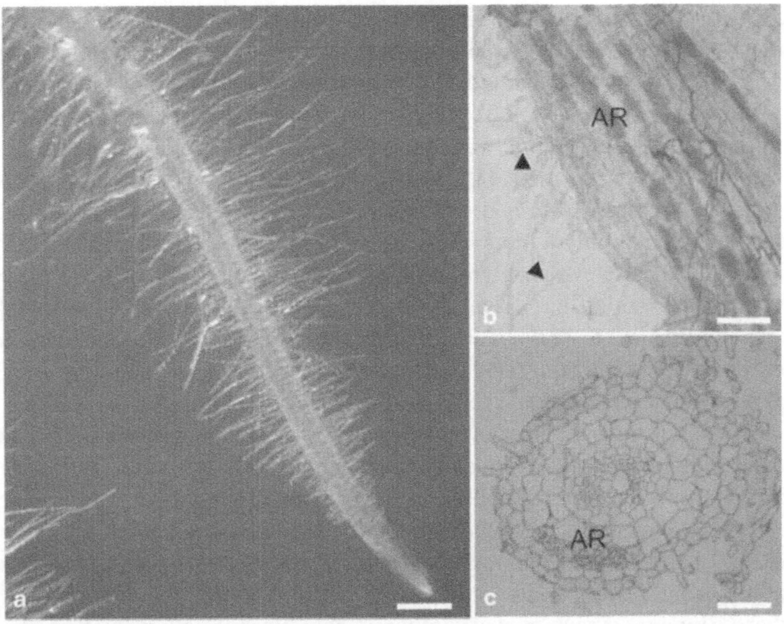

Fig. 3 Transformed tomato root in root organ culture. (**a**) A general view of a nonmycorrhizal root seen under a dissecting microscope shows a huge number of root hairs, suggesting that all the epidermal cells are trichoblasts. Bar corresponds to 500 μm. (**b**) Mycorrhizal root under light microscope. Extra- and intraradical hyphae as well as arbuscules (AR) are stained with cotton blue. Penetrations through the root hairs (*arrowheads*) are not seen. Bar corresponds to 100 μm. (**c**) Transverse semithin section of a mycorrhizal root after toluidine blue staining shows a comparable situation. Bar corresponds to 60 μm. See Appendix for methodological details

4 Lessons from the Mutants

One of the most efficient methods used to understand the colonization process in AM is to dissect it by using genetic approaches, which take advantage of mutant plants affected in their symbiotic capacities. Mutant studies have demonstrated the existence of a genetic control over root compatibility with symbiotic fungi in many legume and nonlegume plants (Duc et al. 1989; Barker et al. 1998; David-Schwartz et al. 2001; Kistner et al. 2005). Isolation of the corresponding genes from mutant backgrounds has been instrumental in the identification of plant gene functions that are essential for the first steps of symbiosis establishment (Oldroyd and Downie 2006). The cellular and molecular characterization of root interactions with AM fungi in mycorrhiza-defective mutants pinpoints a role for symbiosis-related plant genes in sensing and responding to fungal signals, giving clues as to how biologically active root or fungal factors may mediate cell functions linked to a successful symbiosis (Oldroyd and Downie 2006).

Among the legume mutants available so far, a class of nodulation mutants has been identified as hypernodulating when infected with rhizobia. This class includes the *Lotus japonicus har1-1* (hypernodulation aberrant root formation) mutant (Krusell et al. 2002), which also exhibits a hypermycorrhizal phenotype (as described in Solaiman et al. 2000). The line was obtained by EMS mutagenesis (Wopereis et al. 2000) and its characterization led the way to the identification of several *L. japonicus* mutants as suppressors (Murray et al. 2006b). This means that these mutations affect or reduce the plant ability to nodulate, compared to the *har1-1* parental line. After a screening of the suppressors, a group of *L. japonicus* mutants has been identified, with various aberrations in root hair development concomitant with a low nodulation phenotype (Karas et al. 2005). This group is composed of nine independent double mutant lines (e.g., *Ljrhl1/har1-1*). Their microscopical analysis led to the identification of four phenotypic classes: root hairless (*Ljrhl*), petite root hairs (*Ljprh*), short root hairs (*Ljsrh*), and variable root hairs (*Ljvrh*). Since all the root hair double mutant lines were derived from chemically mutagenized *L. japonicus har1-1/har1-1* homozygous mutant seeds, the observed aberrations in growth and development of the root hairs were considered to be the result of secondary mutations. This was confirmed by subsequent genetic analysis and by isolation of the corresponding single mutant lines (e.g. *Ljrhl1*) (Karas et al. 2005).

Other *L. japonicus* mutants with impaired root hair formation, such as, the slippery (*Ljslp*) mutant, have also been identified (Kawaguchi et al. 2002). These are low-nodulating mutants that almost completely lack root hairs, which were obtained from the wild-type ecotype *L. japonicus* Gifu by chemical mutagenesis using EMS (Kawaguchi et al. 2002). Thanks to the availability of such mutants, and to the abundance of root hairs in the wild-type *L. japonicus* (almost all epidermal cells of *L. japonicus* behave as trichoblasts; Karas et al. 2005), *L. japonicus* is currently considered a good model plant to understand the link between root hair abundance and mycorrhization dependency.

Ljrhl, *Ljsrh*, *Ljvrh*, and *Ljslp* mutants (used as single and/or double mutants, with the latter group carrying *har1-1* allele) were inoculated with *Gigaspora margarita* using the "Millipore sandwich method," which, according to Novero et al. (2002), allows controlled-sterile conditions for a bidimensional plant development. When the mycorrhization rate was evaluated 28 days postinoculation using the method by Trouvelot et al. (1986), no significant differences were found, compared with that in wild-type plants (Fig. 4 and Table 1). See Appendix for methodological details. In a parallel set of experiments, two petite root hair mutants (*Ljprh1-1/har1-1* and *Ljprh1-2/har1-1*) were in-pot inoculated with *Glomus intraradices* and the mycorrhization rate was evaluated using the magnified intersections method described by McGonigle et al. (1990). Under these conditions, a reduction in the rate of mycorrhization was seen when the mutants were compared to the *har1-1* parental line. The mutant *Ljprh1-2/har1-1* in fact exhibited fewer extraradical hyphae, hyphopodia, arbuscules, and vesicles than did the parent. In the case of the *Ljprh1-1/har1-1* mutant the reduced colonization was only apparent at the vesicle formation stage. This line had fewer extraradical hyphae, but the number of hyphopodia was unaffected. Therefore, although fewer extraradical hyphae were available to colonize the root, no reduction in

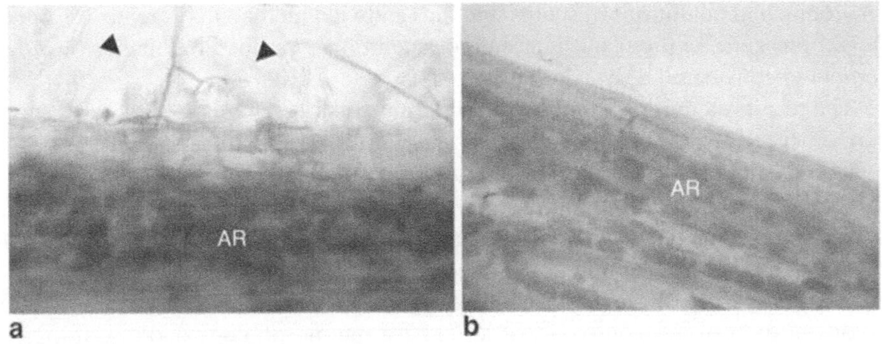

Fig. 4 The presence or the absence of root hairs do not affect the colonization process in *Lotus japonicus* roots stained with cotton blue to visualize the intraradical hyphae. (**a**) In the wild-type roots arbuscules (AR) are visible. (*Arrowheads*: root hair). Bar corresponds to 100 μm. (**b**) In the mutant *Ljrhl1-1/har1-1* root, where root hairs are lacking, the intraradical structures present the same distribution (AR). Bar corresponds to 100 μm. See the Appendix for methodological detalis

Table 1 The presence or the absence of root hairs does not affect the colonization process in *Lotus japonicus* roots, based on the quantification of mycorrhization (Trouvelot et al. 1986)

	F%	*M%*	*a%*	*A%*
Single mutants				
Ljvrh1.1	79.14	26.80	76.29	25.21
Ljvrh1.2	78.31	24.19	88.99	21.79
Slippery	49.69	14.05	81.56	12.77
Ljrhl1.1	54.00	16.59	90.08	15.90
Double mutants				
Ljvar1-1/har1-1	87.16	25.49	80.80	20.19
Ljvar1-2/har1-1	65.04	24.76	73.61	17.18
Ljvar1-3/har1-1	61.86	11.19	68.96	7.27
Ljsrh1/har1-1	80.09	27.40	84.27	22.95
Ljrhl1-1/har1-1	80.95	20.43	87.30	17.33
WT	60.18	21.09	86.72	20.95

Nine mutants of *L. japonicus* are compared with the WT *L. japonicus* genotype
No significant differences are present among the values assumed by the parameters in the different mutants according to the ANOVA test of variance (post hoc Tukey test)
F% frequency of mycorrhization (percentage of segments showing internal colonization), *M%* intensity of mycorrhization (average colonization meant as visible coverage of a root segment), *a%* percentage of arbuscules (average presence of arbuscules within colonized areas), *A%*, percentage of arbuscules in the root system (quantity of arbuscules as referred to the whole root system)

colonization attempts was evident. As a consequence, the parental line was identified as AM^{++} while the *Ljprh* mutants were identified as AM^{low} (Murray et al. 2006a).

These experiments clearly indicate that some phenotypes depend on the fungal species, as already suggested by David-Schwartz et al. (2001), as well as on the environmental conditions. It seems that the sandwich method may create a stronger fungal pressure on the plant than the pot conditions. On the other hand, *G. intraradices* is considered a stronger colonizer than *Gi. margarita*.

The results stemming from the root hair mutants of *L. japonicus* do not seem to support the Baylis hypothesis. Mutants grown in "Millipore sandwich" in sterile conditions and mycorrhized with *Gi. margarita* do not show any difference in mycorrhization intensity when compared with that in wild type plants (Novero et al., unpublished). The results from in-pot grown plants mycorrhized with *G. intraradices* even indicate a reduction in the rate of mycorrhization.

Taken together, these experiments suggest that the genetic dissection of the colonization process requires integration between morphology and physiological data, which is often missing. At the moment, very little is known about the efficiency of the uptake in the root hair mutants, compared to the wild types. Molecular markers for P uptake such as the *Medicago truncatula* transporter MtPt4 (Javot et al. 2007) or the tomato ones (Balestrini et al. 2007) are yet to be identified in *L. japonicus*. It would be interesting to know whether plant P transporters are located to a specific membrane domain of the trichoblast cells: on the hair shaft or at the basis of the trichoblast. It is possible that *L. japonicus* roots, even in the absence of hairs, are equally efficient in absorbing phosphate from the soil, because of a different distribution of P transporters. Jakobsen et al. (2005) measured soil P scavenging by root-hair-less mutants of barley, and found that the extraradical mycelium of AM fungi constitutes an alternative and efficient pathway, which can even overcompensate the reduced P uptake of these plants. Nonetheless, the same authors report that mutant plant growth, in terms of dry weight, is reduced in mycorrhizal conditions, which can be related to the fact that part of the acquired P is used by the fungus itself.

Our data, however, indicate that the absence of root hairs in *L. japonicus* has negligible influence on the plant mycorrhization capacity (Fig. 4 and Table 1).

5 The Input of Confocal Microscopy and In Vivo Imaging

In vivo confocal microscopy and fluorescent tagging of plant subcellular components in living roots have provided new tools to visualize cell responses to AM colonization. This led to the identification of the prepenetration apparatus (PPA), an aggregation of cytoplasm that develops in the epidermal cells of *M. truncatula* under an adhering AM fungal hyphopodium, and predicts the pathway that the penetrating hypha will follow on its way to the internal root tissues (Genre et al. 2005). PPA assembly is closely related to nuclear movements. As soon as the hyphopodium has developed on the root surface, the nucleus of the contacted cell moves to the contact point, and is surrounded by an accumulation of cytoplasm. Later on, a second migration takes the nucleus to the inner tangential wall of the epidermal cell. During this migration, a column of cytoplasm organizes behind the nucleus, and this leads to the complete development of the transcellular apparatus known as PPA. Besides the nucleus, the PPA is made up of a dense aggregation of cytoskeletal elements, endoplasmic reticulum, and membrane vesicles, suggesting the existence of an intense secretory activity concentrated along the core of the PPA. This is believed to be the mechanism by which the interface compartment, which later surrounds each hypha, is built. In fact, fungal penetration occurs only after the PPA has completely developed, and the penetrating hyphae closely follow the PPA path, widening its diameter as they grow through it.

Transformed "hairy" roots, the only experimental material that has so far allowed this type of in vivo approach, see the Appendix for methodological detalis, bear root hairs on most of their surface, possibly due to their juvenile status. Consequently, fungal penetration often occurs through trichoblasts, possibly with a higher frequency than in natural root systems. Transformed roots can therefore be particularly useful to study root hair involvement in AM symbiosis. In our experience, functional hyphopodia developed only on the root surface, especially along the grooves between adjacent epidermal cells, although hyphopodia could often stretch over an area of several cells. Fungal adhesion to the root hair shaft has sometimes been observed, but although this resulted in a local aggregation of cytoplasm under the contact area, complete PPA development and fungal penetration have never been observed.

On the contrary, when a hyphopodium contacted the base of a trichoblast of the transformed root, this cell behaved exactly like atrichoblast cells, mobilising its nucleus and organizing the PPA. It is important to stress here that, due to the particular shape and developmental program of root hair cells, the nucleus usually locates in the vicinity of the hair tip during tip growth (see Emons and Ketelaar 2008) or along the shaft in mature hairs. This requires a much longer movement of the trichoblast nucleus in response to fungal adhesion, compared to the few tens of micrometers that on average separate the nucleus starting position from the contact area, in atrichoblasts. Nonetheless, apart from this initial topological difference, PPAs develop with normal morphology and timing in the root hair cell body (Fig. 5).

Another early response, possibly related to trichoblast morphology, has been observed, which is specific to root hairs. Soon after hyphopodium development at the root epidermis, and by the time the initial phases of PPA development have occurred, the nuclei of root hair cells in the vicinity of the already contacted cells migrate from the hair shaft down into the cell body. This response stretches over an area of a few hundred micrometers around the contact site (Fig. 6). The functional meaning of such a diffused response still has to be demonstrated, although it is possible to speculate that since more hyphopodia may develop later on in that area, trichoblasts anticipate fungal contact by moving their nuclei as close as possible to potential future contact points. This would reasonably shorten PPA development in the case where a hyphopodium actually contacts the hair cell base.

The signal by which root hair cells perceive fungal presence from a distance of several cell lengths is even more obscure. A diffusible molecule of fungal origin, the so far elusive "myc factor," could be a good candidate, although, since this

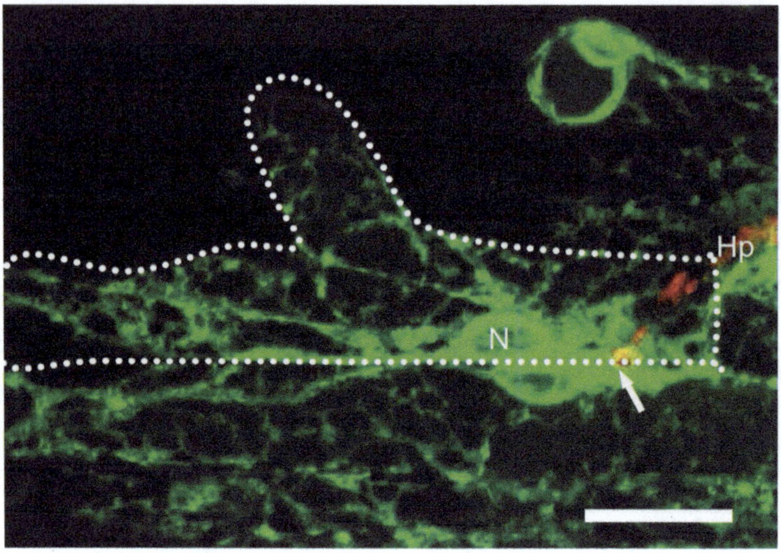

Fig. 5 Confocal micrograph showing an early phase of prepenetration apparatus (PPA) development in the cell body of a trichoblast (*dotted line*) of *Medicago truncatula* contacted by a hyphopodium (Hp) of *Gigaspora gigantea*. GFP:HDEL-tagged endoplasmic reticulum (*green*) accumulates between the nucleus (N) and the site of contact (*arrow*) with the hypha (*red*). An analogous response is present in the neighboring epidermal cell, where nuclear repositioning and endoplasmic reticulum accumulation have led to a very similar pattern. Bar corresponds to 20 μm

repositioning has always been observed in the presence of an already developed hyphopodium, an endogenous plant signal, diffusing from the contacted cells, cannot be ruled out.

Under natural conditions, root penetration by both AM fungi and nodulating bacteria can be far more variable and include the exploitation of occasional opportunities such as cracks in the epidermis, wounds, or secondary root emergence. Having said this, root hair cells, at least in artificial laboratory condition, appear to be as good as nonhair epidermal cells in setting up the prepenetration responses that allow AM establishment and the preservation of plant cell integrity.

The observation of long-distance nuclear movements supports the high sensitivity of trichoblasts to diffusible molecules (possibly as a consequence of their enlarged surface). However, the need for longer nuclear movements might have resulted, especially under natural conditions where root hairs are less abundant, in a preference for non-root hair cell-mediated AM penetration, as the most cost-effective choice for the plant. Both of these considerations contribute to the hypothesis that nitrogen-fixing bacteria may have found a favorable and mostly nonexploited gate to enter the root tissues via root hairs, when coevolution has led

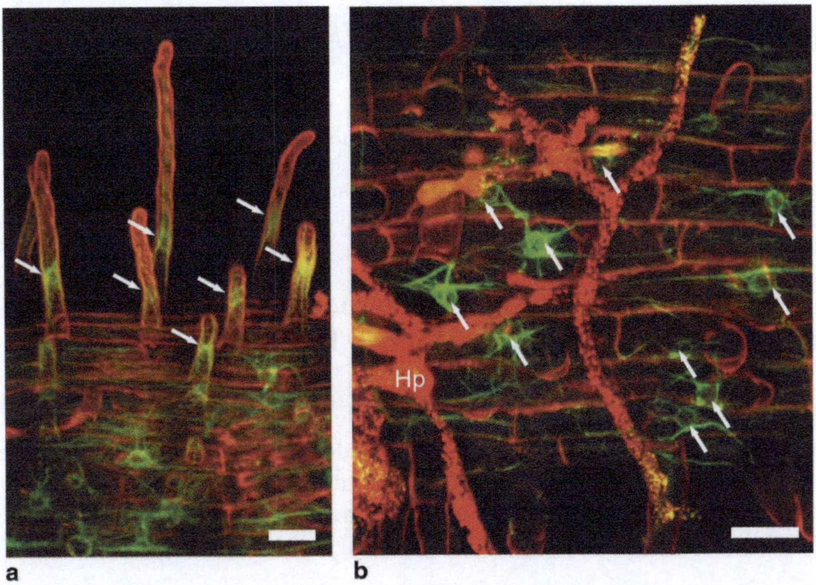

Fig. 6 Confocal micrograph showing nuclear repositioning in *Medicago truncatula* root hairs in the presence of a hyphopodium of *Gigaspora gigantea*. GFP:ABD-tagged actin cables (*green*) allow easy identification of nuclei, thanks to the dense perinuclear actin cytoskeleton; plant cell wall and fungus autofluoresce in *red*. (**a**) Control root with nuclei (*arrows*) positioned within the hair shaft. (**b**) Hyphopodium (Hp) development triggers the movement of root hair nuclei (*arrows*) into the cell body, also in trichoblasts that are not directly contacted by the hyphopodium. Bars correspond to 30 μm

to their ability to interact with legumes several million years after AM fungi had established their own *modus vivendi* with plants.

6 Conclusions

Among the many mycorrhizal types that exist, the presence of root hairs seems to be a constant only in roots that are involved in ectomycorrhizas and AMs. Root hairs in ectomycorrhizal plants seem to offer a first contact point for hyphae exploring the rhizosphere, while the possibility of AM fungi starting their colonization through root hairs is an exception rather than a rule. This observation may have an evolutionary meaning. More and more data suggest that the mechanism of root colonization by rhizobia has evolved partly based on the recruitment of functions used by more ancient AM symbiosis (see Limpens and Bisseling 2008). Choosing root hairs as cells that are able to set up accommodation responses but are often neglected by AM fungi may have been a key to finding a specific anatomical niche for rhizobial colonization as well as to harmonious cooperation between these two major classes of beneficial root symbioses.

Acknowledgments This research was funded by the Italian Project Prin 2006, the EU Marie Curie – Integral project (MRTN-CT-2003-505227) and the local 60% project of Torino University. The results on root hair cell responses to AM colonization were obtained in collaboration with David Barker in LIPM – INRA/CNRS, Castanet-Tolosan, France.

Appendix: How to Study the Interactions Between Root Hairs and Arbuscular Mycorrhizal Fungi: Technical Aspects

The achievement of mycorrhizas under controlled conditions is mandatory to investigate plant–fungal interactions according to a defined time scale. Two major techniques can be followed, one suitable for quantitative and morphological analysis on fixed material, the other for *in vivo* observation.

Morphological and Quantitative Analysis on Fixed Material

Mycorrhizal Synthesis

Ten-day-old seedlings were inoculated with *Gigaspora margarita* Becker and Hall (strain deposited in the Bank of European Glomales as BEG 34) using the Millipore (Bedford, MA) sandwich method (Giovannetti et al. 1993). Two seedlings were placed between two membranes (Millipore, Bedford; pore diameter of 0.45 μm), with 10–15 fungal spores (or without any spores for controls). Membranes containing the seedlings were planted in sterile acid-washed quartz sand in Magenta GA-7 vessels (Sigma, St Louis, MO) and grown in climate chamber at 22°C, 60% humidity, with 14 h of light per day. After 3 weeks, samples from roots were cut after observation under a stereo microscope and processed for quantification and cytological analyses.

Mycorrhizal Intensity Evaluation

Mycorrhized root segments (1 cm long) were incubated overnight at room temperature in 0.1% Cotton blue (w/v) in lactic acid. Segments were then washed in lactic acid, and observed under microscope for quantification assays.

According to the method by Trouvelot et al. (1986), segments were classified into four categories depending on the percentage of segment length occupied by mycelium and by arbuscules. Five parameters were considered:

(1) *F*%, reporting the percentage of segments showing internal colonization (frequency of mycorrhization)
(2) M%, indicating the average percent colonization of root segments (intensity of mycorrhization)
(3) *a*%, quantifying the average presence of arbuscules within the infected areas (percentage of arbuscules)
(4) *A*%, quantifying the presence of arbuscules in the whole root system (percentage of arbuscules in the root system)

Morphological Observation

Root segments were fixed in 2.5% (v/v) glutaraldehyde in 10 mM sodium phosphate buffer (PBS, pH 7.2) for 2 h at 4°C. After rinsing with the same buffer, samples were postfixed in 1% (w/v) osmium tetroxide in double-distilled water for 1 h, washed three times with double-distilled water, and dehydrated in an ethanol series (30, 50, 70, 90, and 100%; each step for 10 min) at room temperature. The root segments were infiltrated in 2:1 (v/v) ethanol–LR White resin (Polysciences, Warrington, PA) for 1 h at room temperature, in 1:2 (v/v) ethanol–LR White for 1 h at room temperature and in 100% LR White overnight at 4°C.

Semithin sections (1 μm) cut from root-embedded samples were stained with 1% toluidine blue for morphological observations.

Thin sections (70 nm) were counterstained with uranyl acetate and lead citrate and observed under a transmission electron microscope.

Cytochemical Localization of Polysaccharides (Thiéry et al. 1967)

Thin sections were treated with periodic acid (1%) for 30 min, washed three times in distilled water, and incubated overnight at 4°C in thiocarbohydrazide (0.2%) in 20% acetic acid. They were then washed in descending concentration of acetic acid and incubated for 30 min in silver proteinate (1%) in the dark. Silver-stained sections were viewed without additional staining under a transmission electron microscope.

In Vivo Analysis

In Vivo *Microscopic Observation of Mycorrhizal Infection*

The targeted AM inoculation technique developed by Chabaud et al. (2002) is currently the most suitable method for studying early stages of the symbiotic association between *Gigaspora* species and *A.-rhizogenes*-transformed root cultures of *M. truncatula*. Axenic spores of *Gi. rosea* or *Gi. gigantea* are gently inserted into the M medium containing 0.5% Phytagel in square Petri dishes, and cultured at 32°C at a slope of ~70° in a 2% CO_2 atmosphere for optimal germination according to Bécard et al. (1992). Under these conditions, spores germinate within 3–6 days. Germinated spores are then transferred within a gel plug to the Petri dish containing the transformed root culture, and positioned underneath a growing secondary root, in such a way that the germination hyphae (with negative geotropism) quickly reach the roots, thereby facilitating the identification of potential infection sites. Hyphal growth and root contacts can be recorded daily on the underside of the dish. For confocal microscopy observations, the root and fungus are covered with 1 ml of sterile water, on top of which is laid a 25-μm gas-permeable plastic film (bioFOLIE 25, Sartorius AG, Vivascience Support Center, Göttingen, Germany). The refractive index of the film is compatible with the use of long-distance water-immersion confocal objectives, thus allowing continuous prolonged microscopic observation, convenient transfer of the dish between the growth chamber and the microscope stage, as well as minimizing potential contamination of the coculture. Initial hyphal contact with the root can be monitored using a stereomicroscope. The potential infection points are then observed and followed in detail under a confocal microscope, using a long distance 40× water-immersion objective (Genre et al. 2005).

References

Akiyama K, Matsuzaki K, Hayashi H (2005) Plant sesquiterpenes induce hyphal branching in arbuscular mycorrhizal fungi. *Nature* 435:824–827

Bago B, Pfeffer PE, Shachar-Hill Y (2000) Carbon metabolism and transport in arbuscular mycorrhizas. *Plant Physiol* 124:949–958

Balestrini R, Bonfante P (2005) The interface compartment in arbuscular mycorrhizae: a special type of plant cell wall? *Plant Biosyst* 139(1):8–15

Balestrini R, Gómez-Ariza J, Lanfranco L, Bonfante P (2007) Laser microdissection reveals that transcripts for five plant and one fungal phosphate transporter genes are contemporaneously present in arbusculated cells. Mol Plant Microbe Interact 20:1055–1062

Barker SJ, Stummer B, Gao L, Dispain I, O'Connor PJ, Smith SE (1998) A mutant in *Lycopersicon esculentum* Mill. with highly reduced VA mycorrhizal colonisation: isolation and preliminary characterisation. *Plant J* 15:791–797

Bates TR, Lynch JP (1996) Stimulation of root hair elongation in *Arabidopsis thaliana* by low phosphorus availability. *Plant Cell Environ* 19:529–538

Bates TR, Lynch JP (2000) The efficiency of *Arabidopsis thaliana* (Brassicaceae) root hairs in phosphorus acquisition. *Am J Bot* 87:964–970

Baylis GTS (1975) The Magnolioid mycorrhiza and mycotrophy in root systems derived from it. In: Sanders FE, Mosse B, Tinker PB (eds) "Endomycorrhizas". Academic, London, pp 373–389

Bécard G, Douds DD, Pfeffer PE (1992) Extensive in vitro hyphal growth of vesicular arbuscular mycorrhizal fungi in the presence of CO_2 and flavonols. *Appl Environ Microbiol* 58:821–825

Béguiristain T, Lapeyrie F (1997) Host plant stimulates hypaphorine accumulation in *Pisolithus tinctorius* hyphae during ectomycorrhizal infection while excreted fungal hypaphorine controls root hair development. *New Phytol* 136:525–532

Bianciotto V, Barbiero G, Bonfante P (1995) Analysis of the cell cycle in an arbuscular mycorrhizal fungus by flow cytometry and bromodeoxyuridine labelling. *Protoplasma* 188:161–169

Bonfante P (1984) Anatomy and morphology of VA Mycorrhizae. In: Mycorrhiza VA, Powell CL, Bagyaraj DJ (eds) CRC Press, Boca Raton, pp 5–33
Bonfante P (2001) At the interface between mycorrhizal fungi and plants: the structural organization of cell wall, plasma membrane and cytoskeleton. In: Hock B (ed) The Mycota IX: fungal associations. Springer, Berlin Heidelberg New York, pp 45–61
Bonfante P, Perotto S (1995) Strategies of arbuscular mycorrhizal fungi when infecting host plants. *New Phytol* 130(1):3–21
Bonfante P, Genre A, Faccio A, Martini I, Schauser L, Stougaard J, Webb J, Parniske M (2000) The *Lotus japonicus* LjSym4 gene is required for the successful symbiotic infection of root epidermal cells. *Mol Plant Microbe Interact* 13(10):1109–1120
Brundrett MC (1991) Mycorrhizas in natural ecosystem. In: Macfayden A, Begon M, Fitter AH (eds) Advances in ecological research, vol 21. Academic, London, pp 171–313
Brundrett MC (2002) Coevolution of roots and mycorrhizas of land plants. *Tansley Rev New Phytol* 154:275–304
Bucher M (2007) Functional biology of plant phosphate uptake at root and mycorrhiza interfaces. *Tansley Rev New Phytol* 173:11–26
Chabaud M, Venard C, Defaux-Petras A, Bécard G, Barker DG (2002) Targeted inoculation of *Medicago truncatula in vitro* root cultures reveals *MtENOD11* expression during early stages of infection by arbuscular mycorrhizal fungi. *New Phytol* 156:265–273
Clarkson DT (1985) Factors affecting mineral nutrient acquisition by plants. *Annu Rev Plant Physiol* 36:77–115
Dauphin A, De Ruijter NCA, Emons AMC, Legué V (2006) Actin organization during *Eucalyptus* root hair development and its response to fungal hypaphorine. *Plant Biol* **8:**204–211
David-Schwartz R, Badani H, Smadar W, Levy AA, Galili G, Kapulnik Y (2001) Identification of a novel genetically controlled step in mycorrhizal colonization: plant resistance to infection by fungal spores but not extra-radical hyphae. *Plant J* 27:561–569
Demchenko K, Winzer T, Stougaard J, Parniske M, Pawlowska K (2004) Distinct roles of *Lotus japonicus* SYMRK and SYM15 in root colonization and arbuscule formation. *New Phytol* 163:381–392
Duc G, Trouvelot A, Gianinazzi-Pearson V, Gianinazzi S (1989) First report of non-mycorrhizal plant mutants (Myc-) obtained in pea (*Pisum sativum* L.) and fava bean (*Vicia faba* L.). *Plant Sci* 60:215–222
Emons AM, Ketelaar T (2008) Intracellular organization: a prerequisite for root hair elongation and cell wall deposition. In: Emons AMC, Ketelaar T (eds) Root hairs: excellent tools for the study of plant molecular cell biology. Springer, Berlin Heidelberg New York. doi:10.1007/7089_2008_4
Föhse D, Jungk A (1983) Influence of phosphate and nitrate supply on root hair formation of rape, spinach and tomato plants. *Plant Soil* 74:359–368
Frank AB (1885) Ueber die auf Wurzelsymbiose beruhende Ernährung gewisser Bäume durch unterirdische Pilze. *Ber Dtsch Bot Ges* 3:128–145
Gahoonia TS, Nielsen NE (2004) Barley genotypes with long root hairs sustain high grain yields in low-P field. *Plant Soil* 262:55–62
Garcia-Romera I, Garcia-Garrido JM, Ocampo JA (1991) Pectolytic enzymes in the vescicular-arbuscular mycorrhizal fungus *Glomus mosseae. FEMS Microbiol Lett* 78:343–346
Genre A, Bonfante P (1998) Actin versus tubulin configuration in arbuscule-containing cells from mycorrhizal tobacco roots. *New Phytol* 140:745–752
Genre A, Bonfante P (2005) Building a mycorrhizal cell: how to reach compatibility between plants and arbuscular mycorrhizal fungi. *J Plant Interact* 1:3–13
Genre A, Bonfante P (2007) Check-in procedures for plant cell entry by biotrophic microbes. Mol Plant Microbe Interact 20:1023–1030
Genre A, Chabaud M, Timmers T, Bonfante P, Barker DG (2005) Arbuscular mycorrhizal fungi elicit. a novel intracellular apparatus in *Medicago truncatula* root epidermal cells before infection. *Plant Cell* 17:3489–3499
Gianinazzi-Pearson V, Brechenmacher L (2004) Functional genomics of arrbscular mycorrhiza: decoding the symbiotic cell programme. *Can J Bot* 82:1228–1234

Gianinazzi-Pearson V, Séjalon-Delmas N, Genre A, Jeandroz S, Bonfante P (2007) Plants and arbuscular mycorrhizal fungi: cues and communication in the early steps of symbiotic interactions. Advances in Botanical Research 46:181–219

Gilroy I, Jones DL (2000) Through form to function: root hair development and nutrient uptake. *Trends Plant Sci* 5:56–60

Giovannetti M, Avio L, Sbrana C, Citernesi AS (1993) Factors affecting appressorium development in the vesicular-arbuscular mycorrhizal fungus *Glomus mosseae* (Nicol. & Gerd.) Gerd. & Trappe. *New Phytol* 123:115–122

Graham JH, Syvertsen JP (1985) Influence of vescicular-arbuscular mycorrhiza on the hydraulic conductivity of roots of two citrus rootstocks. *New Phytol* 97:277–284

Guinel F, Hirsch AM (2000) The involvement of root hairs in mycorrhizal associations. In: Ridge RW and Emos AMC (eds) Cell and molecular biology of plant root hairs. Springer, Berlin Heidelberg New York , 285–310

Hartig T (1840) Vollständige Naturgeschichte der forstlichen Culturpflanzen Deutschlands. Förstner'sche Verlagsbuchhandlung, Berlin

Jakobsen I (1995) Transport of phosphorus and carbon in VA mycorrhizas. In: Varma A, Hock B (eds) Mycorrhiza. Springer, Berlin Heidelberg New York, pp 297–324

Jakobsen I, Chen B, Munkvold L, Lundsgaard T, Zhu YG (2005) Contrasting phosphate acquisition of mycorrhizal fungi with that of root hairs using the root hairless barley mutant. *Plant Cell Environ* 28:928–938

Javot H, Penmetsa VR, Terzaghi N, Cook DR, Harrison MJ (2007) A *Medicago truncatula* phosphate transporter indispensable for the arbuscular mycorrhizal symbiosis. *PNAS* 104:1720–1725

Karas B, Murray J, Gorzelak M, Smith A, Sato S, Tabata S, Szczyglowski K (2005) Invasion of *Lotus japonicus root hairless 1* by *Mesorhizobium loti* involves the nodulation factor-dependent induction of root hairs. *Plant Physiol* 137:1331–1344

Kawaguchi M, Imaizumi-Anraku H, Koiwa H, Niwa S, Ikuta A, Syono K, Akao S (2002) Root, root hair and symbiotic mutants of the model legume *Lotus japonicus. Mol Plant Microbe Interact* 15(1):17–26

Kistner C, Winzer T, Pitzschke A, Mulder L, Sato S, Kaneko T, Tabata S, Sandal N, Stougaard J, Webb JK, Szczyglowski K, Parniske M (2005) Seven *Lotus japonicus* genes required for transcriptional reprogramming of the root during fungal and bacterial symbiosis. *Plant Cell* 17:2217–2229

Kosuta S, Chabaud M, Lougnon G, Gough C, Denarie J, Barker DG, Becard G (2003) A diffusible factor from arbuscular mycorrhizal fungi induces symbiosis-specific MtENOD11 expression in roots of *Medicago truncatula. Plant Physiol* 131:952–962

Krusell L, Madsen LH, Sato S, Aubert G, Genua A, Szczyglowski K, Duc G, Kaneko T, Tabata S, De Bruijn F, Pajuelo E, Sandal N, Stougaard J (2002) Shoot control of root development and nodulation is mediated by a receptor-like kinase. *Nature* 420:422–426

Lagrange H, Jay-Allgmand C, Lapeyrie F (2001) Rutin, the phenolglycoside from *Eucalyptus* root exudates, stimulates *Pisolithus* hyphal growth al picomolar concentration. *New Phytol* 149(2):349–355

Le Quere A, Wright DP, Soderstrom B, Tunlid A, Johansson T (2005) Global patterns of gene regulation associated with the development of ectomycorrhiza between birch (*Betula pendula* Roth.) and *Paxillus involutus* (Batsch) fr. *Mol Plant Microbe Interact* 18(7):659–673

Ligrone R, Carafa A, Lumini E, Bianciotto V, Bonfante P, Duckett JG (2007) Glomeromycotean associations in liverworts: a molecular cellular and taxonomic analysis. Am J Bot 94:1756–1777

Limpens E, Bisseling T (2008) Nod factor signal transduction in the Rhizobium-Legume symbiosis. In: Emons AMC, Ketelaar T (eds) Root hairs: excellent tools for the study of plant molecular cell biology. Springer, Berlin Heidelberg New York. doi:10.1007/7089_2008_10

Lohse S, Schliemann W, Ammer C, Kopka J, Stracke D, Fester T (2005) Organization and metabolism of plastids and mitochondria in arbuscular mycorrhizal roots of *Medicago truncatula. Plant Physiol* 139:329–340

Lopez MF, Manner P, Willmann A, Hampp R, Nehls U (2007) Increased trehalose biosynthesis in the Hartig net hyphae of ectomycorrhizas. *New Phytol* 174:389–398

Ma Z, Bielenberg DG, Brown KM, Lynch JP (2001) Regulation of root hair density by phosphorus availability in *Arabidopsis thaliana. Plant Cell Environ* 24:459–467

Marschner H (1995) Mineral nutrition of higher plant. Academic, London

Martin F, Perotto S, Bonfante P (2007) Mycorrhizal fungi: a fungal community at the interface between soil and roots. In: Pinton R, Varanini Z, Nannipieri P (eds) The Rhizosphere: biochemistry and organic substances at the soil-plant interface, vol 19. CRC Press, pp 201–136

McGonigle TP, Miller MH, Evans DG, Fairchild GL, Swan JA (1990) A new method which gives an objective measure of colonization of roots by vesicular-arbuscular fungi. *New Phytol* 115:495–501

Menge GA, Johnson ELV, Platt RG (1978) Mycorrhizal dependency of several citrus cultivars under three nutrient regimes. *New Phytol* 81:553–559

Murray J, Karas B, Ross L, Brachmann A, Wagg C, Geil R, Perry J, Nowakowski K, MacGillivary M, Held M, Stougaard J, Peterson L, Parniske M, Szczyglowski K (2006a) Genetic suppressors of the *Lotus japonicus har1-1* hypernodulation phenotype. *Mol Plant Microbe Interact* 19:1082–1091

Murray J, Geil R, Wagg C, Bogumil K, Szczyglowski K, Peterson LR (2006b) Genetic supressors of *Lotus japonicus har1-1* hypernodulation show altered interactions with *Glomus intraradices. Funct Plant Biol* 33:749–755

Navazio N, Moscatiello R, Genre A, Novero M, Baldan B, Bonfante P, Mariani P (2007) Diffusible signal from arbuscular mycorrhizal fungi elicits a transient cytosolic calcium elevation in host plant cells. *Plant Physiol* 144:673–681

Novero M, Faccio A, Genre A, Stougaard J, Webb KJ, Mulder L, Parniske M, Bonfante P (2002) Dual requirement of the *LjSym4* gene for the mycorrhizal development in epidermal cells and cortical cells of *Lotus japonicus* roots. *New Phytol* 154:741–749

Olah B, Briere C, Becard G, Denarie J, Gough C (2005) Nod factors and a diffusible factor from arbuscular mycorrhizal fungi stimulate lateral root formation in *Medicago truncatula* via the DMI1/DMI2 signalling pathway. *Plant J* 44:195–207

Oldroyd GE, Downie JA (2006) Nuclear calcium changes at the core of symbiosis signalling. *Curr Opin Plant Biol* 9(4):351–357

Paszkowski U (2006) A journey through signaling in arbuscular mycorrhizal symbioses. *Tansley Rev New Phytol* 172:35–46

Peretto R, Bettini V, Favaron F, Alghisi P, Bonfante P (1995) Polygalacturonase activity and location in arbuscular mycorrhizal roots of *Allium porrum* L. *Mycorrhiza* 5:157–163

Peterson L, Massicotte HB, Melville LH (2004) Mycorrhizas: anatomy and cell biology. NRC Press, Ottawa

Pirozynski KA, Malloch DW (1975) The origin of land plants: a matter of mycotrophism. *Biosystems* 6:153–164

Pühler A, Strack D (2007) Molecular basics of mycorrhizal symbioses. *Phytochemistry* 68(1):6–7

Redecker D, Kodner R, Graham LE (2000) Glomalean fungi from the Ordovician. *Science* 289(5486):1920–1921

Rémy W, Taylor TN, Hass H, Kerp H (1994) Four hundred-million-year-old vesicular arbuscular mycorrhizae. *Proc Natl Acad Sci USA* 91:11841–11843

Scannerini S, Bonfante P (1983) Comparative ultrastructural analysis of mycorrhizal associations. *Can J Bot* 61(3):917–943

Schüßler A, Schwarzott D, Walker C (2001) A new fungal phylum, the Glomeromycota: phylogeny and evolution. *Mycol Res* 105:1413–1421

Selosse MA, Faccio A, Scappaticci G, Bonfante P (2004) Chlorophyllous and achlorophyllous specimens of *Epipactis microphylla* (Neottieae, Orchidaceae) are associated with ectomycorrhizal septomycetes, including truffles. *Microb Ecol* 47:416–426

Selosse MA, Richard F, He X, Simard S (2006) Mycorrhizal networks: les liaisons dangereuses. *Trends Ecol Evol* 21(11):621–628

Sisti D, Giomaro G, Cecchini M, Faccio A, Novero M, Bonfante P (2003) Two genetically related strains of *Tuber borchii* produce *Tilia mycorrhizas* with different morphological traits. *Mycorrhiza* 13:107–115

Smith SE, Read DJ (1997) Mycorrhizal symbiosis. Academic, London, pp 1–605

Solaiman MZ, Senoo K, Kawaguchi M, Imaizumi-Anraku H, Akao S, Tanaka A, Obata H (2000) Characterization of mycorrhizas formed by *Glomus* sp. on roots of hypernodulating mutants of *Lotus japonicus*. *J Plant Res* 113:443–448

Thiéry JP (1967) Mise en evidence des polysaccharides sur coupes fines en microscopie électronique. *J microsc* 6:987–1018

Toth R, Miller RM (1984) Dynamics of arbuscule development and degeneration in a *Zea mays* mycorrhiza. *Am J Bot* 71:449–460

Trouvelot A, Kough JL, Gianinazzi-Pearson V (1986) Mesure du taux de mycorhization VA d'un système radiculaire. recherche de méthodes d'estimation ayant une signification functionnelle. In: Gianninazzi-Pearson V and Gianninazzi S (eds) Les mycorrhizes: physiologie et génétique. ESM/SEM, Dijon, 1–5 July. INRA Press, Paris, pp 217–222

Voegele RT, Mendgen KW (2007) Impact of genomics on fungal biology. *New Phytol* 173(3):458–462

Wopereis J, Pajuelo E, Dazzo FB, Jiang QY, Gresshoff PM, de Bruijn FJ, Stougaard J, Szczyglowski K (2000) Short root mutant of Lotus japonicus with a dramatically altered symbiotic phenotype. *Plant J* 23:97–114

Zobel R (1996) Genetic control of root systems. In: Waisel Y, Eshel A, Kafkafi U (eds) Plant roots: the hidden half, 2nd edn. Marcel Dekker, New York, pp 21–30

Index

GPSR Compliance

The European Union's (EU) General Product Safety Regulation (GPSR) is a set of rules that requires consumer products to be safe and our obligations to ensure this.

If you have any concerns about our products, you can contact us on ProductSafety@springernature.com

In case Publisher is established outside the EU, the EU authorized representative is:

Springer Nature Customer Service Center GmbH
Europaplatz 3
69115 Heidelberg, Germany

Batch number: 10761599

Printed by Printforce, the Netherlands